Charles D. Reese

Occupational Health

and

Safety Management

A Practical Approach

Second Edition

CRC Press
Taylor & Francis Group
Boca Raton   London   New York

CRC Press is an imprint of the
Taylor & Francis Group, an **informa** business

CRC Press
Taylor & Francis Group
6000 Broken Sound Parkway NW, Suite 300
Boca Raton, FL 33487-2742

© 2009 by Taylor & Francis Group, LLC
CRC Press is an imprint of Taylor & Francis Group, an Informa business

No claim to original U.S. Government works
Printed in the United States of America on acid-free paper
10 9 8 7 6 5 4 3 2 1

International Standard Book Number-13: 978-1-4200-5180-3 (Hardcover)

Library of Congress Cataloging-in-Publication Data

Library of Congress Cataloging-in-Publication Data
  Reese, Charles D.
  Occupational health and safety management : a practical approach / Charles D. Reese. -- 2nd ed.
      p. cm.
  "A CRC title."
  Includes bibliographical references and index.
  ISBN 978-1-4200-5180-3 (alk. paper)
  1. Industrial hygiene--Management. 2. Industrial safety--Management. I. Title.

RC967.R447 2008
658.3'82--dc22   2008014778

Visit the Taylor & Francis Web site at
http://www.taylorandfrancis.com

and the CRC Press Web site at
http://www.crcpress.com

Dedication

This book is dedicated to those employers and safety and health professionals who are working diligently to provide a workplace free from the safety and health hazards which could adversely affect their workforce.

Contents

Preface

This book was developed to provide safety and health professionals, students, and employers with the basic tenets for the creation of an occupational safety and health initiative for those responsible or planning to implement and manage safety and health programs for their companies or future employers. The intent of this book is to provide a management blueprint for occupational safety and health for the smallest to the largest companies who are beginning to or have seen the need to develop or improve their safety and health approach in their workplace. This would include the construction industry as well as other industrial groups.

Every chapter provides major requirements that will put you on your way toward building a successful and effective safety and health effort for your company or employer.

This book will encompass a total management approach to the development of written programs, the identification of hazards, the mitigation of hazards by the use of common safety and health tools, the development of a safe workforce through communications, and motivational techniques including behavior-based safety, involvement, and training. It will also address the tracking and acceptable risk from both safety and health hazards. It will address how to work with and within the Occupational Safety and Health Administration (OSHA) compliance approach as well as how to deal with the OSHA regulations.

This is a how-to book not just an information-providing text. It will show you how to write a program, how to identify hazards, how to involve workers and attain their cooperation. It will explain how to use identification and intervention tools such as hazard hunt, audits, and job hazard analysis. It will also provide a listing of potential resources if more detailed information is needed. It will encourage the development of a working relationship with OSHA and how to go about determining which regulations are applicable to the workplace that confronts you and how to budget for safety and how to find assistance and sources that will guide you in compliance with OSHA regulations. It will also address the topic of environmental issues, workplace security/violence, and regulations as related to the workplace. All facets of this book are tools for managing safety and health and are supported using checklists, illustrations, diagrams, figures, photographs, and tables as necessary to clarify or add to a topic.

A major push in the direction of managing occupational safety and health for your company or employer will reap many benefits and rewards. The most recognizable will be "safe production." Other positive effects that will be attained are less human (worker) suffering, better overall morale, a positive commitment to the company's goals and objective, a decrease in overall risk, and less liability.

The professionalism of the safety and health professional and the ethical approaches that need to be followed in order to maintain the integrity of occupational safety and health in the workplace will be discussed in some detail.

Occupational safety and health at any workplace are always dynamic and fluid. You will never have a perfect safety and health program since it will always be evolving as the company and the program matures and changes with time.

The content in this book is your primer for a more effectively managed approach with regard to safety and health for any company or employer.

Charles D. Reese, PhD

Acknowledgments

I would like to express an appreciation to Taylor & Francis acquiring editor, Cindy Carelli, who recommended updating this book. Also, I appreciate the courtesy extended to me by the following organizations and individuals:

National Institute for Occupational Safety and Health
Occupational Safety and Health Administration
United States Department of Energy
Mine Safety and Health Administration

Robert Franko

Author

For 31 years, **Dr. Charles D. Reese** has been involved with occupational safety and health as an educator, manager, and consultant. In his early career, Dr. Reese was an industrial hygienist at the National Mine Health and Safety Academy. He later became manager for the nation's occupational trauma research initiative at the National Institute for Occupational Safety and Health's Division of Safety Research. Dr. Reese has played an integral role in trying to ensure workplace safety and health. As the managing director for the Laborers' Health and Safety Fund of North America, he was responsible for the welfare of the 650,000 members of the laborers' union in the United States and Canada.

Dr. Reese has developed many occupational safety and health training programs, which range from radioactive waste remediation to confined space entry. He has also written numerous articles, pamphlets, and books on related issues.

Dr. Reese, professor emeritus, was a member of the graduate and undergraduate faculty at the University of Connecticut, where he taught courses on Occupational Safety and Health Administration regulations, safety and health management, accident-prevention techniques, industrial hygiene, and ergonomics. As professor of environmental/occupational safety and health, he was instrumental in coordinating the safety and health efforts at the University of Connecticut. He is often invited to consult with industry on safety and health issues and is asked for expert consultation in legal cases.

Dr. Reese is also the principal author of the *Handbook of OSHA Construction Safety and Health (Second Edition)*; *Material Handling Systems: Designing for Safety and Health*; *Annotated Dictionary of Construction Safety and Health*, *Occupational Health and Safety Management: A Practical Approach*; *Office Building Safety and Health*; *Accident/Incident Prevention Techniques*; *Industrial Safety and Health for Goods and Materials Services*; *Industrial Safety and Health for Infrastructure Services*; *Industrial Safety and Health for Administrative Services*; and *Industrial Safety and Health for People-Oriented Services*.

CHAPTER 1

In the Beginning: Introduction

The early silversmith plying his trade.

This book is meant for those who need a blueprint to start a safety and health program or improve upon an existing one. It is designed to guide you in having an effective occupational safety and health effort at your company and will provide you the specific areas for each aspect of incorporating occupational safety and health into your company. It is a management approach using the practical lessons that have been provided by the safety and health community. These techniques, tools, and guidance have been implemented by many companies and have been recommended by others who are proponents of safe and healthy workplaces. They should provide the foundation to build on as you address your unique needs related to occupational safety and health.

The appendices at the end of the book will provide you further assistance in specific areas. Within these pages you will find guidance and directions for applying good safety and health practices to your workplace. Most companies in today's marketplace perceive occupational safety and health as an integral part of doing business. If you have an existing safety and health program, you will hopefully be looking for ways to improve upon it using this guide. If you do not have a program in place, you should glean from these pages a foundation for undertaking an effort to make occupational safety and health part of your business climate. The suggestions, principles, and practices found within this book are those that have stood the test of time. You will most likely want to use other more detailed and technical books to assist you in developing an occupational safety and health approach, which will meet the specific needs of your company.

It was envisioned that this book would be a simple yet practical guide for embracing occupational safety and health as a right and just component of any business that has a workforce. This book is also for those businesses that are concerned about the bottom line and cost containment in the competitive business environment of today.

A short history of occupational safety and health has been provided depicting its development over the years. This may help put things in perspective related to the evolution of occupational safety and health as we know it today.

HISTORY

Historically, the Egyptians were aware of the dangers from gold and silver fumes. They even had a first aid manual for workers as early as 3000 BC. In 2000 BC, Hammurabi placed a value on permanent injuries, such as the loss of an eye, for which the owner paid the worker or paid the doctor's bill. In 1500 BC, Ramses hired a physician for quarry workers. In 400 BC, Hippocrates, the father of medicine, realized stone cutters were having breathing problems. In 100 BC, the Romans were aware of the dangers faced by workers. They would free a slave if he survived the launching of a ship. The Romans even had a goddess of safety and health named *Salus* whose image is often embedded on their coins.

In the Middle Ages, people became more aware of the link between the type of work they did and the types of injuries and illnesses that they sustained. For example, English chimney sweeps in the 1700s were more susceptible to testicular cancer because of the soot. With the advent of the industrial revolution, the use of machinery and the changed work environment saw a rapid rise in the number of injuries, illnesses, and deaths. During this period, the first unions began to be organized to try to protect workers from the hazards of the workplace. The only improvement in the 1800s was fire protection because of pressure from insurance companies. This was soon followed by Massachusetts' requirement for factory inspections. Also, the first acts and regulations pertaining to mining were put forth. Some safety measures were adopted for other industries such as the railroads with the invention of air brakes and automatic couplers, which save many lives and amputations.

During the first part of the 1900s, workers' compensation laws started appearing and were finally deemed constitutional by the Supreme Court in 1916. Before this, most employers blamed their workers and held them responsible for workplace incidents citing what were known as the "the common laws" that stated:

1. Employer is not responsible when a fellow worker negligently causes your injury.
2. Employer is not responsible if the worker is injured due to his/her own negligence.
3. If an employee takes up a risky job knowing fully well the inherent hazards, the employer is not responsible.

Under the workers' compensation laws the employers assumed responsibilities for their workplaces' safety and health. They were required to provide and pay for medical care and lost wages due to on-the-job incidents. Also at this time interest was generated to count the numbers of injuries and deaths; the most famous of these undertakings was the work-related death count for Allegheny County, Pennsylvania, by the Russell Sage Foundation.

It was during this time that mining catastrophes continued to occur and more laws were passed to protect miners. Some catastrophes such as the 1910 Triangle Shirtwaist Co. where 146 young women were killed in a fire because exit doors were locked demonstrated a need to better protect workers. When 2000 workers or 50% of the workforce died from silica exposure at Gauley Bridge, West Virginia, the Walsh–Healey Act was passed that required safety and health measures for any employer receiving a government contract. Some companies began to understand their moral responsibility. The American Match Co. allowed other companies in the match-producing industry to use their process, which substituted a safer substance for phosphorus in match making. This resulted in the decrease of an occupational illness called "Phossy jaw," which caused swelling and pain in the jaw due to phosphorus exposure. A more detailed timeline related to occupational safety and health can be found in Figure 1.1.

Important events in the history of occupational safety and health

1700—Ramazzini publishes his treatise on occupational diseases.
1806—Unions' right to strike curtailed by "conspiracy" decision.
1842—Right to strike upheld by Massachusetts Supreme Court.
1864—Pennsylvania Mine Safety Act passed.
1864—First accident insurance policy in North America issued.
1867—Massachusetts instituted first factory inspection.
1869—Railroad air brake invented.
1877—Massachusetts passed law requiring guarding dangerous machinery and authority for factory inspection enforcement.
1878—First recorded call by labor for federal occupational safety and health law.
1881—American Federation of Labor (AFL) founded.
1886—First U.S. Commission of Labor appointed.
1902—Maryland passed first U.S. workers' compensation law; it was declared unconstitutional in 1904.
1906—Pittsburgh survey conducted by Russell Sage Foundation.
1907—In Monongah, West Virginia, 362 miners died in mine disaster.
1908—Congress passed a workers' compensation law covering government employees.
1909—Bureau of Labor study on Phossy jaw released.
1910—U.S. Bureau of Mines created.
1910—First National Conference on Industrial Diseases called by the American Association for Labor Legislation.
1912—U.S. Public Health Service expanded to include a division of occupational health.
1912—Children's Bureau created by Congress to investigate dangerous occupations, accidents, and employment.
1913—U.S. Department of Labor established.
1916—U.S. Supreme Court declared workers' compensation laws to be constitutional.
1918—American Standards Association founded; it sponsored many safety standards. (It is now called the American National Standards Institute.)
1921—The International Labor Organization (founded in 1919 in Geneva, Switzerland) set up a safety service.
1930–1936—Up to 2000 workers died while constructing a tunnel at Gauley Bridge, West Virginia.
1934—U.S. Bureau of Labor Standards was created.
1936—Frances Perkins, Secretary of Labor, called for a federal occupational safety and health law.
1936—Walsh–Healey (Public Contracts) Act passed.
1937—Congress of Industrial Organizations (CIO) founded.
1943—Publication of Alice Hamilton's *Exploring the Dangerous Trades*.
1952—Coal Mine Safety Act passed.
1955—AFL and CIO reunited.

Figure 1.1 Timeline of events that led to occupational safety and health as we know it today. (Courtesy of the Occupational Safety and Health Administration.)

(continued)

1958—Safety amendments added to Longshoremen's and Harbor Workers' Compensation Act.

1960—Specific safety standards promulgated for the Walsh–Healey Act.

1964—Presidential Conference on Occupational Safety convened.

1965—"Mission Safety-70" begun for federal employees.

1966—Metal and Nonmetallic Mines Safety Act passed.

1968—President Johnson called for federal occupational safety and health law.

1969—Construction Safety Act passed.

1970—President Richard M. Nixon signed the Occupational Safety and Health Act of 1970.

1971—First standards adopted to provide baseline for safety and health protection in American workplaces.

1971—AFL–CIO requested OSHA to adopt emergency standard on asbestos.

1972—OSHA Training Institute established to instruct OSHA inspectors and the public.

1972—First states approved (South Carolina, Montana, and Oregon) to run their own OSHA programs.

1974—OSHA adopted health standards for 14 carcinogens.

1975—Free consultation program created. Nearly 500,000 businesses participated in the past 30 years.

1978—Cotton dust standard promulgated to protect 600,000 workers from byssinosis cases of "brown lung" have declined to 0.1 cases per 10,000 workers.

1978—Supreme Court decision setting staffing benchmarks for state plans to be "at least as effective" as federal OSHA.

1978—New Directions Grants program (now known as the Susan Harwood Training Grants program) to foster development of occupational safety and health training and education for employers and workers. (More than 1.3 million trained since 1978.)

1978—Lead standard published to reduce permissible exposures by three-quarters to protect 835,000 workers from damage to nervous, urinary, and reproductive systems. (Construction standard adopted in 1995.)

1980—AFL–CIO held first Occupational Safety and Health Conference.

1980—Comprehensive Environmental Response, Compensation and Liability Act (CERCLA) passed.

1980—Supreme Court decision on Whirlpool affirming workers' rights to engage in safety- and health-related activities.

1980—Medical and exposure records standard finalized to permit worker and OSHA access to employer-maintained medical and toxic exposure records.

1980—Supreme Court decision vacates OSHA's benzene standard, establishing the principle that OSHA standards must address and reduce "significant risks" to workers.

1980—Fire protection standard updated and rules established for fire brigades responsible for putting out nearly 95% of worksite fires.

1981—Electrical standards updated to simplify compliance and adopt performance approach.

1982—Voluntary protection programs (VPPs) created to recognize worksites with outstanding safety and health programs (nearly 1400 sites currently participating).

1983—Hazard communication standard promulgated to provide information and training and labeling of toxic materials for manufacturing employers and employees. Other industries added August 24, 1987.

1984—First "final approvals" granted to state plans (Virgin Islands, Hawaii, and Alaska) giving them authority to operate with minimal oversight from OSHA.

1986—Superfund Amendments and Reauthorization Act passed.

1986—Reauthorization of 1980 CERCLA passed.

1986—BLS revised the OSHA Record keeping Guidelines.

1986—First instance-by-instance penalties proposed against Union Carbide's plant in Institute, West Virginia, for egregious violations involving respiratory protection and injury and illness record keeping.

1987—Grain handling facilities standard adopted to protect 155,000 workers at nearly 24,000 grain elevators from the risk of fire and explosion from highly combustible grain dust.

1987—OSHA Asbestos Standard for general industry and construction passed.

1987—OSHA Hazard Communication Standard for all U.S. industry (Right-to-Know) passed.

Figure 1.1 (continued)

1989—*Safety and Health Program Management Guidelines*, voluntary guidelines for effective safety and health programs based on VPP experience, published.

1989—Hazardous waste operations and emergency response standard promulgated to protect 1.75 million public and private sector workers exposed to toxic wastes from spills or at hazardous waste sites.

1989—Lockout/tagout of hazardous energy sources standard issued to protect 39 million workers from unexpected energization or start-up of machines or equipment and prevent 120 deaths and 50,000 injuries each year.

1991—Occupational exposure to bloodborne pathogens standards published to prevent more than 9,000 infections and 200 deaths per year, protecting 5.6 million workers against AIDS, hepatitis B, and other diseases.

1992—Education centers created to make OSHA training courses more widely available to employers, workers, and the public. Twenty centers train more than 300,000 students each year—over 370,000 students were trained in financial year 2005 alone.

1992—Process safety management of highly hazardous chemicals standard adopted to reduce fire and explosion risks for 3 million workers at 25,000 workplaces, preventing more than 250 deaths and more than 1,500 injuries each year.

1993—Permit-required confined spaces standard promulgated to prevent more than 50 deaths and more than 5,000 serious injuries annually for 1.6 million workers who enter confined spaces at 240,000 workplaces each year.

1993—Maine 200 program created to promote development of safety and health programs at companies with high numbers of injuries and illnesses.

1994—First expert advisor software, GoCad, issued to assist employers in complying with OSHA's cadmium standard.

1994—Fall protection in construction standard revised to save 79 lives and prevent 56,400 injuries each year.

1994—Asbestos standard updated to cut permissible exposures in half for nearly 4 million workers, preventing 42 cancer deaths annually.

1995—Formal launch of OSHA's expanded Web page to provide OSHA standards and compliance assistance via the Internet.

1996—Phone–fax complaint handling policy adopted to speed resolution of complaints of unsafe or unhealthful working conditions.

1996—Scaffold standard published to protect 2.3 million construction workers and prevent 50 deaths and 4,500 injuries annually.

1998—OSHA Strategic Partnership Program launched to improve workplace safety and health through national and local cooperative, voluntary agreements.

1999—Site-Specific Targeting Program established to focus OSHA resources where most needed on individual worksites with the highest injury and illness rates.

2000—Ergonomics program standard promulgated to prevent 460,000 musculoskeletal disorders among more than 102 million workers at 6.1 million general industry worksites. (This regulation was struck down by Congress in 2001.)

2001—Jersey public employee plan receives final approval.

2001—Steel erection standard, developed in concert with industry and union groups, preventing 30 fatalities and 1142 injuries annually and saving employers nearly $40 million a year, issued. It is the first OSHA safety standard to be developed under the negotiated rulemaking process.

2001—Record keeping rule revised to improve the system that employers use to track and record workplace injuries and illnesses.

2001—As mandated by the Needlestick Safety and Prevention Act, OSHA revised its bloodborne pathogens standard to clarify the need for employers to select safer needle devices.

2001—Under the auspices of the Congressional Review Act the Senate votes 56–44 to repeal ergonomics rule. The House follows suit the next day and votes 223–206 to repeal the rule. It is the first time that Congress exercises its authority under the act to repeal a federal standard.

2001—President signs S.J. Resolution 6, repealing the ergonomics rule.

2001—OSHA celebrates its 30th anniversary; over the past three decades job-related fatalities are cut in half, injuries and illnesses declined by 40%.

Figure 1.1 (continued)

(continued)

2001—OSHA responds to terrorist attacks at the World Trade Center in New York City and the Pentagon outside Washington, DC. More than 1000 OSHA employees from New York and around the country volunteer to help protect workers involved in the cleanup and recovery efforts at both sites.

2002—OSHA instituted a new record keeping guideline and reporting forms.

2002—Agency launches biweekly electronic newsletter *Quick Takes*.

2002—Secretary Chao unveils a comprehensive plan designed to reduce ergonomic injuries through a combination of industry-specific guidelines, strong enforcement, outreach and assistance, and further research.

2002—The recovery phase and cleanup at the World Trade Center disaster site comes to an end. For more than 8 months, 3 million work hours were logged on a worksite like no other, yet only 35 workers missed workdays due to injury and no more lives were lost to work.

2003—OSHA announces an enhanced enforcement policy to focus on those employers who have received "high gravity" citations.

2003—Ergonomics guidelines issued for the nursing home industry.

2003—Final rule establishes criteria for recording work-related hearing loss.

2003—OSHA welcomes 1000th site to achieve "star" status in VPP.

2004—OSHA unveils its National Emergency Management Plan, a new directive that clarifies the agency's policies during responses to national emergencies.

2004—Ergonomics guidelines published for retail grocery stores.

2004—Final rule establishes procedures for handling whistleblower complaints under the Corporate and Criminal Fraud Accountability Act of 2002, also known as the Sarbanes–Oxley Act.

2004—Ergonomics guidelines announced for the poultry processing industry.

2004—Federal agencies required to adopt private sector worker safety and health record keeping and reporting requirements.

2005—Oregon becomes the 17th state to receive final approval to operate their own job safety and health program.

2005—OSHA responds to Hurricane Katrina and offers the full resources of the agency to help protect the safety and health of workers responding to the disaster along the Gulf Coast.

2005—OSHA reaches settlement agreement with BP Products North America Inc. The company pays more than $21 million in fines following a fatal explosion at its Texas City plant. It is the largest penalty ever issued by OSHA.

2006—OSHA publishes final rule on hexavalent chromium, lowering the permissible exposure limit (PEL) from 52 to 5 $\mu g/cm^3$ based on an 8 h workday.

Figure 1.1 (continued)

As pressure mounted from workers and unions to pass some federal laws and the number of injuries, illnesses, and deaths increased, it became more apparent that the state programs for occupational safety and health were not protecting workers effectively. If it were not for unions attempting to protect their members, the Occupational Safety and Health Act of 1970 would probably not have passed and workers would have much less protection from job hazards today. Now most employers have realized that a safe and healthy workplace is more productive and makes good business sense.

In the area of prevention, there are still the three Es of safety:

1. Engineering—this entails awareness of safety issues when designing equipment
2. Education—training employees in safety procedures and how to safely perform their job
3. Enforcement—rules and policies must be strictly enforced if a safe workplace is to be accomplished

There are six good reasons to prevent accidents, injuries, illnesses, and deaths:

1. Destruction of human life is morally unjustified.
2. Failure of employers or workers to take precautions against occupational injuries and illnesses makes them morally responsible for these incidents.
3. Occupational incidents limit efficiency and productivity.

4. Occupational accidents and illnesses produce far-reaching social harm.
5. Safety techniques have produced reduction of accident rates and severity rates.
6. Recent cries and mandates have come forth at the state and federal levels to provide a safe and healthy workplace.

You might ask, "How big is the problem?" The only way to convey this is to provide some of the numbers, which depict the enormity and magnitude of the injuries, illnesses, and deaths, and their impact on the American workforce. Seldom is any company or industry free from being a part of one of these statistics.

NUMBERS

Healthy People 2010 Objectives from the U.S. Department of Health and Human Services (DHHS) has made the facts available relevant to occupational injuries and illnesses. Every 5 s a worker is injured, and every 10 s a worker is temporarily or permanently disabled. Each day, an average of 137 persons die from work-related diseases, and an additional 17 die from workplace injuries on the job. Each year, about 70 youths under 18 years of age die from injuries at work and 70,000 require treatment in a hospital emergency room. In 1996, an estimated 11,000 workers were disabled each day due to work-related injuries. That same year, the National Safety Council estimated that on-the-job injuries cost society $121 billion, including lost wages, lost productivity, administrative expenses, healthcare, and other costs. A study published in July 1997 reports that the 1992 combined U.S. economic burden for occupational illnesses and injuries was an estimated $171 billion.

A number of data systems and estimates exist to describe the nature and magnitude of occupational injuries and illnesses, all of which have advantages as well as limitations. In 1996, information from death certificates and other administrative records indicated that at least 6112 workers died from work-related injuries. No reporting system for national occupational chronic disease and mortality currently exists in this country. Therefore, scientists and policymakers must rely on estimates to understand the magnitude of occupational disease generated from a number of data sources and published epidemiological (or population-based) studies. Estimates generated from these sources are generally thought to underestimate the true extent of occupational disease, but the scientific community recognizes these estimates as the best available information. Such compilations indicate that an estimated 50,000–70,000 workers die each year from work-related diseases.

Current data collection systems are not sufficient to monitor disparities in health-related occupational injuries and illnesses. Efforts will be made over the coming decade to improve surveillance systems and data points that may allow evaluation of health disparities for work-related illnesses, injuries, and deaths. Data from the National Traumatic Occupational Fatalities Surveillance System (NTOF), based on death certificates from across the United States, demonstrate a general decrease in occupational mortality over the 15 year period from 1980 to 1994. However, the number and rates of fatal injuries from 1990 to 1994 remained relatively stable—at over 5,000 deaths per year and about 4.4 deaths per 100,000 workers. Motor vehicle-related fatalities at work, the leading cause of death for U.S. workers since 1980, accounted for 23% of deaths during the 15 year period. Workplace homicides became the second leading cause of death in 1990, surpassing machine-related deaths. While the rankings of individual industry divisions have varied across the years, the largest number of deaths is consistently found in construction, transportation, public utilities, and manufacturing, while those industries with the highest fatality rates per 100,000 workers are mining, agriculture/forestry/fishing, and construction. Data from the Bureau of Labor Statistics (BLS), Department of Labor, indicate that for nonfatal injuries and illnesses, incidence rates have been relatively stable since 1980. The rate in 1980 was 8.7 per 100,000 workers whereas in 1994 it was 8.4 per 100,000 workers. Incidence varied between a low of 7.7 per 100,000 workers (1982) and a high of 8.9 per 100,000 workers (1992) over the 15 year period from 1980 to 1994.

The toll of workplace injuries and illnesses continues to harm our country. Six million workers in the United States are exposed to workplace hazards ranging from falls from elevations to exposures to lead. The hazards vary depending upon the type of industry (e.g., manufacturing) and the types of work being performed by workers (e.g., welding).

The consequences of occupational accidents/incidents have resulted in pain and suffering, equipment damage, exposure of the public to hazards, lost production capacity, and liability. Needless to say these occupationally related accidents/incidents have a direct impact upon profit, which is commonly called the bottom line.

TRUE BOTTOM LINE

Without exception, all industries and companies face safety and health issues, which could have adverse effects upon their workforce and workplace. It does not matter whether you are a service industry, insurance agency, construction operation, or manufacturer of widgets. Your workforce will be exposed to the hazards unique to that worksite. It is definitely beneficial to your bottom line to not have any of your workforces injured or ill from something within your place of employment. Whether you are large or small having anyone in your workforce who has been incapacitated in any way disrupts the work process. Not counting the loss of a potential key employee, the time spent addressing an incident that has caused injuries or illnesses definitely cuts into the bottom line. If you think this is bad, you have no idea about the impact of occupationally related deaths.

In the past, I have gone to investigate an occupational death and could not find the company. It had folded because it could not absorb the impact and cost of a job-related death. The cost alone for such an occurrence exceeds $1.2 million. If you are a small company, this can be a business-ending event.

Taking a reasonable amount of time to address occupational safety and health will have a very positive impact upon your particular operation. Certainly, the magnitude of your safety and health effort will vary depending on whether your workplace is an office environment or a construction jobsite. If you were to just address the key components raised in this book, you will have made great strides forward in making safety and health an integral part of your workplace and in your philosophy regarding your overall view of what encompasses a workplace.

The safety and health of your workforce should be a priority for you since employees spend most of their waking hours in their place of employment. They want, and you want them, to go home the same way they came to work—with all their fingers intact, in their car rather than in an ambulance, and with a desire to return to work the next day. I have always thought that Dan Petersen had the right idea for employers when he said that what employers should want is "safe production."

CONTENT

The content of the chapters that follow provides the multiplicity of factors that go into achieving occupational safety and health excellence. These components of your safety and health program and initiative at your workplace should include:

- Development of a written safety and health program tailored to your business, which includes a statement of management commitment. This commitment is demonstrated by an allocation of funding (budget) as well as an assignment of accountability and responsibility for safety and health.
- Identification of hazards specific to your workplace as well as a way of involving everyone in identifying hazards without any feeling of guilt or it being viewed in a negative manner.
- Everyone has responsibilities for safety and health from top management to supervisors and the workforce.

- There should be consequences for anyone who does not take good safety and health performance seriously. Supervisors, as well as workers, must be a part of safety and health solutions and not the cause of the accidents and incidents that are costly both in terms of expenses and human suffering.
- Special efforts should be undertaken to involve not only management but the workforce in supporting and seeking solutions to safety and health efforts.
- With everyone involved it is important to pay close attention to guide everyone in a positive direction. Are there ways in which you can reinforce good safety and health practices and recognize those who are achieving the safety and health goals that you have set? Your approach to motivating safety and health at your workplace will have a large impact on the success or failure of your safety and health effort.
- You need to recognize why the hazards exist at your workplace, and to use the techniques of safety and health (as well as the interventions and preventive mechanisms that you identify or that are identified by your workforce) to decrease the hazards existing at your place of operations.
- You need to look at the types of injuries and illness that are occurring or have the potential to occur. Injuries and illnesses need to be addressed in different ways since the factors that cause them are very different.
- You need to understand that Occupational Safety and Health Administration (OSHA) regulations exist because workers exposed to hazards were being killed, maimed, or becoming ill. OSHA did not invent the hazards, but for many years has seen their terrible outcomes. In their inspections OSHA personnel have seen the many ways used by industry to address those hazards. These have become the core of their written regulations.
- OSHA must be seen as an asset and helping hand that can provide guidance and resources to assist you in your attempt to provide a safe and healthy workplace for your workforce and yourself. OSHA does not want to have to come to your workplace to investigate a death, address a complaint, or conduct an inspection. OSHA has limited resources and would prefer to answer your questions via the telephone or e-mail. Use them and they will have no reason to abuse you with violations and fines.
- Unless your supervisors and employees are trained in safety and health regulations and how to do their jobs in a safe manner, you cannot expect the other parts of your safety and health program to operate. A trained workforce is a safer workforce, not a perfect workforce.

Each of these is discussed in the following chapters. It is imperative that you address each area in a manner that is appropriate for your type of business to build for yourself and your workers a type of safety and health approach that you can be proud of. The general content addressed in each chapter is as follows:

Chapter 1—In the Beginning: Introduction—The introduction deals with why we need to address safety and health in the workplace. It makes the point that safety and health are not new but have evolved and matured over many years.

Chapter 2—Making a Commitment: Management's Commitment and Involvement—This chapter focuses on the commitment that is needed from the company and its management before starting a safety and health initiative. It details the specifics that management must address to demonstrate that it is serious about safety and health. Without this commitment the effort is doomed from the start.

Chapter 3—Being a Part: Workforce Involvement—This chapter delineates the roles and responsibilities of management, supervisors, workers, and the safety and health professional. It discusses how to organize joint labor/management safety and health committees and how the employer can benefit most from workforce involvement.

Chapter 4—Putting It in Writing: A Written Safety and Health Program—It has long been recognized by the safety community that a written safety and health program is necessary to lay the foundation for an effective safety and health effort. The components will be discussed and a model or example of a safety and health program will be provided. The employer will be encouraged to develop a program that meets their company's needs and not be a paper process, but a program that acts as guidelines for everyone in the company.

Chapter 5—Getting Safe Performance: Motivating Safety and Health—This chapter looks at how motivational techniques can assist in getting safe performance and how such techniques can be used to motivate the entire workforce to follow good safety and health practices. This is often the toughest area that employers attempt to deal with. Many studies have indicated that between 85% and 90% of the accidents are caused by unsafe acts (how people behave related to safety and health). The human issues related to safety and health must be addressed using approaches that have been shown to work by other organizations but need to be tailored for a company's/employer's specific situation.

Chapter 6—How They Act: Behavior-Based Safety—Behavior-based safety has become the new kid on the block. Employers have realized that most accidents/incidents have a human side to them and have begun to address workers' behavior in the workplace especially with regard to occupational safety and health. Attempts are being made by employers and programs to elicit safe and healthy working behaviors by ensuring that these behaviors are recognized and reinforced to reduce or annul incidents of unsafe actions.

Chapter 7—Search for the Culprits: Hazard Identification—Each workplace has its own set of hazards. These hazards must be identified and recognized by the workforce. Since hazards are not static but dynamic, hazard identification is an ongoing process, which needs to involve not only management and supervisors, but the entire workforce. There should be some mechanism for reporting hazards without such reporting being viewed in a negative manner. This mechanism for reporting should foster the addressing of the hazards, which exists so that companies and employers can reduce or eliminate the existing hazards. The how-to of hazard recognition and avoidance will be presented.

Chapter 8—Taking a Serious Look: Analyzing Hazards—It is the intent of this chapter to delve into the causes of incidents and how to analyze them in a practical manner so that some action can be taken to intervene or prevent the occurrence of the situations. An understanding of why accident/incidents are occurring is paramount to determining how one would go about addressing them.

Chapter 9—Hurting: Occupational Injuries—The occurrence of injuries is usually inherently different from the onset of occupational illness. Injuries occur in real time with observable outcomes and the cause is usually easily identifiable since the sequence of events is reasonably easy to reconstruct. Most of the injuries are caused by the release of some sort of energy. This chapter will present the common types of energy, which need to be evaluated to prevent injuries from occurring, as well as events and factors that have more subtle causes, but are at least equal or more important as a causal factor of the injuries. A procedure of how best to evaluate and analyze the injuries being seen will be provided.

Chapter 10—Sick: Occupational Illnesses—Occupational illnesses often do not occur in real time, but have a latency period before their manifestation. When illnesses arise where symptoms occur immediately, the acute results can often follow a cause scenario similar to that of an injury. But, when exposures to chemicals, radiation, noise, biological entities, or environmental extremes transpire, effects are often not immediate. The approach to prevention has to be addressed before the event. In this chapter, illnesses and their preventive approaches will be addressed by following a normal industrial hygiene approach to head off possible occupationally related illnesses. The emphasis is on identifying the potential hazards to health and how to best preclude them or protect your workforce from exposure.

Chapter 11—Bent Too Far: Ergonomics—Ergonomic-related problems arise often from poorly designed work areas or processes due to repetitive movements, the amount of force needed to do an activity, or workers placing their bodies in awkward positions. This chapter will emphasize the detection of workplaces that are not designed to fit the worker, as well as the mitigation of such work areas in an effort to head off ergonomic-related injuries. It will look at practical as well as cost-effective ways to prevent the types of injuries and illnesses that are inherent in poor work design and processes.

Chapter 12—Addressing Illnesses: Industrial Hygiene—At times the potential of environmental factors that can cause both long- and short-term health effects needs to be addressed. This is the time when a business cannot change its product or process but still needs to evaluate and determine the best preventive methods. This chapter will explain when an industrial hygienist (IH) is needed and how to work with the IH. It also explains when an industrial hygiene program might be required as well as the components of such a program.

Chapter 13—Taking Action: Interventions, Controls, and Prevention—This chapter approaches the issue of how to take control and eliminate, decrease, or replace the existing hazards with which we are faced. This entails prioritizing the hazards which have been identified and using the most cost-effective control possible, which will gain the greatest risk reduction for the resources that are applied. Many examples of different types of controls will be provided so that the company will be able to select the most relevant to their situation.

Chapter 14—Using the Tools: Accident Prevention Techniques—The occupational safety and health community has provided many tools to assist us in preventing occupational injuries and illnesses. These tools range from audits to the use of consultants. In this chapter, these tools will be presented along with practical examples of how they can be put to use. Some forms will be provided that can be used by the company.

Chapter 15—Who Knows What: Safety and Health Training—Training is a factor in good safety and health performance. If the supervisor does not know how to direct a safety and health effort or the worker cannot perform their work in a safe manner, training then becomes an issue that the company needs to address. Companies need to know how to determine if the lack of training is indeed a culprit in the injuries and illnesses that they are seeing. This chapter will assist the company and employer in determining if they should invest their resources in training. In this chapter, guidance will be provided related to the determination of the amount, kind, source, and length of training which will be most effective to meet the employers' and the workforce's needs.

Chapter 16—Guiding Light: OSHA Compliance—The entity that provides the impetus for occupational safety and health is the OSHA. In recent years, OSHA has tried to address its mandate of development, promulgation, and enforcement of workplace safety and health regulations in a less heavy-handed fashion. In this chapter, a description of how OSHA works and how its representatives can be of help are described. Since the record keeping guidelines are so new (2002), they will be discussed. Guidance will be given on how to make the most effective use of OSHA in support of companies' or employer's safety and health effort.

Chapter 17—Golden Rules: OSHA Regulations—This chapter explains the reason and intent of OSHA regulations. They have not been developed to punish companies and employers, but to address the hazards that are causing, or are likely to cause, serious physical harm or irreparable health problems to those working in the workplace. This chapter and its tables or figures will demonstrate how to determine if a regulation actually applies to a particular workplace and which regulations have no application. It will also guide the company/employer to the location of sources, which will help them comply with the regulations that are applicable to their workplace.

Chapter 18—All Around: Workplace Environmental Issues—The environmental laws and regulations have application to workplaces. In fact, the fines levied by the U.S. Environmental Protection Agency are substantially higher than those from OSHA. Companies need to be aware that they must be responsive to the environmental requirements specified in the laws. This is the reason that many companies have individuals who have titles like director of environmental health and safety (EHS). This chapter considers what the employer needs to do to comply with the environment component in the workplace.

Chapter 19—Keep Me Safe: Workplace Security and Violence—In this chapter, workplace security is discussed as paramount to having a well-functioning workplace. Workers must feel that they are safe from violation and violence if they are going to be as productive as possible. This chapter looks at the needs and issues faced by companies trying to protect their workforce and their property. It discusses the most favorable approaches to assuring a secure work area for the companies' employees. Also, the mechanism for heading off potential violence within the workplace is discussed along with the best practices for identifying and mitigating the potential for violence.

Chapter 20—Mean Ones: Workplace Bullying—This chapter discusses a relatively new topic to occupational safety and health. With the emphasis placed on violence in the workplace, what was once viewed as a personal issue has now become a potentially dangerous practice since either the bully (when trying to exert his/her control or will) on the target (when defending himself/herself) may resort to violence when pushed to the limit. Application of company policies, nontolerance of violence, and action to prevent bullying can go a long way to putting an end to this type of behavior.

Chapter 21—Let us Find a Way: Safety Communications—This chapter may seem to tackle an odd subject, but many times a company makes huge efforts to develop a safety and health approach only to fail to communicate the intent of its efforts. Much of what they are trying to do or have implemented is not conveyed in a positive and factual manner to those who are going to be impacted by it. In this chapter, the topic of communicating the purpose and the roles served in the safety and health initiative is discussed along with ideas, processes, and procedures to assure that the communications that are being disseminated are effective and relevant to achieving a safe and healthy workplace.

Chapter 22—Everything Costs Dollars: Budgeting for Safety and Health—In this chapter, the emphasis is on having a separate budget for the company's occupational safety and health initiative. This means that a formal budget must be developed to fund personnel, equipments, unforeseen events, and program needs. This budget should be linked to the achieving of realistic measurable safety and health goals. Safety and health should be budgeted for in the same manner as any other company component that requires financial support.

Chapter 23—All's Well That Ends Well: Summary—This chapter summarizes and makes concluding remarks regarding the development of a safety and health effort by companies/employers and their workforces.

Occupational safety and health should be an integral part of the way you do business. You will be surprised how you will be viewed by the community at large as a positive place to work. Even your insurance company that is charged with covering any losses you suffer and those with whom you do business will view you as progressive. After all, it is in their best interest to be affiliated with a company that is concerned with not only its workforce but leads by example. Supervisors whose efforts to provide a safe work environment you support and your workforce who advertise and support your business will be willing to laud your business. You have taken their well-being to be important and that demonstrates that they are an important asset to your company.

Granted everything may not work perfectly and problems will arise. After all, we do not live in a perfect world and neither are the people we work with perfect employees. The problems that you face can be solved and are less severe if you have taken time to implement and plan, whether it is occupational safety and health or any other aspect of your business.

CHAPTER 2

Making a Commitment: Management's Commitment and Involvement

Workplace efforts require management support and involvement.

Before reading this chapter, you should evaluate management's commitment by circling the number that best represents the state of your safety and health initiative at this time.

MANAGEMENT'S COMMITMENT AND INVOLVEMENT

The term commitment is really directed at management since it is solely management's responsibility to provide a safe and healthy workplace for its employees. When occupational injuries and illnesses occur, they are considered to be failings within the management system. Management sets the tone for safety and health within the workplace. It does so by demonstrating a commitment. The first item I look for is a safety and health policy, which expresses the position of the company in relationship to safety and health. This policy must be signed by the president, CEO, or similar official, not the safety director. Your safety and health program must start at the top.

A safety and health policy statement clarifies the policy, standardizes safety within the company, provides support for safety, and supports the enforcement of safety and health within the

Management Leadership

Topic	Circle Answer	Answer Options
Clear worksite safety and health policy	5	We have a safety and health policy and all employees accept it, can explain it, and understand it
	4	We have a safety and health policy and a majority of employees can explain it
	3	We have a safety and health policy and some employees can explain it
	2	We have a written (or oral, where appropriate) policy
	1	We have no policy
Clear goals and desired objectives are set and communicated	5	All employees are involved in developing goals and can explain results and how results are measured
	4	Majority of employees can explain results and measures for achieving them
	3	Some employees can explain results and measures for achieving them
	2	We have written (or oral, where appropriate) goals and objectives
	1	We have no safety and health goals and objectives
Management leadership	5	All employees can give examples of management's commitment to safety and health
	4	Majority of employees can give examples of management's active commitment to safety and health
	3	Some employees can give examples of management's commitment to safety and health
	2	Some evidence exists that top management is committed to safety and health
	1	Safety and health are not top management values or concerns
Management example	5	All employees recognize that managers in this company always follow the rules and address the safety behavior of others
	4	Managers follow the rules and usually address the safety behavior of others
	3	Managers follow the rules and occasionally address the safety behavior of others
	2	Managers generally follow basic safety and health rules
	1	Managers do not follow basic safety and health rules
Assigned safety and health responsibilities	5	All employees can explain what performance is expected of them
	4	Majority of employees can explain what performance is expected of them
	3	Some employees can explain what performance is expected of them
	2	Performance expectations are generally spelled out for all employees
	1	Specific job responsibilities and performance expectations are generally unknown or hard to find
Authority and resources for safety and health	5	All employees believe they have the necessary authority and resources to meet their responsibilities
	4	Majority of employees believe they have the necessary authority and resources to meet their responsibilities
	3	Authority and resources are spelled out for all, but there is often a reluctance to use them
	2	Authority and resources exist, but most are controlled by supervisors
	1	All authority and resources come from supervision and are not delegated

Management Leadership (continued)

Topic	Circle Answer	Answer Options
Accountability	5	Employees are held accountable and all performance is addressed with appropriate consequences
	4	Accountability systems are in place, but consequences used tend to be for negative performance only
	3	Employees are generally held accountable, but consequences and rewards do not always follow performance
	2	There is some accountability, but it is generally hit or miss
	1	There is no effort toward accountability
Program review (quality assurance)	5	In addition to a comprehensive review, a process that drives continuous correction is used
	4	A comprehensive review is conducted at least annually and drives appropriate program modifications
	3	A program review is conducted, but it does not drive all necessary program changes
	2	Changes in programs are driven by events such as accidents or near-misses
	1	There is no program review process

company. It should set forth the purpose and philosophy of the company, delineate the program's goal, assign responsibilities for all company personnel, and be positive in nature. It should be as brief as humanly possible (see Figure 2.1).

It is expected that the top management will sign off on the company's safety and health program. This does not mean that they developed it, but that they support it. This is another example of commitment.

To all employees:

(Name of the Company) has been in business for over _____ years. The company prides itself on the fact that the safety and health of our employees is its first priority.

It has always been our policy to provide a safe and healthy environment for all employees at our facilities. We expect every employee to uphold the standards of the Occupational Safety and Health Act (OSHAct) and the safety and health measures of the company. No priority is to be placed above safety and health at any time.

Preventive measures and the elimination of any potential hazard are of utmost importance for the safety and health of all employees, visitors, and the public in general. It is everyone's responsibility to report any hazard observed to the proper person for correction.

Our safety director, supervisors, and the safety inspectors have the responsibility to immediately report to the managers, superintendents, and senior management any potential hazardous conditions. The superintendent and manager are responsible for immediate actions to avoid injury or illness from the hazard.

It is the responsibility of the corporate safety director to periodically report to senior management the safety and health status of all operations. It is the responsibility of every employee to support and assist in establishing safety and health measures. The safety and health program will be implemented and reviewed annually by the management.

(Signature)
President

Figure 2.1 Example of a safety and health policy statement.

Top management determines to whom safety and health professionals report. It has always been my contention that the higher the better since safety and health often need some teeth when dealing with line management. The president of the company would be ideal. Of course, everyone realizes that due to size or other constraints, this is not always possible.

At one company as I sat with the safety director trying to solve an existing accident problem for the company, he answered the telephone saying personnel. This told me at once what the safety director thought his job was and also gave me an indication as to why they were having problems. The company did not have a true commitment to occupational safety and health. Commitment to occupational safety and health was only a necessary evil to this company. They had no direction in their program.

Management helps to set the goals and objectives of the program and then signs off on them. You cannot accomplish anything in a positive fashion without attainable goals. Goals are the target. You have been asked by others, "What are your goals?" If you cannot answer, others will view you as nonmotivated or at best, lost. If you have not set goals, how will your workforce know what is expected of them in relationship to occupational safety and health? You would not go through a year without setting some production goals, would you?

Goals and objectives are very important and should be directly observable and measurable. They should be reasonable and attainable. The following are some examples of goals and objectives and the issues faced by those using them:

- Ensure zero fatalities or serious injuries. (This is usually a pie in the sky or unreachable goal for most employers. For example, if you had 25 accidents last year, zero is probably not possible).
- Reduce injuries, lost workday accidents, and workers' compensation claims by_____%.
- Prevent damage or destruction to company property or equipment.
- Increase productivity through reduction of injuries by _____%.
- Reduce workers' compensation costs by decreasing the number of claims to _____ or the cost by _____%.
- Enhance company's image by working safely. Can you measure this in some way?
- Keep safety a paramount part of workers' daily activities. What are indicators of this? (They are the number of near-misses, reports of hazards, or the number of observable unsafe acts.)
- Recognize and reward safe work practices. How is this part of a goal? What could be the measurable outcome of this objective?

Management should develop and implement a set of safety and health rules and policies. Since these rules are developed for the good and welfare of everyone, everyone should obey and follow them at all times, even when it is not convenient. A manager should never enter a work area that requires protective eyewear without the required eyewear. No special favors should exist. After all, the manager is a direct reflection of management's commitment to the company's occupational safety and health. Managers are role models who must emulate the company's philosophy and commitment to the safety and health program, to which the top manager has given his support.

You cannot expect a program to run without financial assistance. Thus, it is management's responsibility to support safety and health with an adequate budget. Safety and health should have their own budgetary resources. Actually, safety and health should be managed like any other component of the company whether it is research, development, or production.

Management commitment should include "tough love." This may sound silly to you but what I mean is that if management is committed to enforcing its safety and health program, rules and regulations, and attaining its goals, then it must have some form of a disciplinary policy and procedure that results in negative consequences for failure to follow or abide by the company's safety and health expectations. This is not just for the workforce but should be evenly applied to all of management and the line supervisors, for any events or times of noncompliance.

ROLES AND RESPONSIBILITIES

Everyone is responsible for accident prevention. This statement also means that no one in particular is accountable or responsible. Four entities must be accountable and responsible for accident prevention:

1. All levels of management must demonstrate their commitment to the company's policies and procedures regarding safety and health by their presence, visibility, actions, adherence, and behavior.
2. Person who, by background or experience, has been assigned responsibility and therefore assigned accountability to ensure that the company's safety and health program is adhered to.
3. Supervisor, who models the company's safety personality and is the liaison between management and the worker relevant to the implementation of safety, must be held both responsible and accountable for safety in his/her work area.
4. Employees are responsible for abiding by the company's rules and policies and are accountable for their own behavior, safe or unsafe.

Each of these entities must understand both their responsibilities and accountabilities regarding the safety and health policies and procedures of the company. The sum total of their actual actions and behavior makes up the outward manifestation of the company's culture.

Management

The accountability for workplace safety and health falls on the management. Thus, those who are held accountable are the ones who are responsible for all aspects of the safety and health program. Occupational Safety and Health Administration (OSHA) does not cite workers or the safety and health professional. It cites the company, and the representatives of the company are the management team.

Safety and Health Professional

The individual who is assigned the ultimate responsibility for safety is the safety and health professional. This may be an individual with academic training in safety and health, or an individual who has both experience and understanding of the specific hazards that exist in the company's workplace. This individual may be called the safety director, coordinator, or person. No matter the title, his/her responsibilities are varied and wide-ranging. Some possible performance expectations may be as follows:

- Establishing programs for detecting, correcting, or controlling hazardous conditions, toxic environments, and health hazards.
- Ensuring that proper safeguards and personal protective equipment are available, properly maintained, and properly used.
- Establishing safety procedures for employees, plant design, plant layout, vendors, outside contractors, and visitors.
- Establishing safety procedures for purchasing and installing new equipment and for purchasing and safe storing of hazardous materials.
- Maintaining an accident recording system to measure the organization's safety performance.
- Staying abreast of, and advising management on, the current federal, state, and local laws, codes, and standards related to safety and health in the workplace.
- Carrying out the company's safety obligations as required by law or union contract.
- Conducting investigations of accidents, near-misses, and property damage, and preparing reports with recommended corrective action.

- Conducting safety training for all levels of management, new hires, and current employees.
- Assisting in the formation of both management and union/management safety committees (department heads and superintendents) and attending monthly departmental safety committee meetings.
- Keeping him/herself informed on the latest developments in the field of safety, such as personal protective equipment, new safety standards, workers' compensation legislation, and new literature pertaining to safety, as well as attending safety seminars and conventions.
- Maintaining liaison with national, state, and local safety organizations and taking an active role in the activities of such groups.
- Accompanying OSHA compliance officers during plant inspections and insurance safety professionals on audits and plant surveys. The safety engineer further reviews reports related to these activities with management and initiates action for necessary corrections.
- Distributing the organization's statement of the safety and health policy as outlined in its organizational manual.

If some facets of the safety effort do not proceed well, this individual will usually be held accountable even though he or she may not have the authority to rectify the existing problem. Usually the safety person comes from a staff position, which seldom allows him or her to interfere in any way with the line function of production. Without some authority to influence line function when necessary, the safety professional has little clout as to worksite implementation of the company's safety and health effort. Accountability must go beyond the safety professional. Dan Petersen, a noted safety expert, espouses that what is desired is safe production.

First-Line Supervisor

The line or first-line supervisor is the key component in a working safety and health program. No other person has as much personal contact on a day-to-day basis with the workers. He or she is the communicator of the company's policies and procedures. No one else models safety and health quite like the line supervisor. He or she is the example from which workers draw their behavior relevant to job safety and health. The first-line supervisor sets the tone for the workplace and is the role model for the company by conveying, implementing, supporting, and enforcing all the company's policies and procedures from production to safety.

Just think of all that the first-line supervisor does:

- Hires new employees
- Suspends or discharges
- Reports on probationary employees
- Prepares work schedules
- Trains new employees
- Delegates work to others
- Holds safety meetings
- Prepares vacation schedules
- Coaches employees on the job
- Grants leaves of absence
- Controls quality and quantity
- Lay off others for lack of work
- Stops a job in progress
- Processes grievances
- Takes unsafe tools out of production
- Authorizes maintenance and repairs
- Investigates accidents
- Makes suggestions for improvement

- Inspects the work area
- Discusses problems with management
- Corrects unsafe conditions and unsafe acts
- Reduces waste
- Recommends promotions or demotions
- Prepares budgets
- Transfers employees in and out of the work area
- Approves expenditures
- Fosters employee morale
- Grants pay raises
- Motivates workers
- Issues warnings and administers discipline
- Reduces turnover

Is it any wonder that the success or failure of the safety program is dependent on the first-line supervisor? Certainly, everyone would acknowledge that the first-line supervisor is responsible for safety within the work area but seldom is the first-line supervisor evaluated on his or her safety performance in the same manner as production performance. Until each supervisor is held as accountable for safety as for production, with equal consequences for poor safety performance as for poor production performance, safety will never be a priority with supervisors. Until then, the value the supervisor places on safety will always be far less than the value placed on production. Using a separate evaluation form for each supervisor's safety performance, which compares the safety performance records of one supervisor to another, may make a significant difference.

Worker

Although workers do not have the control that management has over their workplace, they are still responsible for complying with the company's safety and health policies and procedures. Some of the commonly accepted worker responsibilities are to

1. Comply with OSHA regulations and standards
2. Not remove, displace, or interfere with the use of any safeguards
3. Comply with the employer's safety and health rules and policies
4. Report any hazardous conditions to the supervisor or employer
5. Report any job-related injuries and illnesses to the supervisor or employer
6. Report near-miss incidents to the supervisor or employer
7. Cooperate with the OSHA inspector during inspections when requested to do so
8. Report to work on time
9. Wear suitable work clothes
10. Observe good personal hygiene
11. Never sleep, gamble, horseplay, fight, steal, bring fireworks or firearms on the job, or face grounds for immediate dismissal
12. Never use or be under the influence of alcohol, narcotics, or other drugs or intoxicants while on the job
13. Wear personal protective equipment as prescribed for each task
14. Maintain order since housekeeping is everyone's responsibility
15. Observe Danger, Warning, Caution, and No Smoking signs and notices
16. Use and handle equipment, material, and safety devices with care
17. Never leave discharged fire extinguishers in the work areas
18. Never expose yourself to dangerous conditions or actions
19. Never operate any equipment for which training has not been provided
20. Participate in all safety and health training provided
21. Attend all safety and toolbox meetings (mandatory)

A safety and health program is a management function that requires the management's participation in planning, setting objectives, organizing, directing, and controlling the program. The management's commitment to safety and health is an integral part of every decision the company makes and every action this company takes. Therefore, the management of (Name of Company) assumes total responsibility for implementing and ensuring the effectiveness of this safety and health program. The best evidence of our company's commitment to safety and health is this written program, which will be fully implemented at each of the company's facilities.

(Name of Person and Title) is assigned the overall responsibility and authority for implementing this safety and health program. (Company Name) fully supports (Name of Person and Title) and will provide the necessary resources (budget, etc.) and leadership to ensure the effectiveness of this safety and health program.

Figure 2.2 Example of a statement of responsibility.

RESPONSIBILITY

If everybody is responsible for workplace safety and health, then nobody in particular is responsible, and someone needs to be in charge and be assigned the responsibility for the direction and implementation of workplace safety and health. This in no way relieves management of its responsibility or commitment. In addition, this does not relieve supervisors and employees of their responsibility to enforce and adhere to the company's safety and health requirements, rules, policies, or procedures. A statement should be placed in the safety and health program assigning responsibility to some individual (see Figure 2.2).

Employee Responsibility

In dealing with employee responsibility, the expectation of employees should be spelled out as part of the safety and health program. Figure 2.3 is an example of an employee responsibility statement.

Safety and health are the management's responsibility; however, each employee is expected, as a condition of employment, to work in a manner that will not inflict self-injury or cause injury to fellow workers. Each employee must understand that responsibility for his or her own safety is an integral job requirement. Each employee of (Name of Company) will:

- Observe and comply with all safety rules and regulations that apply to his or her trade
- Follow instructions and ask questions of his or her supervisor when in doubt about any phase of his or her operation
- Report all unsafe conditions or situations that are potentially hazardous
- Report all on-the-job accidents and injuries to his or her supervisor immediately
- Report all equipment damage to his or her supervisor immediately
- Help to maintain a safe and clean work area

Figure 2.3 Employee responsibility statement.

The company expects that all workers, including the management, will adhere to the company's safety and health rules as well as applicable state and federal regulations. Whenever a violation of safety rules occurs, the following enforcement policy will be implemented:

First offense—Verbal warning and proper instruction pertaining to the specific safety or health violation will be given the employee. (A notation of the violation may be made and placed in the employee's personnel file.)
Second offense—Written warning with a copy placed in the employee's personnel file.
Third offense—Dismissal from employment *.

*The company reserves the right to terminate immediately any employees who flagrantly endanger themselves or others by their unsafe actions while on (Name of Company) worksites.

Figure 2.4   Safety and health discipline policy.

DISCIPLINE

It goes without saying that unless there are consequences for not performing in a safe and healthy manner you cannot expect supervisors, workers, or even managers to adhere to your intent to have a workplace free from hazards that can cause injury or make your workforce sick. To ensure that you get the safety and health performance that you desire, you need to have a safety and health discipline policy that you strictly enforce (see Figure 2.4). Failure to enforce it in an even-handed and constant manner will result in the failure of your safety and health effort. Some companies place another step in their discipline policy, which results in time off without pay for a certain number of days followed by dismissal on the next offense.

MANAGEMENT ACCOUNTABILITY

All management personnel are expected to follow and enforce the safety and health policies and procedures the company has set forth. They should be held both physically and financially accountable for their safety and health practice performance as well as the performance of those they supervise. Figure 2.5 shows a management accountability statement.

Active participation in and support of safety and health programs are essential. Therefore, all management officials of the (Name of Company) will display their interest in safety and health matters at every opportunity. At least one manager (as designated) will participate in each facility's safety and health meetings, accident investigations, and jobsite inspections. All management personnel are expected to follow the job safety and health rules and enforce them equally. Management personnel's safety performance will comprise a significant portion of their annual merit evaluations.

Figure 2.5   Management accountability statement.

Standard safety and health merit evaluation

Supervisor's name _____

Safety record _____

Setting a good example _____

Compliance with OSHA standards _____

General safety attitude _____

Housekeeping _____

Prompt correction of hazards _____

Accident reporting _____

Hard hat compliance _____

Safety eye wear compliance _____

Injury rate for crew _____

Severity rate for crew _____

Degree of meeting safety goals _____

Safety meeting and toolbox talks _____

Other protective equipment _____

Composite score _____

Previous composite score _____

Change + or – _____

Ranking _____

Scoring: 1 = Excellent
        2 = Above average
        3 = Average
        4 = Fair
        5 = Poor (needs improvement)

Figure 2.6   Standard safety and health merit evaluation.

SUPERVISORY ACCOUNTABILITY

The first-line supervisor is the key to a good occupational safety and health effort. The supervisor must be integrally involved in the safety and health effort. Thus, his or her participation must be mandatory.

The supervisors are to be held accountable for their safety performance and the performance of their crews. A standard safety and health merit evaluation (SSHME) form should be developed for evaluating all supervisory personnel's safety and health performance (see Figure 2.6).

CHAPTER 3

Being a Part: Workforce Involvement

Employee involvement is paramount to an effective safety and health effort.

Before reading this chapter, you should evaluate employee involvement by circling the number that best represents the state of your safety and health initiative at this time.

Employment Involvement

Topic	Circle Answer	Answer Options
Employee involvement	5	All employees have ownership of safety and health and can explain their roles
	4	Majority of employees feel they have a positive impact on identifying and resolving safety and health issues
	3	Some employees feel that they have a positive impact on safety and health
	2	Employees generally feel that their safety and health input will be considered by supervisors
	1	Employee involvement in safety and health issues is neither encouraged nor rewarded

INVOLVEMENT

Management has to make the commitment to involve the workforce in safety and health at the worksite. Since management is in control, it is their prerogative to decide the extent of involvement, depending on what they can endure or on the amount of control they can forgo. However, there are positive outcomes from employees' involvement. It is a real way to gain commitment to safety and health. Safety and health become personal issues for your workforce.

You may wonder why you should involve your workforce. It is the smart thing to do. Some of the reasons for this are

- Your workforce is most in contact with potential safety and health hazards. It has a vested interest in having a safe and healthy workplace.
- Group decisions have the advantage of making use of the wide range of experience each member brings to the table.
- Your workforce is more likely to support and participate in an effort in which it has been involved.

Involvement will only work where employees are encouraged to offer their ideas but foremost their ideas or contributions must be taken seriously. This will result in greater satisfaction and workers who are more productive. There are many ways to elicit employee participation. They are as follows:

- Participating in joint labor/management (L/M) committees or advisory workgroups
- Assisting in conducting inspections
- Identifying/analyzing the common hazards in each step of the job, or job process, and preparing safe work practices or controls to eliminate or reduce the exposure
- Helping in developing or revising the site safety and health rules
- Assisting in training both newly hired and current employees
- Providing programs or presentations at safety and health meetings
- Participating with the team doing accident/incident investigations
- Reporting hazards
- Fixing hazards within their abilities and scope of control
- Supporting their fellow workers by providing feedback on risks and assisting to eliminate hazards
- Conducting a preuse or change analysis for new equipment or processes to determine existing hazards before actual use

Employees want to feel that they have some control over their work lives and safety and health participation is one way to accomplish this. Everyone wants to belong and feel wanted. Employee involvement in safety and health can help accomplish this.

JOINT LABOR/MANAGEMENT SAFETY AND HEALTH COMMITTEES

A joint L/M committee is a formal committee, which is organized to address specific issues such as safety and health or production processes. It is a committee with equal representation, which gives both parties an opportunity to talk directly to each other and educate each other concerning the problems faced by either group. In contrast, labor or management only committees have self-serving goals with no consensus on solving problems. They are the only ones who have the authority or power to make changes. Thus, joint committees are aimed at gaining solutions, ensuring equal participation, and having some degree of authority or power.

Joint L/M committees have a different purpose from committees set up by either labor or management alone. When compared with other committees existing at the worksite, L/M committees are different both in their goals and in methods of operation. In addition, because of the nature of their goals, they are also much more challenging since they require many different skills from all the participants.

Joint committees provide both parties with the opportunity and structure to discuss a wide range of issues challenging them. Neither partner of the L/M committee has enough information, commitment, or power to institute the changes that the joint committee eventually identifies as critical to the success of the business.

Thus, a key purpose of joint committees is to gather, review, analyze, and solve problems that are critical to the success of the business and are not appropriate for the collective bargaining process. Another purpose of these committees is the formation of a level playing field, which has as its ultimate purpose the success of the business. These committees can help build bridges of cooperation, which can lead to increased productivity, quality, efficiency, safety, health, and economic gains shared by all parties.

However, the purpose of joint L/M committees goes beyond quality and productivity. They are builders of true and honest relationships, which help to realize success through focusing on outcomes, using resources more efficiently, fostering real-world flexibility, supporting an information sharing system, opening communications, and fostering a better working relationship.

In the past, L/M relationships were built on confrontation, distrust, acrimony, and the perception of loss or gain of control and power. The ultimate goal, to use an overused phrase, is to attain a win–win situation.

With all the downsizing, right sizing, and reengineering going on within the workplace, individuals believe that they can help and have an impact on their continued employment and the survival of the company, if only asked and given the chance. To do this successfully, it is imperative that they have access to the information needed to solve issues facing them and their employer.

Committee Makeup

The joint L/M committee should be composed of at least as many employee members as employer members. Labor must have the sole right to appoint or select its own representatives, just as employers have the right to appoint theirs. Both parties should clearly understand that the members of this committee must not only be risk-takers, but also be fully capable of making the critical decisions needed to make this process successful.

To cover all facets of the workplace, the labor organization may find it useful to have a broad spectrum of their membership represented. Labor should also allow for turnover in its membership and address this issue by identifying and involving adjunct and alternate members. By doing this you will not compromise the committee's progress by introducing new members into the committee who are unfamiliar with the process and are untrained concerning the subject matter.

The chairperson must be elected by the committee and this position should be a rotating position between labor and management. Each committee member should receive training on the joint committee process and receive other specific instructions that are deemed necessary, such as job-related safety and health training. The labor members should be paid for all committee duties, including attendance at meetings, inspections, training sessions, etc.

Record Keeping

Each participating party (labor and management) should keep its own notes of all meetings and inspections, as well as copies of agendas. This will ensure that agreements and disagreements, time schedules, actions to be taken, etc., are not lost, forgotten, neglected, or misinterpreted. Time has a way of encouraging each one of these things to happen. Good record keeping will also assist in keeping the direction and undertakings of the joint committee in focus. The agenda could include the following:

- Date and time for the meeting to begin and a projected time for adjournment
- To's and from's—who issued the agenda and to whom it is sent

- Location
- Review of any audits, assessments, unsolved problems, or inspections
- Old business
- New business

A formal set of minutes and reports on inspections should be maintained by the joint committee. Minutes should include

1. Employer's name and union for identifiers
2. Date and time of meeting
3. Chairperson
4. Members in attendance
5. Old business
6. Actions taken and dates completed since last meeting
7. New business
8. New actions and proposed dates of completion
9. Other business
   a. Outside or Occupational Safety and Health Administration (OSHA) inspections
   b. Injury and illness incidents
   c. Educational initiatives
   d. Administrative activities
10. Joint representatives' signatures giving approval of minutes

Do's and Don'ts of L/M Committees

Do's:

- Always give an agenda to committee members in advance of a meeting; this allows everyone time for preparation.
- Cancel a meeting only for emergencies; hold meetings on schedule.
- Set timelines for solving problems.
- Focus on the issues involved.
- Do stay on schedule and stick to the starting and ending times in the agenda.
- Decide on a structured approach to recording and drafting minutes, as well as mechanisms for disseminating them.
- Keep the broader workforce informed of the activities of the committee.
- Keep issues on the agenda until they are resolved to everyone's satisfaction.
- Give worker representatives time to meet as a group and prepare for the meeting.
- Be on time for the meetings.
- Make sure that everyone understands the issues and problems to be discussed.

Don'ts:

- Tackle the most difficult problems first since some early successes will build a stronger foundation.
- Work on broadly defined issues, but deal with specific problems and concrete corrective actions.
- Allow the meeting to be a gripe session when problem solving is the result.
- Allow any issue to be viewed as trivial; each issue is important to someone.
- Let individual personalities interfere with neither the meetings nor the intent of the committee.
- Be a know-it-all and assume you know the answer; give everyone an opportunity to participate in solving the problem.
- Neglect to get all the facts before trying to solve an issue or problem.
- Prolong meetings.

- Delay conveying and communicating the solutions to problems and the outcomes or accomplishments that the committee has achieved.
- Expect miraculous successes or results immediately since many of the problems and issues did not occur overnight.

Organizing a Joint Committee

When organizing a joint committee, some specifics should be set forth:

- Set up the ground rules or procedural process
- Have a set place to meet
- Establish, as a group, the goals, objectives, function, and mission of the committee
- Select the frequency of the meetings—at a minimum, monthly—and set parameters for the length of meetings
- Agree to maintain and post minutes of all meetings

Expectations

Anytime new endeavors are undertaken, such as joint L/M committees, there are expectations that accompany them. Some of the expectations are

- Improved workplaces and work environment
- Improved working relationships
- Positive, cooperative approaches
- A compromise of mutual interests, versus self-serving interests
- A true team approach
- Sharing of information, thinking, and substantive decision making
- New/fresh ideas
- Increased participation and involvement

Outcomes

A study done by the Work in America Institute lists some of the outcomes you can expect when you have a functional joint L/M committee. According to this study, both labor and management stand to benefit from joint undertakings. Some of the benefits include

- Economic gains: higher profits, less cost overruns, increased productivity, better quality, greater customer satisfaction, and fewer injuries and illnesses. Working together, workers and supervisors can solve problems, improve product quality, and streamline work processes.
- Improved worker capacities, which more effectively contribute to the improvement of the workplace.
- Human resource benefits.
- Innovations at the bargaining table.
- Committee member growth.
- Workplace democracy.
- Employment security.
- Positive perceptions.

Other outcomes that will arise from joint L/M committees in all likelihood are

- Shared responsibilities
- Increased individual involvement

- Company and labor proactive with each other
- Better communication between company and labor
- Employee ownership of ideas, goals, activities, outcomes, and the company
- Union leadership and members more challenged

Joint L/M Occupational Safety and Health Committees

A joint L/M occupational safety and health committee is a specialized application of the joint L/M committee and is an excellent format, which others can replicate. This type of committee is organized to address specific workplace issues such as the following:

- Monitor safety and health programs.
- Inspect the workplace to identify hazards.
- Conduct and review accident investigations.
- Recommend interventions and prevention initiatives.
- Review injury and illness data for incident trends.
- Act as a sounding board for workers who express health and safety concerns.
- Become involved in designing and planning for a safe and healthy workplace.
- Make recommendations to the company regarding actions, solutions, and program needs for safety and health.
- Participate in and observe workplace exposure monitoring and medical surveillance programs.
- Assure that training and education fully address safety and health issues facing the workplace.

The goals of the joint L/M occupational safety and health committees are:

- To reduce accidents, through a cooperative effort, by eliminating as many workplace hazards as possible
- To reduce the number of safety- and health-related complaints filed with regulatory agencies without infringing on the workers' federal and state rights
- To promote worker participation in all safety and health programs
- To promote training in the areas of recognition, avoidance, and the prevention of occupational hazards
- To establish another line of communication whereby the workers can voice their concerns regarding potential hazards and then receive feedback on the status or action taken

Summary

What can joint L/M committees accomplish?

- Increased commitment to achieving the organization's goals or mission.
- Improved productivity, safety and health, customer service, and product quality.
- Joint resolution of problems and issues facing the organization.
- Shared responsibility toward and accountability for results and outcomes.
- Better and more constructive relationship between labor and management.
- Enhanced employee morale and job satisfaction.
- Heightened communication and information sharing, which brings all employees into the decision-making process. This helps them understand the mission, goals, and objectives of the organization and fosters employee support of the organization's undertakings.
- Increased job security and compensation.

To make joint L/M committees work, certain actions must occur and certain procedures must be followed:

- Ensure that upper management supports the joint effort and that this is conveyed both to the union and other company representatives.
- Acknowledge that reservations exist on both sides and try to gradually build trust.
- Keep the committee focused on its goals and mission.
- Strive for a good balance of employee and management representatives who are willing to invest in the process.
- Keep the committee structured; do not allow it to turn into a bull session.
- Remember that the committee is designed to serve all workplace constituencies, not just workers and management.
- Assure that committee leadership is elected or selected by consensus to fill various roles.
- Make decisions fairly and use the consensus process.
- Know and work within the guidelines of federal and state regulations.
- Do not raise issues that really must be addressed at the collective bargaining table; it will only undermine the viability and success of this process.

POLICY ESTABLISHING JOINT COMMITTEES

This partnership can only succeed when problems are identified, goals exist, priorities are set forth, and trust is the foundation upon which it is built. Thus, everyone must be willing to work on two-way communications with a forthright exchange of information. Joint solutions and real action will become the visible products of these joint committees. A sample written statement for your safety and health program related to joint safety and health committees can be found in Figure 3.1.

(Name of Company) jobsite will establish a safety and health committee to assist with the implementation of this program and the control of identified hazards. The safety and health committee will be composed of employees and management representatives. The committee should meet regularly but not less than once a month. Written minutes from safety and health committee meetings will be available and posted on the project bulletin board for all employees to see.

The safety and health committee will participate in periodic inspections to review the effectiveness of the safety program and make recommendations for improvement of unsafe and unhealthy conditions. This committee will be responsible for monitoring the effectiveness of this program. The committee will review safety inspection and accident investigation reports and, where appropriate, submit suggestions for further action.

Figure 3.1   Sample statement for safety and health committees.

CHAPTER 4

Put It in Writing: A Written Safety and Health Program

Developing the written safety and health program.

The need for written safety and health programs has been an area of controversy for some time. Many in the industry feel that written safety and health programs are just more paperwork, a deterrent to productivity, and nothing more than another bureaucratic way of mandating safety and health on the job. However, over a period of years, data and information have been mounting in support of the need to develop and implement written safety and health programs for all workplaces.

This perceived need for written programs must be tempered with a view to their practical development and implementation. A very small employer who employs one to four employees and no supervisors in all likelihood needs only a very basic written plan, along with any other written programs that are required as part of an Occupational Safety and Health Administration (OSHA) regulation. However, as the size of the company and the number of employees increase, the employer becomes more removed from the hands-on aspects of what now may be multiple facilities or worksites.

Now companies must find a way to convey support for safety to all those who work with and for the company. As with all other aspects of business, the employer must plan, set the policies, apply management principles, and assure adherence to the company's goals to facilitate efficient and effective completion of projects. Again, job safety and health should be managed the same as any other part of the company's business.

In other words, to effectively manage safety and health, a company must pay attention to some critical factors. These factors are the essence in managing safety and health on worksites. The questions that need to be answered regarding managing safety and health are

1. What is the policy of the employer regarding safety and health on his or her projects?
2. What are the safety and health goals for the company?
3. Who is responsible for occupational safety and health?
4. How are supervisors and employees held accountable for job safety and health?
5. What are the safety and health rules for their type of industry or work?
6. What are the consequences of not following the safety rules?
7. Are there set procedures for addressing safety and health issues that arise in the workplace?
8. How are hazards identified?
9. How are hazards controlled or prevented?
10. What type of safety and health training occurs? And, who is trained?

Specific actions can be taken to address each of the previous questions. The written safety and health program is of primary importance in addressing these items. Have you ever wondered how your company is doing in comparison with a company without a safety and health professional and a viable safety and health program? Well, wonder no more.

In research conducted by the Lincoln Nebraska Safety Council in 1981, the following conclusions were reached based on a comparison of responses from a survey of 143 national companies. All conclusions have a 95% confidence level or more (see Table 4.1). Even though this study has not been replicated in a number of years, it is most likely still relevant and a good picture of reality for companies today.

It seems apparent from this research that to have an effective safety program, at a minimum, an employer must satisfy the following conditions:

• Have a demonstrated commitment to job safety and health
• Commit budgetary resources
• Train new personnel

Table 4.1   Effectiveness of Safety and Health Program Findings

Fact	Statement	Findings
1	Do not have separate budget for safety	43% more accidents
2	No training for new hires	52% more accidents
3	No outside sources for safety training	59% more accidents
4	No specific training for supervisors	62% more accidents
5	Do not conduct safety inspections	40% more accidents
6	No written safety program compared with companies that have written programs	106% more accidents
7	Those using canned programs, not self-generated	43% more accidents
8	No written safety program	130% more accidents
9	No employee safety committees	74% more accidents
10	No membership in professional safety organizations	64% more accidents
11	No established system to recognize safety accomplishments	81% more accidents
12	Did not document/review accident reports and reviewers did not have safety as part of their job responsibility	122% more accidents
13	Did not hold supervisor accountable for safety through merit salary reviews	39% more accidents
14	Top management did not actively promote safety awareness	470% more accidents

- Ensure that supervisors are trained
- Have a written safety and health program
- Hold supervisors accountable for safety and health
- Respond to safety complaints and investigate accidents
- Conduct safety audits

Other refinements, which will help in reducing workplace injuries and illnesses, can always be part of the safety and health program. They are as follows: (1) more worker involvement (e.g., joint labor/management committees); (2) incentive or recognition programs; (3) getting outside help from a consultant or safety association; and (4) setting safety and health goals.

A decrease in occupational incidents that result in injury, illness, or damage to property is enough reason to develop and implement a written safety and health program.

REASONS FOR A COMPREHENSIVE SAFETY PROGRAM

The three major considerations involved in the development of a safety program are

1. Humanitarian—Safe operation of workplaces is a moral obligation imposed by modern society. This obligation includes consideration for loss of life, human pain and suffering, family suffering and hardships, etc.
2. Legal obligation—Federal and state governments have laws charging the employer with the responsibility for safe working conditions and adequate supervision of work practices. Employers are also responsible for paying the costs incurred for injuries suffered by their employees during their work activities.
3. Economic—Prevention costs less than accidents. This fact has been proven consistently by the experience of thousands of industrial operations. The direct cost is represented by medical care, compensation, etc. The indirect cost of 4 to 10 times the direct cost must be calculated, as well as the loss of wages to employees and the reflection of these losses on the entire community.

All three of these are good reasons to have a health and safety program. It is also important that these programs be formalized in writing, since a written program sets the foundation and provides a consistent approach to occupational health and safety for the company. There are other logical reasons for a written safety and health program. Some of them are

- It provides standard directions, policies, and procedures for all company personnel.
- It states specifics regarding safety and health and clarifies misconceptions.
- It delineates the goals and objectives for workplace safety and health.
- It forces the company to define its view of safety and health.
- It sets out in black and white the rules and procedures for safety and health that everyone in the company must follow.
- It is a plan that shows how all aspects of the company's safety and health initiative work together.
- It is a primary tool for communicating the standards set by the company regarding safety and health.

Written safety and health programs have a real place in modern safety and health practices not to mention the potential benefits. If a decrease in occupational incidents that result in injury, illness, or damage to property is not reason enough to develop and implement a written safety and health program, the other benefits from having a formal safety and health program seem well worth the investment of time and resources. Some of these are

- Reduction of industrial insurance premiums and costs
- Reduction of indirect costs of accidents
- Fewer compliance inspections and penalties

- Avoidance of adverse publicity from deaths or major accidents
- Less litigation and fewer legal settlements
- Lower employee payroll deductions for industrial insurance
- Less pain and suffering by injured workers
- Fewer long-term or permanent disability cases
- Increased potential for retrospective rating refunds
- Increased acceptance of bids on more jobs
- Improved morale and loyalty from individual workers
- Increased productivity from work crews
- Increased pride in company personnel
- Greater potential of success for incentive programs

BUILDING A SAFETY AND HEALTH PROGRAM

The length of a written safety and health plan is not as important as the content. It should be tailored to the company's needs and the health and safety of its workforce. It could be one or two pages or a multiple-page document. However, it is suggested you adhere as much as possible with the keep it simple, stupid (KISS) principle. To ensure a successful safety program, three conditions must exist, and are as follows: (1) management leadership, (2) safe working conditions, and (3) safe work habits by all employees. The employer must do the following:

- Let the employees know that you are interested in safety on the job by consistently enforcing and reinforcing safety regulations.
- Provide a safe working place for all employees; it pays dividends.
- Be familiar with federal and state laws applying to your operation.
- Investigate and report all OSHA-recordable accidents and injuries. This information may be useful in determining areas where more work is needed to prevent such accidents in the future.
- Make training and information available to the employees, especially in areas such as first aid, equipment operation, and common safety policies.
- Develop a prescribed set of safety rules to follow, and ensure that all employees are aware of the rules.

OTHER REQUIRED WRITTEN PROGRAMS

Many of the OSHA regulations have requirements for written programs that coincide with the regulations. This may become a bothersome requirement to many within industry, but the failure to have these programs in place and written is a violation of the regulations and will result in a citation for the company. At times, it is difficult to determine which regulations require a written program but, in most cases, the requirements are well known. Some of the other OSHA regulations that require written programs are

- Process safety management of highly hazardous chemicals
- Bloodborne pathogens/exposure control plan
- Emergency action plan
- Fire prevention plan
- Respirator program
- Lockout/tagout/energy control program
- Hazard communications program
- Hazardous waste and emergency response/site-specific safety and health program, training program, and personal protective equipment program
- Fall protection plan
- Confined space permit entry plan

The specific requirements for the content of written programs vary with the regulation. The respirator regulation requires that the following exist:

- Written standard operating procedures
- Program evaluation procedures
- Respirator selection procedures
- Training program
- Fit testing requirements and procedures
- Inspection, cleaning, maintenance, and storage procedures
- Provision for medical examinations
- Process for work area surveillance
- Acceptable air quality standards
- Use of approved respirators

REQUIREMENTS AND ELEMENTS OF OSHA GUIDELINES FOR A SAFETY AND HEALTH PROGRAM

Although federal regulations do not currently require employers to have a written safety and health program, the best way to satisfy OSHA requirements and reduce accidents is for employers to produce one. In addition, distributing a written safety and health program to employees can increase employee awareness of safety and health hazards while, at the same time, reducing the costs and risks associated with workplace injuries, illnesses, and fatalities.

Federal guidelines for safety and health programs suggest that an effective occupational safety and health program must include evidence for the following:

1. Management commitment and employee involvement are complementary. Management commitment provides the motivation force and the resources for organizing and controlling activities within an organization. In an effective program, management regards worker safety and health as a fundamental value of the organization and applies its commitment to safety and health protection with as much vigor as to other organizational purposes. Employee involvement provides the means through which workers develop or express their own commitment to safety and health protection, for themselves and for their fellow workers.
2. Worksite analysis involves a variety of worksite examinations, to identify not only existing hazards but also conditions and operations in which changes might occur to create hazards. Unawareness of a hazard that stems from failure to examine the worksite is a sure sign that safety and health policies and practices are ineffective. Effective management actively analyzes the work and worksite to anticipate and prevent harmful occurrences.
3. Hazard prevention and control are triggered by a determination that a hazard or potential hazard exists. Where feasible, hazards are prevented by effective design of the job site or job. Where it is not feasible to eliminate them, they are controlled to prevent unsafe or unhealthful exposure. Elimination or control is accomplished in a timely manner once a hazard or potential hazard is recognized.
4. Safety and health training addresses the safety and health responsibilities of all personnel concerned with the site, whether salaried or hourly. It is most effective when incorporated into other training about performance requirements and job practices. Its complexity depends on the size and complexity of the worksite and the nature of the hazards and potential hazards at the site.

If a representative from OSHA visits a jobsite, he or she will evaluate the safety program using the elements listed here. The compliance officer will review the previous items to assess the effectiveness of the safety and health program. You are not confined to only these elements of your safety and health program. You might want to address accountability and responsibility, emergency procedures, program evaluation, firefighting, or first aid and medical care. This is your program; design it to meet your specific needs. These can be addressed in add-on sections. You will find that your fines for OSHA violations can be reduced if you have a viable written safety and health program that meets the minimum OSHA guidelines for safety and health programs.

The composition or components of your safety and health program may vary depending on the complexity of your operations. They should at least include

- Management's commitment and safety and health policy
- Hazard identification and evaluation
- Hazard control and prevention
- Training

Of course, each of these may have many subparts that address the four elements in more detail. The safety and health program that you develop should be tailored to meet your specific needs. It is now up to you to develop and implement your own effective safety and health program. You can build a more comprehensive program or pare down the model to meet your specific needs.

In summary, management commitment and leadership includes a policy statement that should be developed and signed by the top official in the company. Safety and health goals and objectives should be included to assist with establishing workplace goals and objectives that demonstrate the company's commitment to safety. An enforcement policy is provided to outline disciplinary procedures for violations of the company's safety and health program. This safety and health plan, as well as the enforcement policy, should be communicated to everyone on the jobsite. Some of the key aspects found under the heading, management commitment and leadership, are

- Policy statement: goals established, issued, and communicated to employees
- Program should be revised annually
- Participation in safety meetings, inspections, and safety items addressed in meetings
- Commitment of resources adequate in the form of budgeted dollars
- Safety rules and procedures incorporated into jobsite operations
- Procedure for enforcement of the safety rules and procedures
- Statement that management is bound to adhere to safety rules

Sample written program statements related to management commitment, leadership, and employee involvement can be found in Chapters 2 and 3.

Identification and assessment of hazards include those items that can assist you with identifying workplace hazards and determining what corrective action is necessary to control them. Actions include jobsite safety inspections, accident investigations, and meetings of safety and health committees and project safety meetings. To accomplish the identification of hazards, the following should be reviewed:

- Periodic site safety inspections involving supervisors
- Preventative controls in place (personal protective equipment [PPE], maintenance, engineering controls)
- Action taken to address hazards
- Establishing safety committee, where appropriate
- Documenting technical references available
- Enforcement procedures implemented by management

The employer must carry out an initial assessment, and reassess thereafter as often as necessary to ensure compliance. Worksite assessments involve a variety of worksite examinations to identify not only existing hazards, but also conditions and operations where changes might occur and create hazards. Becoming aware of a hazard that stems from failure to examine the worksite is a sure sign that safety and health policies and practices are inadequate. Effective management actively analyzes the work and worksite to anticipate and prevent harmful occurrences. Worksite analysis is intended to ensure that all hazards are identified. This can be accomplished by

1. Conducting comprehensive baseline worksite surveys for safety and health and periodically doing a comprehensive updated survey
2. Analyzing planned and new facilities, processes, materials, and equipment
3. Performing routine job hazard analyses

Sample written program statements can be found in Chapters 6 and 7 for your use and revision.

Hazard prevention and controls are triggered by a determination that a hazard or potential hazard exists. Where feasible, hazards are prevented by effective design of the jobsite or job. Where it is not feasible to eliminate them, they are controlled to prevent unsafe and unhealthful exposure. Elimination of controls is to be accomplished in a timely manner, once a hazard or potential hazard is recognized. To ensure that all current and potential hazards, however detected, are corrected or controlled in a timely manner, procedures should be established using the following measures:

1. Engineering techniques where feasible and appropriate.
2. Procedures for safe work that are understood and followed by all affected parties, as a result of training, positive reinforcement, correction of unsafe performance, and, if necessary, enforcement through a clearly communicated disciplinary system.
3. Provision of PPE.
4. Administrative controls, such as reducing the duration of exposure.

More details on this element of a written safety and health program can be found in Chapter 13.

The employer must ensure that each employee is provided information and training on the safety and health program. Each employee exposed to a hazard must be provided information and training in that hazard. *Note*: Some OSHA standards impose additional, more specific requirements for information and training. The employer must provide general information and training on the following subjects:

- Nature of the hazards to which the employee is exposed and how to recognize them
- What is being done to control these hazards
- What protective measures the employee must follow to prevent or minimize exposure to these hazards
- Provisions of applicable standards

The employer must provide specific information and training on the following subjects:

- New employees must be informed and properly trained, before their initial assignment to a job involving exposure to a hazard.
- Employer is not required to provide initial information and training in cases where the employer can demonstrate that the employee has already been adequately trained.
- Employer must provide periodic information and training as often as necessary to ensure that employees are adequately informed and trained; and to ensure that safety and health information and changes in workplace conditions, such as when a new or increased hazard exists, are communicated.

Safety and health training addresses the safety and health responsibilities of all personnel concerned with the site, whether salaried or hourly. The employer must provide all employees who have program responsibilities with the information and training necessary to carry out their safety and health responsibilities. More information on safety and health training can be found in Chapter 15.

EMERGENCY AND MEDICAL PLANNING

Emergency and medical planning should be a part of your written safety and health program. This may not be the ideal location for this topic, but there seemed to be no other area within this book where it fit any better. Make sure that you make emergency and fire response and medical care for your workers a part of any safety and health initiative.

Before reading this section, you should evaluate your emergency and medical planning by circling the number that best represents the state of your safety and health effort.

Emergency and Medical Planning

Topic	Circle Answer	Answer Options
Emergency planning and preparation	5	There is an effective emergency response plan and employees know how to respond immediately as a result of effective planning, training, and drills
	4	There is an effective emergency response plan and employees have a good understanding of responsibilities as a result of planning, training, and drills
	3	There is an effective emergency response plan and team, but other employees may be uncertain of their responsibilities
	2	There is an effective emergency response plan, but training and drills are weak and roles may be unclear
	1	Little effort is made to prepare for emergencies
Emergency equipment	5	Facility is fully equipped for emergencies; all systems and equipment are in place and regularly tested; all personnel know how to use equipment and communicate during emergencies
	3	Emergency phones, directions, and equipment are in place, but only emergency teams know what to do
	2	Emergency phones, directions, and equipment are in place, but employees show little awareness
	1	There is little or no effort made to provide emergency equipment and information
Medical program (health providers)	5	Occupational health providers are regularly on-site and fully involved
	4	Occupational health providers are involved in hazard assessment and training
	3	Occupational health providers are consulted about significant health concerns in addition to accidents
	2	Occupational health providers are available, but normally concentrate on employees who get hurt
	1	Occupational health assistance is rarely requested or provided
Medical program (emergency care)	5	Personnel fully trained in emergency medicine are always available on-site
	4	Personnel with basic first-aid skills are always available on-site, all shifts.
	3	Either on-site or nearby community aid is always available on day shift
	2	Personnel with basic first-aid skills are usually available, with community assistance nearby
	1	Neither on-site nor community aid can be ensured at all times

All employees will be informed by posted notice of the existence, location, and availability of medical or exposure records at the time of initial employment and at least annually thereafter. (Name/Title of Individual) is responsible for maintaining and providing access to these records.

Each (Name of Company) facility/workplace will have adequate first-aid supplies and certified, trained personnel or readily available medical assistance in case of injury. It is also imperative that all treatments be documented in the first-aid log (see Figure 4.3 for an example). Each (Name of Company) facility/workplace will have medical services available either on the worksite or at a location nearby. Emergency phone numbers will be posted on the worksite for employees to call in the event of an injury or accident on the worksite. Nurses will be available from _____ a.m. until _____ p.m. to respond to medical emergencies. First aid will be available from the (Name) Fire Department at all other times.

Figure 4.1 Sample policy statement for medical and emergency situations.

FIRST AID AND MEDICAL AVAILABILITY

You should have some set guidelines related to how workers are to obtain first-aid treatment for any injuries or illnesses. In addition, you should have made arrangements with medical providers such as hospitals, emergency transport services, and health care providers (i.e., physicians) when more extensive medical care is needed for workers at your worksite. This should be spelled out, posted, and available to all employees at your workplace. An example of a written policy that could be placed in your written program can be found in Figure 4.1.

It is always advisable to maintain a first-aid log, which provides a record of injuries that could have been potentially worse or recordable for OSHA record keeping. This is not required but is recommended as good business practice. A sample first-aid log is found in Figure 4.2.

EMERGENCY PROCEDURES AND RESPONSE

OSHA regulations require that you have a written emergency response plan as well as a fire prevention plan, which could be all in the same plan if designed that way. Workers should know what you expect them to do in case of an emergency or fire. It is your job to provide that guidance. A sample written statement of a policy related to emergency response and fire is found in Figure 4.3.

An example of an emergency evacuation plan can be found in Figure 4.4 (see Appendix A).

A model written safety and health program can be found in Appendix B. This model should be taken and adapted to fit the needs of your company.

First aid log form (Company Name)

Injured worker's name _____

Trade of injured worker_____

Immediate supervisor _____

Date _____ Time of injury_____

Cause of injury_____

Body part injured_____

Nature of injury_____

Treatment rendered_____

_____Returned to work _____ Sent home _____ Sent to hospital

Injured worker's name _____

Trade of injured worker_____

Immediate supervisor_____

Date_____Time of injury_____

Cause of injury_____

Body part injured_____

Nature of injury_____

Treatment rendered_____

_____Returned to work _____ Sent home _____ Sent to hospital

Injured worker's name _____

Trade of injured worker_____

Immediate supervisor _____

Date_____Time of injury_____

Cause of injury_____

Body part injured_____

Nature of injury_____

Treatment rendered _____

_____Returned to work _____ Sent home _____ Sent to hospital

Injured worker's name _____

Trade of injured worker_____

Immediate supervisor_____

Date_____Time of injury_____

Cause of injury_____

Body part injured _____

Nature of injury_____

Treatment rendered _____

_____Returned to work _____Sent home _____ Sent to hospital

Figure 4.2 First-aid log form.

Fire is one of the most hazardous situations encountered on a facility/workplace because of the potential for large losses. Prompt reaction to and rapid control of any fire are essential. (Name of Company) is responsible to provide fire protection procedures for each worksite to assure that they are followed. It is the supervisor /foreperson's responsibility to review all aspects of the firefighting and fire prevention program with his or her workers. The program should provide for effective firefighting equipment to be available without delay and be designed to effectively meet all fire hazards as they occur.

Some emergencies may require company personnel to evacuate the workplace. In the event of an emergency that requires evacuation from the workplace, the signal will be a (describe the actual sound that will be used). All employees are required to go to the area adjacent to the worksite, which has been designated as the safe area. The safe area for this worksite is located at (Description of Location).

Figure 4.3 Sample policy for emergency procedures.

Sample emergency evacuation plan

1. In the event of an emergency necessitating the evacuation of the office, facility areas, or any portions thereof, the supervisor in charge will immediately make repeated announcements over the public address system that an emergency exists and that all personnel in the danger area will evacuate the building in an orderly manner.
2. As soon as an evacuation signal is given (unique signal) any and all supervisors will assume station in the vicinity of the exit doors to receive reports regarding the completion of the evacuation of the building or buildings.
3. When orders are given to evacuate, all supervisors will render assistance to persons evacuating the building and shall begin an immediate check of each room or office, if at all possible, to make sure that everyone has left the building. If fire or emergency wardens have been trained and assigned to duty, they can perform this function.
4. After making sure that a building or work areas have been completely evacuated, the supervisor shall report the same to the manager.
   a. If an evacuation occurs at night, the supervisor on duty will perform these assignments and at the first opportunity notify the manager.
   b. Shop evacuation will be performed in essentially the same manner with one exception. Notification of personnel in the shop will be handled by voice. Shop forepersons, or their assistants or any employee, day or night, will declare an emergency and give the order to evacuate.
5. Emergency telephone numbers of the fire department, rescue, etc., are posted in the dispatch office, dock office, supervisor's office, and other operating areas. It is essential that the appropriate emergency service be called immediately. If the emergency occurs during office hours, the manager or shop foreperson will make the emergency telephone call. When an emergency exists after office hours, the supervisor in charge will make the call.
6. Emergency escape route chart will be posted in the office and basement.
7. Major workplace fire hazards and control procedure will be discussed with the local fire department's authorized personnel to ensure better control of risks. This information along with the aforementioned procedure will be discussed with all personnel in the safety meeting.
8. Emergency escape procedures from the dock and shop will be verbally discussed with all workers and all new personnel before assignment.
9. These procedures will be revised when there are any physical changes to the facility or changes in evacuation personnel or evacuation routes.

Figure 4.4 Sample emergency evacuation plan.

CHAPTER 5

Getting Safe Performance: Motivating Safety and Health

The task of motivating safe and healthy work performance.

It is always interesting to me that while working with employers as well as safety and health professionals, there is such an investment of time and effort in developing a safety and health initiative. At times, the cost is sizable. However, in motivating employees, the employer is looking for a quick fix. Seldom do employers take time to plan, apply principles of human behavior, or invest reasonable amounts of money to motivate employees to work in a safe manner. This seems even odder when one considers that all workplaces are made up of employees and the safety and health effort is aimed at these individuals. Many times I hear employers and others say, "What can we give them?" Often, when given a reward or incentive employees do not even understand why they are receiving it. What most employers and others are asking is "What is the quick fix?" They think that they can easily address such a complex subject as human behavior by throwing some money at it or giving employees a trinket. I contend that much more effort should be invested in Getting Safety than usually occurs.

As stated earlier, employers take little, if any, time to think about motivating safety and health. This certainly seems like an oversight when data indicate that 85%–90% of accident causes are likely the result of unsafe behavior (acts). With this realization, it seems beyond logic that employers do not pay more attention to motivation in the workplace, especially related to occupational safety and health performance.

Let me caution you that paying attention to developing motivational approaches to safety and health will be to no avail unless all of the other components discussed in this book are addressed first.

We cannot expect workers to be motivated toward safety and health without the foundation of a safety and health program. You cannot be motivated without knowing what you are expected to be motivated about (or even why you should care about being motivated) if the company has not put forth the effort to define and direct the performance desired regarding safety and health. Much of the development and planning for implementing a motivational approach will have already been completed if the guidelines provided in the other chapters are implemented. You will have the needed directions, goals, policies, and procedures in place. I would caution you that there is no foolproof motivation plan that is fail-safe and assures you of achieving the results that you desire. As we all know, the most difficult tasks faced by all of us in the workplace are those in which people are involved.

The reason that most of us fail to get the type of motivation we desire is that we try to change people's values, which are set in early life, or change their attitudes, which are an integral part of their personalities. Both values and attitudes are not measurable or easily observable and are accepted or rejected based on our own set of values and attitudes. The best that we can hope to accomplish is to change an individual's behavior, which is observable and measurable. Over time, the workers' attitudes may change or their values may be altered by your motivational attempt, but that is not as important as obtaining safe and healthy work behavior. It is imperative that we motivate workers to exhibit a behavior that will keep them safe and healthy in the performance of their jobs.

SETTING THE STAGE

The aim is to provide a basis that will allow you to motivate yourself and your employees. When you discuss motivating yourself or others, you are always in search of a blueprint. To most of us, the magic formula for motivation is perceived to be composed of plans, tricks, gimmicks, or inducements. This may be the case in some instances, or for some individuals, but this is not the panacea.

The human aspect of safety and health is often brushed aside in industrial settings. After all, each workplace is different and the fundamental principles of motivating and dealing with construction workers or office workers are, therefore, different. Granted, people are unique, but all workers have basic needs to be addressed and fulfilled in order for them to work effectively and productively. The principles and examples in this chapter have been used for miners as well as office workers and are applicable to any group of workers. It is up to you to use your ingenuity and creativity to develop an effective approach that will motivate the workers you have to deal with.

The management of people is directly related to understanding the basic principles of motivation. Obtaining good safety and health behavior and work practices can be directly attributable to how effectively you apply good principles of motivation. The intention of this chapter is to provide a practical and insightful view of the workings of motivation in the work or occupational environment. One must remember that many theories related to motivation exist. Using the principles discussed in this chapter is not a sure-fire guarantee that you will be able to motivate your workers, but if you do not do something, you certainly will not attain the motivation you desire in them. Most of the time managers, contractors, and safety and health professionals want a band-aid solution to their motivational problems in the workplace.

Motivation is a somewhat imprecise science and undertaking. No guarantees for success exist. Each workplace needs to pay close attention to the motivational needs of the individuals who work in it or there will be no progress in attaining the desired safe behavior, and you will probably see a decrease in morale. Since morale is as difficult to define as is motivation, for the context here, let us use motivation to attain good morale.

There are many legitimate reasons for trying to motivate workers and employees. These reasons may be as simple as trying to get an employee to work safely, or as complex as fostering safe work teams. It is very challenging to instill motivation where motivation does not exist. While you should hesitate to use workers as behavioral guinea pigs, workers are no different from anyone else when it comes to motivational issues. Motivating yourself and others is usually tackled because you care about someone or some group and want to see them accomplish a goal or conquer more than they ever thought was possible. It is rewarding to see a group of workers attain the goal of working 1,000,000 h without an accident and go on from there.

Although all the principles espoused here can apply to families, relationships, friendships, coworkers, teams, or employees, most of the successes discussed are those relevant to the workplace. This is, of course, because the majority of our adult life is spent in the workplace.

DEFINING MOTIVATION

Some people believe motivation has the potential to answer everyone's problems. You may have heard such statements as "If I were only motivated!"; "You should motivate me!"; "You should motivate him or her!"; "You are not motivated!"; "You had better get motivated!"; or, "All you have to do is find his/her 'hot button'!" However, these statements, do not tell us what motivation really is nor do they tell us how we can measure or even understand motivation.

Motivation, in the broadest sense, is self-motivation, complex, and either need- or value-driven. Someone once stated that he believed hope was the secret ingredient to a person being motivated (the hope to accomplish a goal, a dream, or attain a need); there is reason to support this theory. However, possibly a better definition is "motivation presumes valuing, and values are learned behaviors; thus, motivation, at least in part, is learned and can be taught" (Frymier, 1985). This definition provides us with the encouragement we need to go forward and achieve motivation for ourselves and others.

If we want to be successful, we must believe that we can teach someone to be motivated toward specific outcomes (goals) or, at the least, be able to alter some unwanted behavior. On the other hand, we should not want to completely manipulate an individual to the point that he responds blindly to our motivational efforts. If that occurred, we would lose that most important part of a person, the unique human will.

Thus, motivation is internal. We cannot directly observe or measure it but a glimpse of the results may be seen when we observe a positive change take place in behavior. Such a change might be something as simple as a worker wearing protective eyewear or something as far reaching as going a full year without an accident or injury. By observing these outwardly manifested behaviors, you can then be encouraged when you see even the smallest of successes that are related to your motivational techniques.

PRINCIPLES OF MOTIVATION

Goals are an integral part of the motivational process and tend to structure the environment in which motivation takes place. The environment in which we find ourselves is many times the springboard to the overall motivational process. You may be fortunate enough to accidentally step into a high-energy motivational environment. On the other hand, you may find yourself in an environment that is not at all conducive to motivating others and, thereby, it is very difficult to attain your desired goals. If this is the case, you may need to make a change in the physical environment or possibly even make a change in the work atmosphere (i.e., allowing more independence of individuals in the decision-making process).

The changing of the environment may not affect all individuals in the same way. It has been said that there are three certainties that can be stated regarding people. Those certainties are "People are Unique, People are Unique, and People are Unique!" Since people are unique, what motivates one person may be demotivating to another.

If you are that person with the responsibility of trying to motivate an individual or group, you will need to address their motivational needs. Employees fall along a continuum—some need little motivation from you and others need constant attention. It is unrealistic to expect all of them to achieve your level of expectations. The quality of your leadership will be the determining factor to your success with these people.

Why are some leaders more motivational than others are? What are the unique talents that these dynamic leaders possess? Some people believe that these individuals were born to be leaders. Most of us do not believe that they are just born leaders but individuals who possess a set of talents and have chosen to develop those talents to the maximum. These talents are developed because they have the burning desire (goal) to become leaders and those desires motivate them to learn the necessary skills.

To be a successful motivational leader you must have some sort of a plan that will get you from point A to point B. This plan should include your desired goals and objectives, levels of expectations, mechanisms for communication, valuative procedures and techniques for reinforcement, feedback, rewards, and incentives. Any motivational plan is a dynamic tool, which must be flexible enough to address changes that may occur over a period of time and take into consideration the universality of people and situations. These plans can use a variety of techniques and gadgetry to facilitate the final desired outcomes or performances, which lead to a safer and healthier workplace.

MOTIVATIONAL ENVIRONMENT

Everything that surrounds us is part of our motivational environment. Depending on our environment, we are motivated differently at a given point in time. We actually exist in what we call micromotivational environments. These microenvironments make up the sum total of our motivational environment and are composed of our work environment, family environment, social environment, team environment, peer environment, or even a nonfunctional environment.

Any one or all of these microenvironments can have an impact on the other. The negative impact of one of a person's microenvironments may cause that person to also react negatively in another one of his or her environments; this can happen even when the environment is, in itself, a positive one. For example, if an individual has problems at home, it may, and many times does, cause that highly motivated employee to become less safety conscious or productive at work.

In illustrating the complexity of this issue, let us think for a moment about problem employees. Many times these employees ask to be moved to a different job because they are either dissatisfied or are performing poorly in their current job. Amazingly, once the reassignment is made, their performance vastly improves. It is almost as if this worker becomes a different individual. When they are put into a new and different environment, they get a new spark and the new environment becomes their positive motivator; they have been revitalized! Many individuals do not like change but all of us react and are energized both negatively and positively by change. Therefore, you will need to make changes in your motivational approach when you see motivation waning.

Structuring the Motivational Environment

It takes some degree of organization and commitment to structure an environment where workers will be motivated to perform their work in a safe and healthy manner. The safety and health environment must have a foundation. A written safety and health program is the key component in providing and structuring that foundation. This written program should set the tone for safety and health within the work environment.

There are some keys to motivating safety and health. A listing of them is as follows:

- Explanations
- Goals
- Reinforcement
- Involvement
- Self-monitoring
- Rewards

First, you must explain and clarify the safety and health performance expectations. You cannot assume that workers know what is expected of them unless you tell them. You must make sure that your expectations are concise and consistent. If you want your workplace to be the safest in your industry, then you must not deviate from what you expect. Tell people how you hope to achieve your expectations and do not fail to ask for advice on how to attain these expectations from everyone in your workforce. Always remember that people have a definite need to know. They certainly like to know what is going on.

Second, a way to keep your expectations out front regarding safety and health is to establish attainable and reachable safety and health goals. Goals that are understandable by all are much better motivators than ones that workers do not understand. It is good to involve those who will be impacted by the goal in the development of that goal. A goal to reduce our injury rate by 20% may sound fine to you but many workers do not know what an injury rate is or the components that go into calculating it. A better goal would be to keep the number of injuries below 10 per month. This can be easily tracked and counted on a monthly basis. The progress toward this goal can be posted regularly. All of us are goal-driven, whether we recognize it or not. I venture to say that most of what an organization, team, or individual accomplishes is the result of a goal set.

A third key to motivation is providing feedback. If you post the number of injuries each month, workers are being given feedback on the progress toward that goal. We need to know how we are doing to maintain our focus and motivation toward an outcome. Providing feedback is vitally important. How many times have you heard someone say, "I wish they would tell me how I am doing. I don't care if it is bad or good. I just want to know." That person is saying, "Talk to me, please give me feedback."

Once you have explained your expectation and set the goals, some employees will be on board immediately while others will see no value in adhering to the safety expectation and goals. Thus, if all of your motivational efforts fail with them, what do you have left? Actually, you have nothing unless there are consequences to their failure to be motivated by your previous efforts. For example, you have warned a worker to wear protective eyewear and even given him a written warning. If you do nothing the next time he fails to comply with your safety rule, then the worker has no consequence. You have reinforced his negative safety performance. However, if you give him 3 days off without pay or dismiss him, this says to others that you have placed a value on wearing safety eyewear. With that, value becomes a consequence for this unsafe behavior. You send a strong message that you are serious about this rule. For without consequence, many individuals will not be motivated to perform as you would like them to. Although a very negative approach, without it you have lost a very critical motivational key. It is possible for the consequence to be positive. For example, a commendation for safe performance is a positive consequence.

Reinforcement is an important key in the motivational process. How you reinforce safety performance will determine whether it is strengthened or weakened. Reinforcers can be verbal feedback, a reward, or a consequence. Reaching the safety and health goal depends on how the reinforcement is used to drive home the message of accomplishment or failure. Telling a worker that you really appreciate the safe way he is performing his job is feedback, which reinforces the type of behavior you desire and fosters motivation in that individual. Reinforcement for safety and health needs to be more frequent than once a year. Monthly would be best but quarterly is also adequate. Unless the reward is very large, a year is too long a period to have to wait for reinforcement.

The key, which has been discussed earlier, is involvement. We should make every effort to involve workers who have a vested interest and invest their energy toward the outcome of a safe and healthy workplace. There are many ways to involve workers in the safety and health initiative. These range from participation in a safety and health committee to conducting inspections. This involvement needs to be nurtured and recognized. When our contributions are supported and recognized, it has a positive effect on our behavior and thus we are more motivated. You may need to be highly creative in finding ways to get workers involved.

Workers like to be able to monitor their own progress toward a goal or expectation. If you have a chalkboard, workers themselves can mark the board each time that they or a fellow worker has gone without an injury for an agreed upon period. Even if somewhat inaccurate, workers need to sense that they are involved in the process of safety and health at their workplace. This can lead to more teamwork and more motivation in the workplace. Self-monitoring may not always be an option but do not overlook its effect when it can be used.

One of the most debated keys to motivation in the safety arena is rewards. In all aspects of life, we tend to focus and perform better when rewards are involved. Rewards are not a quick fix to problems with your company's safety and health performance. Rewards are a complement to the safety and health initiative. If you do not have all the nuts and bolts of a safety and health program in place, then rewards are not a replacement for failure to effectively manage the safety and health effort. You can consider rewards as the icing on the cake. If your company is not performing well to reach your safety and health goals, then rewards may be used to keep workers focused, reinforce performance, or recognize the attainment of goals and expectations. Employers often say, "I already pay my workforce to work safely," and I say to them, "Then why in the world do you give them production bonuses when you already pay them to produce." Of course, the answer is that it gives them a little more motivation. I do hate to think that a person's safety can be bought and it would not be in their best interest to become injured or ill. However, on the other hand, it is my sense that using rewards as motivators, reinforcers, and reminders for a higher purpose as that of trying to provide the best approaches to obtain a safe and healthy workplace is appropriate.

One caution with rewards that I would make you aware of is that money is a onetime occurrence and the reason that one receives it is quickly forgotten when it is spent. In addition, any reward must be of value to the person receiving it. A gift certificate to Fashion Bug may not be viewed as valuable to your male employees. Therefore, if gifts of some sort are to be given out you might want a variety available through a catalog that would appeal to a wider range of your workforce. You could provide catalog dollars that could increase in numbers as progress toward your goals is attained.

Rewards need to be tangible so that when employees see it in their home or workplace it serves as a reminder of why they received it. Appropriate rewards might include a bond, a plaque, a pin with a one on it (one year without an accident), a certificate, an embossed hat, jersey, jacket, or other assorted items that are reminders to the individual to stay motivated toward safety and health performance goals.

The organized approach to safety and health should address each of the previous eight keys. Within these paragraphs you find these keys to motivation, which should be an integral part of the occupational safety and health prevention program initiative.

Many tangible and intangible factors comprise the motivational work environment. Something that is tangible could be something as simple as moving a piece of equipment to create a more desirable environment or granting a request. Something intangible could be your ability to change the way someone feels about you.

When it comes to developing a safe and productive motivational work atmosphere, the intangible motivational issues are just as important as the tangible ones, but they are also the most time consuming. These challenges run a broad spectrum and, just to list a few, could be some of the following examples: changing the way a person is treated by peers, colleagues, or supervisors; helping individuals gain a positive perception of their contribution; or developing a new and positive attitude toward workplace safety and health.

You should develop an environment where the majority will be positively motivated to perform but this should not prohibit you from making adjustments, when possible, to address individual needs. Furthermore, be sure that you do not allow too much flexibility (i.e., favoritism, etc.) or it could destroy a good situation for the majority.

To assure equal treatment, require all to abide by the rules. For instance, when there is a set group of safety and health rules for your workplace, you should never allow one person to abuse these rules while holding others rigidly to them; this will cause disenchantment with safety and health issues. Top management, supervisors, foremen, and workers should be treated equally and fairly by requiring them to comply with the safety and health rules and policies. As an example of this, while I was working with one company the safety director wore moccasins while everyone else was required to wear hard-toed shoes. This type of behavior should not be acceptable for one person and not others. Although you may be the head of an organization, or the boss, you should never consider yourself so lofty that you do not adhere to your own safety and health rules and policies. It is very important that management and supervisors set the tone of the work environment with regard to safety and health on the job.

When setting up an environment where you want those involved to be motivated, you should first address the physical needs. For example, in the work environment there may be the need to provide the proper tools and personal protective equipment in order for the workers to do their work safely.

Your ability to structure an environment that provides individual needs and adequate stimulus to motivate each person to his or her full capacity is desirable but not usually possible. In fact, you actually have little chance of setting up the perfect environment for every person. Just too many other environments and factors compete with you and what you desire each individual to accomplish. However, do the best you can for each person and then each individual will make a conscious decision as to whether he or she wants to perform safely in the workplace. This is the reason that each worker should know the consequences of any unsafe performance. You should develop mechanisms to assist these individuals to perform safely, but also have disciplinary procedures for those who elect not to comply with the safety and health rules.

As part of setting the environment, be assured that each worker understands the expectations regarding working safely. It is also useful to involve them in setting the safety and health rules and goals and to know the expected outcomes. Each worker needs to understand that there will be consequences or penalties for disregarding or violating the safety and health requirements of his or her work. Therefore, goals are important in setting performance objectives for the company's safety and health program.

Track the progress of the safety and health goals and provide feedback. This allows the workers to monitor their own accomplishments in their work area. Recognize the workers who are progressing toward the safety and health goals and reinforce safe work behaviors.

Reacting to the Motivational Environment

You can provide all of the bells and whistles, but if you do not pay attention to some fundamental characteristics of people, you will not be successful in developing a good motivational environment. Some of the fundamental principles you need to be aware of when working with people are

1. Individuals view themselves as very special. Thus, praise, respect, responsibility, delegated authority, promotions, recognitions, bonuses, and raises add to their feelings of high self-esteem and need to be considered when structuring a motivational environment.
2. Instead of criticism, use positive approaches and ask for corrected behavior. Individuals usually react in a positive manner when this approach is used.
3. Verbally attacking (disciplining) individuals tends to illicit a very defensive response (even a mouse, when cornered, will fight back in defense of itself). Therefore, it is better to give praise in public and, when necessary, criticize in private.
4. Individuals are unique and given the proper environment, will astound you with their accomplishments and creativity (even those individuals whom you consider noncreative).

Remember, the final outcome lies with the employees; they will decide whether or not to perform safely. However, if the employer has done his or her part, workers will not be able to hold you responsible for the decisions they have chosen to make.

There will be people who elect to work unsafely even though the environment may be very motivational to the majority. Thus, when discussing work, you will need to pay close attention to the motivational environment, and work at making it the very best! However, when there are those who fail to perform safely, there should be consequences and discipline administered quickly and fairly. If no one enforces the speed limit, then who will abide by it? Either enforce the rules or lose the effectiveness of your motivational effort.

The key to a successful motivational environment is to pique the interest of people. Let them know you want them to succeed; give them responsibility; and leave them alone to accomplish those goals and succeed. If the above principles are not taken into consideration when setting up your motivational environment, you will be more likely to encounter problems with your success rate. As an illustration, a supervisor noticed that his workers were not giving him the performance he expected. He was having difficulty receiving top-quality written reports from them and, therefore, had been rewriting each report. When the supervisor was asked if his employees were aware that he was rewriting their reports and, if so, did he think they were putting forth their best effort, he answered, Oh yes. However, after thinking this question through he decided to go back to his work area and ask his workers the same question that was asked to him. Their reply was, as expected, that they were only giving a half-hearted effort since they knew the report would be rewritten. As you can see from this example, you need to be cautious that you do not set yourself up for this type of response.

The way you structure the motivational environment will allow individuals within the work groups to accomplish safety and health goals and assure that they are free from injury and illness. What you need to do is set up an environment where people can be successful. And, in order for that environment and the people within it to succeed, you must demonstrate that you genuinely care about them and the purpose of the mission (goals) they are trying to attain, which in this case is a safe and healthy workplace.

Next, you need to be open to learning from your own experiences, as well as from others. This will facilitate flexibility in your encounters and give you the ability to make the necessary changes. You need to be honestly perceived by everyone as working diligently to prevent workplace incidents and be willing to work at motivating those who are not in tune with your safety and health attempts. It does take an added effort to motivate others.

It is imperative that you realize when you have reached a point where you have accomplished as much as you can and have lost the effectiveness of the safety and health environment that you have structured. This may be an indicator that you need to change your approach. As an example of this type of situation, think for a moment about coaches, especially those who are in the professional ranks or at larger collegiate institutions. When they become ineffective, they are forced to move onto other coaching positions. However, once they are in their new position, and even though they had become ineffective in their previous one, they often are able to rejuvenate a program, which, until their arrival, was unsuccessful. In these cases, we realize that the coaches are still the same people but they become ineffective in their previous jobs because the environment changed and they were unable to adapt to those changes. Nevertheless, when they were introduced into a new position, they once again became successful.

I believe that psychologist Frederick Herzberg was on target with his concept of successful motivators. He said that in order for individuals to be satisfied with their jobs and remain motivated, they need competent supervision, job security, adequate salary, adequate benefits, and good working conditions; but, more importantly, they need the satisfaction of achievement, recognition, responsibility, and challenge. These are the real motivators, the internal ones; the ones that truly satisfy the individual's specific needs. Thus, when structuring your motivational environment, be sure to load it with as many of these true motivators as possible. They are the most successful incentives and encourage consistent and improved safe performance. Some motivational environment examples are described in the following paragraphs.

A company installed a new air conditioner for the workers of their appliance assembly in an effort to improve their physical work environment. The new air conditioner did make the work environment more comfortable but, for some unknown reason, the production of the workers actually decreased, and accidents and rejects increased. In an attempt to increase production and decrease rejects and workplace injuries, some new incentives were introduced but, to the company's dismay, the production remained low and work-related incidents continued to occur. Finally, the discovery was made that the new air-conditioning system was so loud that the workers could no longer talk or be heard by their fellow workers during the assembly process. Since the workers were doing a repetitive task and were also receiving low wages, their job satisfaction depended on their social interaction with fellow workers. Once the air conditioner was shut down and replaced with a quieter unit, productivity improved and rejects and injuries also decreased.

In the past, companies have tried to motivate people by reducing the hours worked, giving longer vacations, increasing wages, increasing benefit packages, providing career counseling services, training supervisors in communications, and organizing interactive groups. However, these incentives have not proven to be highly effective in increasing productivity. Therefore, it is important that we determine what affects the satisfaction or dissatisfaction on a job or, for that matter, anywhere else.

In structuring a motivational environment, it is important to help people grow and learn through the task that they are asked to perform. Prepare them to stretch their abilities to new and more difficult tasks and help them advance to higher levels of achievement. Help them use and recognize their unique abilities and make sure they can see the results of their efforts. Be sure to recognize when a task is well done; give a promotion or award and provide or reinforce performance with constructive feedback. This is not only applicable at the workplace but is also standard for life situations whether it be school, sports, home, social groups, or peer groups.

In recent years companies such as Ford, Volvo, and General Motors, as well as many others, have found that team approaches to the work environment are very effective. They have found that an increase in quality and overall job satisfaction transpires when a work group is assigned a task and then given control over such decisions as who does what tasks, how the tasks will be accomplished, and who has the authority to stop the process if quality is in question.

With this type of system in force, the supervisor is no longer responsible for completion of the task; the group has that responsibility and control. The supervisor's main duty becomes one of advising, providing feedback, and assuring that all materials and tools are available to accomplish the job. This approach has also been very successful with quality circles but may not work in all environments since the end product is not the same for all individuals and in all situations.

When there are barriers that keep you from being able to set up a good motivational setting, put an even greater emphasis on the nontangibles (recognition, achievement, responsibility, and challenge). As an example, let us consider the MASH television series. The physical setting was terrible, the wounded were disheartening, and the tools needed to accomplish their mission were often missing. However, discipline was not stringent, protocol was lax, individuality and recognition were endeared, and this made the mission not only challenging but also rewarding.

As you can see from these examples, you cannot always predict the way in which individuals will react to a motivational environment, but you can predict with some certainty that if there is no attempt to set up a good motivational environment, an integral part of motivation will be lost. Thus, with this piece of the puzzle missing, the other facets of the motivational plan cannot be effectively applied.

SELF-MOTIVATION

The question that arises is "Who motivates you?" Is it a person, is it peer groups, is it incentives, or is it the environment? It is the contention of this chapter that it is you who motivates you. Excuses, blame, and alibiing will not negate this fact. Nobody can motivate you. You must assume the responsibility of motivating yourself within the environment in which you find yourself. Some individuals are motivated by positive happenings within their lives, while others succeed through adversity. Certainly, an employer may work very hard to set up a motivational environment, but the individual decides whether he will be motivated by that environment.

People Are Amazing

One person who was motivated by his failures was Dan Jensen, Olympic speed skater. After failing to receive the gold medal in three previous Olympic Games, he went on to become a gold medal winner in the 1994 Winter Olympic Games. He had been expected to win the gold medal in previous Olympic Games but through disastrous falls or unexpected losses he was unable to accomplish that goal. He was determined to make his failure lead him to his success and ultimate goal, a gold medal in the 1994 Winter Olympics' 1000 m race. Failures can bring success!

On the other hand, what would have happened to Bonnie Blair if she had experienced the same fate as Dan Jensen? She culminated her career with five gold medals and had at least one medal in each of the previous three Olympics. No one can say, or even guess the answer since she was motivated by her successes each time, instead of her failures. Mistakes can either have a positive or negative effect on the motivational environment, but tend to be demotivational; we need to realize that people who do something are going to make mistakes.

In a recent article, Bill Gates, Chairman and CEO of Microsoft, stated, "Reacting calmly and constructively to a mistake is not the same as taking it lightly. Every employee must understand that management cares about mistakes and is on top of fixing problems. But setbacks are normal, especially among people and companies trying new things."

Regardless of all the efforts made to assure that no accidents, injuries, or work-related illnesses occur, there will still be, at times, mistakes made and negative outcomes that occur. When it happens, this should be an incentive to try even harder; do not trash the safety and health effort over a setback.

Self in Motivation

The basis for motivation seems to be in our perceptions of ourselves. These perceptions govern our behavior and support the concept of self-motivation. In order for people to motivate themselves, there must be meaning in what they are doing. If they do not perceive that the goal set before them will satisfy their needs, they cannot possibly motivate themselves to accomplish it. You must realize that no matter how unrealistic a perception may seem to us, it is a reality to the person who holds it. No matter how we try to debunk a perception, there is always some degree of truth and reality within it and, therefore, it is very real to that person.

Individuals will not be motivated to work safely unless they have internalized the goals and expectations of the company. It is not enough for them to know that they will be fired for violation of a safety rule, they need to be motivated to perform their work safely even when there is no one watching them.

People must be inspired to be accountable to themselves. If they put their goals and plans down on paper, then they take possession of their own behavior to a greater extent. This motivates them to do something and gives them the time, direction, and a reason to find new or better ways to accomplish their goals and plans. As many experts will tell you, you should put your goals or plans in writing. If you can not write them down, then you probably will never achieve them.

The three most important things to remember about people are people are different, people are different, and people are different! With this in mind, you will need to view each person on a continuum. When trying to figure out what motivates him or her and how you can begin to get a change, look at every aspect concerning that person's life, and try to evaluate what is and is not of importance to him or her.

Some individuals are superstars. These individuals are self-motivated and all you have to do is give them support and minimal guidance and then just step back and watch them go. Others, on the other hand, seem to lack any motivation at all. These individuals need to have things structured for them, know exactly what is expected of them, know what happens if they do not perform, and know what the reward or outcome of their performance will be.

You will also find individuals who need to be around other people; they perform best when they are in a social environment and, therefore, are more affected by peer group pressures. Finally, there are people who prefer to work alone. Frequently these individuals are achievement oriented and all they want is your recognition and reinforcement. All it may take to motivate them is to grant their request for something as simple as a tool or piece of equipment that will help them do their job in a safer manner.

NEEDS MOVE MOUNTAINS AND PEOPLE

Dr. Abraham Maslow of Brandeis University believes that people are motivated not only by their unique personalities and by how they want to fit into their world, but they are also motivated by their own individual needs. The premise that runs through Dr. Maslow's book is "motivation is internal—thus, self-perpetuating."

Dr. Maslow identified five needs. They run the gamut from the basic animal needs to the highly intellectual needs of modern man. They are the physiological, safety, social, ego, and self-fulfillment needs. In order for you to understand the relationship between these needs and the motivational process, a description of each one follows:

- Physiological needs are the requirements we have for our survival. They encompass the basic needs that are necessary for the body to sustain life or physical well-being. These needs are such things as the food we eat, the clothing we wear, and the shelter we live in. Each of these must be satisfied before other needs can be dealt with. The physiological needs appear in all of the actions each of us take to ensure our survival and physical well-being. Individuals who are motivated primarily by these needs will do anything that you ask them to do, no matter how unsafe it might be.

- Safety needs include the requirements for our security. If first the physiological needs are reasonably well satisfied, then people become aware of and start to act to satisfy their safety needs. These needs are such things as having freedom from fear, anxiety, threat, danger, and violence and being able to have stability in their lives. Striving to satisfy these needs might show up in such actions as (for their safety) avoiding people or situations that are threatening and may be of danger to them, or (for stability) lobbying for a pension plan or putting money into an agency's credit union. These individuals are concerned about whether you are taking all the precautions to protect them from workplace injury and illness.

- Social needs include the requirements for feeling loved and wanted, and the sense of belonging and being cared for. If the safety needs can be reasonably satisfied, social needs begin to emerge. Some behaviors that take place and indicate a social need of acceptance are asking the opinion of the group before acting, following group preferences instead of personal preferences, or joining job-related interest groups. These individuals will follow the safety and health pattern set by their workgroup.

- Ego needs include the requirements for self-identity, self-worth, status, and recognition. When social needs are reasonably satisfied, individuals are able to explore the dimensions of who they are and consider how they wish to sell/market themselves. Some examples of ego (esteem) needs are self-respect, esteem of others, self-confidence, mastery, competence, independence, freedom, reputation, prestige, status, fame, glory, dominance, attention, importance, dignity, and appreciation. These individuals will want to be involved in and be a part of the ongoing safety and health effort set by the company.

- Self-fulfillment needs are composed of the requirements it takes to become all that one is capable of becoming and to fulfill oneself as completely as possible. The self-fulfillment needs are so complex that people never reach a point where they are completely fulfilled. They do what they do, not because they want others to notice or reward them, but because they feel a need to be creative, to grow, to achieve, and to become all that they are capable of becoming. These individuals understand the true importance of safety and health on the job; it is a part of them. They will follow the safety and health rules because they have internalized the true function of the safety and health program; they realize that it is a vital component of the whole operation. These individuals have a sense of needing to help others reach an understanding of the safety and health issue.

Maslow was right when he suggested that needs are motivators for people. As a motivator, you cannot motivate another person by depending on elements that you deem as important. What you must do is be sensitive to the needs and wants of the people you are trying to motivate. It is sometimes difficult for many of us to remember where we came from and to relate to someone who has basic needs (physical and security) that are far below our own needs. If you are to be a real motivator, you will need to spend time understanding the real needs of those around you.

There are many ways to motivate people but I have found that the simplest way has always been asking the individual what he wants and what presses his button. What makes him go? Involvement is probably the best motivator available to most of us. No matter where a person is along the continuum, he is ego-centered enough to want to be part of the decision-making process that affects his life.

In summary, this means that each individual you are trying to motivate will need individualized attention. You will need to tailor, as best you can, a motivational plan that will meet his needs and, thus, causes him to be motivated toward the goals that have been developed. A person has his own reasons, based on his own values, needs, and desires, that determine how he applies his energies. How you accomplish this is not scientific. It may be accomplished by trial and error or, at best, by small successes followed by bigger successes until the goal is reached. It seems safe to say that what works well for one person may fail miserably for another or with modification, may also be successful.

MOTIVATIONAL LEADERSHIP

Describing Leadership

Leaders are not born, but some individual personalities are more suited for leadership positions. There may be several potential leaders within an organization, company, or team, but only one of them may meet the criteria or have the leadership skills that are needed for a position at a particular point in time. This does not mean that the other potential leaders are not qualified leaders, but that their unique leadership traits are not appropriate at that time.

There are two types of leaders: those who lead by coercion and those who lead by example. The question that arises is, "Which is the right type of leadership?" It seems that there are occasions where leadership by power is appropriate (or the only way), but this type of leadership seems more appropriate, for example, in the military. Please keep in mind that power does not, in itself, have to be bad. Many people who have leadership responsibility use power in a responsible manner to help others. Many good leaders use their leadership position and power not as a divisive tool, but to help people get things done.

Role Models

To be a good leader, one must be a good role model and must be willing to sacrifice his or her own wants, desires, needs, and ego. Unfortunately, many leaders are not willing to do this and are, therefore, not good role models. Supervisors and employers should not be privileged to be what they want to be and then expect employees and workers to do as they say, not as they do. The leaders must set the example that they want their employees to follow because most of us follow the models that are set before us.

For example, a safety expert should never go on a worksite without wearing the proper safety equipment for that particular jobsite. He or she should always ask, before going to a jobsite, what the requirements are for that particular jobsite. Although it may be a hassle, a good and responsible safety and health professional or advocate will always carry his or her bulky safety equipment to the jobsite, even if it means carrying it clear across the country on an airplane. How can the safety professional expect others to wear the appropriate safety equipment if he or she does not wear it?

Leadership Characteristics

Leaders should lead by example, but there are additional things leaders can do to facilitate their leadership, motivate others, and achieve their goals. The motivational leader is capable of building on the strengths of the people he or she deals with by developing (coaching) confidence in others, depending on goodwill, inspiring enthusiasm, and saying things like WE and LET US DO. As a role model for motivation, he or she must build trust, recognize abilities, gain commitment, ensure rewards, and always expect the best of people. Leaders like this are facilitators of the development of human potential and inspire others to adhere to prudent safety and health practices at work. This is in stark comparison to the old style leader who had all the answers and told people what to do. These leaders were quick to point out weaknesses, inspire fear, use authority, push people, and use words like I and DO THIS or DO THAT.

Effective leadership is critical to the structuring of a motivational environment. Managers, supervisors, and others achieve results through the efforts of working with other people. While planning, directing, and controlling receive most of our attention, motivating people is also a critical part of everyone's responsibilities and must be given the right attention.

Thus, it is imperative that we train managers, supervisors, forepersons, and safety and health professionals on how to apply and make use of motivational techniques. Individuals in leadership positions can do much to undermine your motivation techniques and need to be held accountable to assure that they support this facet of your occupational safety and health initiative.

Applying Leadership

Motivating people is an ongoing process. It requires a continuing commitment, an objective view of our own style and abilities, and an understanding of the effect our behavior has on others. Recent studies show that a majority of individuals are still motivated by traditional incentives; however, money no longer has the same clout it once did. A significant portion of people today place greater value on positive reinforcers that are related to their accomplishments. They look for more control, responsibility, and meaningful accomplishments that are worthy of their talents and skills.

A good leader is one who is willing to listen. Nothing is more rewarding than to see someone who was considered a lost cause or less than average blossom just because someone took the time to listen. People need someone to listen to their problems and to help solve them.

In today's world of diversity, everyone brings a variety of personal experiences to each situation. Failure to show sensitivity to the feelings of others can result in misunderstanding, resentment, anxiety, communication gaps, wasted time, unnecessary work, lower productivity, poor morale, and other negative effects.

Generally, the golden rule "Do unto others as you would have them do unto you," is a good rule for guidance. This is effective in all situations. The following are some specifics that good leaders need to be cognizant of and apply to the motivation of safety and health:

- Communications—always keep people informed of what is going on within their organization. They like to feel that they can be trusted with information when it becomes available. Make your expectations clear and follow the old adage, "Say what you mean and mean what you say." Make time available to meet with people and make that time unhurried and without interruption. Actively listen to what they say to you. Get to know the people you are trying to motivate and find out what their goals and aspirations really are.
- Involve people—allow people the flexibility of involving in the decisions that directly affect them. This will increase their personal commitment and their feelings of exercising some control over what affects them. Include individuals in goal setting; this increases their stake in accomplishing the established goals. Let them know what part they play in accomplishing these goals and how they can contribute. It is critical to get everyone involved.
- Respond to others—frequently provide feedback to others. Whether the comments are positive or negative, do not wait until a specific time or until something is finished. Feedback is most effective when you let people know how they are doing immediately following their performance.
- Support others—help others reach their goals by offering advice and guidance, and recognize and reward good performance. In the work environment, help others get the rewards they deserve and make every effort to get a raise or a promotion for individuals who warrant it.
- Demonstrate respect for others—when meeting with someone, do not disrupt a meeting by answering the telephone; their time is also valuable to them. Avoid canceling or scheduling meetings at the last minute; this is indicative of poor preparation and the lack, again, of consideration for others' schedules. Do not reprimand another person in front of siblings, peers, or fellow workers; this will certainly result in ill will.
- Be a role model—no matter what sacrifices you make, you cannot be an effective motivator unless you demonstrate and live what you expect of others. You need to take the lead by being prompt, conscientious, and consistent if you expect others to mimic your leadership.

Key Person

In maintaining communications, fostering good morale, attaining production goals, and assuring that workers are working safely, no other person is as important as the first-line supervisor. All that affects workers comes directly from the first-line supervisor. This includes all training, job communications, enforcement of safety and health rules, the company line, and feedback on the overall function of the company. Thus, the supervisor sets the tone for the motivational environment and is the role model upon which workers base their own degree of motivation. The supervisor's role is critical to the function of the jobsite and the efficient and effective accomplishment of all work activities. No other person has more control over the workforce than this individual. If the supervisor emphasizes production above safety and health, then his or her workforce will tend to be motivated in that direction. This is the reason that the supervisor's skills as a facilitator of people are more critical than his or her expertise related to the work performed. These individuals need more training and support than anyone else on the jobsite.

EFFECTS OF OTHER FACTORS

Peer Pressure

Peer pressure is a very powerful motivator and can be either rewarding or punishing. Peer groups who are doing just enough to get by tend to draw or attract less motivated individuals. The less motivated individuals tend to identify with the peer group and are governed more by their peer group than are the highly motivated individuals. Normally, highly motivated individuals do not succumb easily to peer pressure. At certain times, peer pressure can result in an individual being bullied, which should not be the intent of positive peer pressure.

Social and family pressure has a role in motivating individuals, but most people sense that peer groups are the prime movers in the workplace motivational arena; thus, to motivate an individual who is under the influence of peers, one must spend time trying to change the peer group behavior. This is the only way to achieve the motivation of an individual within that group.

The following is given as an illustration of this type of peer pressure. On one occasion, a company needed to reduce the number of accidents occurring on a project. The management decided that an incentive program would be installed. In this program, the workers were told that if their group went a certain length of time without an accident, the group would receive an award, and each individual in the work group would also receive an additional reward. This was visualized as creating a peer pressure situation because everyone began looking out for each other to keep anyone from getting injured. This incentive program did seem to work and have a positive effect on the workers. This was best illustrated when one of the miners in this group slipped and broke an ankle. As they were loading him into the ambulance, one of his fellow workers yelled to him saying, "Tell them you fell and broke it at home." As you can see from this example, there was pressure from the peer group to avoid accidents so they could receive their rewards. This type of pressure can be viewed by some as positive and others as negative since it would encourage individuals to not report their injuries or illnesses.

Family Pressure

The following is an example of how family pressure can be used as a positive reinforcement or motivator. In this particular situation, the workers in a company were told that if they worked safely at their jobsite they would accumulate points, which could be used to purchase items from their company catalog. These items could be purchased for either themselves or their family members. One worker said his son reminded him daily to work safely. He did this because his father had promised him a bicycle when he accumulated enough points. The father was extra cautious at work because he had made this promise to his son and did not want to let him down. He said this had been a true incentive for him and had motivated him to work safely. Therefore, as you can see by this example, there are times when family pressure can also be a true motivator.

Incentives

Using an incentive as a reward can be a positive motivator but unless you are able to achieve a behavioral change, it may only be a temporary motivator. Therefore, to get the behavioral change you desire, you need to be aware that workers hold a more positive attitude toward their work when their supervisor provides them with a reward that they desire and is most meaningful to them. In addition, it has been found that employees are very receptive to rewards that are given to them and that had been expected. Rewards of this nature seem to be even more satisfying and are received more enthusiastically than the rewards they knew they were going to receive. Many of us try to use rewards or incentives to achieve the behavior we would like to see, but, as you must realize, there is a potential to backfire on us. Therefore, you can see how important it is that your incentives and rewards be well thought out and planned so that this does not happen to you.

Incentive Program

There is no need to try to implement an incentive program if you do not have an implemented safety and health program. All other aspects of safety and health must be in place before using an incentive program. In other words, an incentive program is a component to reinforce what presently exists. If you have not structured an environment in which workers sense the importance of safety, feel involved in the process, have safety conscious leaders and supervisors as role models, and are directed by goals to prevent incidents that can cause workplace injuries and illnesses, then an incentive program is of no use. An example of an incentive program that tries to address many of the pitfalls faced by employers is presented in Figure 5.1.

This is an incentive program that tries to address individual injuries, supervisor commitment, crew peer pressure, absenteeism, and nonreporting of incidents. You can use anything deemed valuable by the workforce for a reward. It does not have to be personal days off (PDOs). The reward

---

Incentive program policy

This program is started to eliminate accidents and reduce absenteeism at the _____ (Company).
This program will provide personal days off (PDOs) for quarters without accidents.
Individual awards (includes supervisors):

- Initial two quarters without an accident = 1 PDO
- Third quarter without an accident = 1 PDO
- Fourth quarter without an accident = 1 PDO

Supervisory awards:

- 1 PDO for every 10 workers who earn a PDO

Crew awards:

- Crews without accidents for two successive quarters will receive a free dinner and an extra PDO per member.

Figure 5.1   Sample incentive program.

Company award:

- Banquet for all employees will be provided if the incident rate, severity rate, and average lost-work days are below the national average for our specific industry for the calendar year.

Penalties:

- After a lost-time accident, workers and crews must have one quarter of accident-free work to earn a PDO.
- For any days missed without authorized approval, the worker and his or her crew will not be eligible to earn PDOs for one successive quarter.
- Failure to report an injury by a worker or supervisor will result in loss of ability to earn PDOs for two quarters.

Figure 5.1 (continued)

should increase as the length of time after an accident increases. This is only an example. You should create what will work for you.

Contests

Some organizations use contests for incentives but generally, they are found to be unsatisfactory. In many instances, too many negative response factors come into play. Contests are a type of competition and not everyone likes to compete. Many times, there is only one winner in the contest and therefore, many receive no type of reward even though they have worked very hard and to the best of their ability. This can affect an individual's status and, at times, even his or her morale. Contests do not increase total productivity, do not increase cooperation, do not motivate everyone, are self-serving, decrease group problem solving, and may cause suspicion and hostility. After reading the previous list of negatives, you can understand why contests generally are not a recommended practice. As with anything, though, there are always exceptions to the rule.

If you do decide that a contest is good for your particular situation, design the contest so that it will involve all individuals, foster status and pride, involve group competition, and involve the management. Be sure that the same individual does not win all the time; this can be very discouraging to the other participants. If varied skill levels exist, use handicaps. Make sure winning and losing are distributed and give prizes to first, second, and third place winners. Contests can be somewhat effective when used properly, but do not "bet the farm" on them as the answer to your motivational issues.

Gimmicks and Gadgets

Gimmicks and gadgets are novel or unconventional ideas, gifts, or devices, which call attention to a desired response and maintain motivation in an unusual manner. Some examples are

- Presenting a crew with monogrammed jackets for their achievement or performance. This type of gift increases group morale, peer pressure, and recognition of safety and health performance.
- Using an Olympic weight lifter to demonstrate the right way to lift.
- Using other gadgets such as knives, belt buckles, caps, pens, key chains, patches, rings, t-shirts, or trophies. Supervisors can be given a number of items and he can distribute these as reinforcers when he or she witnesses a worker performing in a safe manner (these items should not be given

to everyone since they would not be effective as motivators). Since these gadgets often have the company logo on them or due to their value, they can become collector items and constant reminders of why they received them, such as for safe performance or not having an accident.

All of these ideas are short-term motivators and should only be used to supplement an existing comprehensive program; they should not be the entire program.

Visuals

Visuals, such as posters and bulletin boards, can be used as motivational tools. They are beneficial in that they serve as a constant reminder of the desired goal you are trying to reach. Bulletin boards and posters need to be changed often and kept updated. You can even get your employees who have graphic or artistic talents to develop posters, which you can have reproduced and displayed. This way the employees get some recognition and you do not have the cost of commercially made posters. Videos are another excellent way of motivating individuals or groups especially if they are made by other workers or work groups and are played repeatedly in cafeterias, break rooms, or common areas.

Another type of visual is electronic signs with streaming messages regarding successes either as numbers or accomplishments in safety and health initiatives or goals. The use of visuals during a talk normally increases its effectiveness. Using personalized information, written literature, and statistics tends to more readily hold the participants' attention.

Conferences and Seminars

Conferences and seminars can be used as good motivational tools. In these meetings all members of the group are focused on one topic, problem, or activity and everyone can respond to the same information and materials. In this setting the group can be motivated to act as one entity. Therefore, do not overlook the possibility of using this type of method; it can be very beneficial. When returning from a conference or seminar, individuals or groups are often rejuvenated because of the new ideas they received during the meetings.

Nonfinancial Incentives

Nonfinancial incentives can be such things as the use of praise, knowledge of results (output), competition, experience of progress, experience of achievement, or granting a request. Nonfinancial incentives such as "safety and health employees of the month awards" may result in a special parking spot, picture of the employee, and recognition in the company newsletter, at a meeting, or impromptu pizza party, etc. Some of the most powerful motivators are achievement, recognition of a person's work or task, responsibility, and growth potential.

These motivators can become functional by giving someone more control, but at the same time, holding him or her accountable. You can make him or her more accountable by making him or her responsible for a discrete outcome, or allowing him or her additional authority. You will also be more successful if you keep the people you are trying to motivate informed by direct communication, instead of through someone else. Challenge them with tasks that are more difficult and allow them to become specialized in a certain area.

Communication including interpersonal relations, employers/supervisors, promotion/recognition, work conditions, and status are some external factors that motivate people. Personal or internal motivators give more freedom of choice of activity, freedom from criticism, work environment, choice of peers, fewer status factors, less supervisor or employee conflict, and more opportunity to be oneself.

SUMMARY

It is evident that we spend a large portion of our lives either motivating ourselves or trying to motivate someone else. Thus, it is hoped that you have gained some insights on ways in which you can be more effective at motivating yourself and others. In summary, some of the key traits that I believe are critical to understanding how to motivate people are

1. People are self-motivated
2. What people do seems logical and rational to them
3. People are influenced by what is expected of them
4. To each individual, the most important person is one's self
5. People support what they create or are involved in
6. Conflict is natural (normal) and can be used positively
7. People prefer to keep things the way they are rather than to make a change
8. People are underutilized

With these thoughts in mind, people can be motivated by

- Allowing them involvement and participation
- Delegating responsibility with authority to them
- Effectively communicating with them
- Demonstrating concern and assisting them with counseling and coaching
- Being a good role model to them
- Having high expectations of them
- Providing rewards and promotions based on their achievements

Workers need to know what is expected of them, what happens if they do not perform, and what the rewards, outcomes, or consequences will be for not meeting expectations. We are motivated by what we think the consequences of our actions will be. Those consequences should be immediate, certain, and positive if we expect them to be motivational.

Practically all our motivational attempts are geared toward peers or employees and this is accomplished through the employer, fellow workers, or supervisors. These motivators are, among other things, a funnel that directs all materials and information to those who need to be motivated. The motivator also directs or carries out the vast majority of learning.

People have many abilities and talents that they are unaware of or just do not use. As a motivator, it is your responsibility to bring out those hidden abilities and talents and channel them toward the goals, outcomes, behaviors, and objectives you desire. If this is done as discussed in previous sections, it will give them a new sense of enthusiasm and self-esteem.

Motivation takes a lot of nurturing and caring for both the people involved to benefit and for the goals to be attained. Many organizations say that people are their most important asset but fail to exhibit that principle by the manner in which they treat their employees.

Motivation is not something that you can schedule for a Thursday at 2:00 p.m. It is a process that requires your continuing commitment and your ability to have an objective view of your own self. You must also have an understanding of your effect on others.

The essence of motivation is to find meaning in what you are doing. Motivation is the predisposition of doing something to satisfy a need. In real life, most people rarely have just one need; they have several needs at any one given time and consequently, they are moved to do something about them. Unfortunately, if they have too many needs facing them at one time, they may become indecisive, highly aggressive, negative, or even irrational.

Motivation is internal and can be stimulated by leadership and incentives. However, unless you know something about the needs, desires, and drives of the other person, your leadership and incentives may be completely ineffective. When a person's task or job does not permit him to satisfy his own personal needs, he is less likely to work as hard at accomplishing the task you have chosen for him. It seems safe to say that people do things well if they are excited about their assigned tasks. When their external environment assures that their own needs, wants, and desires will be met, it further enhances their desire to do a good job. You, as a leader and motivator, are also responsible for helping others meet the demands of their world to the level of their capabilities. When each of these aspects is fulfilled, you will have an excellent motivational situation.

REFERENCE

Frymier, J.R., *Motivation to Learn*, Kappa Delta Pi Press, West Lafayette, IN, 1985.

CHAPTER 6

How They Act: Behavior-Based Safety

A worker demonstrating safe behavior as a part of his job function.

BEHAVIOR-BASED SAFETY

Behavior-based safety (BBS) is not a new concept since its preceptors have been around since the early work of H.W. Heinrich, who in the 1930s and 1940s suggested that unsafe acts by workers were the primary cause of accidents/incidents that resulted in occupationally related injuries and illnesses. Over the years, the figure attributed to these unsafe acts has ranged from 75% to 95%.

In fact, many safety researchers have viewed these figures as resulting in fault finding rather than fact finding as well as causing workers to underreport accident/incident events to prevent themselves from becoming the target of scrutiny, enforcement, discipline, or reprisal.

The major complaint with this concept is that it overlooks the importance or contribution to prevention of accidents that unsafe conditions attribute to the numbers of accidents/incidents occurring. When addressing accident/incident prevention, it has never been my personal opinion or contention that addressing only unsafe acts was the cure-all in the accident prevention puzzle.

Figure 6.1   Five factors in the accident sequence. (Courtesy of Mine Safety and Health Administration.)

HEINRICH'S DOMINOS CONCEPT

H.W. Heinrich suggested that an accident/incident sequence was composed of five components. These components are

1. Ancestry and social environment
2. Fault of the person
3. Unsafe act and mechanical or physical hazard
4. Accident
5. Injury (see Figure 6.1)

Ancestry and social environment were of the idea that recklessness, stubbornness, avariciousness, and other undesirable traits of character may be passed along through inheritance. In addition, the environment may develop undesirable traits of character or may interfere with education and training. Both inheritance and environment cause faults of the person.

Faults of the person are due to inherited or acquired faults such as recklessness, violent temper, nervousness, excitability, inconsiderateness, ignorance of safe practices, etc., and constitute proximate reasons for committing unsafe acts or for the existence of mechanical or physical hazards.

Unsafe acts or performance of workers, such as standing under a suspended load, starting machinery without warning, horseplay, and removal of safeguards; mechanical or physical hazards such as unguarded gears, unguarded points of operation, absence of rail guards and insufficient light, result in accidents.

Accident events such as falls of persons, striking of persons by flying objects, etc., are typical accidents/incidents that can cause injuries. Injuries are fractures, lacerations, etc. that result directly from accidents/incidents.

Heinrich suggests that if any one of these dominos falls a chain of events that results in injury will occur (see Figure 6.2), but if the unsafe act domino could be removed then no accident/incident or injury could occur (see Figure 6.3).

Figure 6.2 Injury is caused by the action of preceding factors. (Courtesy of Mine Safety and Health Administration.)

Figure 6.3 Removal of the unsafe act and mechanical hazard results in no injury. (Courtesy of Mine Safety and Health Administration.)

As can be seen this places the onus almost entirely upon the worker or person to prevent an accident/incident and its resulting injury.

BBS TODAY

BBS cannot be viewed as the panacea or end-all solution for the prevention of accidents/incidents, but only as one tool in the arsenal of tools, and it does not supplant a complete and organized overall approach in addressing occupational safety and health issues of today. All the components discussed in this book must be in place, such as training, safety and health program, accident/incident analysis, safety engineering, controls, interventions, etc. It is only then that BBS can become an integral part of the occupational safety and health initiative.

It has always been a goal to motivate all employees to perform their tasks in a safe and healthy manner, but this goal is achievable only when all safeguards are in place, all feasible protections are provided, all hazards are eliminated or controlled, safety and health is managed effectively, etc., and the workplace has been structured to protect the workforce as best as possible. Then behavioral approaches can be implemented to elicit changed behavioral pattern and attitudes toward the standard practice of all the workforce to self and groups, taking responsibility and becoming involved with the prevention process by adhering to policies, safe procedures/practices, and rules with regard to working and performance of all aspects of their jobs in a safe manner, without any thought of circumvention of standards of practice or best practices.

This chapter is not a how-to chapter but an explanation BBS, the pros and cons of using a BBS approach, and why it is used. BBS is not a cure-all for safety and health issues, nor is it meant to replace the existing safety and health program. It is often viewed as a quick fix for safety and health problems, but is not easy to implement. It is an organized approach and not common sense or a new way to blame employees.

Every BBS approach must be designed to fit the needs and culture of the organization or business. It is based upon the notion that safety and health is a shared responsibility and not just a personal matter. It is a way the employers provide the tools to optimize safety performance in the employees' unique work environment by developing methods to measure successes in safety performance, rather than using the traditional failure rates.

BBS Described

BBS is a process used to identify at-risk behaviors that are likely to cause injury to workers and is dependent on the involvement of workers in this process so that they become a willing participant and buy into the concepts and purpose of BBS. Finally, they will be asked to observe each other and their coworkers to determine if decreases in at-risk behavior have resulted in a reduction in these unsafe behaviors.

This is a very simplistic description of BBS. Although there are an infinite number of variations to BBS programs, they all share common characteristics. These are

- Identifying critical behaviors
- Gather data and information
- Encouraging two-way feedback
- Stimulating continuous improvement

Identifying Critical Behaviors

In this first step of BBS an analysis of at-risk behaviors needs to be completed, which will be used to make safe behavior more probable and at-risk behavior less probable. This is to be done by a steering committee made up of operational personnel. This includes front-line individuals such as supervisors, experienced operators, and other interested workers. They will need to use resources at their disposal to identify the major contributors to incidents, accidents, injuries, near-misses, and property damage. This group is often called a steering committee and is the first step in gaining support for behavior-based programs. They will identify a cadre of at-risk behaviors that will act as the foundation for this approach. By consulting the workers doing the jobs or tasks, a more complete set of the types of behaviors that contribute to incidents can be used to help develop a data sheet that can be used by trained observers to assess the amount of safe behaviors occurring. The iceberg found in Figure 6.4 attempts to indicate that there are more at-risk behaviors and near-misses than

Figure 6.4 Usually the tip of the iceberg is observed.

actual fatalities, lost time accidents, Occupational Safety and Health Administration (OSHA) recordables, or first-aid events.

Gathering Data

This step is completed by trained observers who are in most cases fellow workers. These observers are not intent on finding fault or blaming workers for their safety behaviors, but document the rate at which workers perform tasks in a safe manner or in an at-risk fashion. The observations usually take 10–15 min. The data sheets developed by the steering committee should be the guide and specify the expected behaviors. These observations are strictly conducted under the conditions that no names are used and no blame is placed. These observations are best when they are conducted peer-to-peer and feedback given immediately. Observation data are entered into a database for analysis and problem solving. This approach builds a sense of ownership.

Two-Way Feedback

The observer can provide the person performing the task with immediate feedback on at-risk behaviors and provide reinforcement on safe work performance. If the observer asks the worker why he or she used at-risk behaviors to perform a task, the observer can learn if there are roadblocks that prevent the worker from performing the task in a safe manner. The observer many find that the protective equipment is not available or is no longer usable, which becomes a follow-up item for the steering committee or management.

This process is founded on the premise that for every accident there are hundreds or sometimes thousands of at-risk behaviors. When at-risk behaviors are reduced, the likelihood of injuries is also reduced. A successful approach does not look for blame, but provides two-way feedback that promotes the idea that workers are indeed the solution. Any positive reinforcement that can be provided will help to strengthen the safe performance and cause it to become the norm.

Continuous Improvement

Using the comment data will allow site personnel to target areas of improvement and demonstrate to workers that their input is critical and an important component of the program. If at-risk behaviors occur in certain areas, it is an accident waiting to happen. If there are barriers to safe performance, continuous improvement is deterred. The most common barriers to safe performance are

- Hazard recognition: If workers do not realize that they are performing an at-risk behavior, then they can never perform the task in a safe manner.
- Business systems: At-risk behavior is the result of an organization system that is unreliable due to inefficiency. If this occurs workers will avoid using the system; they will find a way around it.
- Disagreement on safe practices: There can be legitimate disagreement as to what constitutes safe performance and this needs to be reviewed and addressed in some manner. This is best accomplished by working toward an agreed upon consensus.
- Culture: The way that something was always done may be at odds with what is a safe practice. It is hard to teach old dogs new tricks.
- Inappropriate rewards: Rewards for achieving production may be at odds with safety and reinforce that at-risk shortcuts are more of a benefit than safe performance.
- Facilities and equipment: Outdated facilities or processes, and rigged, missing, or damaged equipment may cause workers to act in an unsafe manner.

- Personal factors: The at-risk behavior results from personal characteristics of the worker, which results in him or her deliberately taking risks or refusing to work safely as a result of factors such as fatigue, medication, stress, or illness.
- Personal choice: A worker with adequate skill, knowledge, and resources chooses to work at risk to save, time, effort, or something similar.

These barriers can be removed with a successful approach by observing and talking with employees and never, in any manner, implying that workers are the problem. It is important that the workers are viewed as the solution.

The idea that consequences control behavior is the conceptual foundation of BBS. Thus, the majority of behaviors rely on applying previous experience of consequences (both negative and positive) as the reinforcing factor. A picture of an amputated finger portrays visually the consequences of at-risk behavior and this reminder of a negative consequence may be enough to cause workers to alter their behavior before a similar incident.

Even though the previous barriers are addressed there are no guarantees that BBS will work in your situation, but the principles are applicable to any situation when designed and implemented to meet your needs.

HINDRANCES TO IMPLEMENTING BBS

It is never easy to implement anything new and BBS is no different. If you cannot overcome the hindrances to implementing a program such as BBS, then failure is guaranteed. Some of these deterrents are as follows:

- Failure to ensure the participation of all those expected to participate in the basic principles and philosophies behind the program and in its success will result in BBS becoming just the newest kid on the block or as some call it the flavor of the month. To be fully successful everyone needs to be a participant in order for BBS to become part of the company's culture.
- If the participants understand the principles and honest intent of the BBS program, then they will feel more at ease in taking ownership. If they do not have ownership then they will not incorporate it into the culture of the company or organization.
- When it is realized that the workers performing the tasks are the true safety experts, they must be enticed to share responsibility for its implementation. Without the support of the front-line supervisors and workers, failure is assured. Techniques to elicit bottom-up involvement are critical.
- Resources and lip service of BBS are important since no program functions without top-down support, but the real commitment is the physical presence from the top since what management does speaks louder than their words. Physical/visible presence is more powerful than other forms of support.
- Every program needs its cheerleaders. This means that management, supervisors, and labor leaders need to support BBS by becoming role models and extending themselves to commit time and effort. To be leader for BBS the most valued gift is to give of yourself and your time to it. A real champion is committed. Management must give these individuals the education and training as well as resource support in time and materials to facilitate the program.
- The mission of the organization or company is the high ground and usually very lofty such as an accident or injury-free workplace while the goal may by a percent reduction of accidents or injuries over a given period, which is a possible, realistic, and attainable goal. Progress toward this goal is measurable and accountability can be obtained if the individual or group has direct control and resources to achieve it. Remember that purpose and goals are separate animals.
- It is difficult to effectively measure the outcomes of BBS since the reduction of injury rates can be manipulated and may not be a true indicator of the effectiveness of this program. The completion of observations will indicate, the number of individuals actually participating with deeds, or the

refining of at-risk behavior identification as denoted by extinction of some behavior, the reporting of more at-risk behaviors or near-misses, or first aid case. Measures that indicate changes in culture and successes in the program may be slow to evolve, but are better positive reinforcers than just counting bodies.

By addressing these potential hindrances to the BBS the company or organization can take steps toward an effective implementation of a successful program.

SUMMARY

First, BBS is based on the general principles that behavior causes the majority of accidents, but this does not excuse employers from providing a safely engineered workplace with all controls in place to prevent the occurrence of incidents. Second, accountability inspires behavior and accountability facilitates accomplishments. Third, feedback that fosters good communications is the key to continuous improvement, and excellence in safety needs to be established as the underlying culture desired in the organization or company.

These premises are driven by the following strategies:

- Obtaining of objective evidence of at-risk behaviors
- Defining barriers to safe behavior
- Teaching ways to substitute safe behavior for at-risk practices
- Holding employees accountable to improve their safety-related behaviors and help others do the same
- Demonstrating the effectiveness by measurement that garners continued management support

BBS is used to increase safety awareness and to decrease accidents/incidents, by focusing on identification and elimination of unsafe behaviors. Workers are trained to conduct safety observation and give guidance on specific behaviors while collecting the information in a readily available format for providing immediate feedback. Observations are structured to have minimal impact on the workload and data are shared with the entire workforce.

For such an approach to be successful, a good organizational safety culture (see Chapter 21) and people's participation and involvement are required. Since real-time safety analysis is an integral part of BBS using operational personnel involvement to identify hazards and risks that are the key to effective BBS, the behavior-based process allows an organization or company to create and maintain a positive safety culture that continually reinforces safe behaviors over unsafe behaviors, which ultimately results in a reduction of risk.

The organization or company's purpose must be to continuously improve with the ultimate goal being a workplace that is free of injuries and illnesses. While attitudes are not addressed directly, it is a deep-seated intention to have employees accept safety as a value over time.

CHAPTER 7

Search for the Culprits: Hazard Identification

Hazard identification is a major step in prevention.

Before you start reading this chapter, please take a moment to circle the number of the answer option that best fits the state of your safety and health initiative at this time.

Hazard Identification

Topic	Circle Answer	Answer Options
Hazard identification (expert survey)	5	Comprehensive expert surveys are conducted regularly and result in corrective action and updated hazard inventories
	4	Comprehensive expert surveys are conducted periodically and drive appropriate corrective action
	3	Comprehensive expert surveys are conducted, but corrective action sometimes lags
	2	Expert surveys in response to accidents, complaints, or compliance activity only
	1	No comprehensive surveys have been conducted

(continued)

Hazard Identification (continued)

Topic	Circle Answer	Answer Options
Hazard reporting system	5	A system exists for hazard reporting, employees feel comfortable using it, and employees feel comfortable correcting hazards on their own initiative
	4	A system exists for hazard reporting and employees feel comfortable using it
	3	A system exists for hazard reporting and employees feel they can use it, but the system is slow to respond
	2	A system exists for hazard reporting but employees find it unresponsive or are unclear how to use it
	1	There is no hazard reporting system or employees are not comfortable reporting hazards
Hazard identification (inspection)	5	Employees and supervisors are trained, conduct routine joint inspections, and all items are corrected
	4	Inspections are conducted and all items are corrected; repeat hazards are seldom found
	3	Inspections are conducted and most items are corrected, but some hazards are still uncorrected
	2	An inspection program exists, but corrective action is not complete; hazards remain uncorrected
	1	There is no routine inspection program in place and many hazards can be found
Accident/incident investigation	5	All loss-producing incidents and near-misses are investigated for root cause with effective prevention
	4	All Occupational Safety and Health Administration (OSHA) reportable incidents are investigated and effective prevention is implemented
	3	OSHA-reportable incidents are generally investigated; accident cause and correction may be inadequate
	2	Some investigation of incidents takes place, but root cause is seldom identified and correction is spotty
	1	Injuries are either not investigated or investigation is limited to report writing required for compliance

HAZARD IDENTIFICATION

Hazard identification is a process controlled by management. You must assess the outcome of the hazard identification process and determine whether immediate action is necessary or whether, in fact, there is an actual hazard involved. When you do not view a reported hazard as an actual hazard, it is critical to the ongoing process to inform the worker that you do not view it as a true hazard and explain why. This will ensure the continued cooperation of workers in hazard identification.

It is important to remember that a worker may perceive something as a hazard, when in fact it may not be a true hazard; the risk may not match the ranking that the worker places on it. Also, even if hazards exist, you need to prioritize them according to the ones that can be handled quickly, which may take time, or which will cost money above your budget. If the correction will cause a large capital expense and the risk is real but does not exhibit an extreme danger to life and health, you might need to wait until the next year's budget cycle. An example of this would be when workers complain of smell and dust created by a chemical process. If the dust is not above accepted exposure limits and the smell is not overwhelming, then the company may elect to install a new ventilation system, but not until the next year because of budgetary constraints. The use of personal protective equipment (PPE) until hazards can be removed may be required.

The expected benefits of hazard identification are a decrease in incidents of injuries, a decrease in lost workdays and absenteeism, a decrease in workers' compensation costs, increased productivity, and better cooperation and communication. The baseline for determining the benefit of the hazard identification can be formulated from existing company data on occupational injuries/illnesses, workers' compensation, attendance, profit, and production.

Hazard identification includes those items that can assist you with identifying workplace hazards and determining the corrective action necessary to control them. These items include jobsite safety inspections, accident investigations, safety and health committees, and project safety meetings. Identification and control of hazards should include periodic site safety inspection programs that involve supervisors and, if you have them, joint labor management committees. Safety inspections should ensure that preventive controls are in place (PPE, guards, maintenance, engineering controls), that action is taken to quickly address hazards, that technical resources such as OSHA, state agencies, professional organizations, and consultants are used, and that safety and health rules are enforced.

Many workplaces have high accident incidence and severity rates because they are hazardous. Hazards are dangerous situations or conditions that can lead to accidents. The more the hazards present, the greater the chance that there will be accidents. Most hazards are from the direct release of some form of energy (i.e., mechanical, electrical, thermal, etc.). The control of hazardous energy sources is the main avenue for prevention of incidents that could result in injury, illnesses, or death. Unless safety procedures are followed, a direct relationship will exist between the number of hazards in the workplace and the number of accidents that will occur there.

As in most industries, people work together with machines in an environment that causes employees to face hazards, which can lead to injury, disability, or even death. To prevent industrial accidents, the people, machines, and other factors that can cause accidents, including the energies associated with them, must be controlled. This can be done through education and training, good safety engineering, and enforcement.

The core of an effective safety and health program is hazard identification and control. Periodic inspections and procedures for correction and control provide methods of identifying existing or potential hazards in the workplace and eliminating or controlling them. The hazard control system provides a basis for developing safe work procedures and injury and illness prevention training. Hazards occurring or recurring reflect a breakdown in the hazard control system.

The written safety and health program establishes procedures and responsibilities for the identification and correction of workplace hazards. The following activities can be used to identify and control workplace hazards: hazard reporting system, jobsite inspections, accident investigation, and expert audits.

After all basic steps of the operation of a piece of equipment or job procedure have been listed, we need to examine each job step to identify hazards associated with it. The purpose is to identify and list the possible hazards in each step of the job. Some hazards are more likely to occur than others, and some are more likely to produce serious injuries than others. Consider all reasonable possibilities when identifying hazards.

To make this task manageable you should work with basic types of accidents. The question to ask oneself is "Can any of these accident types or hazards inflict injury to a worker?" There are 11 basic types of accidents:

1. Struck-against
2. Struck-by
3. Contact-with
4. Contacted-by

  5. Caught-in
  6. Caught-on
  7. Caught-between
  8. Fall-same-level
  9. Fall-to-below
 10. Overexertion
 11. Exposure

You should look at each of these basic accident types to identify procedures, processes, occupations and tasks that present a hazard to cause one of the accident types in the following section.

ACCIDENT TYPES

Struck-Against Type of Accidents

Consider the first four basic accident types (struck-against, struck-by, contact-with, and contacted-by) in more detail, with the job step walk-around inspection in mind. Can the worker strike against anything while doing the job step? Think of the worker moving and contacting something forcefully and unexpectedly (an object capable of causing injury). Can he or she forcefully contact anything that will cause injury? This forceful contact may be with machinery, timber or bolts, protruding objects, or sharp, jagged edges. Identify not only what the worker can strike against, but how the contact can come about. This does not mean that every object around the worker must be listed.

Struck-By Type of Accidents

Can the worker be struck by anything while doing the job step? The phrase struck-by means that something moves and strikes the worker abruptly with force. Study the work environment for what is moving in the vicinity of the worker, what is about to move, or what will move as a result of what the worker does. Is unexpected movement possible from normally stationary objects? Examples are ladders, tools, containers, and supplies.

Contact-With and Contacted-By Types of Accidents

The subtle difference between contact-with and contacted-by injuries is that in the first, the agent moves to the victim, while in the second, the victim moves to the agent.

Can the worker be contacted by anything while doing the job step? The contacted-by accident is one in which the worker could be contacted by some object or agent. This object or agent is capable of injuring by nonforceful contact. Examples of items capable of causing injury are chemicals, hot solutions, fire, electrical flashes, and steam.

Can the worker come in contact with some agent that will injure without forceful contact? Any type of work that involves materials or equipment that may be harmful without forceful contact is a source of contact-with accidents. Two kinds of work situations account for most of the contact-with accidents. One situation is working on or near electrically charged equipment and the other is working with chemicals or handling chemical containers.

Caught-In and Caught-On Types of Accidents

The next three accident types involve caught accidents. Can the person be caught in, caught on, or caught between objects? A caught-in accident is one in which the person, or some part of his or her body, is caught in an enclosure or opening of some kind. Can the worker be caught on anything while doing the job step? Most caught-on accidents involve worker's clothing being caught on some projection of a moving object. This moving object pulls the worker into an injury contact. Alternatively, the worker may be caught on a stationary protruding object, causing a fall.

Caught-Between Type of Accidents

Can the worker be caught between any objects while doing the job step? Caught-between accidents involve a part of the body caught between something moving and something stationary, or between two moving objects. Always look for pinch points.

Fall-Same-Level and Fall-to-Below Types of Accidents

Slip, trip, and fall accident types are some of the most common accidents occurring in the workplace. Can the worker fall while doing a job step? Falls are such frequent accidents that we need to look thoroughly for slip, trip, and fall hazards. Consider whether the worker can fall from something above ground level, or fall to the same level. Two hazards account for most fall-to-same level accidents: slipping hazards and tripping hazards. The fall-to-below accidents occur in situations where employees work above ground or above floor level, and the results are usually more severe.

Overexertion and Exposure Types of Accidents

The next two accident types are overexertion and exposure. Can the worker be injured by overexertion; that is, can he or she be injured while lifting, pulling, or pushing? Can awkward body positioning while doing a job step cause a sprain or strain? Can the repetitive nature of a task result in injury to the body? An example of this is excessive flexing of the wrist, which can cause carpal tunnel syndrome (which is abnormal pressure on the tendons and nerves in the wrist).

Finally, can exposure to the work environment cause injury to the worker? Environmental conditions such as noise, extreme temperatures, poor air, toxic gases and chemicals, or harmful fumes from work operations should also be listed as hazards.

HAZARD REPORTING SYSTEM

Hazard identification is a technique used to examine the workplace for hazards with the potential to cause accidents. Hazard identification, as envisioned in this section, is a worker-oriented process. Workers are trained in hazard identification and asked to recognize and report hazards for evaluation and assessment. Management is not as close to the actual work performed as are those performing the work. Even supervisors can use extra pairs of eyes looking for areas of concern.

Workers already have hazard concerns and have often devised ways to mitigate the hazards, thus preventing injuries and accidents. This type of information is invaluable when removing and reducing workplace hazards.

This approach to hazard identification does not require that someone with special training conduct it. It can usually be accomplished by the use of a short fill-in-the-blank questionnaire (see Appendix C). This hazard identification technique works well where management is open and genuinely concerned about the safety and health of its workforce. The most time-consuming portion of this process is analyzing the assessment and response regarding potential hazards identified. Empowering workers to identify hazards, make recommendations on abatement of the hazards, and then suggest how management can respond to these potential hazards is essential. Only three responses are required:

Hazard identification form

Worker's name (optional) _____ Date _____

Jobsite_____ Job titles_____

1. Describe the hazard that exists.

_____

2. What are your recommendations for reducing or removing the hazard?

_____

3. What suggestions do you have for management for handling the hazard?

_____

4. Manager's or supervisor's response to hazard concern identified.

_____

Supervisor _____Date _____Time _____

_____

Note: Use a separate form for each hazard identified.

Figure 7.1   Hazard identification form.

1. Identify the hazard.
2. Explain how the hazards could be abated.
3. Suggest what the company could do.

Use a form similar to the one found in Figure 7.1. A sample statement to place in your written safety and health program is found in Figure 7.2.

WORKPLACE INSPECTIONS OR AUDITS

Workplace audits are inspections that are conducted to evaluate certain aspects of the work environment regarding occupational safety and health. The use of safety and health audits has been shown to have a positive effect on a company's loss control initiative. In fact, companies that perform safety and health audits have fewer accidents/incidents than companies that do not.

Safety and health audits (inspections), which are often conducted in workplaces, serve a number of evaluative purposes. Audits or inspections can be performed to

---

The (Company Name) is committed to identifying and removing or controlling hazards. This can only be accomplished by full cooperation of the workforce in the process. Employees are to report any perceived hazard using the company's standard form and giving it to the supervisor. The supervisor will respond to your observation by the next shift. If this does not occur, you are to inform your supervisor's immediate supervisor or proceed up the chain of management until you receive a response. A response may be that no hazard exists, that it will be fixed or removed, or it may be an explanation for the delay in fixing or removing the hazard.

Supervisors are to respond in writing on the company form detailing action or nonaction that will be taken and are to return a copy to the employee submitting the hazard identification form.

Figure 7.2   Sample written statement on hazard identification reporting.

1. Identify the existence of hazards
2. Check compliance with company rules and regulations
3. Check compliance with OSHA rules
4. Determine the safety and health conditions of the workplace
5. Determine the safe condition of equipment and machinery
6. Evaluate supervisors' safety and health performance
7. Evaluate workers' safety and health performance
8. Evaluate progress regarding safety and health issues and problems
9. Determine the effectiveness of new processes or procedural changes

NEED FOR AN AUDIT

First, determine what needs to be audited. You might want to audit specific occupations (e.g., machinist), tasks (e.g., welding), topics (e.g., electrical), teams (e.g., rescue), operators (e.g., crane operator), part of the worksite (e.g., loading/unloading), compliance with an OSHA regulation (e.g., hazard communication standard), or the complete worksite. You may want to perform an audit if any of the previous lists or activities have unique identifiable hazards, new tasks involved, increased risk potential, changes in job procedures, areas with unique operations, or areas where comparison can be made regarding safety and health factors.

In the process of performing audits, you may discover hazards that are in a new process, hazards once the process has been instituted, a need to modify or change processes or procedures, or situational hazards that may not exist at all times. These audits may verify that job procedures are followed and identify work practices that are both positive and negative. They may also detect exposure factors both chemical and physical and determine monitoring and maintenance methods and needs.

At times audits are driven by the frequency of injury; potential for injury; the severity of injuries; new or altered equipment, processes, and operations; and excessive waste or damaged equipment. These audits may be continuous, ongoing, planned, periodic, intermittent, or dependent on specific needs. Audits may also determine employee comprehension of procedures and rules and the effectiveness of workers' training, assess the work climate or perceptions held by workers and others, and evaluate the effectiveness of a supervisor in his or her commitment to safety and health.

At many active workplaces, daily site inspections are performed by the supervisor or foremen to detect hazardous conditions, equipment, materials, or unsafe work practices. At other times, periodic site inspections are conducted by the site safety and health officer. The frequency of inspections is established in the workplace safety and health program. The supervisor, in conjunction with the safety and health officer, determines the required frequency of these inspections, based on the level and complexity of the anticipated activities and on the hazards associated with these activities. When audits address site hazards or protect site workers as well as review worksite conditions and activities, inspections should include an evaluation of the effectiveness of the company's safety and health program. The safety and health officer should revise the company's safety and health program as necessary to ensure the program's continued effectiveness.

Before the start of each shift or new activity, a workplace and equipment inspection should take place. This should be done by the workers, crews, supervisor, and other qualified employees. At a minimum, they should check the equipment and materials that they will be using during the operation or shift for damage or defects that could present a safety hazard. In addition, they should check the work area for new or changing site conditions or activities that could also present a safety hazard. All employees should immediately report any identified hazards to their supervisors. All identified hazardous conditions should be eliminated or controlled immediately. When this is not possible, the following should be done:

- Interim control measures should be implemented immediately to protect workers.
- Warning signs should be posted at the location of the hazard.
- All affected employees should be informed of the location of the hazard and the required interim controls.
- Permanent control measures should be implemented as soon as possible.

When a supervisor is not sure how to correct an identified hazard, or is not sure if a specific condition presents a hazard, he or she should seek technical assistance from a competent person, a site safety and health officer, or from other supervisors or managers.

When to Audit

The supervisor or project inspector must perform daily inspections of active worksites to detect hazardous conditions resulting from equipment or materials or unsafe work practices. The supervisor, inspector, or site safety and health officer must perform periodic inspections of the workplace at a frequency established in the worksite's specific safety and health program. The supervisor, in conjunction with the site safety and health officer, should determine the required frequency of these inspections based on the level and complexity of anticipated work activities and on the hazards associated with these activities. In addition to a review of worksite conditions and activities, these inspections must include an evaluation of the effectiveness of the worksite safety and health program in addressing site hazards and in protecting site workers. The safety and health officer may need to revise the safety and health program as necessary to ensure the program's continued effectiveness. Work crew supervisors, foremen, and employees need to inspect their workplace before the start of each work shift or new activity. At a minimum, supervisors and employees should

1. Check the equipment and materials that they will use during the operation or work shift for damage or defects that could present a safety hazard
2. Check the work area for new or changing site conditions or activities that could present a safety hazard
3. Report identified hazards immediately to their supervisors

What to Audit

The complexity of the worksite and the myriad areas, equipment, tasks, materials, and requirements can make the content of most audits overwhelming. As seen in Figure 7.3, the audit topics that could be targeted on a worksite are expansive.

Safety and health audits should be an integral part of your safety and health effort. Anyone conducting a safety and health audit must know the workplace, the procedures or processes that are audited, the previous accident history, and the company's policies and operations. This person should also be trained in hazard recognition and interventions for safety and health.

The use of workplace inspections makes it necessary to have or develop written safety and health audit instruments. These instruments need to be tailored to meet the specific needs and intended purposes. Those using audit instruments need to be trained on their use and application. This means that they should be able to identify the hazards that are recorded on the audit form and must understand the workings of the workplace, process, equipment, etc., that they are to audit. These completed instruments are a record and documentation of one facet of your safety and health effort. A written sample safety and health audit policy is found in Figure 7.4. In Figure 7.5, you can find a sample jobsite inspection that can be altered and tailored to meet the unique needs of a particular company.

Acids	Fire protection	Personal services and
Aisles	Flammables	first aid
Alarms	Fork lifts	Power sources
Atmosphere	Fumes	Power tools
Automobiles	Gas cylinders	Radiation
Barrels	Gas engines	Railroad cars
Barriers	Gases	Respirators
Boilers	Generators	Safety devices
Buildings	Hand tools	Signs
Cabinets	Hard hats	Scaffolds
Catwalks	Hazardous chemical	Shafts
Caustics	processes	Shapers
Chemicals	Heavy equipment	Shelves
Compressed-gas cylinders	Hoists	Solvents
Containers	Horns and signals	Stairways
Controls	Hoses	Steam systems
Conveyors	Housekeeping	Storage facilities
Cranes	Jacks	Tanks
Confined spaces	Ladders	Transportation equipment
Docks	Lifting	Trucks
Doors	Lighting	Ventilation
Dusts	Loads	Walkways
Electrical equipment	Lockout/tagout	Walls and floor openings
Elevators	Machines	Warning devices
Emergency procedures	Materials	Welding and cutting
Environmental factors	Mists	Work permit
Explosives	Noise	Working surfaces
Extinguishers	Piping	Unsafe conditions
Fall protection	Platforms	Unsafe acts
Fibers	Personal protective	X-rays
Fire extinguishers	equipment	

Figure 7.3 Audit topics.

ACCIDENT INVESTIGATIONS

Introduction

Although accident investigation is an after-the-fact approach to hazard identification, it is still an important part of this process. At times hazards exist, which no one seems to recognize until they result in an accident or incident. Complicated accidents require an investigation to determine the

---

Jobsite inspections

Safety audits/inspections of the jobsite will be conducted, usually on a monthly basis, when conditions change, or when a new process or procedure is implemented. The inspections are to identify and correct potential safety and health hazards. A standard site evaluation worksheet (see examples in Figure 7.5 and Appendix D) will be used to conduct these jobsite safety inspections. Safe operating procedures will be used to determine the effectiveness of safety and health precautions. These audits/inspections are to be used to improve jobsite safety and health.

Figure 7.4 Sample written statement regarding jobsite inspections.

Sample jobsite inspection form*

Check if no unsafe act/conditions exist. Otherwise denote the extent of the problem.

Jobsite _____ Date _____

_____Housekeeping. Explain _____

_____No protruding nails exist. Explain _____

_____Adequate Illumination. Explain _____

_____Floor openings are covered or guarded. Explain _____

_____All stairways are in good condition. Explain _____

_____Ventilation is adequate. Explain _____

_____Fire extinguishers present and accessible. Explain _____

_____All equipment guards are in place. Explain_____

_____All ladders are in good condition. Explain _____

_____All gas cylinders are secured. Explain_____

_____No open access to energized electrical circuits. Explain _____

_____GFCIs are being used. Explain _____

_____Guardrail systems are in place. Explain _____

_____Hard hats are being worn. Explain _____

_____All chemical containers are labeled. Explain _____

_____Trenches and excavations are inspected by competent person. Explain _____

_____Workers are following safe lifting practices. Explain _____

_____Compressed air is below 30 psi. Explain _____

_____First aid supplies exist and are stocked. Explain _____

*This is only a short, noncomprehensive example of a jobsite inspection or audit instrument. A more detailed description of job inspection instruments, as well as recommendations for designing your own site evaluation instrument for your company, can be found in Appendix D.

Figure 7.5   Sample jobsite inspection form.

cause of the accident. This is especially true in cases where death results and few or no witnesses exist. An accident investigation is a fact-finding process and not a faultfinding process with the purpose of affixing blame. The result of an accident investigation should be assuring that the type of hazard or accident does not exist or will not occur in the future.

Your company should have a formalized accident investigation procedure, which is followed by everyone. It should be set down in writing and end with a written report, using as a foundation your standard company accident investigation form. This may be a form developed by you (such as the example provided in Figure 7.6). It may be your workers' compensation form or an equivalent from your insurance carrier.

Accidents and even near-misses should be investigated by your company if you are intent on identifying and preventing hazards in your workplace. Thousands of accidents occur throughout the United States every day. The failure of people, equipment, supplies, or surroundings to behave or react as expected causes most of the accidents. Accident investigations determine how and why these failures occur. Using the information gained through an investigation, a similar or perhaps more disastrous accident may be prevented. Accident investigations should be conducted with accident prevention in mind. Investigations are *not* to place blame.

An accident is any unplanned event that results in personal injury or in property damage. When the personal injury requires little or no treatment, it is minor. If it results in a fatality or in a permanent total, permanent partial, or temporary total (lost time) disability, it is serious. Similarly, property damage may be minor or serious. Investigate all accidents regardless of the extent of injury or damage.

Accidents are part of a broad group of events that adversely affect the completion of a task. These events are incidents. For simplicity, the procedures discussed in later sections refer only to

---

Accident investigation report form

Accident number _____

Company _____ Address _____

Department or location of accident if different from above

_____

Who was injured, ill, or died?

Name of injured _____ social security number _____

Sex _____ Age _____ Date of birth _____ Date of accident _____

Home address _____ Telephone # _____

Employee's usual occupation _____

Occupation at time of accident _____

Length of employment _____ Time in occup. at time of accident _____

Employment category _____ (i.e., full time)

Type of injury _____ Part of body _____

Severity of the injury _____

Names of others injured in same accident

_____

_____

_____

When did the accident occur?

Date of accident _____ Time of accident _____

Shift _____

Where did the accident occur?

Location of accident _____

On employer's premise? _____

Activity at time of accident _____

Supervisor in charge _____

What happened or caused the accident?

Describe how the accident occurred

_____

_____

Figure 7.6   Accident investigation report form.

accidents. They are, however, also applicable to incidents. This discussion introduces the reader to basic accident investigation procedures and will be followed in the next chapter by descriptions of accident analysis techniques.

Goals and Uses of Accident Investigation

The purpose or goals of accident investigation are to

- Prevent the recurrence of the same or similar event
- Act as a window to view safety and health performance
- Provide a mechanism that can be used to prevent recurrence

- Be an organized method of determining the cause of incidents
- Identify the root causes
- Be an integral part of the safety and health program

The use of accident investigation as a safety and health prevention tool is important as an integral part of prevention since it provides

- An understanding of what happened
- An organized mechanism for gathering information and data for present and future use
- Answers regarding the effectiveness of interventions and prevention approaches
- A documentation of the circumstances and chronology for legal and workers' compensation issues
- Identification of hazards and potential risks
- A tool for cooperation by allowing labor and management the opportunity to work together while conducting the investigation
- A tool as a vital component of the safety and health program

Accident investigation is an important part of a viable prevention effort that is aimed at helping companies not to repeat the past.

Accident Prevention

Accidents are usually complex. An accident may have 10 or more events that can be causes. A detailed analysis of an accident will normally reveal three cause levels: basic, indirect, and direct. At the lowest level, an accident results only when a person or object receives an amount of energy or hazardous material that cannot be absorbed safely. This energy or hazardous material is the *direct cause* of the accident. The direct cause is usually the result of one or more unsafe acts or unsafe conditions, or both. Unsafe acts and conditions are the *indirect causes* or symptoms. In turn, indirect causes are usually traceable to poor management policies and decisions, or to personal or environmental factors. These are the *basic causes*.

In spite of their complexity, most accidents are preventable by eliminating one or more causes. Accident investigations determine not only what happened, but also how and why. The information gained from these investigations can prevent recurrence of similar or perhaps more disastrous accidents. Accident investigators are interested in each event as well as in the sequence of events that led to an accident. The accident type is also important to the investigator. The recurrence of accidents of a particular type or those with common causes shows areas needing special accident prevention emphasis.

Investigative Procedures

The actual procedures used in a particular investigation depend on the nature and results of the accident. The agency with jurisdiction over the location determines the administrative procedures. In general, responsible officials appoint an individual to be in charge of the investigation. An accident investigator should use most of the following steps:

1. Define the scope of the investigation.
2. Select the investigators. Assign specific tasks to each (preferably in writing).
3. Present a preliminary briefing to the investigating team, including
   a. Description of the accident
   b. Normal operating procedures
   c. Maps (local and general)
   d. Location of the accident site
   e. List of witnesses
   f. Events that preceded the accident

4. Visit the accident site to get updated information.
5. Inspect the accident site.
   a. Secure the area. Do not disturb the scene unless a hazard exists.
   b. Prepare the necessary sketches and photographs. Label each carefully and keep accurate records.
6. Interview each victim and witness. Also, interview individuals who were present before the accident and those who arrived at the site shortly after the accident. Keep accurate records of each interview. Use a tape recorder if desired and if approved.
7. Determine the following:
   a. What was not normal before the accident?
   b. Where the abnormality occurred?
   c. When it was first noted?
   d. How it occurred?
8. Analyze the data obtained in step 7. Repeat any of the previous steps, if necessary.
9. Determine the following:
   a. Why the accident occurred?
   b. A likely sequence of events and probable causes (direct, indirect, and basic)?
   c. Alternative sequences.
10. Check each sequence against the data from step 7.
11. Determine the most likely sequence of events and the most probable causes.
12. Conduct a postinvestigation briefing.
13. Prepare a summary report, including the recommended actions to prevent a recurrence. Distribute the report according to applicable instructions.

An investigation is not complete until all data are analyzed and a final report is completed. In practice, the investigative work, data analysis, and report preparation proceed simultaneously over much of the time spent on the investigation.

Fact-Finding

Gather evidence from many sources during an investigation. Get information from witnesses and reports as well as by observation. Interview witnesses as soon as possible after an accident. Inspect the accident site before any changes occur. Take photographs and make sketches of the accident scene. Record all pertinent data on maps. Get copies of all reports. Documents containing normal operating procedures, flow diagrams, maintenance charts, or reports of difficulties or abnormalities are particularly useful. Keep complete and accurate notes in a bound notebook. The following should be carried out such as recording preaccident conditions, the accident sequence, and postaccident conditions. In addition, document the location of victims, witnesses, machinery, energy sources, and hazardous materials.

In some investigations, a particular physical or chemical law, principle, or property may explain a sequence of events. Include laws in the notes taken during the investigation or in the later analysis of data. In addition, gather data during the investigation that may lend itself to analysis by these laws, principles, or properties. An appendix in the final report can include an extended discussion.

Interviews

In general, experienced personnel should conduct interviews. If possible, the team assigned to this task should include an individual with a legal background. In conducting interviews, the team should

1. Appoint a speaker for the group.
2. Get preliminary statements as soon as possible from all witnesses.
3. Locate the position of each witness on a master chart (including the direction of view).
4. Arrange for a convenient time and place to talk to each witness.
5. Explain the purpose of the investigation (accident prevention) and put each witness at ease.
6. Listen, let each witness speak freely, and be courteous and considerate.

7. Take notes without distracting the witness. Use a tape recorder only with the consent of the witness.
8. Use sketches and diagrams to help the witness.
9. Emphasize areas of direct observation. Label hearsay accordingly.
10. Be sincere and do not argue with the witness.
11. Record the exact words used by the witness to describe each observation. Do not put words into a witness' mouth.
12. Word each question carefully and be sure the witness understands.
13. Identify the qualifications of each witness (name, address, occupation, years of experience, etc.).
14. Supply each witness with a copy of his or her statements. Signed statements are desirable.

After interviewing all witnesses, the team should analyze each witness' statement. They may wish to reinterview one or more witnesses to confirm or clarify key points. While there may be inconsistencies in witnesses' statements, investigators should assemble the available testimony into a logical order. Analyze this information along with data from the accident site. Not all people react in the same manner to a particular stimulus. For example, a witness within proximity to the accident may have an entirely different story from one who saw it at a distance. Some witnesses may also change their stories after they have discussed it with others. The reason for the change may be additional clues. A witness who has had a traumatic experience may not be able to recall the details of the accident. A witness who has a vested interest in the results of the investigation may offer biased testimony. Finally, eyesight, hearing, reaction time, and the general condition of each witness may affect his or her powers of observation. A witness may omit entire sequences because of a failure to observe them or because their importance was not realized.

Problem-Solving Techniques

Accidents represent problems that must be solved through investigations. Several formal procedures solve problems of any degree of complexity. These types of analysis will be discussed in Chapter 8.

Report of Investigation

As noted earlier, an accident investigation is not complete until a report is prepared and submitted to proper authorities. Special report forms are available in many cases. Other instances may require a more extended report. Such reports are often very elaborate and may include a cover page, a title page, an abstract, a table of contents, a commentary or narrative section, a discussion of probable causes, and a section on conclusions and recommendations. The following outline has been found especially useful in developing the information to be included in the formal report:

1. Background information
   a. Where and when the accident occurred
   b. Who and what were involved
   c. Operating personnel and other witnesses.
2. Account of the accident (what happened)
   a. Sequence of events
   b. Extent of damage
   c. Accident type
   d. Agency or source (of energy or hazardous material)
3. Discussion (analysis of the accident—How? Why?)
   a. Direct causes (energy sources; hazardous materials)
   b. Indirect causes (unsafe acts and conditions)
   c. Basic causes (management policies; personal or environmental factors)

4. Recommendations (to prevent a recurrence) for immediate and long-range action to remedy
   a. Basic causes
   b. Indirect causes
   c. Direct causes (such as reduced quantities or protective equipment or structures)

Accident investigation should be an integral part of your written safety and health program. It should be a formal procedure. A successful accident investigation determines not only what happened, but also finds how and why the accident occurred. Investigations are an effort to prevent a similar or perhaps more disastrous sequence of events. You can then use the resulting information and recommendations to prevent future accidents.

A sample of a written statement for a written safety and health program for accident investigation can be found in Figure 7.7, which also includes a sample statement related to the record keeping policies of the company regarding accidents, incidents, injuries, and illnesses.

Keeping records is also very important in recognizing and reducing hazards. A review of accident and injury records over a period of time can help pinpoint the cause of some accidents. If a certain worker shows up several times on the record as being injured, it may indicate that the person is physically unsuited for the job, is not properly trained, or needs better supervision. If one or two occupations experience a high percentage of the accidents in a workplace, they should be carefully analyzed and countermeasures should be taken to eliminate the cause. If there are multiple accidents involving one machine or process, it is possible that work procedures must be changed or that maintenance is needed. Records that show many accidents during a short period of time would suggest an environmental problem.

Once the hazards have been identified then the information and sources must be analyzed to determine their origin and the potential to remove or mitigate their effects on the workplace. Analysis of hazards forces us to take a serious look at them.

Accident investigations

Supervisors/forepersons will conduct an investigation of any accident/incident that results in death, injury, illness, or equipment damage. The supervisor will use the company's standard investigation form (see example, Figure 7.6). The completed accident investigation report will be submitted to the individual assigned responsibility for occupational safety and health.

Record keeping

OSHA requires (Name of Company) to record and maintain injury and illness records. These records are used by management to evaluate the effectiveness of this safety and health program. A summary of all recordable injuries and illnesses will be posted during the months of February through April on an OSHA 300A annual summary form for all employees to see.

Figure 7.7 Sample accident investigation and record keeping statements.

CHAPTER 8

Taking a Serious Look: Analyzing Hazards

How could this accident have been prevented? Effective analysis is a key factor. (Courtesy of Mine Health and Safety Administration.)

Before reading this chapter, you should evaluate each of the following topics by circling the answer that most closely represents the current state of hazard analysis in your company.

Hazard Analysis

Topic	Circle Answer	Answer Options
Hazard analysis	5	All workers and supervisors are involved in assessing hazards and deriving solutions
	4	Only supervisors are involved in analyzing hazards and addressing interventions
	3	Only serious hazards are analyzed and controls recommended
	2	Hazards are analyzed after accidents/incidents have occurred
	1	No routine hazard analysis takes place
Root cause analysis	5	There is a system in place to evaluate the root cause of all accidents/ incidents, even near misses
	4	Supervisors and safety professional determine the root causes of accidents/incidents
	3	Someone looks at the root causes of very serious accidents/incidents
	2	There are times when accidents/incidents are evaluated further than fixing blame on the injured worker
	1	There is no root cause analysis

(continued)

Hazard Analysis (continued)

Topic	Circle Answer	Answer Options
Change analysis	5	Every planned or new facility, process, material, or equipment is fully reviewed by a competent team, along with affected workers
	4	Every planned or new facility, process, material, or equipment is fully reviewed by a competent team
	3	High-hazard planned or new facilities, processes, materials, or equipment are reviewed
	2	Hazard reviews of planned or new facilities, processes, materials, or equipment are problem driven
	1	No system for hazard review of planned or new facilities exists
Hazard identification (job and process analysis)	5	A current hazard analysis exists for all jobs, processes, and material; it is understood by all employees; and employees have contributed inputs into the analysis for their jobs
	4	A current hazard analysis exists for all jobs, processes, and material and it is understood by all employees
	3	A current hazard analysis exists for all jobs, processes, or phases and it is understood by many employees
	2	A hazard analysis program exists, but few are aware of it
	1	There is no routine hazard analysis system in place
Injury/illnesses analysis	5	Data trends are fully analyzed and displayed, common causes are communicated, management ensures prevention, and employees are fully aware of trends, causes, and means of prevention
	4	Data trends are fully analyzed and displayed, common causes are communicated, and management ensures prevention
	3	Data are centrally collected and analyzed and common causes are communicated to supervisors for prevention
	2	Data are centrally collected and analyzed but not widely communicated for prevention
	1	Little or no effort is made to analyze data for trends, causes, and prevention

HAZARD ANALYSIS

Hazard analysis is a technique used to examine the workplace for hazards with the potential to cause accidents. The information obtained by the hazard identification process provides the foundation for making decisions upon which jobs should be altered in order for the worker to perform the work safely and expeditiously. In addition, this process allows workers to become more involved in their own destiny. For some time, involvement has been recognized as a key motivator of people. This is also a positive mechanism in fostering labor/management cooperation. This is especially true if everyone in the workplace is continuously looking for potential hazards that can result in injury, illness, or even death.

Hazards analysis can get sophisticated and go into much detail. Where the potential hazards are significant and the possibility for trouble is quite real, such detail may well be essential. However, for many processes and operations—both real and proposed—a solid look at the operation or plans by a variety of affected people may be sufficient.

Analysis often implies mathematics, but calculating math equations is not the major emphasis when attempting to address hazards or accidents/incidents that occur within the industry. Analysis in the context of this module means taking time to examine systematically the worksite's existing or potential hazards. This can be accomplished in a variety of ways.

However, if you are faced with fairly sophisticated and complex risks with a reasonable probability of disaster if things go wrong, you may want some help with some of the other hazard analysis methodologies. What follows is a very brief look at the common ones. If you decide to try one of the approaches, check with your local Occupational Safety and Health Administration (OSHA) consultation office or call an engineering firm that specializes in hazard analysis.

ROOT CAUSE ANALYSIS

Accidents are rarely simple and almost never result from a single cause. Accidents may develop from a sequence of events involving performance errors, changes in procedures, oversights, and omissions. Events and conditions must be identified and examined to find the cause of the accident and a way to prevent that accident and similar accidents from occurring again. To prevent the recurrence of accidents one must identify the accident's causal factors. The higher the level in the management and oversight chain in which the root cause is found, the more diffused the problem can be.

Root cause analysis aids in the development of evidence, by collecting information and arranging it in a logical sequence so that it can be easily examined. This will lead to the causal factors of the accident and then to a development of new methods to help eliminate that accident or similar accidents from recurring in the future. By creating an event in the causal factor chain, multiple causes can be visually illustrated and a visual relationship between the direct and contributing causes can be shown. Event causal charting also visually delineates the interactions and relationships of all involved groups or individuals. By using root cause analysis, one can develop an event causal chain to examine the accident in a step-by-step manner by looking at the events, conditions, and causal factors chronologically, to prevent future accidents.

Root cause analysis is used when there are multiple problems with a number of causes of an accident. A root cause analysis is a sequence of events that shows, step by step, the events that took place in order for the accident to occur. Root cause analysis puts all the necessary and sufficient events and causal factors for an accident in a logical, chronological sequence. It analyzes the accident and evaluates evidence during an investigation. It is also used to help prevent similar accidents in the future and to validate the accuracy of preaccidental system analysis. It is used to help identify an accident's causal factors, which, once identified, can be fixed to eliminate future accidents of the same or of similar nature.

On the downside, root cause analysis is a time-consuming process and requires the investigator to be familiar with the process for it to be effective. As you will see later in this chapter, you may need to revisit an accident scene multiple times and look at areas that are not directly related to the accident to have a complete event and causal factor chain. Analysis requires a broad perspective of the accident to identify any hidden problems that would have caused the accident.

One of the simplest root cause analysis techniques is to determine the causes of accidents/incidents at different levels. During any hazard analysis we are always trying to determine the root cause of any accident or incident. Experts who study accidents often do a breakdown or analysis of the causes. They analyze them at three different levels:

1. Direct causes (unplanned release of energy or hazardous material)
2. Indirect causes (unsafe acts and unsafe conditions)
3. Basic (root) causes (management safety policies and decisions, and personal factors)

Direct Causes

When making a detailed analysis of an accident or incident, consider the release of energy or hazardous material as a direct cause. Energy or hazardous material is considered to be the force that results in injury or other damage at the time of contact. It is important to identify the direct causes. To prevent injury, it is often possible to redesign equipment or facilities and provide personal protection against energy release or release of/contact with hazardous materials. Some examples of direct causes in the form of energy or hazardous material sources are found in Table 8.1 and Figure 8.1.

Table 8.1   Sources of Direct Causal Agents

Energy Sources	Hazardous Materials
1. Mechanical	1. Compressed or liquefied gas
Machinery	Flammable
Tools	Nonflammable
Noise	2. Corrosive material
Explosives	3. Flammable material
Moving objects	Solid
Strain (self)	Liquid
2. Electrical	Gas
Uninsulated conductors	4. Poison
High-voltage sources	5. Oxidizing material
3. Thermal	6. Dust
Flames	
Hot surfaces	
Molten metals	
4. Chemical	
Acids	
Bases	
Fuels	
Explosives	
5. Radiation	
Lasers	
X-rays	
Microwave	
Radiation sources	
Welding	

Indirect Causes

Unsafe acts (behavior) or unsafe conditions comprise indirect causes of accidents or incidents. These indirect causes can inflict injury, property damage, or equipment failure. They allow the energy or hazardous material to be released. Unsafe acts can lead to unsafe conditions and vice versa. Examples of unsafe acts and unsafe conditions are found in Table 8.2.

Basic Causes

Some accident investigations result only in the identification and correction of indirect causes, but indirect causes of accidents are symptoms that some underlying causes exist, which are often termed basic causes. By going one step further, accidents can best be prevented by identifying and correcting the basic or root causes. Basic causes are grouped into policies and decisions, personal factors, and environmental factors, as found in Tables 8.3 through 8.5.

When basic causes are eliminated, unsafe acts and unsafe conditions may not occur. (For example, Millie Samuels used a broken ladder because no unbroken ladder existed on the jobsite.) In Millie's case, the basic cause, lack of an unbroken ladder, set up her subsequent unsafe act.

Hazardous energy sources

Chemical energy	Electrical energy	Thermal energy
Corrosive materials Flammable materials Toxic materials Reactive materials Oxygen deficiency Carcinogens	Capacitors Transformers Batteries Exposed conductors Static electricity	Steam Fire Solar Friction Chemical reactions Spontaneous combustion Cryogenic materials Ice, snow, wind, rain
Radiant energy	Kinetic energy	Pressure energy
Intense light Lasers Ultraviolet X-rays, gamma rays Infrared sources Electron beams Magnetic fields RF fields Nuclear criticality High energy particles	Pulleys, belts, gears Shears, sharp edges Pinch points Vehicles Mass in motion	Confined gases Explosives Noise
	Potential energy	Biological energy
	Falling Falling objects Lifting Tripping, slipping Earthquakes	Allergens Pathogens (virus, bacteria, etc.)

Figure 8.1   Hazard energy sources. (Courtesy of Lawrence Livermore National Laboratory.)

Table 8.2   Unsafe Acts and Conditions

Unsafe Acts	Unsafe Conditions
1. Failure to warn coworkers or to secure equipment	1. Congested work areas
2. Failure to warn coworkers or to secure equipment	2. Defective machinery/tools
3. Ignoring equipment/tool defects	3. Improperly stored explosive or hazardous materials
4. Improper lifting	4. Poor illumination
5. Improper working position	5. Poor ventilation

6. Improper use of equipment
  At excessive speeds
  Using defective equipment
  Servicing moving equipment
	6. Inadequate supports/guards
	7. Poor housekeeping

7. Operating equipment without authority	8. Radiation exposure
8. Horseplay	9. Excessive noise
9. Making safety devices inoperable	10. Hazardous atmospheric conditions
10. Drug misuse	11. Dangerous soil conditions
11. Alcohol use	12. No firefighting equipment
12. Violation of safety and health rules	13. Unstable work areas/platforms
13. Failure to wear assigned personal protective equipment (PPE)	

Table 8.3   Basic Causes of Accidents from Policies and Procedures

1. Safety policy is not
   a. In writing
   b. Signed by top management
   c. Distributed to each employee
   d. Reviewed periodically
2. Safety procedures do not provide for
   a. Written manuals
   b. Safety meetings
   c. JSA
   d. Housekeeping
   e. Medical surveillance
   f. Accident investigations
   g. Preventive maintenance
   h. Reports
   i. Safety audits/inspections

3. Safety is not considered in the procurement of
   a. Supplies
   b. Equipment
   c. Services
4. Safety is not considered in the personnel practices of
   a. Selection
   b. Authority
   c. Responsibility
   d. Accountability
   e. Communications
   f. Training
   g. Job observations

Table 8.4   Basic Causes of Accidents from Personal Factors

1. Physical
   a. Inadequate size
   b. Inadequate strength
   c. Inadequate stamina
2. Experiential
   a. Insufficient knowledge
   b. Insufficient skills
   c. Accident records
   d. Unsafe work practices
3. Motivational
   a. Needs
   b. Capabilities

4. Attitudinal
   a. Toward others
      i. People
      ii. Company
      iii. Job
   b. Toward self
      i. Alcoholism
      ii. Drug use
      iii. Emotional upset
5. Behavioral
   a. Risk taking
   b. Lack of hazard awareness

Table 8.5   Basic Causes of Accidents due to Environmental Factors

1. Unsafe facility design
   a. Poor mechanical layout
   b. Inadequate electrical system
   c. Inadequate hydraulic system
   d. Crowded limited access ways
   e. Insufficient illumination
   f. Insufficient ventilation
   g. Lack of noise control

   h. Normal
   i. Emergency
2. Unsafe operating procedures
   a. Normal
   b. Emergency
3. Weather
4. Geographical area

Thus, accidents have many causes. Basic (root) causes lead to unsafe acts and unsafe conditions (indirect causes). Indirect causes may result in a release of energy or hazardous material (direct causes). The direct cause may allow for contact, resulting in personal injury or property damage or equipment failure (accident). You can use the accident report form found in Figure 8.2 to identify and analyze these three causes.

Root (Basic) Cause Analysis

Root causes are those that, if corrected, would eliminate the accident from occurring again or similar accidents from occurring. They may include several contributing causes. They are a higher order of causes that address multiple problems rather than focusing on the single direct cause. An example would be, "Management failed to implement the principles and core functions of a safety and health program. It is management's responsibility to ensure that the workplace has an effective safety and health program and that the workplace is safe for employees to work."

A root cause analysis is not a search for the obvious but an in-depth look at the basic or underlying causes of occupational accidents or incidents. The purpose of investigating and reporting the causes of occurrences is to enable the identification of corrective actions adequate to prevent recurrence and thereby protect the health and safety of the public, the workers, and the environment. Every root cause investigation and reporting process should include five phases. While there may be some overlap between phases, every effort should be made to keep them separate and distinct. The phases of a root cause analysis are

- Phase I—data collection
- Phase II—assessment
- Phase III—corrective actions
- Phase IV—informing
- Phase V—follow-up

The objective of investigating and reporting the cause of occurrences is to enable the identification of corrective actions adequate to prevent recurrence and thereby protect the health and safety of the public, the workers, and the environment. Programs can then be improved and managed more efficiently and safely.

Root Cause Analysis Methods

The most common root cause analysis methods are as follows:

1. Events and causal factor analysis identifies the time sequence of a series of tasks and actions and the surrounding conditions leading to an occurrence.
2. Change analysis is used when the problem is obscure. It is a systematic process that is generally used for a single occurrence and focuses on elements that have changed.
3. Barrier analysis is a systematic process that can be used to identify physical, administrative, and procedural barriers or controls that should have prevented the occurrence.
4. Management oversight and risk tree (MORT) analysis is used to identify inadequacies in barriers and controls, specific barrier and support functions, and management functions. It identifies specific factors relating to an occurrence and identifies the management factors that permitted these risk factors to exist. MORT/Mini-MORT is used to prevent oversight in the identification of causal factors. It lists on the left side of the tree specific factors relating to the occurrence; and on the right side of the tree, it lists the management deficiencies that permit specific risk factors to exist. Management factors support each of the specific barrier and control factors. Included is a set of questions to be asked for each of the barrier and control factors on the tree. They are useful in preventing oversight and ensuring that all potential causal factors are considered. They are especially useful when there

Accident report form

Department:	Employee name:	Location of accident:
Date of accident:	Employee age:	Accident type:
Time of accident:	Employee occupation:	Accident classification:

Description of accident/incident:

Itemize personal injury involved:
Itemize property damage involved:
Itemize tools/equipment involved:

	Direct		Indirect		Basic	
	Energy sources	Hazardous materials	Unsafe acts	Unsafe conditions	Inadequate policy and/or decisions	Environment and/or personal factors
Causes						
Recommendations	Direct level		Indirect level		Basic level	

Foreman's signature: _____   Department head's signature: _____

Figure 8.2 Accident report form. (Courtesy of Mine Safety and Health Administration.)

is a shortage of experts of whom to ask the right questions. However, because each management oversight factor may apply to specific barrier and control factors, the direct linkage or relationship is not shown but is left to the analyst. For this reason, events and causal factor analysis and MORT should be used together for serious occurrences: one to show the relationship, the other to prevent oversight. A number of condensed versions of MORT, called Mini-MORT, have been produced. For a major occurrence justifying a comprehensive investigation, a full MORT analysis could be performed while Mini-MORT would be used for most other occurrences.

5. Human performance evaluation identifies factors that influence task performance. The focus of this analysis method is on operability, work environment, and management factors. User–system interface studies are frequently done to improve performance. This takes precedence over disciplinary measures. Human performance evaluation is used to identify factors that influence task performance. It is most frequently used for user–system interface studies. Its focus is on operability and work environment, rather than training of operators to compensate for bad conditions. Human performance evaluations may be used to analyze most occurrences, since many conditions and situations leading to an occurrence have ultimately originated from some task performance problem that results from management planning, scheduling, task assignment, maintenance, and inspections. Training in ergonomics and human factors is needed to perform adequate human performance evaluations, especially in user–system interface situations.

6. Kepner–Tregoe problem solving and decision making provides a systematic framework for gathering, organizing, and evaluating information and applies to all phases of the occurrence investigation process. Its focus on each phase helps keep them separate and distinct. The root cause phase is similar to change analysis. Kepner–Tregoe is used when a comprehensive analysis is needed for all phases of the occurrence investigation process. Its strength lies in providing an efficient, systematic framework for gathering, organizing, and evaluating information and consists of four basic steps:

a. Situation appraisal to identify concerns, set priorities, and plan the next step.

b. Problem analysis to precisely describe the problem, identify and evaluate the causes, and confirm the true cause. (This step is similar to change analysis.)

c. Decision analysis to clarify purpose, evaluate alternatives, assess the risks of each option, and to make a final decision.

d. Potential problem analysis to identify safety degradation that might be introduced by the corrective actions and identify the likely causes of those problems, take preventive action, and plan contingent action. This final step provides assurance that the safety of no other system is degraded by changes introduced by proposed corrective actions.

These four steps cover all phases of the occurrence investigation process. Thus, Kepner–Tregoe can be used for more than causal factor analysis. Separate worksheets (provided by Kepner–Tregoe) provide a specific focus on each of the four basic steps and consist of step-by-step procedures to aid in the analyses. This systematic approach prevents overlooking any aspect of concern. As formal Kepner–Tregoe training is needed for those using this method, a further description is not included in this book.

The use of different methods to conduct root cause analysis has been widely accepted over a period of years. There have been many creative adaptations and permutations using the root cause analysis approach, but the foundation for it has stood the test of time. Certain methods are used for different circumstances (see Table 8.6 and Figure 8.3).

The analysis of an accident does not stop with the identification of the direct, indirect, and basic (root) causes of the accident or incident. To make positive gains from the event, changes should be made in the interaction of users, systems, materials, methods, and physical and social environments. These changes should result from the recommendations that are derived from the causes identified during the investigation. The goal of these changes is the prevention of future accidents and incidents similar to the one investigated.

Let us consider one type of root cause analysis before leaving the topic of root cause analysis completely. Most workplaces are dynamic and subject to change. Thus, the use of change analysis is often an appropriate root cause analysis to apply.

Table 8.6   Summary of Root Cause Methods

Method	When to Use	Advantages	Disadvantages	Remarks
Events and causal factor analysis	Use for multifaceted problems with long or complex causal factor chain	Provides visual display of analysis process. Identifies probable contributors to the condition	Time consuming and requires familiarity with process to be effective	Requires a broad perspective of the event to identify unrelated problems. Helps to identify where deviations occurred from acceptable methods
Change analysis	Use when cause is obscure. Especially useful in evaluating equipment failures	Simple six-step process	Limited value because of the danger of accepting wrong, "obvious" answer	A singular problem technique that can be used in support of a larger investigation. All root causes may not be identified
Barrier analysis	Use to identify barrier and equipment failures and procedural or administrative problems	Provides systematic approach	Requires familiarity with process to be effective	This process is based on the MORT Hazard/Target Concept
MORT/Mini-MORT	Use when there is a shortage of experts to ask the right questions and whenever the problem is a recurring one. Helpful in solving programmatic problems	Can be used with limited prior training. Provides a list of questions for specific control and management factors	May only identify area of cause, not specific causes	If this process fails to identify problem areas, seek additional help or use cause and-effect analysis
Human performance evaluations (HPE)	Use whenever people have been identified as being involved in the problem cause	Thorough analysis	None if process is closely followed	Requires HPE training
Kepner–Tregoe	Use for major concerns where all spects need thorough analysis	Highly structured approach focuses on all aspects of the occurrence and problem resolution	More comprehensive than may be needed	Requires Kepner–Tregoe training

Source: Courtesy of the Department of Energy.

CHANGE ANALYSIS

As its name implies, this technique emphasizes change. To solve a problem, an investigator must look for deviations from the norm. Consider that all problems result from some unanticipated change. Make an analysis of the change to determine its causes. Use the following steps in this method:

1. Define the problem (what happened).
2. Establish the norm (what should have happened).
3. Identify, locate, and describe the change (what, where, when, to what extent).
4. Specify what was and what was not affected.
5. Identify the distinctive features of the change.
6. List the possible causes.
7. Select the most likely causes.

Change analysis is used when the problem is obscure. It is a systematic process that is generally used for a single occurrence and focuses on elements that have changed. It compares the previous trouble-free

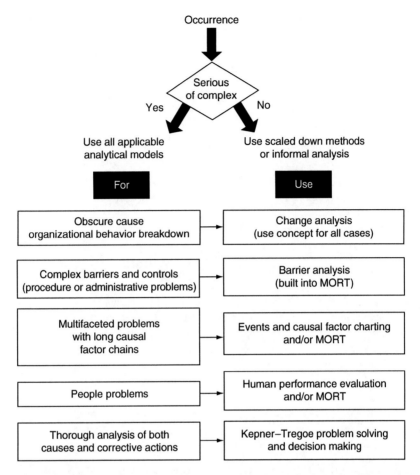

Figure 8.3 Flowchart of root cause analysis methods. (Courtesy of the Department of Energy.)

activity with the occurrence to identify differences. These differences are subsequently evaluated to determine how they contributed to the occurrence.

Change analysis studies a problem by analyzing the deviation between what is expected and what actually happened. The evaluator essentially asks what differences occurred to make the outcome of this task or activity different from all the other times this task or activity was successfully completed. This technique consists of asking the questions: What? When? Where? Who? and How? Answering these questions should lead to answering the root cause question: Why? Primary and secondary questions included within each category provide the prompting necessary to thoroughly answer the overall question. Some of the questions are not applicable to any of the existing conditions. Some amount of redundancy exists in the questions to ensure that all items are addressed. Several key elements for addressing any change in the standard operating process include:

- Consider the event containing the undesirable consequences.
- Consider a comparable activity that did not have the undesirable consequences.
- Compare the condition containing the undesirable consequences with the reference activity.
- Set down all known differences whether they appear to be relevant or not.
- Analyze the differences for their effects in producing the undesirable consequences. This must be done with careful attention to detail, ensuring that obscure and indirect relationships are identified (e.g., a change in color or finish may change the heat transfer parameters and consequently affect system temperature).
- Integrate information into the investigative process relevant to the causes of, or the contributors to, the undesirable consequences.

Change analysis is a good technique to use whenever the causes of the condition are obscure, you do not know where to start, or you suspect a change may have contributed to the condition. Not recognizing the compounding of change (e.g., a change made 5 years previously combined with a change made recently) is a potential shortcoming of change analysis. Not recognizing the introduction of gradual change as compared with immediate change is also possible. This technique may be adequate to determine the root cause of a relatively simple condition. In general, though, it is not thorough enough to determine all the causes of more complex conditions. Problem/change analysis involves six steps.

Change analysis in accident investigation helps you to systematically analyze problems. It builds new skills and reinforces existing ones. The practical application will enable you to focus on causes rather than symptoms and aid you in becoming more perceptive to the contributing factors of an accident. It provides a step-by-step process for getting to the core of any problem.

This type of analysis may trigger the need to more closely analyze a job or task that has been identified as presenting a high risk of producing hazards or injuries.

JOB HAZARD ANALYSIS

The easiest and possibly most effective method is the step-by-step process of the job hazard analysis (JHA). JHA is sometimes referred to as job safety analysis (JSA). The hazard analysis process looks at jobs or processes. Done for every job, a JHA or JSA ensures safe steps, teaches new workers, eliminates or controls hazardous materials, and much more. Some companies have work teams complete JHAs or JSAs on every job or process and then use them as a guide to do the job. The JHA is a hazard identification tool, an analysis tool, a training tool, and an accident prevention tool all rolled into one.

Job safety/hazard analysis is a basic approach to developing improved accident prevention procedures by documenting the firsthand experience of workers and supervisors, and at the same time, it tends to instill acceptance through worker participation. Job safety/hazard analysis can be a central element in a safety program; and the most effective safety programs are those that involve employees. Each worker, supervisor, and manager should be prepared to assist in the recognition, evaluation, and control of hazards. Worker participation is important to efficiency, safety, and increased productivity. Through the process of job safety/hazard analysis, these benefits are fully realized.

JSA, also called JHA, is a process used to determine hazards of, and safe procedures for, each step of a job. A specific job, or work assignment, can be separated into a series of relatively simple steps. The hazards associated with each step can be identified and solutions can be developed to control each hazard. A simple form can be used to carry out a JHA (see Table 8.7). You will find more information on JHAs in Chapter 17.

Table 8.7   Sample JHA Form

Key Job Steps	Tools, Equipment or Materials Used	Potential Health or Injury Hazard to Worker	Potential Hazard to System	Recommendations for Worker Protection	Systems Counter Measures

Source: Courtesy of Occupational Safety and Health Administration.

ANALYZING ACCIDENT DATA

Many companies conduct accident investigations and keep accident records and other data on the company's safety and health initiatives. If a company has a sufficient number of accidents/incidents and enough detail in their occupational injury/illness investigation data, the company can begin to examine trends or emerging issues relevant to their safety and health intervention/prevention effort. The analysis of these data can be used to evaluate the effectiveness of safety and health on various projects and jobsites or groups of workers. The safety and health data can be used to compare your company to companies that engage in similar work or to employees with a comparable occupation in their workforce, or to bid on the same size of projects on a state, regional, or national basis.

By analyzing your accidents/incidents, you are in a better position to compare apples to apples rather than apples to oranges. You will be able to identify not only the types of injuries and types of accidents and causes, but you will also be able to intervene and provide recommendations for preventing these accidents/incidents in the future. You will be able to say with confidence that "I do" or "I do not" have a safety and health problem. If you find that you have a problem, your analysis and data will be essential if you try to elicit advice on how to address your health and safety needs.

Gathering and analyzing accidents/incidents data do not comprise your entire safety and health program, but are single elements that provide feedback and evaluative information as you proceed toward accomplishing your safety and health goals; thus, they are important elements.

The two most frequent statistical pieces of information, which are designed to allow you to compare your company's safety and health performance with others, are the incident rate and the severity rate. These two rates, respectively, answer the questions of "How often or frequently do accidents occur?" and "How bad are the injuries/illnesses that occur?" The number of times that occupational injuries/illnesses happen is the determinant of the incident rate whereas the number of days (lost workdays) is the prime indicator of the severity rate. Both of these rates provide unique information regarding your safety and health effort.

How can you compare your company of 15 workers with a company that has 250 workers? This can be accomplished only by using a statistic or formula that allows a standardized numerical value to be developed, which takes into account the differences and places your company on the same playing field so that each company's front line appears to be the same weight.

To find the incident rate, count the number of distinct events that resulted in injuries/illness. To compare your incident rate with that of other companies, you must normalize your data. This is accomplished by using a constant of 200,000 work hours, which was established by the Bureau of Labor Statistics. The 200,000 work hours are the number of hours that 100 full-time workers would work during 50 weeks at 40 h per week. Thus, you can calculate your incident rate in the following manner:

$$\text{Incident rate} = \frac{\text{Number of your OSHA recordable injuries/illnesses} \times 200,000 \text{ (work hour constant)}}{\text{Total number of hours which your employees worked during the year}}$$

The incident rate could be a rate calculated for recordable (combined) injuries and illnesses, recordable injuries, recordable illnesses, all injuries with lost workdays, all illnesses with lost workdays, injuries requiring only medical treatment, or first-aid injuries. These calculated rates would not normally be calculated on a national basis, but could be used to compare your progress on a yearly basis, or between jobsites or facilities.

The severity rate, which is often called the lost-time workday rate, is used to determine how serious the injuries and illnesses are. A company may have a low incident rate or few injuries and illnesses but, if the injuries and illnesses that occur result in many days away from work, the lost workday cases can be as costly or more costly than having a large number of no lost workday injuries, which have only medical costs associated with them. Lost workday cases can definitely have a greater impact on your worker compensation costs and premiums.

Calculation of the severity rate is similar to the incident rate except that the total number of lost-time workdays is used in place of the number of OSHA-recordable injuries/illnesses. Thus, you can calculate your severity rate in the following manner:

$$\text{Severity rate} = \frac{\text{Number of your lost-time workdays} \times 200,000 \text{ (work hour constant)}}{\text{Total number of hours which your employees worked during the year}}$$

The incident and severity rates are both expressed as a rate per 100 full-time workers. This provides a standard comparison value for a company whether it has 20 or 1000 workers. Thus, both the 20 employee company and a 1000 employee company can compare their safety and health performances with those of each other. If a company has different departments, shifts, facilities, operations, or locations, rates can be calculated to compare each type of situation or operations within the company. This can allow a company to identify problem areas or locations.

Sometimes the temptation exists to focus on only the lost-time workday cases, but how can you identify which injury or illness is going to result in a lost-time workday? At times, the difference between a medical treatment injury and a lost-time workday injury may only be a matter of inches or chance. Thus, it is more logical to address your total injury problem.

The analysis of industrial-related hazards and the accident/incidents that they cause is an important step in the overall process of reducing construction-related injuries, illnesses, and deaths. Only after a systematic look at the hazards and accidents can you hope to integrate the accident prevention techniques and tools that can have an impact upon your company's safety and health initiative.

RISK VERSUS COST

Workplaces have hazards that present a risk of injury or illness from the dangers that exist. At times, the hazards cannot be removed and the dangers exist and can result in an accident. Risk is the probability of an accident occurring. The amount of risk you deem as acceptable will do much to define the extent of your injury prevention effort. Risk related to safety and health is often a judgment call. However, even a judgment call can be quantified if you develop criteria and place value upon them. W. Fine has provided a mathematical model for conducting a risk assessment that results in a numerical value, which can be used to compare potential risks from accidents as well as determine if the amount of fix justifies the cost involved to fix or remove the hazard. Fine's approach for this book has been simplified and updated. Most of his basic components are present, which will allow you to assess the risk of a hazard as well as make decisions on whether it is logical and economically feasible to fix the hazard.

In determining risk we will need to assess a consequence value for the existing hazard, which indicates its effect on workers if contacted, as well as a value for the exposure potential that denotes how many times during any period workers could come into contact with the hazard. We also need to assess the probability that workers would be injured, become ill, or be killed if they contacted the hazard.

The following is a way to calculate and interpret risk.

$$\text{Risk assessment factor} = \text{Consequence} \times \text{exposure} \times \text{probability}$$

Select the consequence that is most likely for the hazard, based on experience and injury/illness data.

Consequence	
Death	10
Multiple worker injury or illness (two or more)	8
Very serious injury or illness (permanent disabling)	6
Serious injury or illness (lung damage, broken bone, amputation, temporary disabling), lost workdays greater than 1 week	4
Other injuries or illnesses requiring first aid (cuts, sprains, headaches)	2

Select the exposure that best depicts the frequency of the workforce's normal contact with the hazard.

Exposure	
Every hour of the day	10
Every day	8
Every week	6
Every month	4
Once a year	2

Select the probability that best represents the chance that exposure to the hazard would result in injury or illness.

Probability	
100%	10
75%	8
50%	6
25%	4
0%	2

Multiply the values that you have selected for consequence, exposure, and probability together to determine this risk assessment factor and find the value on the following scale.

Risk Assessment Factor	
801–1000	Highest risk
601–800	Higher risk
401–600	High risk
201–400	Lower risk
1–200	Lowest risk

Now you can decide how much risk is acceptable to your operation. Another question that arises from this score is "Should I invest the money to fix or remove the hazard and what will be my gain?" Another way to represent this is "How much fix will I get for the money I invest?" Companies are always looking at the bottom line.

The two factors in determining the justification factor for fixing or removing a hazard are dependent on the cost and the amount of fix or removal of the hazard. This justification factor is obtained by

$$\text{Justification factor} = \text{Amount of fix or removal} \times \text{Cost to fix or remove}$$

Select the value that best represents the amount or degree of fix or removal of the hazard.

Amount of Fix or Removal	
100%	10
75%	8
50%	6
25%	4
0%	2

Select the value that represents what it would cost to fix or remove the hazard.

Cost to Fix or Remove	
>$100,000	2
>$50,000	4
>$10,000	6
>$1,000	8
>$100	10

To determine whether the cost justifies the degree of fix or removal of the hazard, multiply the amount of fix or removal and cost to fix or remove values together. Apply this product to the following chart to determine if the cost is worth the amount of fix or removal of the hazard.

Cost Justification Factor	
81–100	Very definitely fix or remove
61–80	Definitely fix or remove
41–60	Fix
21–40	Probably fix
0–20	Do not fix

This is only a process that you could amend, not use, or use as one tool to help you prioritize your hazards. It will allow you to generate numbers to compare your hazards and make informed decisions on addressing the existing hazard.

It has been my experience that when you are trying to make a case to management to fund a safety-related item, quantitative approaches are better received than ones based on opinions or worded facts. Statistical data always help to bolster the case for doing almost anything.

In summary, many aspects of hazards must be analyzed for a variety of reasons. Thus, you will need to learn to use a myriad of tools to achieve your goals related to hazards. If your goal is to analyze hazards, there are all kinds of tools available from the simple to the complex. If you are attempting to determine the effects of hazards on your accident record, you must analyze your accident data. If you are trying to determine risk and the risk acceptable to your company based on the potential cost incurred, then you must analyze the risk versus the cost. As you can see, analysis is an integral part of your safety and health initiative.

Cost Avoidance

At times even using the preceding risk assessment and cost justification model does not find acceptance when you are trying to convince others of the need to fix or invest heavily in safety and health in the workplace. In the past, I have used the cost avoidance work-up, which usually results in support of my proposal.

For example, suppose that you have a worker who received a back injury from lifting 150 lb, a job that he or she performs two times a week. For a cost of $10,000, the injury would never occur again. Would you think that the cost to fix the situation would be worth the prevention of an injury? First of all what employer would allow an employee to lift such a large load? Simply allowing this practice is a risk in itself. With that said, let us consider our options.

If you applied the risk/cost formula found above, you would find that the risk assessment factor is not very high principally because the lift is only performed twice a week and not regularly. You will also find that the risk should definitely be fixed or removed using the justification factor. Your analysis provides a mixed message to management. Therefore, your case to have the problem fixed may not be very strong in this situation. A cost avoidance approach may be a better avenue. First, let us assume that a worker is injured. What are the costs incurred?

- Lost-time workday: Normal back strains require 7 to 15 days bed rest for the injured. Let us suppose an employee makes $10.00 an hour × 40 h work week = $400.00 week.
- Replacement worker for the injured = $400.00 week.
- Environmental, Health and Safety (EH&S) department conducting the incident investigation: Average wage $15.00/h × 8 h of investigation, assessment and interviewing = $120.00.
- Employees/supervisor involved in the investigation and interviewing process: Worker/witness $10.00 × 1 h = $10.00 and supervisor at average pay of $13.00 × 2 h = $26.00.
- Finance personnel to fill workers' compensation forms and insurance forms. It takes about 3–4 h of work to collect the data. Assume that the finance employee makes a salary of $10.00 h × 4 h = $40.00.
- Take into account loss of production or equipment damage. Process shut down for 2 h resulting in two pieces of equipment not manufactured, each worth $2000 × 2 = $4000.00.
- In addition, do not forget the injured worker who now has to start therapy. Normally a two-week process is required for strains. This will cost close to, or over, $5000.00 depending on the medical services needed.
- Company will need to retrain employees on proper lifting and back safety and again, looking at wages of employees' training time, lost production for training, and the EH&S person's time to do the training, this cost could be well over $5000.00.
- This does not include the rise in premium for your workers' compensation assessed by your insurance carrier or your workers' compensation commissioner.

This is well over the cost of $10,000.00 to prevent the accident from occurring again and ensuring the safety of an employee who can return home and enjoy life. You might have other costs that I have missed, but this is a way of opening eyes to the cost if you do not take action. Another take on this is that the employee can file a complaint with OSHA and the company now has more help than they ever wanted from a government agency. This may greatly increase the cost from violation or citations. There are many points to consider when we look at the safety of the employees or the environment. Using varied approaches to analyze incidents or the possibility of incidents can certainly improve your safety and health approach.

CHAPTER 9

Hurting: Occupational Injuries

Death by the release of energy. (Courtesy of Mine Safety and Health Administration.)

Errant or unplanned release of energy results in injury to workers or even death. Recently, there were three workers killed in my state from failure to lock out (block) energy from being released. They failed to block or support the raised bed of a dump truck. They had positioned themselves between the frame of the truck and the raised bed. The bed's energy was released and came down on them, crushing life from them. This type of accident continues to happen every year. This is an example of the release of energy.

There are such a myriad of safety hazards facing employers and workers within the workplace. It is difficult to select the most important hazards and thus err by leaving out ones that others believe to be important. Although great detail cannot be provided in one chapter, it is the intent of this chapter to provide enough information to set the tone. If detailed information is needed, then other references can be sought by the reader.

Trauma is by definition an injury produced by a force (violence, thermal, chemical, or an extrinsic agent). Occupational trauma transpires from the contact with or the unplanned release of varied sources of energy intrinsic within the workplace. Most workplaces have a plethora of energy sources from potential (stored) energy to kinetic (energy in motion) energy sources. These sources may be stacked materials (potential) or a jackhammer (kinetic). The sources of energy are the primary causes of trauma deaths and injuries to workers.

Trauma-related events are a lot easier to observe and evaluate than those related to occupational illness. The reasons for the validity of this statement are as follows:

- Trauma occurs in real time with no latency period.
- Trauma cases have an immediate sequence of events.
- Outcomes are readily observable (only have to reconstruct a few minutes or hours)
- Root or basic causes are more clearly identified.
- Events allow for easy detection of cause-and-effect relationships.
- Deaths and injuries are not difficult to diagnose.
- Deaths and injuries are highly preventable.

OCCUPATIONAL TRAUMA DEATHS

In 1984, the National Institute for Occupation Safety and Health (NIOSH) estimated that the number of trauma-related deaths were 10,000 per year. As the years progressed, the ability to count these deaths became a more precise process. As more attention was focused on occupational deaths, industries were better able to look at the specific types of deaths that were most prevalent in their industry. You need to know the types of fatalities that occur in your industry.

For example, in the construction industry there are four leading causes of construction-related deaths. They are falls (from elevation)—33%, struck by (a vehicle, falling material, etc.)—22%, caught-in/between (excavation collapse)—18%, shock (electrical)—17%, while all other types make up 10% of these deaths. At present Occupational Safety and Health Administration (OSHA) conducts focused inspections of construction worksites, where it looks specifically for these hazards or for the lack of controls for each of these four causes of construction deaths.

In 2006, there were 5703 fatal work injuries; this was a decline by 31 deaths from 2005. This would be equivalent to a 16 passenger commuter aircraft crashing each day of the year. How many of us would continue to fly if one commuter aircraft crashed each day of the year? I would venture to say that few of us would want to fly. However, there is no uproar over the number of worker deaths. The reason is that they do not happen at one time or in one place. Sixteen deaths a day do not equate to a catastrophe since they are all spread out geographically and not in a cluster. Remember that this is a count of the number of aircraft per day that includes Saturdays and Sundays, which are light workdays. If we factored out holidays and days off, we would need bigger commuter planes to get all the dead into them.

The number of fatalities (5703) and the fatal rate per 100,000 (3.9) workers of the total charged to each industry can be found in Table 9.1.

As you can see mining has the worst rate. Mining has very few workers compared with other industries and 190 deaths are too many deaths for a small population. Agriculture also has a small workforce but suffered 646 deaths, but the danger to the worker in that industry is higher than any other except mining. Note that construction had the most deaths (1226). Only 5% of the workers in the United States are construction workers but they suffer 19.5% of all workplace trauma deaths. As you might see from this example, just counting the number of deaths in an industry would not give us an accurate indication of the danger of death to its workers.

From practical experience over 20 years, I went out to investigate the death of a worker and could not find the company. I soon discovered that the company had gone out of business. They were not financially sound enough to absorb the cost of an occupational death. So much is lost not only in the loss of a skilled worker, but the downtime, slowed production, lower morale, supervisor time, time with OSHA, cost of repair of equipment, lost credibility with customers, time spent with news media, and let us not forget the legal ramifications that abound. I have always used the figure that an occupationally related fatality cost to a company is in excess of a million dollars. Many companies cannot afford to continue in business faced with these types of costs.

It is definitely a good business practice to target the prevention of occupational deaths if you know what events or exposures cause them. You can identify the potential if the event or exposure could exist or occur in your workplace. The types of incidents that caused the 5703 trauma deaths are presented in Table 9.2.

Table 9.1 Number of Deaths and Rate of Fatal Occupational Injuries in the Major Industrial Sectors by Industry Sector, 2006

Major Industry	Number of Deaths	Fatal Rate
Construction	1226	10.8
Transportation and warehousing	832	16.3
Agriculture (forestry and fishing)	646	29.6
Government	501	2.3
Professional and business services	449	3.1
Manufacturing	447	2.7
Retail trade	351	2.1
Leisure and hospitality	252	2.2
Wholesale trade	217	4.8
Mining	190	27.8
Other services	175	2.5
Educational and health services	172	0.9
Financial activities	122	1.2
Information	64	1.9
Utilities	52	6.2

Source: Courtesy of the BLS National Census of Fatal Occupational Injuries in 2006.

You may have different occupations in your workplace. You will want to pay special attention to any of those occupations that carry the highest risk of death. Realize that some work occupations will not have the same exposure as others. Hence, you need to understand the types of events or exposures that are most likely to affect the occupations in your facility or worksite. A summary of the varied occupations and number of deaths occurring in them is found in Table 9.3.

As you identify the sources of energy whose release could be injurious or deadly in your work area, you will need to ensure that these sources are prevented from releasing their force on unsuspecting workers. By becoming knowledgeable regarding the types, causes, occupations, and industries that have fatalities, you can take steps to ensure that occupational deaths do not take place at your company.

OCCUPATIONAL INJURIES

Of the injuries and illnesses reported to the Bureau of Labor Statistics (BLS), 4.1 million were injuries in 2006. Since the early 1990s there has been a gradual decrease in the incident rates, which depict the number of reported cases per 100 full-time workers, falling from 8.4 in 1994 to 4.4 in 2006. The 1999 rate was the lowest since the BLS started reporting injury rates in the 1970s. The injury rate was higher for goods-producing operations (8.9 per 100 full-time workers) than for service-producing operations at 5.3 per 100 full-time workers.

According to the BLS approximately 2.1 million of the 4.1 million were injuries that resulted in days away from work or restricted work activity. Now keep in mind that the causes of trauma injuries are usually the unplanned release of energy or the contact with an energy source. These injuries may result in the need for simple to complex medical care. Each of these events comes with an expense involved. There is direct medical care and potential for the need of workers' compensation

Table 9.2   Causes of Fatal Occupational Injuries by Event or Exposure, 2006

Event or Exposure	Number	Percent
Transportation incidents	2413	42
Highway	1329	23
Nonhighway (farm and industrial)	342	6
Worker struck by vehicle	372	7
Aircraft	215	4
Water vehicle	89	2
Rail vehicle	65	1
Assaults and violent acts	754	13
Homicides	516	9
Self-inflicted	199	3
Contact with objects and equipment	983	17
Struck by object (flying or falling)	583	10
Caught in or compressed by equipment or object	281	5
Caught in or crushed in collapsing material	107	2
Falls	809	14
Fall to lower level	728	13
Fall on the same level	59	1
Exposure to harmful substances or environments	525	9
Contact with electrical current	247	4
Contact with temperature extremes	53	1
Exposure to caustic, noxious, or allergic substances	153	3
Oxygen deficiency (e.g., drowning)	64	1
Fire and explosion	201	4
Other events or exposures	1001	18

Source: Courtesy of the BLS National Census of Fatal Occupational Injuries in 2006.

benefits. Thus, the prevention of these injuries is in the best interest of both the employer and the worker. The release of this energy can come from many sources as has been noted in an earlier chapter. The energy sources are shown in Table 9.4.

There are approximately 6 million workplaces in the United States and each one of them has unique sources of energy that are an integral part of the industry or the occupation in that workplace. To take a critical and nonbiased look at the injuries that you are experiencing, other pieces of information are necessary. You should gather information regarding injuries to evaluate them against national trends, industry trends, and your own prevention effort. Injuries involving lost workdays from the 1999 BLS Annual Survey of Work-Related Injuries and Illnesses will be used as an example.

Injuries that result in days away from work are considered to be the most severe. It is not enough just to count the number of injuries. We must make an effort to decrease not only the number of injuries but the ones that are most costly in time lost and medical treatment. Using lost workdays provides a standard way of looking at these most severe types of injuries. One of the questions that you ask first is "Do I have employees in occupations that have historically been at risk for severe injuries?" Table 9.5 shows the numbers of lost-workday injuries that have occurred in high-risk occupations.

Once you have identified the different occupations and their propensity for injury, you can target those who are at most risk. However, if you already have injuries you need to gather some specific

Table 9.3   Fatal Occupational Injuries by Occupation and Selected Event or Exposure, 2006[a]

Occupations	Number of Deaths	(Event or Exposure [Percent])				
		Percent	Highway	Homicide	Struck by Object	Fall to Lower Level
Management occupations	543	10	11	9	13	9
Business and financial operation occupations	38	1	26	13	—	18
Computer and mathematical occupations	16	—	19	—	—	25
Architecture and engineering occupations	54	1	26	—	6	19
Life, physical, and social science occupations	25	—	16	—	—	20
Community and social services occupations	31	1	55	23	—	10
Legal occupations	11	—	—	45	—	—
Education, training, and library occupations	23	—	30	—	—	22
Arts, design, entertainment, sports, and media occupations	56	1	9	—	—	11
Healthcare practitioners and technical occupations	54	1	28	19	—	—
Healthcare support occupations	20	—	40	15	—	15
Protective service occupations	274	5	27	33	2	1
Food preparation and serving-related occupations	78	1	9	54	10	—
Building and grounds cleaning and maintenance occupations	277	5	11	5	14	29
Personal care and service occupations	56	1	11	21	5	11
Sales and related occupations	308	5	20	48	3	6
Office and administrative support occupations	62	1	33	29	5	10
Farming, fishing, and forestry occupations	289	5	11	—	23	4
Construction and extraction occupations	1258	22	10	1	11	33
Installation, maintenance, and repair occupations	415	7	12	3	20	18
Production occupations	282	5	6	4	18	9
Transportation and material moving occupations	1463	26	50	4	7	4
Military occupations	51	1	25	—	14	—

Source: Courtesy of the BLS National Census of Fatal Occupational Injuries in 2006.
[a] This is an example of the specific information that can be obtained for occupations under the major headings.

Table 9.4   Source of Potential Energy Release

Mechanical	Compressed or liquefied gas
Machinery	Flammable gases
Tools	Nonflammable gases
Noise	Corrosive material
Explosives	Flammable material
Moving objects	Flammable solids
Strain (self)	Flammable liquids
Stored/stacked materials, liquid	Poisons
Electrical	Oxidizing materials
Uninsulated conductors	Dust
High-voltage sources	Explosives
Thermal	Radiation
Flames	Ionizing radiation
Hot surfaces	Nonionizing radiation
Molten metals	Lasers
Chemical	Microwaves
Acids	X-rays
Bases	Radiation sources
Fuels	Welding
Moving vehicles	Bullets
Sharp object	Bombs
Caving soil	Potential for a fall
Biological	Cold temperatures
Ignitions	Fatigued structure or equipment
Pressure vessels	Sharp edges
Vibration	Hydraulic pressure
Lightning strikes	Wind
High heat source	Flooding

kinds of information. First, you must know what the nature of the injuries is. This means that you need to determine if the injury resulted in a fracture, amputation, burn, etc. Another piece of information you would need is the body part that was injured such as head, trunk, foot, etc. A third area of information would be to determine the source of the injury, for example, machinery, stacked materials, tools, etc. The final information would be to determine the event or exposure that occurred as contact with electricity, fall to same level, assault, chemical exposure, etc. These are the must-know types of information. Others that may be of interest to you regarding the injury might be location in your facility, time, shift, day of week, supervisor, etc.

Table 9.6 shows how these criteria are used to compile statistical information on injuries. This kind of information will allow you to evaluate the potential for mitigation or prevention of injuries within your workplace. This decrease in injuries should be highly beneficial to your company's bottom line.

The information and data provided in this chapter were presented for the purpose of awareness as well as to show you the types of data and information that are available to you from different sources. The best and most current source of occupational safety and health injury and illness data

Table 9.5 Number of Nonfatal Occupational Injuries and Illnesses Involving Days Away from Work[a] by Major Occupational Group and Major Industry Sector, 2006

Occupation	Private Industry[b,c,d]	Goods Producing				Service Providing							
		Total Goods Producing	Natural Resources and Mining[b,c]	Construction	Manufacturing	Total Service Providing	Trade, Transportation and Utilities[d]	Information	Financial Activities	Professional and Business Services	Education and Health Services	Leisure and Hospitality	Other Services
Total cases	1,183,500	380,440	26,290	153,180	200,970	803,060	354,510	18,560	33,300	89,940	182,210	96,910	27,640
Management occupations	19,800	3,880	390	2,100	1,390	15,930	3,300	490	2,660	1,210	3,650	3,540	1,070
Business and financial operations occupations	6,000	520	20	80	410	5,490	1,130	120	1,780	1,430	770	110	150
Computer and mathematical occupations	2,790	360	—	60	290	2,430	260	590	350	880	250	80	—
Architecture and engineering occupations	5,150	1,840	130	340	1,380	3,300	550	440	30	2,140	40	50	50
Life, physical, and social science occupations	2,940	770	140	20	600	2,170	100	—	30	1,520	320	30	160
Community and social services occupations	8,190	20	—	20	—	8,170	—	—	60	310	7,400	—	390
Legal occupations	970	—	—	—	—	960	—	—	250	550	130	—	—
Education, training, and library occupations	6,900	—	—	—	—	6,890	90	30	20	260	5,970	140	370

(continued)

Table 9.5 (continued)   Number of Nonfatal Occupational Injuries and Illnesses Involving Days Away from Work[a] by Major Occupational Group and Major Industry Sector, 2006

Occupation	Private Industry[b,c,d]	Goods Producing				Service Providing							
		Total Goods Producing	Natural Resources and Mining[b,c]	Construction	Manufacturing	Total Service Providing	Trade, Transportation and Utilities[d]	Information	Financial Activities	Professional and Business Services	Education and Health Services	Leisure and Hospitality	Other Services
Arts, design, entertainment, sports, and media occupations	6,240	170	—	40	130	6,070	1,220	1,080	40	610	330	2,670	120
Healthcare practitioners and technical occupations	48,020	140	50	—	80	47,880	980	—	590	1,780	44,360	30	140
Healthcare support occupations	67,790	20	—	—	20	67,770	220	—	60	1,700	65,360	130	300
Protective service occupations	9,520	360	70	90	200	9,160	780	110	340	5,280	1,280	1,190	180
Food preparation and serving related occupations	77,820	670	90	50	520	77,150	10,460	140	270	1,260	8,210	56,450	360
Building and grounds cleaning and maintenance occupations	67,270	4,560	420	1,100	3,040	62,710	3,810	380	5,600	23,290	13,950	13,610	2,060
Personal care and service occupations	22,750	120	100	—	—	22,640	7,040	150	250	1,100	8,350	3,200	2,540

Occupation												
Sales and related occupations	76,210	1,850	600	1,220	74,360	63,840	970	2,010	2,830	570	3,140	990
Office and administrative support occupations	83,320	8,710	1,100	7,310	74,610	38,120	3,670	8,160	9,510	11,560	1,940	1,660
Farming, fishing, and forestry occupations	14,330	12,340	30	620	1,340	800	—	—	270	20	240	—
Construction and extraction occupations	149,910	5,160	124,470	7,510	12,770	6,160	150	830	3,640	850	700	430
Installation, maintenance, and repair occupations	94,890	26,730	11,570	13,620	68,160	36,260	6,480	4,780	6,620	3,200	2,580	8,230
Production occupations	171,030	1,950	5,820	126,670	36,590	19,430	1,820	510	6,420	1,750	2,740	3,930
Transportation and material moving occupations	239,710	44,630	3,530	5,550	35,550	195,080	159,390	1,890	4,490	17,070	3,600	4,290	4,350

Sources: From Bureau of Labor Statistics, U.S. Department of Labor, Survey of Occupational Injuries and Illnesses in cooperation with participating State agencies; Courtesy of the BLS Lost Work Time Injuries and Illnesses Report, 2006.

Note: Dash indicates data do not meet publication guidelines. Because of rounding and data exclusion of nonclassifiable responses, data may not sum to the totals.

[a] Days-away-from-work cases include those that resulted in days away from work, some of which also included job transfer or restriction.

[b] Excludes farms with fewer than 11 employees.

[c] Data for Mining (Sector 21 in the North American Industry Classification System—United States, 2002) include establishments not governed by the Mine Safety and Health Administration rules and reporting, such as those in Oil and Gas Extraction and related support activities. Data for mining operators in coal, metal, and nonmetal mining are provided to BLS by the Mine Safety and Health Administration, U.S. Department of Labor. Independent mining contractors are excluded from the coal, metal, and nonmetal mining industries. These data do not reflect the changes the Occupational Safety and Health Administration made to its record keeping requirements effective January 1, 2002; therefore, estimates for these industries are not comparable to estimates in other industries.

[d] Data for employers in rail transportation are provided to BLS by the Federal Railroad Administration, U.S. Department of Transportation.

Table 9.6 Percent Distribution of Nonfatal Occupational Injuries and Illnesses Involving Days Away from Work^e by Selected Injury or Illness Characteristics and Major Industry Sector, 2006

Characteristic	Private Industry[b,c,d]	Goods Producing				Service Providing							
		Total Goods Producing	Natural Resources and Mining[b,c]	Constr-uction	Manufac-turing	Total Service Providing	Trade, Transportation and Utilities[d]	Inform-ation	Financial Activities	Professional and Business Services	Education and Health Services	Leisure and Hospitality	Other Services
Total [1,183,500 cases]	100.0	100.0	100.0	100.0	100.0	100.0	100.0	100.0	100.0	100.0	100.0	100.0	100.0
Nature of injury or illness													
Sprains, strains	39.9	33.9	31.3	34.5	33.7	42.8	44.4	42.3	39.7	36.6	49.4	32.9	38.6
Bruises, contusions	8.6	7.7	11.3	6.2	8.4	9.0	9.5	8.5	9.4	8.4	8.6	8.6	6.7
Cuts, lacerations	8.4	10.9	7.9	12.1	10.4	7.2	7.2	4.1	4.6	8.7	2.6	15.6	8.8
Fractures	8.0	9.9	13.0	11.5	8.3	7.0	7.6	8.7	6.8	7.1	5.8	6.7	7.7
Heat burns	1.5	1.4	1.7	1.0	1.7	1.5	0.7	0.3	0.6	0.8	1.2	6.3	1.8
Carpal tunnel syndrome	1.1	1.6	0.2	0.6	2.5	0.9	0.8	2.3	2.5	0.9	0.7	0.7	1.3
Tendonitis	0.4	0.5	0.3	0.1	0.8	0.3	0.3	0.5	0.3	0.5	0.4	0.3	0.5
Chemical burns	0.6	0.8	0.6	0.7	0.9	0.6	0.5	0.1	0.3	0.8	0.4	0.9	0.3
Amputations	0.7	1.4	1.2	0.8	1.9	0.3	0.4	0.2	0.9	0.5	e	0.5	0.3
Multiple traumatic injuries	3.9	3.7	4.2	3.7	3.5	4.0	3.8	4.4	6.5	4.2	4.3	3.1	3.7
Part of body affected by the injury or illness													
Head	7.0	8.1	10.1	8.2	7.7	6.4	6.5	5.9	7.1	7.6	5.5	6.6	6.7
Eye	3.0	4.7	4.4	4.6	4.8	2.3	2.3	1.7	2.2	3.2	1.9	2.0	3.1
Neck	1.5	1.2	1.6	1.4	1.0	1.6	1.7	2.2	1.9	1.2	2.0	1.2	0.9
Trunk	34.0	30.3	31.0	29.1	31.2	35.7	37.1	31.2	32.8	30.0	41.1	28.0	34.2

Shoulder	6.4	5.9	4.7	5.0	6.7	6.7	7.6	6.5	4.9	4.9	6.8	4.9	7.9
Back	21.2	17.8	18.0	18.3	17.4	22.8	22.7	19.8	21.6	18.4	28.9	17.8	19.1
Upper extremities	23.2	29.0	22.5	25.0	32.9	20.4	19.6	19.9	15.3	20.5	15.8	32.2	25.1
Wrist	4.1	4.5	3.1	3.2	5.7	3.9	3.8	5.0	4.7	3.3	4.0	4.3	4.2
Hand, except finger	4.2	5.2	4.6	5.2	5.3	3.7	3.4	3.0	1.9	3.7	2.5	7.5	3.7
Finger	9.0	13.0	9.4	10.7	15.1	7.1	6.9	5.4	4.7	7.8	4.2	13.1	9.1
Lower extremities	22.2	22.0	23.0	26.2	18.6	22.2	23.7	23.7	25.9	24.4	19.0	20.1	20.1
Knee	8.1	7.8	7.9	9.5	6.6	8.2	8.5	9.9	8.0	8.2	8.0	7.2	7.3
Foot, except toe	3.7	3.7	3.5	4.0	3.4	3.7	4.6	2.7	6.0	3.8	2.2	2.9	3.5
Toe	1.1	1.2	1.0	1.2	1.3	1.1	1.4	1.4	1.1	1.1	0.8	0.5	0.8
Body systems	1.5	1.2	1.8	1.1	1.2	1.7	1.1	2.2	3.2	2.2	2.3	1.7	1.7
Multiple parts	9.8	7.3	8.6	8.1	6.6	11.0	9.3	14.1	13.2	12.8	13.5	9.4	9.7
Source of injury or illness													
Chemicals and chemical products	1.6	1.8	2.7	1.1	2.2	1.6	1.3	0.8	2.5	2.1	1.8	1.7	1.0
Containers	12.4	8.5	5.6	4.9	11.6	14.3	20.6	8.3	9.0	11.2	5.8	14.7	8.8
Furniture and fixtures	3.8	2.2	.9	1.6	2.9	4.6	4.0	3.3	7.8	3.8	4.9	6.3	3.9
Machinery	6.5	10.3	8.6	7.1	13.0	4.8	5.0	5.5	5.5	5.8	2.2	6.7	7.0
Parts and materials	10.5	19.5	10.6	22.8	18.2	6.3	9.4	5.7	5.1	7.1	1.1	2.7	12.5
Worker motion or position	13.8	14.7	9.2	12.9	16.8	13.4	13.6	22.1	16.7	13.2	12.2	11.8	14.4
Floors, walkways, ground surfaces	18.1	14.9	19.1	18.3	11.8	19.6	16.7	25.3	25.8	19.0	21.6	24.9	17.1

(continued)

Table 9.6 (continued)   Percent Distribution of Nonfatal Occupational Injuries and Illnesses Involving Days away from Work[a] by Selected Injury or Illness Characteristics and Major Industry Sector, 2006

Characteristic	Private Industry[b,c,d]	Goods Producing				Service Providing							
		Total Goods Producing	Natural Resources and Mining[b,c]	Construction	Manufacturing	Total Service Providing	Trade, Transportation and Utilities[d]	Information	Financial Activities	Professional and Business Services	Education and Health Services	Leisure and Hospitality	Other Services
Tools, instruments, and equipment	6.9	9.1	7.6	12.0	7.1	5.9	5.1	6.4	5.5	6.6	4.7	9.8	9.4
Vehicles	8.6	5.6	9.5	5.5	5.2	10.0	13.7	10.0	9.5	10.6	5.3	4.5	9.7
Health care patient	4.4	—	—	—	—	6.6	(e)	—	.4	.8	28.3	—	.7
Event or exposure leading to injury or illness													
Contact with objects and equipment	28.3	38.0	37.5	38.2	38.0	23.8	26.4	18.4	23.1	28.1	13.2	29.7	28.9
Struck by object	13.9	17.8	18.4	19.7	16.2	12.1	13.5	7.6	10.8	13.0	6.9	15.7	16.9
Struck against object	7.2	8.3	8.0	9.5	7.5	6.7	6.9	6.0	7.8	8.6	4.2	9.2	5.8
Caught in equipment or object	5.0	8.6	8.6	5.2	11.1	3.3	4.0	3.8	2.3	4.1	1.4	3.3	3.3
Fall to lower level	6.3	7.4	7.6	11.9	3.9	5.8	6.4	9.5	8.1	8.1	3.5	4.4	4.2

Fall on same level	12.8	8.7	11.6	8.0	8.9	14.8	11.6	17.2	17.0	11.9	18.4	21.4	13.2
Slip, trip, loss of balance without fall	3.0	2.5	2.5	2.8	2.3	3.2	2.8	3.8	3.6	3.3	3.1	4.6	2.6
Overexertion	24.1	19.7	14.7	17.4	22.1	26.1	28.2	16.5	17.3	17.9	34.8	15.4	24.4
Overexertion in lifting	12.8	10.6	6.4	10.1	11.6	13.8	16.0	7.9	9.2	9.4	15.8	9.1	11.7
Repetitive motion	3.2	4.6	1.3	1.7	7.3	2.6	2.5	6.2	4.9	2.8	2.1	1.8	2.8
Exposure to harmful substances	4.8	4.5	5.4	3.3	5.3	4.9	2.9	4.1	5.1	6.3	5.5	10.0	4.1
Transportation accidents	4.7	3.1	5.5	4.1	2.0	5.5	6.8	7.4	6.7	7.2	3.4	2.6	5.0
Fires and explosions	.2	.3	.4	.4	.2	.1	.1	—	.1	.4	.1	.2	—
Assaults and violent acts by person	1.3	.1	.4	.1	.1	1.9	.5	.5	2.6	1.0	5.7	1.4	1.2

Sources: From Bureau of Labor Statistics, U.S. Department of Labor, Survey of Occupational Injuries and Illnesses in cooperation with participating State agencies; Courtesy of the BLS Lost Work Time Injuries and Illnesses Report, 2006.

Note: Dash indicates data do not meet publication guidelines. Because of rounding and data exclusion of nonclassifiable responses, percentages may not add to 100.

a Days-away-from-work cases include those that resulted in days away from work, some of which also included job transfer or restriction.

b Excludes farms with fewer than 11 employees.

c Data for Mining (Sector 21 in the *North American Industry Classification System*–United States, 2002) include establishments not governed by the Mine Safety and Health Administration rules and reporting, such as those in Oil and Gas Extraction and related support activities. Data for mining operators in coal, metal, and nonmetal mining are provided to BLS by the Mine Safety and Health Administration, U.S. Department of Labor. Independent mining contractors are excluded from the coal, metal, and nonmetal mining industries. These data do not reflect the changes the Occupational Safety and Health Administration made to its recordkeeping requirements effective January 1, 2002; therefore, estimates for these industries are not comparable to estimates in other industries.

d Data for employers in rail transportation are provided to BLS by the Federal Railroad Administration, U.S. Department of Transportation.

e Data too small to be displayed.

is the BLS (http:// www.bls.gov). Accessing this information and data will allow you to compare your injury and illness data profile with the national trends as a benchmark for improvement of your safety and health effort.

INJURY AND DEATH COST

According to the National Safety Council, the true cost of occupational injuries and deaths to the nation far surpasses the cost incurred from workers' compensation alone. In 2005, the cost was $160.4 billion, which includes wage and productivity losses of $80.0 billion, medical cost of $31.3 billion, and administrative cost of $34.4 billion. It also includes $10.7 billion for such uninsured costs as the money value of time lost by workers other than those with disabling injuries who are directly or indirectly involved in the injuries and the cost to investigate injuries, write up reports, etc. In addition, it includes damages to motor vehicles in work injuries of $1.7 billion and fire losses of $2.3 billion.

Each and every worker would have to generate goods and service equal to $1100 to offset the cost of work injuries. If a worker dies on the job, the cost of the death is $1,190,000 while the cost of a disabling injury is $38,000.

Accidents are more expensive than most realize because of hidden costs, for example, workers' compensation covers direct costs such as medical and indemnity payment for an injured or ill worker. However, the cost to train and compensate a replacement worker, repair damaged property, investigate the accident and implement corrective action, and to maintain insurance coverage will not be covered. Even less apparent are the costs related to schedule delays, added administrative time, lower morale, increased absenteeism, and poorer customer relations. These are all examples of indirect cost.

Studies have shown that the ratio of indirect cost to direct cost varies widely from 20:1 to 1:1. OSHA has shown that the smaller the cost of the accident the greater the direct to indirect cost ratio (see Table 9.7). Over many years I have always used a ballpark figure of around 5–10 for indirect cost to direct cost. There is little doubt that indirect cost can mount up quickly.

Let us use some common figures that may help to put this into perspective. To pay for an accident with a total cost of $1000:

- Soft drink bottler would have to bottle and sell over 122,000 cans of soda.
- Food packer would have to can and sell over 470,000 cans of corn.
- Bakery would have to bake and sell over 470,000 donuts.
- Contractor would have to pour and finish 6000 sq. ft. of concrete.
- Ready-mix company would have to deliver 40 truckloads of concrete.
- Paving contractor must lay 1800 ft. of a two-lane asphalt road.

Table 9.7   OSHA's Ratio of Indirect to Direct Cost

Direct Cost of Claim	Ratio of Indirect to Direct Cost
$0–$2,000	4.5:1
$3,000–$4,999	1.6:1
$5,000–$9,999	1.2:1
$10,000 or more	1:1

Source: Courtesy of Occupational Safety and Health Administration.

Some examples of savings from real companies that made an effort to improve the safety and health performance are presented.

- After focusing on its safety and health program, an Atlanta company reported that from 1994 to 1996, their annual workers' compensation claim cost fell from $591,536 to $91,536, a savings of $500,000.
- After implementing a 100% fall protection program and supervisor accountability, Horizon Steel Erectors, Inc., had a 96% reduction in its accident cost per hour, from $4.26 to $0.18.

One study estimated that implementing a safety and health program provides a gain of $6.00 for every $1.00 invested in workplace safety and health. Companies that are in OSHA's Voluntary Protection Program (180 sites) are estimated to have saved more than $130,000,000.

As can be seen, it is to the best interest of a company to be able to have and use numbers to obtain a profile of the deaths and injuries in their place of business in order to develop and implement a prevention program that influences the company's bottom line.

CHAPTER 10

Sick: Occupational Illnesses

I really do not feel well. Does it have something to do with my job?

Occupational illnesses are not as easily identified as are injuries. According to the Bureau of Labor Statistics there were 4.0 million injuries and illnesses reported in 2004. Of this number only 249,000 cases of occupational illnesses were reported. The reporting by industry can be seen in Table 10.1.

The 249,000 occupational illnesses included repeat trauma such as carpal tunnel syndrome, noise-induced hearing loss, and poisonings. It is my professional opinion that many occupational illnesses go unreported when the employer or worker is not able to link exposure with the symptoms the employees exhibit. In addition, physicians fail to ask the right questions regarding the patient's employment history, which can lead to the commonest diagnosis of a cold or flu. This has become very apparent with the recent occupational exposure to anthrax where a physician sent a worker home with anthrax without addressing potential occupational exposure hazards. Unless the physician is trained in occupational medicine, he or she seldom addresses work as the potential exposure source.

This is not entirely a physician problem by any means since the symptoms that are seen by the physician are often those of flu and other common illnesses suffered by the public. It is often up to the employee to make the physician aware of on-the-job exposure. If you notice, I have continuously

Table 10.1   Number and Percent of Occupational Illnesses by Major Industrial Groups

Industry Division	Total Number of Cases	Total Cases (%)
Agriculture, forestry, fishing	2,988	1.2
Mining	747	0.3
Construction	8,715	3.5
Manufacturing	105,078	42.2
Transportation and warehousing	10,209	4.1
Retail	17,181	6.9
Wholesale	7,221	2.9
Utilities	3,237	1.3
Information	4,233	1.7
Financial activities	8,466	3.4
Professional and business services	16,434	6.6
Education services	1,743	0.7
Health care and social assistance	45,816	18.4
Leisure and hospitality	13,197	5.3
Other services	3,486	1.4

Source: From Bureau of Labor Statistics Annual Report.

used the term exposure since, unlike trauma injuries and deaths, which are usually caused by the release of some source of energy, occupational illnesses are often due to both short- and long-term exposures. If the results of an exposure lead to immediate symptoms, it is said to be acute. If the symptoms come later, it is termed as chronic exposure. The time between exposure and the onset of symptoms is called the latency period. It could be days, weeks, months, or even years, as in the case of asbestos, where asbestosis or lung cancer appears 20 to 30 years after exposure. Looking at a large number of death certificates (20,000) from specific groups of workers, you often see a significant number of cases of specific types of cancer such as liver, thyroid, or pancreatic cancer that do not appear in the same number in the normal adult population. This leads one to believe that something the workers were exposed to in the work environment may have caused their demise.

It is often very difficult to get employers, supervisors, and employees to take seriously the exposures in the workplace as a potential risk to the workforce both short- and long-term, especially long-term. "It can't be too bad if I feel all right now." This false sense of security is illustrated by the 90,000 occupational illness deaths that are estimated by the Bureau of Labor Statistics to occur each year. This far surpasses the 6000 occupation trauma deaths a year. If both trauma and illness deaths are added together it would be equivalent to the lives lost to a jumbo jet crashing every day of the year. Would an aviation record like this be acceptable to you? If not, I doubt that you would be flying. It is time for employers and the workforce to take on-the-job exposures as a potentially serious threat.

IDENTIFYING HEALTH HAZARDS

Health-related hazards must be identified (recognized), evaluated, and controlled to prevent occupational illnesses that come from exposure to them. Health-related hazards come in a variety of forms, such as chemical, physical, ergonomic, or biological:

1. Chemical hazards arise from excessive airborne concentrations of mists, vapors, gases, or solids that are in the form of dusts or fumes. In addition to the hazard of inhalation, many of these materials may act as skin irritants or may be toxic by absorption through the skin. Chemicals can also be ingested, although this is not usually the principal route of entry into the body.
2. Physical hazards include excessive levels of nonionizing and ionizing radiations, noise, vibration, and extremes of temperature and pressure.
3. Ergonomic hazards include improperly designed tools or work areas. Improper lifting or reaching, poor visual conditions, or repeated motions in an awkward position can result in accidents or illnesses in the occupational environment. Designing the tools and the job to be done to fit the worker should be of prime importance. Intelligent application of engineering and biomechanical principles is required to eliminate hazards of this kind.
4. Biological hazards include insects, molds, fungi, viruses, vermin (birds, rats, mice, etc.) and bacterial contaminants (sanitation and housekeeping items such as potable water, removal of industrial waste and sewage, food handling, and personal cleanliness can contribute to the effects from biological hazards). Biological and chemical hazards can overlap.

These health-related hazards can often be difficult and elusive to identify. A common example of this is a contaminant in a building that has caused symptoms of illness. Even the evaluation process may not be able to detect the contaminant, which has dissipated before a sample can be collected. This leaves nothing to control and possibly no answer to what caused the illnesses.

You might want to know the most common reported illnesses in the workplace. This can also assist you in deciding where to put your resources toward prevention of occupational illnesses in your facility or worksite. In Table 10.2 you can find a listing of the most commonly reported occupational illnesses. The cost of compensation should also make you look carefully at the types of illnesses that are most costly and yet are preventable. Most employers look at trauma injuries only and seldom pay attention to the potential cost of occupationally related illnesses.

As can be seen repeat trauma illnesses (ergonomic-related incidents) are most prevalent in the workplace today; these types of illnesses often go unreported until they have reached a serious level of impairment. The medical cost and lost work time can be very large. This is why a specific chapter is reserved for this topic.

Most skin disorders can be prevented with the proper use of personal protective equipment (PPE) and good personal hygiene (washing hands, etc.). Usually skin disorders are caused by exposure to chemical and result in nothing more than a rash that is cured by proper PPE or removal or substitution (using a safer chemical) of the chemical. Some skin disorders can exacerbate into serious conditions when not tended to. At times, a worker's skin disorder may be an allergic reaction

Table 10.2   Percent Values of Illness Reported for 2004[a]

Type of Illness	Total Illnesses Reported (%)
Disorder associated with repeat trauma	66.6
Skin disease or disorders	15.6
Hearing loss	11.4
Respiratory conditions	7.1
Poisoning	1.3
All other diseases	8.8

Source: From Bureau of Labor Statistics Annual Report.
[a]   There were 249,000 total cases of nonfatal occupational illnesses reported.

that may not be solvable unless the worker is removed from that type of work. If the worker continues to do the same work, this could result in a costly illness.

Physical agents, of which noise is the most common in the workplace, can lead to nonreparable hearing loss, which becomes very compensable and degrades the value of that employee to you since he or she may not be able to hear warning signals or cannot communicate effectively with other workers. Although radiation (both ionizing and nonionizing) can be found in the workplace, it is not as common as noise, vibration, or temperature extremes.

What appears to present the most problems within the workplace are chemicals and the effects on workers who are exposed to them. Thus, you will note the emphasis in that direction in this chapter. The major Occupational Safety and Health Administration (OSHA) general standard that affects most workplaces will be discussed. That standard is the hazard communication standard (HCS).

TEMPERATURE EXTREMES

Cold Stress

Temperature is measured in degrees Fahrenheit (°F) or Celsius (°C). Most people feel comfortable when the air temperature ranges from 66°F to 79°F and the relative humidity is about 45%. Under these circumstances, heat production inside the body equals the heat loss from the body, and the internal body temperature is kept around 98.6°F. For constant body temperature, even under changing environmental conditions, rates of heat gain and heat loss should balance. Every living organism produces heat. In cold weather, the only source of heat gain is the body's own internal heat production, which increases with physical activity. Hot drinks and food are also a source of heat.

The body loses heat to its surroundings in several different ways. Heat loss is greatest if the body is in direct contact with cold water. The body can lose 25–30 times more heat when in contact with cold wet objects than under dry conditions or with dry clothing. The higher the temperature differences between the body surface and cold objects, the faster the heat loss. Heat is also lost from the skin by contact with cold air. The rate of loss depends on the air speed and the temperature difference between the skin and the surrounding air. At a given air temperature, heat loss increases with air speed. Sweat production and its evaporation from the skin also cause heat loss. This is important when performing hard work. Small amounts of heat are lost when cold food and drink are consumed. As well, heat is lost during breathing by inhaling cold air and through evaporation of water from the lungs.

The body maintains heat balance by reducing the amount of blood circulating through the skin and outer body parts. This minimizes cooling of the blood by shrinking the diameter of blood vessels. At extremely low temperatures, loss of blood flow to the extremities may cause an excessive drop in tissue temperature resulting in damage such as frostbite, and by shivering, which increases the body's heat production. This provides a temporary tolerance for cold but it cannot be maintained for long periods.

Overexposure to cold causes discomfort and a variety of health problems. Cold stress impairs performance of both manual and complex mental tasks. Sensitivity and dexterity of fingers lessen in cold. At lower temperatures still, cold affects deeper muscles, resulting in reduced muscular strength and stiffened joints. Mental alertness is reduced due to cold-related discomfort. For all these reasons accidents are more likely to occur in very cold working conditions.

The main cold injuries are frostnip, frostbite, immersion foot, and trench foot, which occur in localized areas of the body. Frostnip is the mildest form of cold injury and occurs when ear lobes, noses, cheeks, fingers, or toes are exposed to cold. The skin of the affected area turns white. Frostnip can be prevented by warm clothing and is treated by simple rewarming.

Immersion foot occurs in individuals whose feet have been wet, but not freezing cold, for days or weeks. The primary injury is to nerve and muscle tissue. Symptoms are numbness, swelling, or even superficial gangrene. Trench foot is wet cold disease resulting from exposure to moisture at or near the freezing point for one to several days. Symptoms are similar to immersion foot (swelling and tissue damage).

Hypothermia can occur in moderately cold environments. The body's core temperature does not usually fall more than 2°F–3°F below the normal 98.6°F because of the body's ability to adapt. However, in intense cold without adequate clothing, the body is unable to compensate for the heat loss, and the body's core temperature starts to fall. The sensation of cold followed by pain in exposed parts of the body is the first sign of cold stress. The most dangerous situation occurs when the body is immersed in cold water. As the cold worsens or the exposure time increases, the feeling of cold and pain starts to diminish because of increasing numbness (loss of sensation). If no pain can be felt, serious injury can occur without the victim noticing. Next, muscular weakness and drowsiness are experienced. This condition is called hypothermia and usually occurs when body temperature falls below 92°F. Additional symptoms of hypothermia include interruption of shivering, diminished consciousness, and dilated pupils. When body temperature reaches 80°F, coma (profound unconsciousness) sets in. Heart activity stops around 68°F and the brain stops functioning around 63°F. The hypothermia victim should be immediately warmed, either by moving to a warm room or by the use of blankets. Rewarming in water at 104°F–108°F has been recommended in cases where hypothermia occurs after the body was immersed in cold water.

Although people easily adapt to hot environments, they do not acclimatize well to cold. However, frequently exposed body parts can develop some degree of tolerance to cold. Blood flow in the hands, for example, is maintained in conditions that would cause extreme discomfort and loss of dexterity in unacclimatized persons. This is noticeable among fishermen who are able to work with barehands in extremely cold weather.

In the United States there are no OSHA exposure limits for cold working environments. It is often recommended that work warm-up schedules be developed. In most normal cold conditions, a warm-up break every 2 h is recommended, but, as temperatures fall and wind increases, more warm-up breaks are needed.

Protective clothing is needed for work at or below 40°F. Clothing should be selected to suit the cold, level of activity, and job design. Clothing should be worn in multiple layers, which provide better protection than a single thick garment. The layer of air between clothing provides better insulation than the clothing itself. In extremely cold conditions, where face protection is used, eye protection must be separated from respiratory channels (nose and mouth) to prevent exhaled moisture from fogging and frosting eye shields.

Heat Stress

Operations involving high air temperatures, radiant heat sources, high humidity, direct physical contact with hot objects, or strenuous physical activities have a high potential for inducing heat stress in employees engaged in such operations. Such places include iron and steel foundries, nonferrous foundries, brick-firing and ceramic plants, glass products facilities, rubber products factories, electrical utilities (particularly boiler rooms), bakeries, confectioneries, commercial kitchens, laundries, food canneries, chemical plants, mining sites, smelters, and steam tunnels. Outdoor operations, conducted in hot weather, such as construction, refining, asbestos removal, and hazardous waste site activities, especially those that require workers to wear semipermeable or impermeable protective clothing, are also likely to cause heat stress among exposed workers.

Age, weight, degree of physical fitness, degree of acclimatization, metabolism, use of alcohol or drugs, and a variety of medical conditions, such as hypertension, all affect a person's sensitivity to heat. However, even the type of clothing worn must be considered. Prior heat injury predisposes an individual to additional injury. It is difficult to predict just who will be affected and when, because individual susceptibility varies. In addition, environmental factors include more than the ambient air temperature. Radiant heat, air movement, conduction, and relative humidity all affect an individual's response to heat. There is no OSHA regulation for heat stress. The American Conference of Governmental Industrial Hygienists (ACGIH, 1992) stated that workers should not be permitted to work when their deep body temperature exceeds 38°C (100.4°F).

Complications transpire when workers suffer from heat exposure. The main anomalies are

1. Heat stroke
2. Heat exhaustion
3. Heat cramps
4. Fainting
5. Heat rash

The human body can adapt to heat exposure to some extent. This physiological adaptation is called acclimatization. After a period of acclimatization, the same activity will produce fewer cardiovascular demands. The worker will sweat more efficiently (causing better evaporative cooling) and thus will more easily be able to maintain normal body temperatures. A properly designed and applied acclimatization program decreases the risk of heat-related illnesses. Such a program involves exposing employees to work in a hot environment for progressively longer periods. The National Institute for Occupational Safety and Health (NIOSH, 1986) says that for workers who have had previous experience in jobs where heat levels are high enough to produce heat stress, the regimen should be 50% exposure on day 1, 60% on day 2, 80% on day 3, and 100% on day 4. For new workers who will be similarly exposed, the regimen should be 20% on day 1, with a 20% increase in exposure each additional day.

IONIZING RADIATION

Ionizing radiation has always been a mystery to most people. Actually, much more is known about ionizing radiation than the hazardous chemicals that constantly bombard the workplace. After all, there are only four types of radiation (alpha particles, beta particles, gamma rays, and neutrons) rather than thousands of chemicals. There are instruments that can detect each type of radiation and provide an accurate dose-received value. This is not so for chemicals, where the best that we could hope for in a real-time situation is a detection of the presence of a chemical and not what the chemical is. With radiation detection instruments the boundaries of contamination can be detected and set, while detecting such boundaries for chemicals is near to impossible except for a solid.

It is possible to maintain a lifetime dose for individuals exposed to radiation. Most workers wear personal dosimetry, which provides reduced levels of exposure. The same is impossible for chemicals where no standard unit of measurement, such as the roentgen equivalent in man (rem), exists for radioactive chemicals. The health effects of specific doses are well known such as 20–50 rems—minor changes in blood occur, 60–120 rems—vomiting occurs but no long-term illness, or 5,000–10,000 rems—certain death within 48 h. Certainly, radiation can be dangerous, but one or a combination of three factors, distance, time, and shielding, can usually be used to control exposure. Distance is the best since the amount of radiation from a source drops off quickly as a factor of the inverse square of the distance; for instance, at 8 ft. away the exposure is 1/64th of the radiation emanating from the source. As for time, many radiation workers are allowed to stay in a radiation area only for a certain length of time, and then they must leave that area. Shielding often conjures up lead plating or lead suits (similar to when x-rays are taken by a physician or dentist). Wearing a lead suit may seem appropriate but the weight alone can be prohibitive. Lead shielding can be used to protect workers from gamma rays (similar to x-rays). Once they are emitted, they could pass through anything in their path and continue on their way, unless a lead shield is thick enough to protect the worker.

Aluminum foil will stop the penetration of beta particles. Thus, a protective suit will prevent beta particles from reaching the skin, where they can burn and cause surface contamination. Alpha particles can enter the lungs and cause the tissue to become electrically charged (ionized). Protection for alpha particles can be obtained with the use of air-purifying respirators with proper cartridges to filter out radioactive particles. Neutrons are found around the core of a nuclear reactor and are absorbed by both water and the material in the control rods of the reactor. If a worker is not close to the core of the reactor, then no exposure can occur.

Ionizing radiation is a potential health hazard. The area where potential exposure can occur is usually highly regulated, posted, and monitored on a continuous basis. A maximum yearly exposure is permitted. Once it has been reached, a worker can have no more exposure. For general use 5 rems/year is the exposure of accepted level for workers. This is 50 times higher than what the U.S. Environmental Protection Agency (EPA) recommends for the public on a yearly basis. The average public exposure is supposed to be no more than 0.1 rems/year. A standard of 5 rems has been employed for many years and seems to reasonably protect workers. Exposure to radiation should be considered serious since overexposure can lead to serious health problems or even death.

NOISE-INDUCED HEARING LOSS

Occupational exposure to noise levels in excess of the current OSHA standards places hundreds of thousands of workers at risk of developing material hearing impairment, hypertension, and elevated hormone levels. Workers in some industries (i.e., construction, oil and gas well drilling and servicing) are not fully covered by the current OSHA standards and lack the protection of an adequate hearing conservation program. Occupationally induced hearing loss continues to be one of the leading occupational illnesses in the United States. OSHA is designating this issue as a priority for rulemaking action to extend hearing conservation protection, provided in the general industry standard, to the construction industry and other uncovered industries.

According to the U.S. Bureau of the Census, Statistical Abstract of the United States, there are over 7.2 million workers employed in the construction industry (6% of all employment). NIOSH and National Occupational Exposure Survey (NOES) estimate that 421,000 construction workers are exposed to noise above 85 dBA. NIOSH estimates that 15% of workers exposed to noise levels of 85 dBA or higher will develop material hearing impairment.

Research demonstrates that construction workers are regularly overexposed to noise. The extent of the daily exposure to noise in the construction industry depends on the nature and duration of the work. For example: rock drilling (up to 115 dBA); chain saw (up to 125 dBA); abrasive blasting (105–112 dBA); heavy equipment operation (95–110 dBA); demolition (up to 117 dBA); and needle guns (up to 112 dBA). Exposure to 115 dBA is permitted for a maximum of 15 min for an 8 h workday. No exposure above 115 dBA is permitted. Traditional dosimetry measurement may substantially underestimate noise exposure levels for construction workers since short-term peak exposures that may be responsible for acute and chronic effects can be lost in lower, full-shift time-weighted average (TWA) measurements.

There are a variety of control techniques documented in the literature to reduce the overall worker exposure to noise. Such controls reduce the amount of sound energy released by the noise source, divert the flow of sound energy away from the receiver, or protect the receiver from the sound energy reaching him or her. For example, types of noise controls include proper maintenance of equipment, revised operating procedures, equipment replacements, acoustical shields and barriers, equipment redesign, enclosures, administrative controls, and PPE.

Table 10.3   OSHA Permissible Noise Exposure Levels

Duration Per Day (h)	Sound Level dBA Slow Response
8	90
6	92
4	95
3	97
2	100
1½	102
1	105
½	110
¼ or less	115

Source: Courtesy of Occupational Safety and Health Administration.

Under OSHA's general industry standard, feasible administrative and engineering controls must be implemented whenever employee noise exposures exceed 90 dBA (8 h TWA). In addition, an effective hearing conservation program (including specific requirements for monitoring noise exposure, audiometric testing, audiogram evaluation, hearing protection for employees with a standard threshold shift, training, education, and record keeping) must be made available whenever employee exposures equal or exceed an 8 h TWA sound level of 85 dBA (29 CFR 1910.95). Similarly, under the construction industry standard, the maximum permissible occupational noise exposure is 90 dBA (8 h TWA), and noise levels in excess of 90 dBA must be reduced through feasible administrative and engineering controls (see Table 10.3). However, the construction industry standard (29 CFR 1926.52) includes only a general minimum requirement for hearing conservation and lacks the specific requirements for an effective hearing conservation program included in the general industry standard (29 CFR 1910.95). NIOSH and ACGIH have also recommended exposure limits (NIOSH: 85 dBA TWA, 115 dBA ceiling; ACGIH: 85 dBA).

Noise, or unwanted sound, is one of the most pervasive occupational health problems. It is a by-product of many industrial processes. Sound consists of pressure changes in a medium (usually air), caused by vibration or turbulence. These pressure changes produce waves emanating away from the turbulent or vibrating source. Exposure to high levels of noise causes hearing loss and may cause other harmful health effects as well. The extent of damage depends primarily on the intensity of the noise and the duration of the exposure. Noise-induced hearing loss can be temporary or permanent. Temporary hearing loss results from short-term exposures to noise, with normal hearing returning after a period of rest. Generally, prolonged exposure to high noise levels over a period of time gradually causes permanent damage.

Sometimes the loss of hearing due to industrial noise is called the silent epidemic. Since this type of hearing loss is not correctable by either surgery or the use of hearing aids, it is certainly a monumental loss to the worker. It distorts communication both at work and socially. It may cause the worker to lose his or her job if acute hearing is required to perform effectively. The loss of hearing is definitely a handicap to the worker.

NONIONIZING RADIATION

Nonionizing radiation is a form of electromagnetic radiation, and it has varying effects on the body, depending largely on the particular wavelength of the radiation involved. In the following paragraphs, in approximate order of decreasing wavelength and increasing frequency, are some hazards associated with different regions of the nonionizing electromagnetic radiation spectrum. Nonionizing radiation is covered in detail by 29 CFR 1910.97.

Low frequency, with longer wavelengths, includes power line transmission frequencies, broadcast radio, and shortwave radio. Each of these can produce general heating of the body. The health hazard from these radiations is very small, however, since it is unlikely that they would be found in intensities great enough to cause significant effect. An exception can be found very close to powerful radio transmitter aerials.

Microwaves (MW) have wavelengths of 3 m to 3 mm (100–100,000 MHz). They are found in radar, communications, some types of cooking, and diathermy applications. MW intensities may be sufficient to cause significant heating of tissues. The effect is related to wavelength, power intensity, and time of exposure. Generally, longer wavelengths produce greater penetration and temperature rise in deeper tissues than shorter wavelengths. However, for a given power intensity, there is less subjective awareness to the heat from longer wavelengths than there is to the heat from shorter wavelengths because absorption of longer wavelength radiation takes place beneath the body's surface.

An intolerable rise in body temperature, as well as localized damage to specific organs can result from an exposure of sufficient intensity and time. In addition, flammable gases and vapors may ignite when they are inside metallic objects located in a MW beam. Power intensities for MWs are given in units of milliwatts per square centimeter (mW/cm^2), and areas that have a power intensity of over $10 mW/cm^2$ for a period of 0.1 h or longer should be avoided.

Radio frequency (RF) and MW radiation is electromagnetic radiation in the frequency range 3 kHz to 300 GHz. Usually MW radiation is considered a subset of RF radiation, although an alternative convention treats RF and MW radiation as two spectral regions. MWs occupy the spectral region between 300 GHz and 300 MHz, while RF or radio waves include 300 MHz to 3 kHz. RF/MW radiation is nonionizing in that there is insufficient energy (less than 10 eV) to ionize biologically important atoms. The primary health effects of RF/MW energy are considered to be thermal. The absorption of RF/MW energy varies with frequency. MW frequencies produce a skin effect (you can literally sense your skin starting to feel warm). RF radiation may penetrate the body and be absorbed in deep body organs without the skin effect, which can warn an individual of danger. A great deal of research has discovered other nonthermal effects. All the standards of western countries have so far based their exposure limits solely on preventing thermal problems. In the meantime, research continues. Use of RF/MW radiation includes aeronautical radios, citizen's band (CB) radios, cellular phones, processing and cooking of foods, heat sealers, vinyl welders, high-frequency welders, induction heaters, flow solder machines, communications transmitters, radar transmitters, ion implant equipment, MW drying equipment, sputtering equipment, glue curing, power amplifiers, and metrology.

Infrared radiation does not penetrate below the superficial layer of the skin so that its only effect is to heat the skin and the tissues immediately below it. Except for thermal burns, the health hazard on exposure to low level conventional infrared radiation sources is negligible. Visible radiation, which is about midway in the electromagnetic spectrum, is important because it can affect both the quality and accuracy of work. Good lighting conditions generally result in increased product quality with less spoilage and increased production. Lighting should be bright enough for easy seeing and directed so that it does not create glare. The light should be bright enough to permit efficient seeing.

Ultraviolet radiation in industry may be found around electrical arcs, and such arcs should be shielded by materials opaque to the ultraviolet. The fact that a material may be opaque to ultraviolet has no relation to its opacity to other parts of the spectrum. Ordinary window glass, for instance, is almost completely opaque to the ultraviolet in sunlight; at the same time, it is transparent to the visible light waves. A piece of plastic, dyed a deep red-violet, may be almost entirely opaque in the visible part of the spectrum and transparent in the near-ultraviolet. Electric welding arcs and germicidal lamps are the most common strong producers of ultraviolet in industry. The ordinary fluorescent lamp generates a good deal of ultraviolet inside the bulb, but it is essentially all absorbed by the bulb and its coating.

The most common exposure to ultraviolet radiation is from direct sunlight, and a familiar result of overexposure (one that is known to all sunbathers) is sunburn. Almost everyone is familiar with certain compounds and lotions that reduce the effects of the sun's rays, but many are unaware that some industrial materials, such as cresols, make the skin especially sensitive to ultraviolet rays. After having been exposed to cresols, even a short exposure in the sun usually results in severe sunburn. Nonionizing radiation, although perceived not to be as dangerous as ionizing radiation, does have its share of adverse health effects accompanying it.

VIBRATION

Vibrating tools and equipment at frequencies between 40 and 90 Hz can cause damage to the circulatory and nervous systems. Care must be taken with low frequencies, which have the potential to put workers at risk for vibration injuries. One of the most common cumulative trauma disorders (CTDs) resulting from vibration is Raynaud's syndrome. Its most common symptoms are intermittent numbness and tingling in the fingers, skin that turns pale, ashen, and cold, and eventual loss of sensation and control in the fingers and hands. Raynaud's syndrome comes about by the use of vibrating hand tools such as palm sanders, planers, jackhammers, grinders, and buffers. When such tools are required for a job, an assessment should be made to determine if any other methods can be used to accomplish the desired task. If not, other techniques, such as time/use limitations, alternating workers, or other such administrative actions, should be considered to help reduce the potential for a vibration-induced CTD. The damage caused by vibrating tools can be reduced by

- Using vibration dampening gloves
- Purchasing low-vibration tools and equipment
- Putting antivibration material on handles of existing tools
- Reducing length of exposure
- Changing the actual work procedure if possible
- Using balanced and dampening tools and equipment
- Rotating workers to decrease exposure time
- Decreasing the pace of the job as well as the speed of tools or equipment

Individuals subject to whole-body vibration have experienced visual problems, vertebral degeneration, breathing problems, motion sickness, pains in the abdomen, chest, and jaw, backache, joint problems, muscle strain, and problems with their speech. Although there are still many questions regarding vibration, it is definite that physical problems can transpire from exposure to vibration.

HEALTH HAZARDS

Health hazards are caused by any chemical or biological exposure that interacts adversely with organs within our body causing illnesses or injuries. The majority of chemical exposures result from inhaling chemical contaminants in the form of vapors, gases, dusts, fumes, and mists, or by skin absorption of these materials. The degree of the hazard depends on the length of exposure time and the amount or quantity of the chemical agent. This is considered to be the dose of a substance. Thus, a chemical is considered a poison when it causes harmful effects or interferes with biological reactions in the body. Only those chemicals that are associated with a great risk of harmful effects are designated as poisons.

Dose is the most important factor determining whether or not you will have an adverse effect from a chemical exposure. The longer you work at a job and the more chemical agent that gets into the air or on your skin, the higher the dose potential. Two components that make up dose are

- Length of exposure, or how long you are exposed—1 h, 1 day, 1 year, 10 years, etc.
- Quantity of substance in the air (concentration), how much you get on your skin, and the amount eaten or ingested.

Another important factor to consider about the dose is the relationship of two or more chemicals acting together that cause an increased risk to the body. This interaction of chemicals that multiply the chance of harmful effects is called a synergistic effect. Many chemicals can interact and although the dose of any one chemical may be too low to affect you, the combination of doses from different chemicals may be harmful. For example, the combination of chemical exposures and a personal habit such as cigarette smoking may be more harmful than just an exposure to one chemical. Smoking and exposure to asbestos increase the chance of lung cancer by as much as 50 times.

The type and severity of the body's response are related to the dose and the nature of the specific contaminant present. Air that looks dirty or has an offensive odor may, in fact, pose no threat whatsoever to the tissues of the respiratory system. In contrast, some gases that are odorless, or at least not offensive, can cause severe tissue damage. Particles that normally cause lung damage cannot even be seen. Many times, however, large visible clouds of dust are a good indicator that smaller particles may also be present.

The body is a complicated collection of cells, tissues, and organs with special ways of protecting itself against harm. These are usually called the body's defense systems. The body's defense system can be broken down, overcome, or bypassed. When this happens, injury or illness can result. Sometimes job-related injuries or illness are temporary, and you can recover completely. Other times, as in the case of chronic lung diseases like silicosis or cancer, these are permanent changes that may lead to death.

Acute Health Effects

Chemicals can cause acute (short-term) or chronic (long-term) effects. Whether or not a chemical causes an acute or chronic reaction depends on both the chemical and the dose you are exposed to. Acute effects are seen quickly, usually after exposures to high concentrations of a hazardous material. For example, the dry cleaning solvent perchloroethylene can immediately cause dizziness, nausea and at higher levels, coma and death. Most acute effects are temporary and reverse shortly after the individual is removed from the exposure. However, at high enough exposures permanent damage may occur. For most substances, neither the presence nor absence of acute effects can be used to predict whether chronic effects will occur. Dose is the determining factor. Exposures to cancer-causing substances (carcinogens) and sensitizers may lead to both acute and chronic effects.

An acute exposure may occur, for example, when we are exposed to ammonia while using another cleaning agent. Acute exposure may have both immediate and delayed effects on the body. Nitrogen dioxide poisoning can be followed by signs of brain impairment (such as confusion, lack of coordination, and behavioral changes) days or weeks after recovery.

Chemicals can cause acute effects on breathing. Some chemicals irritate the lungs and some sensitize the lungs. Fluorides, sulfides, and chlorides are all found in various welding and soldering fluxes. During welding and soldering, these materials combine with the moisture in the air to form hydrofluoric, sulfuric, and hydrochloric acid. All three can severely burn the skin, eyes, and respiratory tract. High levels can overwhelm the lungs, burning and blistering them, and causing pulmonary edema (fluid buildup in the lungs that will cause shortness of breath and if severe enough can cause death).

In addition, chemicals can have acute effects on the brain. When inhaled, solvent vapors enter the bloodstream and travel to other parts of the body, particularly the nervous system. Most solvents have a narcotic effect. This means they affect the nervous system by causing dizziness, headaches, feelings of drunkenness, and tiredness. One result of these symptoms may be poor coordination, which can contribute to falls and other accidents on a worksite. Exposure to some solvents may increase the effects of alcoholic beverages.

Chronic Health Effects

A chronic exposure occurs during longer or repeated periods of contact, sometimes over years and often at relatively low concentrations of exposure. Perchloroethylene or alcohol, for example, may cause liver damage or other cancers 10–40 years after first exposure. This period between first exposure and the development of the disease is called the latency period. An exposure to a substance may cause adverse health effects many years from now with little or no effects at the time of exposure. It is important to avoid or eliminate all exposures to chemicals that are not part of normal ambient breathing air. For many chemical agents, the toxic effects following a single exposure are quite different from those produced by repeated exposures. For example, the primary acute toxic effect of benzene is central nervous system damage, while chronic exposures can result in leukemia.

There are two ways to determine if a chemical causes cancer—studies conducted on people and studies on animals. Studies on humans are expensive, hard to conduct, and very often not even possible. This type of long-term research is called epidemiology. Studies on animals are less expensive and easier to do. This type of research is sometimes referred to as toxicology. Results showing increased occurrences of cancer in animals are generally accepted to indicate that the same chemical causes cancer in humans. The alternative to not accepting animal studies means we would have a lot less knowledge about the health effects of chemicals. We would never be able to determine the health effects of the more than 100,000 chemicals used by industry.

No level of exposure to cancer-causing chemicals is safe. Lower levels are considered safer. One procedure for setting health standard limits is called risk assessment. Risk assessment on the surface appears very scientific yet the actual results are based on many assumptions. The differences in these assumptions allow scientists to come up with very different results when determining an acceptable exposure standard. Following are major questions that assumptions are based on:

- Is there a level of exposure below which a substance would not cause cancer or other chronic diseases? (Is there a threshold level?)
- Can the body's defense mechanisms inactivate or break down chemicals?
- Does the chemical need to be at a high enough level to cause damage to a body organ before it causes cancer?
- How much cancer should we allow? (One case of cancer among 1,000,000 people, or one case of cancer among 100,000 people, or one case of cancer among 10 people?)

For exposures at the current permissible exposure limit (PEL), the risk of developing cancer from vinyl chloride is about 700 cases of cancer for every million workers exposed. The risk for asbestos is about 6400 cases of cancer for each million workers exposed. The risk for coal tar pitch is about 13,000 cases for each million workers exposed. PELs set for current federal standards differ because of these different risks.

The dose of a chemical causing cancer in human or animal studies is then used to set a standard PEL below which only a certain number of people will develop illness or cancer. This standard is not an absolute safe level of exposure to cancer-causing agents, so exposure should always be minimized even when levels of exposure are below the standard. Just as the asbestos standard has been lowered in the past from 5 fibers/cm^3 to 0.2 fibers/cm^3, and now to 0.1 fibers/cm^3 (50 times lower), it is possible that other standards will be lowered in the future as new technology for analysis is discovered and public outrage insists on fewer deaths for a particular type of exposure. If a chemical is suspected of causing cancer, it is best to minimize exposure, even if the exposure is below accepted levels.

Chronic Diseases

Chronic disease is not always cancer. Many other types of chronic diseases can be as serious as cancer. These chronic diseases affect the function of different organs of the body. For example, chronic exposure to asbestos or silica dust (fine sand) causes scarring of the lung. Exposure to gases such as nitrogen oxides or ozone may lead to destruction of parts of the lung. No matter what the cause, chronic disease of the lungs will make the individual feel short of breath and limit their activity. Depending on the extent of disease, chronic lung disease can kill. In fact, it is one of the top 10 causes of death in the United States.

Scarring of the liver (cirrhosis) is another example of chronic disease. It is also one of the top 10 causes of death in the United States. The liver is important in making certain essential substances in the body and cleaning certain waste products. Chronic liver disease can cause an individual to be tired all the time, have muscles waste away, and cause swelling of the stomach from fluid accumulation. Many chemicals such as carbon tetrachloride, chloroform, and alcohol can cause cirrhosis of the liver.

The brain is also affected by chronic exposure. Chemicals such as lead can decrease IQ, decrease the ability to remember things, and make someone more irritable. Many times these changes are small and can only be found with special medical tests. Workers exposed to solvents, such as toluene or xylene in oil-based paints, may develop neurological changes over a period of time.

Scarring of the kidney is another example of a chronic disease. Individuals with severe scarring must be placed on dialysis to remove the harmful waste products or have a kidney transplant. Chronic kidney disease can cause an individual to be tired all the time, have high blood pressure and swollen feet, as well as many other symptoms. Lead, mercury, and solvents are suspect causes of chronic kidney disease.

Birth Defects/Infertility

The ability to have a healthy child can be affected by chemicals in many different ways. A woman may be unable to conceive because a man is infertile. The production of sperm may be abnormal, reduced, or stopped by chemicals that enter the body. Men working in an insecticide plant manufacturing 1,3-dibromo-3-chloropropane (DBCP) realized after talking among themselves that none of their wives had been able to become pregnant. When tested, all the men were found to be sterile.

A woman may be unable to conceive or may have frequent early miscarriages because of mutagenic or embryotoxic effects. Changes in genes in the woman's ovaries or the man's sperm from exposure to chemicals may cause the developing embryo to die. A woman may give birth to a child with a birth defect because of a chemical with mutagenic or teratogenic effects. When a chemical causes a teratogenic effect, the damage is caused by the woman's direct exposure to the chemical. When a chemical causes a mutagenic effect, changes in genes from either the man or woman have occurred.

Many chemicals used in the workplace can damage the body. Effects range from skin irritation and dermatitis to chronic lung diseases such as silicosis and asbestosis or even cancer. The body may be harmed at the point where a chemical touches or enters it. This is called a local effect. When the solvent benzene touches the skin, it can cause drying and irritation (local effect).

A systemic effect develops at some place other than the point of contact. Benzene can be absorbed through the skin, breathed into the lungs, or ingested. Once in the body, benzene can affect the bone marrow, leading to anemia and leukemia. (Leukemia is a kind of cancer affecting the bone marrow and blood.) Adverse health effects may take years to develop from a small exposure or may occur very quickly from large concentrations.

HAZARDOUS CHEMICALS

Hazardous and toxic (poisonous) substances can be defined as chemicals present in the workplace that are capable of causing harm. In this definition, the term chemicals includes dusts, mixtures, and common materials such as paints, fuels, and solvents. OSHA currently regulates exposure to approximately 400 substances. The OSHA Chemical Sampling Information file contains a listing for approximately 1500 substances. The EPAs Toxic Substance Chemical Act: Chemical Substances Inventory lists information on more than 62,000 chemicals or chemical substances. Some libraries maintain files of material safety data sheets (MSDSs) for more than 100,000 substances. It is not possible to address the hazards associated with each of these chemicals (see Figure 10.1).

Since there is no evaluation instrument that can identify the chemical or the amount of chemical contaminant present, it is not possible to be able to make a real-time assessment of a worker's exposure to potentially hazardous chemicals. Additionally, threshold limit values (TLVs) provided by the ACGIH in 1968 are the basis of OSHA's PELs. In the early 2000s, workers are provided protection with chemical exposure standards that are 32 years old. The ACGIH regularly updates and changes its TLVs based on new scientific information and research.

The U.S. EPA allows for one death or one cancer case per million people exposed to a hazardous chemical. Certainly, the public needs these kinds of protections. Using the existing OSHA PELs risk factor is only as protective as one death due to exposure in 1000 workers. This indicates that there exists a fence line mentality, which suggests that workers can tolerate higher exposures than what the public would be subjected to. As one illustration of this, the exposure to sulfur dioxide for

Figure 10.1   Toxic chemical exposure is a problem for workers.

the public is set by the EPA at 0.14 ppm average over 24 h, while the OSHA PEL is 5 ppm average over 8 h. Certainly, there is a wide margin between what the public can be subjected to and what a worker is supposed to be able to tolerate. The question is, "Is there a difference between humans in the public arena and those in the work arena?" Maybe workers are assumed to be more immune to the effects of chemicals when they are in the workplace than when they are at home, because of workplace regulations and precautions.

A more significant issue is that regarding mixtures. Information does not exist to show the risk of illnesses, long-term illnesses, or the toxicity of combining these hazardous chemicals. At present, it is assumed that the most dangerous chemical of the mixture has the most potential to cause serious health-related problems than the next most hazardous and so on. However, little consideration is given regarding the increase in toxicity, long-term health problems, or present hazards. Since most chemicals used in industry are mixtures, formulated by manufacturers, it makes it even more critical to have access to the MSDSs and take a conservative approach to the potential for exposure. This means that any signs or symptoms of exposure should be addressed immediately; worker complaints should be addressed with sincerity and true concern; and employers should take precautions beyond those called for by MSDSs if questions prevail.

Actually, the amount of information that exists on dose/response for chemicals and chemical mixtures is limited. This is especially true for long-range effects. If a chemical kills or makes a person sick within minutes or hours, the dose/response is easily understood. However, if chemical exposure over a long period results in an individual's death or illness, then the dose needed to do this is, at best, a guess. It most certainly does not take into account other chemicals the worker was exposed to during his or her work life and whether they exacerbated the effects or played no role in the individual's death or illness. This is why it is critical for individual workers to keep their exposure to chemicals as low as possible. Even then, there are no guarantees that they may not come down with an occupational disease related to chemical exposure.

Many employers and workers as well as physicians are not quick or trained to identify the symptoms of occupational exposure to chemicals. In one case, two men painted for 8 h with a paint containing 2-nitropropane in an enclosed environment. At the end of their shift one of the workers felt ill and stopped at the emergency center at the hospital. After examination, he was told to go home and rest and would probably be better in the morning. Later that evening, he returned to the hospital and died of liver failure from 2-nitropropane exposure. The other worker suffered irreparable liver damage but survived. No one asked the right questions regarding occupational exposure. The symptoms were probably similar to a common cold or flu, which is often the case unless some detective work is done. Often those with chemical poisoning go home and off-gas or excrete the contaminant during the 16 h where they have no exposure. They feel better the next day and return to work and are reexposed. Thus, the worker does not truly recognize this as a poisoning process. Being aware of the chemicals used, reviewing the MSDSs, and following the recommended precautions are important to the safe use of hazardous chemicals.

With this point made, it becomes critical that employers know the dangers that chemicals in use by them present to their workforce. Employers need to get and review material safety data sheets (MSDSs) for all chemicals in use on their worksite and take the proper precautions recommended by the MSDSs. In addition, it behooves workers to get copies of MSDSs for chemicals they use.

MSDSs can also provide information for training employees in the safe use of materials. These data sheets, developed by chemical manufacturers and importers, are supplied with manufacturing or construction materials and describe the ingredients of a product, its hazards, protective equipment to be used, safe handling procedures, and emergency first-aid responses. The information contained in these sheets can help employers identify employees in need of training (i.e., workers handling substances described in the sheets) and train employees in safe use of the substances (see Appendix E). MSDSs are generally available from suppliers, manufacturers of the substance, large employers who use the substance on a regular basis, or they may be developed by employers or trade associations.

MSDS are particularly useful for employers who are developing training in safe chemical use as required by OSHA's HCS.

Carcinogens

Carcinogens are any substances or agents that have the potential to cause cancer. Whether these chemicals or agents have been shown to only cause cancer in animals should make little difference to employers and their workers. Employers and their workers should consider these as cancer causing on a precautionary basis since all is not known regarding their effects upon humans on a long-term basis. Since most scientists say that there is no known safe level of a carcinogen, then zero exposure should be the goal of workplace health and safety. Do not let the label suspect carcinogen or agent put your mind at ease. This chemical or agent can cause cancer. OSHA has identified 13 chemicals as carcinogens. They are as follows:

1. 4-Nitrobiphenyl, Chemical Abstracts Service Register Number (CAS No.) 92933
2. α-Naphthylamine, CAS No. 134327
3. Methyl chloromethyl ether, CAS No. 107302
4. 3,3′-Dichlorobenzidine (and its salts), CAS No. 91941
5. *bis*-Chloromethyl ether, CAS No. 542881
6. β-Naphthylamine, CAS No. 91598
7. Benzidine, CAS No. 92875
8. 4-Aminodiphenyl, CAS No. 92671
9. Ethyleneimine, CAS No. 151564
10. β-Propiolactone, CAS No. 57578
11. 2-Acetylaminofluorene, CAS No. 53963
12. 4-Dimethylaminoazo-benzene, CAS No. 60117
13. *N*-Nitrosodimethylamine, CAS No. 62759

Many other chemicals probably should be identified as carcinogens, but have not come under the scrutiny of the regulatory process to make them such. In many cases, this is probably due to special interests of manufacturers and other groups.

The OSHA regulation 29 CFR 1910.1003 pertains to solid or liquid mixtures containing less than 0.1% by weight or volume of 4-nitrobiphenyl, methyl chloromethyl ether, *bis*-chloromethyl ether, β-naphthylamine, benzidine, or 4-aminodiphenyl and solid or liquid mixtures containing less than 1.0% by weight or volume of α-naphthylamine, 3,3′-dichlorobenzidine (and its salts), ethyleneimine, β-propiolactone, 2-acetylaminofluorene, 4-dimethylaminoazo-benzene, or *N*-nitrosodimethylamine.

The specific nature of the previous requirements is an indicator of the danger presented by exposure to, or work with, carcinogens that are regulated by OSHA. There are other carcinogens that OSHA regulates (not part of the original 13). These carcinogens are

- Vinyl chloride (1910.1017)
- Inorganic arsenic (1910.1018)
- Cadmium (1910.1027 and 1926.1127)
- Benzene (1910.1028)
- Coke oven emissions (1910.1029)
- 1,2-Dibromo-3-chloropropane (1910.1044)
- Acrylonitrile (1910.1045)
- Ethylene oxide (1910.1047)
- Formaldehyde (1910.1048)
- Methylenedianiline (1910.1050)
- 1,3-Butadiene (1910.1051)
- Methylene chloride (1910.1052)

Recently, OSHA has reduced the PEL for methylene chloride from 400 to 25 ppm. This is a huge reduction in the PEL, equal to a 15 times decrease in what workers can be exposed to. This reduction indicates the potential of methylene chloride to cause cancer in workers and should raise the flag that chemicals that are believed to cause cancer are not to be taken lightly. Information and research are continuously evolving and providing new insight into the dangers of these chemicals and agents.

BIOLOGICAL MONITORING

Biological monitoring is the analysis of body systems such as blood, urine, fingernails, teeth, etc., which provide a baseline level of contaminants in the body. Medical testing can have several different purposes, depending on why the worker is visiting a doctor. If it is a preemployment examination, it is usually considered a baseline to use as a reference for future medical testing. Baselines are a valuable tool to measure the amount of toxic substances in the body and often give an indication of the effectiveness of PPE.

OSHA regulations allow the examining physician to determine most of the content reviewed in the examination. Benefits received from an examination will vary with the content of the examination. No matter what tests are included in the examination, there are certain important limitations of medical testing:

- Medical testing cannot prevent cancer. Cancer from exposure to chemicals or asbestos can only be prevented by reducing or eliminating an exposure.
- For many conditions, there are no medical tests for early diagnosis. For example, the routine blood tests conducted by doctors for kidney functions do not become abnormal until half the kidney function is lost. Nine of 10 people with lung cancer die within 5 years because chest x-rays do not diagnose lung cancer in time to save the individual.
- No medical test is perfect. Some tests are falsely abnormal and some falsely normal.

Medical Questionnaire

A medical and work history, despite common perceptions, is probably the most important part of an examination. Most diagnoses of disease in medicine are made by the history. Laboratory tests are used to confirm past illnesses and injuries. Doctors are interested in the history of lung, heart, kidney, liver, and other chronic diseases for the individual and family. Doctors will also be concerned about symptoms indicating heart or lung disease and smoking habits.

A physical examination is very beneficial for routine screening. Good results are important but an individual may have a serious medical problem while physical examination results seem perfectly normal. Blood is taken to check for blood cell production (anemia), liver function, kidney function, and if taken while fasting, for increased sugar, cholesterol, and fat in the blood. Urine tests are obtained to check for kidney function and diabetes (sugar in the urine). It is possible to measure in the blood and urine chemicals that get into the body from exposures on a job site. This type of testing is called biological monitoring.

Pulmonary Function Test

When an individual breathes into a spirometer, it measures how much air volume is present in his lungs and how quickly he can breathe in and out. This is called pulmonary function testing. This is useful for diagnosing diseases that cause scarring of the lungs, which affects the expandability (asbestosis). Emphysema or asthma may also be diagnosed with pulmonary function testing. It is vital for evaluating the ability of an individual to wear a respirator without additional health risk.

Electrocardiogram

An electrocardiogram is a test used to measure heart injury or irregular heartbeats. Work can be extremely strenuous, particularly when wearing protective equipment in hot environments. A stress test using an electrocardiogram while exercising is sometimes a help in determining fitness, especially if there are indications from the questionnaire that an individual has a high risk of heart disease.

Chest X-Ray

X-rays are useful in determining the cause of breathing problems or they may be used as a baseline to determine future problems. A chest x-ray is used to screen for scarring of the lungs from exposure to asbestos or silica. It should not be performed routinely, unless the history indicates a potential lung or heart problem and the physician thinks a chest x-ray would be necessary. Some OSHA regulations require chest x-rays as part of the medical surveillance program. Unnecessary x-ray screening should be eliminated. Five year intervals are plenty for work-related biological monitoring.

HAZARD COMMUNICATIONS STANDARD (HCS)

HCS is one of the most often cited standards by OSHA. It is found in 29 CFR 1910.1200 and 29 CFR 1926.59. The basic goal of a hazard communication program is to be sure employers and employees know about work hazards and how to protect themselves; this should help to reduce the incidence of chemical source illness and injuries. The following information will walk you through the steps and requirements for complying with the HCS and act as a template for planning how to comply with other OSHA standards. OSHA has estimated that more than 32 million workers are exposed to 650,000 hazardous chemical products in more than 3,000,000 American workplaces. This poses a serious problem for exposed workers and their employers.

Chemicals pose a wide range of health hazards (such as irritation, sensitization, and carcinogenicity) and physical hazards (such as flammability, corrosion, and reactivity). OSHA's HCS is designed to ensure that information about these hazards and associated protective measures is disseminated to workers and employers. This is accomplished by requiring chemical manufacturers and importers to evaluate the hazards of the chemicals they produce or import, and to provide information about them through labels on shipped containers and more detailed information sheets called MSDSs. All employers with hazardous chemicals in their workplaces must prepare and implement a written hazard communication program and must ensure that all containers are labeled, employees are provided access to MSDSs, and an effective training program is conducted for all potentially exposed employees (see Appendix E).

The HCS provides workers the right to know the hazards and identities of the chemicals they are exposed to in the workplace. When workers have this information they can effectively participate in their employers' protective programs and take steps to protect themselves. In addition, the standard gives employers the information they need to design and implement an effective protective program for employees potentially exposed to hazardous chemicals. Together these actions will result in a reduction of chemical source illnesses and injuries in American workplaces.

Protection under OSHA's HCS includes all workers exposed to hazardous chemicals in all industrial sectors. This standard is based on a simple concept—that employees have both a need and a right to know the hazards and the identities of the chemicals they are exposed to when working. They also need to know what protective measures are available to prevent adverse effects from occurring.

The HCS covers both physical hazards (such as flammability or the potential for explosions) and health hazards (including both acute and chronic effects). By making information available to employers and employees about these hazards, and recommended precautions for safe use, proper implementation of the HCS will result in a reduction of illnesses and injuries caused by chemicals. Employers will have the information they need to design an appropriate protective program. Employees will be better able to participate in these programs effectively when they understand the hazards involved, and to take steps to protect themselves. Together, these employer and employee actions will prevent the occurrence of adverse effects caused by the use of chemicals in the workplace.

The HCS established uniform requirements to make sure that the hazards of all chemicals imported into, produced, or used in United States workplaces are evaluated and that this hazard information is transmitted to affected employers and exposed employees.

Chemical manufacturers and importers must convey the hazard information they learn from their evaluations to downstream employers by means of labels on containers and MSDSs. In addition, all covered employers must have a hazard communication program to get this information to their employees through labels on containers, MSDSs, and training.

This program ensures that all employers receive the information they need to inform and train their employees properly and to design and put in place employee protection programs. It also provides necessary hazard information to employees so that they can participate in, and support, the protective measures in place at their workplaces.

All employers in addition to those in manufacturing and importing are responsible for informing and training workers about the hazards in their workplaces, retaining warning labels, and making available MSDSs with hazardous chemicals.

Some employees deal with chemicals in sealed containers under normal conditions of use (such as in the retail trades, warehousing and truck and marine cargo handling). Employers of these employees must assure that labels affixed to incoming containers of hazardous chemicals are kept in place. They must maintain and provide access to MSDSs received or obtain MSDSs if requested by an employee. Moreover, they must train workers on what to do in the event of a spill or leak. However, written hazard communication programs will not be required for this type of operation.

All workplaces where employees are exposed to hazardous chemicals must have a written plan that describes how the standard will be implemented in that facility. The only work operations that do not have to comply with the written plan requirements are laboratories and work operations where employees handle chemicals only in sealed containers.

The written program must reflect what employees are doing in a particular workplace. For example, the written plan must list the chemicals present at the site, indicate who is responsible for the various aspects of the program in that facility, and where written materials will be made available to employees. The written program must describe how the requirements for labels and other forms of warning, MSDSs, and employee information and training will be met in the facility.

Guidelines for Employer Compliance

The HCS is based on a simple concept—that employees have both a need and a right to know the hazards and identities of the chemicals they are exposed to when working. They also need to know what protective measures are available to prevent adverse effects from occurring. The HCS is designed to provide employees with the information they need (see Figure 10.2).

Knowledge acquired under the HCS will help employers provide safer workplaces for their employees. When employers have information about the chemicals used, they can take steps to reduce exposures, substitute less hazardous materials, and establish proper work practices. These efforts will help prevent the occurrence of work-related illnesses and injuries caused by chemicals.

The HCS addresses the issues of evaluating and communicating hazards to workers. Evaluation of chemical hazards involves a number of technical concepts and is a process that requires the

Figure 10.2   Informing workers about chemicals is the law.

professional judgment of experienced experts. Therefore, the HCS is designed so that employers who simply use chemicals, rather than produce or import them, are not required to evaluate the hazards of those chemicals. Hazard determination is the responsibility of the producers and importers of the materials. Producers and importers of chemicals are then required to provide the hazard information to employers who purchase their products.

Employers who do not produce or import chemicals need only focus on parts of the rule that deal with establishing a workplace program and communicating information to their workers. This following is a general guide for such employers to help them determine what is required under the rule. It does not supplant or substitute for the regulatory provisions, but rather provides a simplified outline of the steps an average employer would follow to meet those requirements.

Becoming Familiar with the Rule

OSHA has provided a simple summary of the HCS in a pamphlet entitled *Chemical Hazard Communication*, OSHA Publication No. 3084. Some employers prefer to begin to familiarize themselves with the rule's requirements by reading this pamphlet. A copy may be obtained from your local OSHA Area Office, by contacting the OSHA Publications Office at (202) 693-1888 or via the OSHA Web site under publications or search by title.

The standard is long, and some parts of it are technical, but the basic concepts are simple. In fact, the requirements reflect what many employers have been doing for years. You may find that you are already largely in compliance with many of the provisions and will simply have to slightly modify your existing programs. If you are operating in an OSHA-approved State Plan, you must comply with the state's requirements, which may be different from those of the federal rule. Many of the State Plan states had hazard communication or right-to-know laws before promulgation of the Federal rule. Employers in State Plan states should contact their state OSHA offices for more information regarding applicable requirements.

The HCS requires information to be prepared and transmitted regarding all hazardous chemicals. The HCS covers both physical hazards (such as flammability) and health hazards (such as irritation, lung damage, and cancer). Most chemicals used in the workplace have some hazard potential, and thus will be covered by the rule.

One difference between this rule and many others adopted by OSHA is that this one is performance oriented. This means that you have the flexibility to adapt the rule to the needs of your workplace, rather than follow specific, rigid requirements. It also means that you have to exercise more judgment to implement an appropriate and effective program.

The standard's design is simple. Chemical manufacturers and importers must evaluate the hazards of the chemicals they produce or import. Using that information, they must then prepare labels for containers, and more detailed technical bulletins called MSDSs.

Chemical manufacturers, importers, and distributors of hazardous chemicals are all required to provide the appropriate labels and MSDSs to the employers to whom they ship the chemicals. This information should be provided automatically. Every container of hazardous chemicals you receive must be labeled, tagged, or marked with the required information. Your suppliers must also send you a properly completed MSDS at the time of the first shipment of the chemical and with the next shipment after the MSDS is updated with new and significant information about the hazards. You can rely on the information received from your suppliers. You have no duty to analyze the chemical or evaluate its hazards.

Employers who use hazardous chemicals must have a program to ensure that information is provided to exposed employees. Use means to package, handle, react, or transfer. This is an intentionally broad scope and includes any situation where a chemical is present in such a way that employees may be exposed under normal conditions of use or in a foreseeable emergency.

The requirements of the rule that deal specifically with the hazard communication program are found in the standard in paragraphs (e), written hazard communication program; (f), labels and other forms of warning; (g), MSDSs; and (h), employee information and training. The requirements of these paragraphs should be the focus of your attention. Concentrate on becoming familiar with them, using paragraphs (b), scope and application, and (c), definitions, as references when needed, to help explain the provisions.

There are two types of work operations where the coverage of the rule is limited. These are laboratories and operations where chemicals are handled only in sealed containers (e.g., a warehouse). The limited provisions for these workplaces can be found in paragraph (b), scope and application. Employers who have these types of work operations need only keep labels on containers as they are received; maintain MSDSs that are received, and give employees access to them; and provide information and training for employees. Employers do not have to have written hazard communication programs and lists of chemicals for these types of operations.

The limited coverage of laboratories and sealed container operations addresses the obligation of an employer to the workers in the operations involved and does not affect the employer's duties as a distributor of chemicals. For example, a distributor may have warehouse operations where employees would be protected under the limited sealed container provisions. In this situation, requirements for obtaining and maintaining MSDSs are limited to providing access to those received with containers while the substance is in the workplace and requesting MSDSs when employees request access for those not received with the containers. However, as a distributor of hazardous chemicals, that employer will still have responsibilities for providing MSDSs to downstream customers at the time of the first shipment and when the MSDS is updated. Therefore, although MSDSs may not be required for the employees in the work operation, the distributor may, nevertheless, have to have them to satisfy other requirements of the rule.

Identifying Responsible Staff

Hazard communication will be a continuing program in your facility. Compliance with the HCS is not a one shot deal. To have a successful program, it will be necessary to assign responsibility for both the initial and ongoing activities that have to be undertaken to comply with the rule. In some cases, these activities may already be part of current job assignments. For example, site supervisors are frequently responsible for on-the-job training sessions. Early identification of the responsible employees, and involvement of them in the development of your plan of action, will result in a more effective program design. Evaluation of the effectiveness of your program will also be enhanced by involvement of affected employees.

For any safety and health program, success depends on commitment at every level of the organization. This is particularly true for hazard communication, where success requires a change in behavior. This will only occur if employers understand the program, and are committed to its success, and if employees are motivated by the people presenting the information to them.

Identifying Hazardous Chemicals in the Workplace

This standard requires a list of hazardous chemicals in the workplace as part of the written hazard communication program. This list will eventually serve as an inventory of everything for which an MSDS must be maintained. At this point, however, preparing the list will help you complete the rest of the program since it will give you some idea of the scope of the program required for compliance in your facility.

The best way to prepare a comprehensive list is to survey the workplace. Purchasing records may also help, and certainly employers should establish procedures to ensure that in the future purchasing procedures result in MSDSs being received before a material is used in the workplace.

The broadest possible perspective should be taken when conducting the survey. Sometimes people think of chemicals as only liquids in containers. The HCS covers chemicals in all physical forms—liquids, solids, gases, vapors, fumes, and mists—whether they are contained or not. The hazardous nature of the chemical and the potential for exposure are the factors that determine whether a chemical is covered. If it is not hazardous, it is not covered. If there is no potential for exposure (e.g., the chemical is inextricably bound and cannot be released), the rule does not cover the chemical.

Look around. Identify chemicals in containers, including pipes, but also think about chemicals generated in the work operations. For example, welding fumes, dusts, and exhaust fumes are all sources of chemical exposures. Read labels provided by suppliers for hazard information. Make a list of all chemicals in the workplace that are potentially hazardous. For your own information and planning, you may also want to note on the list the location(s) of the products within the workplace and an indication of the hazards as found on the label. This will help you as you prepare the rest of your program.

Paragraph (b) of the standard, scope and application, includes exemptions for various chemicals or workplace situations. After compiling the complete list of chemicals, you should review paragraph (b) to determine if any of the items can be eliminated from the list because they are exempted materials. For example, food, drugs, and cosmetics brought into the workplace for employee consumption are exempt. Therefore, rubbing alcohol in the first-aid kit would not be covered.

Once you have compiled as complete a list as possible of the potentially hazardous chemicals in the workplace, the next step is to determine if you have received MSDSs for all of them. Check your files against the inventory you have just compiled. If any are missing, contact your supplier and request one. It is a good idea to document these requests, either by copy of a letter or a note regarding telephone conversations. If you have MSDSs for chemicals that are not on your list, figure out why. Maybe you do not use the chemical anymore. Alternatively, maybe you missed it in your survey. Some suppliers do provide MSDSs for products that are not hazardous. These do not have to be maintained by you. You should not allow employees to use any chemicals for which you have not received an MSDS. The MSDS provides information you need to ensure that proper protective measures are implemented before exposure.

Preparing and Implementing a Hazard Communication Program

All workplaces where employees are exposed to hazardous chemicals must have a written plan that describes how the standard will be implemented in that facility. Preparation of a plan is not just a paper exercise—all of the elements must be implemented in the workplace to be in compliance with the rule. See paragraph (e) of the standard for the specific requirements regarding written hazard communication programs. The only work operations that do not have to comply with the written plan requirements are laboratories and work operations where employees handle chemicals only in sealed containers. See paragraph (b), scope and application, for the specific requirements for these two types of workplaces.

The plan does not have to be lengthy or complicated. It is intended to be a blueprint for implementation of your program—an assurance that all aspects of the requirements have been addressed. Many trade associations and other professional groups have provided sample programs and other assistance materials to affected employers. These have been very helpful to many employers since they tend to be tailored to the particular industry involved. You may wish to investigate whether your industry trade groups have developed such materials.

Although such general guidance may be helpful, you must remember that the written program has to reflect what you are doing in your workplace. Therefore, if you use a generic program it must be adapted to address the facility it covers. For example, the written plan must list the chemicals present at the site, indicate who is to be responsible for the various aspects of the program in your facility, and indicate where written materials will be made available to employees.

If OSHA inspects your workplace for compliance with the HCS, the OSHA compliance officer will ask to see your written plan at the outset of the inspection. In general, the following items will be considered in evaluating your program.

The written program must describe how the requirements for labels and other forms of warning, MSDSs, and employee information and training, are going to be met in your facility. The following discussion provides the type of information compliance officers will be looking for to decide whether these elements of the hazard communication program have been properly addressed.

Labels and Other Forms of Warning

In-plant containers of hazardous chemicals must be labeled, tagged, or marked with the identity of the material and appropriate hazard warnings. Chemical manufacturers, importers, and distributors are required to ensure that every container of hazardous chemicals they ship is appropriately labeled with such information and with the name and address of the producer or other responsible party. Employers purchasing chemicals can rely on the labels provided by their suppliers. If the material is subsequently transferred by the employer from a labeled container to another container, the employer will have to label that container unless it is subject to the portable container exemption. See paragraph (f) for specific labeling requirements.

The primary information to be obtained from an OSHA-required label is an identity for the material and appropriate hazard warnings. The identity is any term that appears on the label, the MSDS, and the list of chemicals, and thus links these three sources of information. The identity used by the supplier may be a common or trade name (Black Magic Formula), or a chemical name (1,1,1,-trichloroethane). The hazard warning is a brief statement of the hazardous effects of the chemical (i.e., "flammable," "causes lung damage"). Labels frequently contain other information, such as precautionary measures (do not use near open flame), but this information is provided voluntarily and is not required by the rule. Labels must be legible and prominently displayed. There are no specific requirements for size or color, or any specified text.

With these requirements in mind, the compliance officer will be looking for the following types of information to ensure that labeling is properly implemented in your facility:

1. Designation of person(s) responsible for ensuring labeling of in-plant containers
2. Designation of person(s) responsible for ensuring labeling of any shipped containers
3. Description of labeling system(s) used

4. Description of written alternatives to labeling of in-plant containers (if used)

5. Procedures to review and update label information when necessary

Employers who purchase and use hazardous chemicals—rather than produce or distribute them—will primarily be concerned with ensuring that every purchased container is labeled. If materials are transferred into other containers, the employer must ensure that these are labeled as well, unless they fall under the portable container exemption [paragraph (f)(7)]. In terms of labeling systems, you can simply choose to use the labels provided by your suppliers on the containers. These will generally be verbal text labels and do not usually include numerical rating systems or symbols that require special training. The most important thing to remember is that this is a continuing duty—all in-plant containers of hazardous chemicals must always be labeled. Therefore, it is important to designate someone to be responsible for ensuring that the labels are maintained as required on the containers in your facility, and that newly purchased materials are checked for labels before use.

Material Safety Data Sheets

Chemical manufacturers and importers are required to obtain or develop an MSDS for each hazardous chemical they produce or import. Distributors are responsible for ensuring that their customers are provided a copy of these MSDSs. Employers must have an MSDS for each hazardous chemical that they use. Employers may rely on the information received from their suppliers. The specific requirements for MSDSs are presented in paragraph (g) of the standard.

There is no specified format for the MSDS under the rule, although there are specific information requirements. OSHA has developed a nonmandatory format, OSHA form 174, which may be used by chemical manufacturers and importers to comply with the rule. The MSDS must be in English. You are entitled to receive from your supplier a data sheet that includes all the information required under the rule. If you do not receive one automatically, you should request one. If you receive one that is obviously inadequate, with, for example, blank spaces that are not completed, you should request an appropriately completed one. If your request for a data sheet or for a corrected data sheet does not produce the information needed, you should contact your local OSHA Area Office for assistance in obtaining the MSDS.

The role of MSDSs under the rule is to provide detailed information on each hazardous chemical, including its potential hazardous effects, its physical and chemical characteristics, and recommendations for appropriate protective measures. This information should be useful to you as the employer responsible for designing protective programs, as well as to the workers. If you are not familiar with MSDSs and with chemical terminology, you may need to learn to use them yourself. A glossary of MSDS terms may be helpful in this regard. Generally, most employers using hazardous chemicals will primarily be concerned with MSDS information regarding hazardous effects and recommended protective measures. Focus on the sections of the MSDS that are applicable to your situation.

MSDSs must be readily accessible to employees when they are in their work areas during their workshifts. This may be accomplished in many different ways. You must decide what is appropriate for your particular workplace. Some employers keep the MSDSs in a binder in a central location (e.g., in the pickup truck on a construction site). Others, particularly in workplaces with large numbers of chemicals, computerize the information and provide access through terminals. As long as employees can get the information when they need it, any approach may be used. The employees must have access to the MSDSs themselves—simply having a system where the information can be read to them over the phone is permitted only under the mobile worksite provision, paragraph (g)(9), when employees must travel between workplaces during the shift. In this situation, they have access to the MSDSs before leaving the primary worksite and upon return, and so the telephone system is simply an emergency arrangement.

To ensure that you have a current MSDS for each chemical in the plant as required, and that employee access is provided, the compliance officers will be looking for the following types of information in your written program:

1. Designation of person(s) responsible for obtaining and maintaining the MSDSs
2. How such sheets are to be maintained in the workplace (e.g., in notebooks in the work area(s) or in a computer with terminal access), and how employees can obtain access to them when they are in their work area during the workshift
3. Procedures to follow when the MSDS is not received at the time of the first shipment
4. For producers, procedures to update the MSDS when new and significant health information is found
5. Description of alternatives to actual data sheets in the workplace, if used

For employers using hazardous chemicals, the most important aspect of the written program in terms of MSDSs is to ensure that someone is responsible for obtaining and maintaining the MSDSs for every hazardous chemical in the workplace. The list of hazardous chemicals required to be maintained as part of the written program will serve as an inventory. As new chemicals are purchased, the list should be updated. Many companies have found it convenient to include on their purchase orders the name and address of the person designated in their company to receive MSDSs.

Employee Information and Training

Each employee who may be exposed to hazardous chemicals when working must be provided with information and trained before the initial assignment to work with a hazardous chemical and whenever the hazard changes. Exposure or exposed under the rule means that "an employee is subjected to a hazardous chemical in the course of employment through any route of entry (inhalation, ingestion, skin contact or absorption, etc.) and includes potential (e.g., accidental or possible) exposure." See paragraph (h) of the standard for specific requirements. Information and training may be done either by individual chemical or by categories of hazards (such as flammability or carcinogenicity). If there are only a few chemicals in the workplace, then you may want to discuss each one individually. Where there are large numbers of chemicals, or the chemicals change frequently, you will probably want to train generally based on the hazard categories (e.g., flammable liquids, corrosive materials, carcinogens). Employees will have access to the substance-specific information on the labels and MSDSs.

Information and training are a critical part of the hazard communication program. Information regarding hazards and protective measures are provided to workers through written labels and MSDSs. However, through effective information and training, workers will learn to read and understand such information, determine how it can be obtained and used in their own workplaces, and understand the risks of exposure to the chemicals in their workplaces as well as the ways to protect themselves. A properly conducted training program will ensure comprehension and understanding. It is not sufficient to either just read material to the workers, or simply hand them material to read. You must create an environment where workers feel free to ask questions. This will help you to ensure that the information is understood. You must always remember that the underlying purpose of the HCS is to reduce the incidence of chemical source illnesses and injuries. This will be accomplished by modifying behavior through the provision of hazard information and information about protective measures. If your program works, you and your workers will better understand the chemical hazards within the workplace. The procedures you establish regarding, for example, purchasing, storage, and handling of these chemicals will improve, and thereby reduce the risks posed to employees exposed to the chemical hazards involved. Furthermore, your workers' comprehension will also be increased and proper work practices will be followed in your workplace.

If you are going to conduct the training yourself, you will have to understand the material and be prepared to motivate the workers to learn. This is not always an easy task, but the benefits are worth the effort. More information regarding appropriate training can be found in OSHA Publication No. 2254, which contains voluntary training guidelines prepared by OSHA's Office of Training and Education. A copy of this document is available from OSHA's Publications Office at (202) 693-1888.

In reviewing your written program with regard to information and training, the following items need to be considered:

1. Designation of person(s) responsible for conducting training
2. Format of the program to be used (audiovisuals, classroom instruction, etc.)
3. Elements of the training program (should be consistent with the elements in paragraph (h) of the HCS)
4. Procedure to train new employees at the time of their initial assignment to work with a hazardous chemical, and to train employees when a new hazard is introduced into the workplace

The written program should provide enough details about the employer's plans in this area to assess whether or not a good faith effort is being made to train employees. OSHA does not expect that every worker will be able to recite all of the information about each chemical in the workplace. In general, the most important aspects of training under the HCS are to ensure that employees are aware that they are exposed to hazardous chemicals, that they know how to read and use labels and MSDSs, and that, as a consequence of learning this information, they are following the appropriate protective measures established by the employer. OSHA compliance officers will be talking to employees to determine if they have received training, if they know they are exposed to hazardous chemicals, and if they know where to obtain substance-specific information on labels and MSDSs.

The rule does not require employers to maintain records of employee training, but many employers choose to do so. This may help you monitor your own program to ensure that all employees are appropriately trained. If you already have a training program, you may simply have to supplement it with whatever additional information is required under the HCS. For example, construction employers who are already in compliance with the construction training standard (29 CFR 1926.21) will have little extra training to do.

An employer can provide employees with information and training through whatever means found appropriate and protective. Although there would always have to be some training on-site (such as informing employees of the location and availability of the written program and MSDSs), employee training may be satisfied in part by general training about the requirements of the HCS and about chemical hazards on the job, which is provided by, for example, trade associations, unions, colleges and professional schools. In addition, previous training, education, and experience of a worker may relieve the employer of some of the burdens of informing and training that worker. Regardless of the method relied upon, however, the employer is always ultimately responsible for ensuring that employees are adequately trained. If the compliance officer finds that the training is deficient, the employer will be cited for the deficiency regardless of who actually provided the training on behalf of the employer.

Other Requirements

In addition to these specific items, compliance officers will also be asking the following questions in assessing the adequacy of the program:

1. Does a list of the hazardous chemicals exist in each work area or at a central location?
2. Are the methods the employer will use to inform employees of the hazards of nonroutine tasks outlined?

3. Are employees informed of the hazards associated with chemicals contained in unlabeled pipes in their work areas?
4. On multiemployer worksites, has the employer provided other employers with information about labeling systems and precautionary measures where the other employers have employees exposed to the initial employer's chemicals?
5. Is the written program made available to employees and their designated representatives?

If your program adequately addresses the means of communicating information to employees in your workplace and provides answers to the basic questions outlined above, it will be found to be in compliance with the rule.

Checklist for Compliance

The following checklist will help to ensure that you are in compliance with the HCS (see Figure 10.3).

HCS compliance checklist	
Obtained a copy of the rule	_____
Read and understood the requirements	_____
Assigned responsibility for tasks	_____
Prepared an inventory of chemicals	_____
Ensured containers are labeled	_____
Obtained MSDS for each chemical	_____
Prepared written program	_____
Made MSDSs available to workers	_____
Conducted training for workers	_____
Established procedures to maintain current program	_____
Established procedures to evaluate effectiveness	_____

Figure 10.3   HCS compliance checklist.

CHAPTER 11

Bent Too Far: Ergonomics

Designing the workplace to fit the workers so that they can work free of injuries.

ERGONOMICS

When the word ergonomics is mentioned most employers and many others will tend to make derogatory comments and act as though ergonomics is some kind of contrived problem. However, from my experience of visiting many types of industries and workplaces, it is not at all unusual for me to talk to a person who has had five surgeries related to musculoskeletal disorders (MSDs), cumulative trauma disorders (CTDs), or repetitive motion injuries (RMIs). In this chapter these terms will be used interchangeably.

I see ergonomics-related conditions in the shipyards, on construction sites, in manufacturing, in the service industry, and the office environment. When ergonomics is mentioned, many individuals immediately think of computer workstations, which are a small part of this issue and in most cases the problems with them are easily fixable.

Ergonomics is by definition fitting the workplace to the worker. It means more than changing a workstation. It means that the whole environment is designed to fit workers including directions, controls, printed material, warning signals, mental stress, work schedules, the work climate, fatigue and boredom, material handling, noise, vibration, lighting, mental capacity, the worker/machine interface, and the list could go on.

Ergonomics brings to bear many different academic disciplines. This is especially true of the more complex workplace problems. For the most part many solutions can be achieved simply and with little cost involved. To solve most of the problems that we face with ergonomic implications, we do not need a rocket scientist. The workers themselves often have very viable solutions. This is why Occupational Safety and Health Administration (OSHA) was requiring worker involvement in their now defunct ergonomics regulations. This is not to say that some of the existing ergonomic issues in the workplace will not require some time and cost investment by the employer. In most cases, this investment in solving workplace ergonomic problems decreases injuries and improves efficiency and morale.

EXTENT OF THE PROBLEM

In 2004, the Bureau of Labor Statistics (BLS) reported 402,700 MSDs. Of these, sprains, strains, and tears accounted for more than three-fourths of the MSDs that resulted in days away from work. Service-providing industries reported the most MSDs, accounting for 69% of all cases. Within these industries, health care and social assistance reported the most cases. Goods-producing industries reported 31% of all MSD cases, led by manufacturing with about 20%. The three main occupations with the most MSDs were laborers and materials movers; nursing aides, orderlies, and attendants; and heavy and tractor-trailer truck drivers. In Figure 11.1, a chart depicts the type of MSD and its percent of occurrence of the 402,700 MSDs reported by the BLS.

The signs and symptoms that are indicated by workers suffering an MSD are found in Table 11.1.

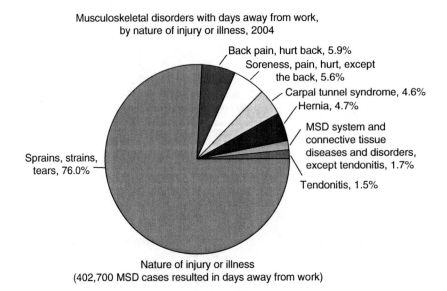

Musculoskeletal disorders with days away from work,
by nature of injury or illness, 2004

Back pain, hurt back, 5.9%
Soreness, pain, hurt, except
the back, 5.6%
Carpal tunnel syndrome, 4.6%
Hernia, 4.7%
MSD system and
connective tissue
diseases and disorders,
except tendonitis, 1.7%
Tendonitis, 1.5%
Sprains, strains,
tears, 76.0%

Nature of injury or illness
(402,700 MSD cases resulted in days away from work)

Figure 11.1 MSDs with days away from work by nature of injury or illness, 2004. (Courtesy of the U.S. Bureau of Labor Statistics, U.S. Department of Labor, November 2005.)

Table 11.1 Examples of MSD Signs and Symptoms

MSD Signs	MSD Symptoms
Deformity	Numbness
Decreased grip strength	Tingling
Decreased range of motion	Pain
Loss of function	Burning
	Stiffness
	Cramping

Source: Courtesy of Occupational Safety and Health Administration.

DEVELOPING AN ERGONOMIC PROGRAM

When you have ergonomic-related issues, I would suggest that the first undertaking be to develop a written ergonomic incident prevention program. This program should contain the elements that will lead to prevention. This would include management commitment and employee involvement (employee involvement is critical in solving ergonomic-related problems, the second element should be hazard identification and assessment, the third element would be hazard control and prevention, and lastly, education and training to name four).

Developing an ergonomics program should consist of these six program elements:

1. Management leadership and employee participation
2. Hazard information and reporting
3. Job hazard analysis and control
4. Training
5. MSD management
6. Program evaluation

An ergonomics program that contains well-recognized program elements is necessary. All of these core elements are essential to the effective functioning of ergonomics programs. These elements have been endorsed by both private industry and OSHA as keys to ergonomic program effectiveness.

The core elements will allow employers to manage all aspects of the process of protecting workers from MSDs and are a way of organizing that process into parts that can be meaningfully understood and implemented. All of the elements are important although many safety and health professionals believe that management leadership and employee participation are the keystone of an effective ergonomics program. OSHA believes that all the elements are necessary to achieve the overall goal of managing MSDs and ensuring that MSD hazards are systematically and routinely prevented, eliminated, or controlled.

In identifying ergonomic hazards you might want to use hazard identification and analysis as the heart of your ergonomics program because it is the first step in eliminating or materially reducing MSD hazards. Through hazard identification and analysis, you can identify and assess where and how employees' physical capabilities have been exceeded in a given job. This can be done by identifying aspects of the physical work activities and conditions of the job and ergonomics risk factors that may be causing or contributing to the MSD hazards. Some of what you will gain from this process is as follows:

- Obtain information about the specific tasks or actions the job involves.
- Obtain information about the job and problems in it from employees who perform the job.
- Observe the job.

- Identify specific job factors.
- Evaluate those factors (e.g., duration, frequency, and magnitude) to determine whether they cause or contribute to the problem.

Once MSD hazards have been identified, the next step is to eliminate or control them. An effective hazard control process involves identifying and implementing control measures to obtain an adequate balance between worker capabilities and work requirements so that MSDs are not reasonably likely to occur.

During the identification and analysis of hazards, you should do the following:

- Include in the hazard identification and analysis all of the employees in the problem job or those who represent the range of physical capabilities of employees in the job.
- Ask the employees whether performing the job poses physical difficulties, and, if so, which physical work activities or conditions of the job they associate with the difficulties.

Ergonomics hazard identification and analysis are processes for pinpointing work-related hazards or causes of MSDs and involve examining the workplace conditions and individual elements or tasks of a job to identify and assess the ergonomic risk factors that are reasonably likely to be causing or contributing to the reported MSDs. They can also be preventive measures used to identify jobs and job tasks where MSDs and MSD hazards are reasonably likely to develop in the future. Job hazard analysis is an essential element in the effective control of MSD hazards. In many situations, the causes of MSD hazards are apparent after discussions with the employee and observation of the job, but in other jobs the causes may not be readily apparent. In part, this is because most MSD hazards involve exposure to a combination of risk factors (i.e., multifactorial hazard). For example, it may not be clear in a repetitive motion job whether exposure to repetition, force, or awkward postures is the risk factor that is causing the problem.

Hazard identification and analysis are also important to pinpoint where the risk of harm exists and to rule out aspects of the job that do not put employees at risk. In this sense, this process is an efficient way to help you focus their resources on the most likely causes of the problem so that the control strategy they select has a reasonable expectation of eliminating or materially reducing the MSD hazards. They also provide you with the information needed to target their efforts to those jobs or tasks that may pose the most severe problems. This is an important step for those of you whose ergonomics programs include early intervention when employees report MSDs. For example, many workplaces provide MSD management first aid (i.e., immediate restricted work activity whenever an employee reports MSD signs or symptoms) and afterward look to see whether they need to take action to fix the job.

Some specific workers need to be evaluated since they may not be indicative of your average worker. This may be especially true of workers performing the same task as others. It is imperative that you look at sizes of workers or handicaps such as the following:

- Shortest employees in the job, because they are likely to have to make the longest reaches or to have a working surface that is too high
- Tallest employees because they may have to maintain the most excessive awkward postures (e.g., leaning over the assembly line, reaching down with the arms) while performing tasks
- Employees with the smallest hands because they may have to exert considerably more force to grip and operate hand and power tools
- Employees who work in the coldest areas of the workplace because they may have to exert more force to perform repetitive motions
- Employees who wear bifocals because they may be exposed to awkward postures (e.g., bending the neck back to see)

An assessment tool such as the one found in Figure 11.2 can be used to evaluate workers in these categories.

Ergonomic hazard identification checklist

Work area _____ Employees _____ Date _____
Conducted by _____ Reviewed by _____ Date _____

Answer the following questions based on the primary job activities of the worker at this particular task.

Use the following responses to describe how frequently the worker is exposed to the job conditions described below:

> **Never**—Worker is never exposed to the condition
> **Sometimes**—Worker is exposed to the condition less than three times daily
> **Usually**—Worker is exposed to the condition three times or more daily

	Never	Sometimes	Usually	If usually, list jobs to which answer applies here
Does worker perform tasks that are externally paced?				
Is the worker required to exert force with his hands (e.g., gripping, pulling, pinching)?				
Does the worker stand continuously for periods of more than 30 min?				
Does the worker sit for periods of more than 30 min without the opportunity to stand or move around freely?				
Does the worker have to stretch to reach the parts, tools, or work area?				
Does the worker use electronic input devices (e.g., keyboards, mouse, joysticks, track balls) for continuous periods of more than 30 min?				
Does the worker kneel (one or both knees)?				
Does the worker perform activities with hands raised above shoulder height?				
Does worker perform activities while bending or twisting at the waist?				
Is the worker exposed to vibration?				
Is the worker required to work in unnatural body positions?				
Does worker lift or lower objects between the floor and waist height or above the shoulder?				
Does the worker lift, lower, or carry objects weighing more than 50 lb?				

TERMS

Primary job activities—Job activities that make up a significant part of the work or are required for safety or contingency. Activities are not considered to be primary job activities if they make up a small percentage of the job (i.e., takes up less than 10% of the worker's time) are not essential for safety or contingency, and can be readily accomplished in other ways (e.g., using equipment already available in the facility).

Externally paced activities—Work activities for which the worker does not have direct control of the rate of work. Externally paced work activities include activities which (1) the worker must keep up with an assembly line or an independently operating machine, (2) the worker must respond to a continuous queue (e.g., customers standing in line, phone calls at a switch).

Figure 11.2 Ergonomic hazard identification checklist. (Courtesy of Occupational Safety and Health Administration.)

It is also a good idea to conduct a symptom or comfort survey (see Figure 11.3). This allows the worker to tell you where he or she is experiencing pain or discomfort. He or she also can tell you what would make it easier to accomplish the work and often can suggest very cost-effective solutions.

You must remember that there are likely to be situations in which the physical work activities or conditions only pose a risk to the reporting employee. For example, an employee in a commercial bakery may report a back or shoulder MSD related to extended reaches involved in sorting rolls. However, other employees who have performed the job for several years do not have (and never have had) difficulties performing the physical work activities of the job. In this case, it might be concluded that the problem is limited to the injured employee. In this situation, you would limit the response (e.g., analysis, control, or training) to physical work activities and conditions confronting that injured employee.

Another example might involve a manufacturing assembly line job where an employee is much shorter than other employees. The employee reports persistent shoulder and elbow pain, which the employer observes is caused by reaching higher than the other employees to perform the job tasks. This may also be an appropriate case for you to focus the analysis and control efforts on the employee who reported the problem.

These efforts may include job task breakdown, videotaping or photographing the job, job or hazard checklists, employee questionnaires, use of measuring tools, or biomechanical calculations. Checklists, together with other screening methods such as walk-through observational surveys, and worker and supervisory interviews, employee symptom or discomfort surveys, are recognized ergonomic evaluation methods.

Videotaping the job is a common practice for observing jobs. A number of employers, especially in situations where the work activities are complex or the causes of the problem may not be easily identifiable, do videotape or photograph the job.

These employers find it helpful to be able to refer to a record of the job while evaluating the ergonomic risk factors or identifying and assessing possible control measures.

Job task analysis is another job hazard analysis process that is widely used. This process involves breaking the job down into its various discrete elements or actions and then identifying and evaluating or measuring the extent to which the risk factors that are present in the physical work activities and conditions are reasonably likely to be contributing to the MSD hazard. To do a job task breakdown, a number of individuals look at the job as a series of individual, distinct tasks or steps. Focusing on each task allows for easier identification of the physical activities required to complete the job. While observing the job, employers record a description of each task for use in later risk factor analysis as well as other information that is helpful in completing the analysis:

- Tools or equipment used to perform task
- Materials used in task
- Amount of time spent doing each task
- Workstation dimensions and layout
- Weight of items handled
- Environmental conditions (cold, glare, blowing air)
- Vibration and its source
- Personal protective equipment worn

In addition, if the controls are likely to be the same for all the employees in a particular job, continuing to conduct job hazard analyses after a certain point may have diminishing returns. Doing hazard identification and analysis for all employees also may be difficult in jobs that do not have fixed workstations (e.g., beverage delivery, package delivery, furniture moving, appliance delivery, home repair, visiting nurse, home health aide). Some of these jobs may have constantly changing work conditions, all of which it may not be possible to analyze.

Symptoms Survey Form: #2

Symptoms Survey: Ergonomics Program

Date _____

Plant	Department	Job #	Job name
Shift	Supervisor	Hours worked/week	Time on this job __ years __ months

Other jobs you have done in the last year (for more than 2 weeks)

Department	Job #	Job name	Time on this job __ months __ weeks
Department	Job #	Job name	Time on this job __ months __ weeks

(If more than 2 jobs, include those you worked on the most)

Have you had any pain or discomfort during the last year?

☐ Yes   ☐ No (If NO stop here)

If YES, check off area of the drawing which bothers you the most.

Front   Back

Figure 11.3   Symptom survey. (Courtesy of the National Institute for Occupational Safety and Health.)

(*continued*)

Symptoms Survey Form (continued)

(Complete a separate page for each area that bothers you)

Check area: ☐ Neck   ☐ Shoulder   ☐ Elbow/forearm   ☐ Hand/wrist   ☐ Fingers
☐ Upper back   ☐ Low back   ☐ Thigh/knee   ☐ Low leg   ☐ Ankle/foot

1. Please check the word(s) that best describe your problem
☐ Aching   ☐ Numbness (asleep)   ☐ Tingling
☐ Burning   ☐ Pain   ☐ Weakness
☐ Cramping   ☐ Swelling   ☐ Other
☐ Loss of Color ☐ Stiffness

2. When did you first notice the problem?   ___ (month)   ___ (year)

3. How long does each episode last?   ☐ 1 hour   ☐ 1 day   ☐ 1 week   ☐ 1 month   ☐ 6 months

4. How many separate episodes have you had in the last year? ___

5. What do you think caused the problem? ___

6. Have you had this problem in the last 7 days?   ☐ Yes   ☐ No

7. How would you rate this problem?

NOW

☐ ☐ ☐ ☐ ☐ ☐ ☐ ☐ ☐
None   **Unbearable**

When it is the WORST

☐ ☐ ☐ ☐ ☐ ☐ ☐ ☐ ☐
None   **Unbearable**

8. Have you had medical treatment for this problem?   ☐ Yes   ☐ No
   8a. If NO, why not? _____
   8a. If YES, where did you receive treatment?
   ☐ 1. Company medical   Times in past year
   ☐ 2. Personal doctor   Times in past year
   ☐ 3. Other   Times in past year
   Did treatment help?   ☐ Yes   ☐ No _____

9. How much time have you lost in the last year because of this problem? ____ days

10. How many days in the last year were you on restricted or light duty because of this problem? ____ days

11. Please comment on what you think would improve your symptoms ____

Figure 11.3 (continued)

Hazards cannot be addressed efficiently without an accurate evaluation of the situation. Line employees are one of the best sources of this information. They are local process experts. Employees need to be involved in the identification, analysis, and control process because no one knows the job better than the person who does it. Employees have the best understanding of what it takes to perform each task in a job, and thus know what parts of the job are the hardest to perform or pose the biggest difficulties. Workers can best tell what conditions cause them pain, discomfort, and injuries. They often have easy and practical suggestions on how such problems can be alleviated. Involving workers, in addition to helping to ensure identification, analysis and control are correct, can make the job process more efficient. Employees can help pinpoint the causes of problems more quickly. Some of the ergonomics risk factors that will likely come to light are found in Table 11.2.

From this point on some more specific information and some of the key elements will be discussed as stand-alone items such as risk factors, exposure limits, ergonomic controls, program evaluation, and training.

Table 11.2 Tasks and Their Risk Factors: Physical Work Activities and Conditions

Ergonomic risk factors that may be present

1. Exerting considerable physical effort to complete a motion
   a. Force
   b. Awkward postures
   c. Contact stress
2. Doing same motion over and over again
   a. Repetition
   b. Force
   c. Awkward postures
   d. Cold temperatures
3. Performing motions constantly without short pauses or breaks in between
   a. Repetition
   b. Force
   c. Awkward postures
   d. Static postures
   e. Contact stress
   f. Vibration
4. Performing tasks that involve long reaches
   a. Awkward postures
   b. Static postures
   c. Force
5. Working surfaces are too high or too low
   a. Awkward postures
   b. Static postures
   c. Force
   d. Contact stress
6. Maintaining same position or posture while performing tasks
   a. Awkward postures
   b. Static postures
   c. Force
   d. Cold temperatures
7. Sitting for a long time
   a. Awkward postures
   b. Static postures
   c. Contact stress
8. Using hand and power tools
   a. Force
   b. Awkward postures
   c. Static postures
   d. Contact stress
   e. Vibration
   f. Cold temperatures

9. Vibrating working surfaces, machinery, or vehicles
   a. Vibration
   b. Force
   c. Cold temperatures
10. Workstation edges or objects press hard into muscles or tendons
    a. Contact stress
11. Using hand as a hammer
    a. Contact stress
    b. Force
12. Using hands or body as a clamp to hold object while performing tasks
    a. Force
    b. Static postures
    c. Awkward postures
    d. Contact stress
13. Gloves are bulky, too large, or too small
    a. Force
    b. Contact stress

Manual material handling (lifting/lowering, pushing/pulling, and carrying)
14. Objects or people moved are heavy
    a. Force
    b. Repetition
    c. Awkward postures
    d. Static postures
    e. Contact stress
15. Horizontal reach is long (distance of hands from body to grasp object to be handled)
    a. Force
    b. Repetition
    c. Awkward postures
    d. Static postures
    e. Contact stress
16. Vertical reach is below knees or above the shoulders (distance of hands above the ground when the object is grasped or released)
    a. Force
    b. Repetition
    c. Awkward postures
    d. Static postures
    e. Contact stress

(continued)

Table 11.2 (continued)

17. Objects or people are moved significant distance	19. Object is slippery or has no handles
a. Force	a. Force
b. Repetition	b. Repetition
c. Awkward postures	c. Awkward postures
d. Static postures	d. Static postures
e. Contact stress	20. Floor surfaces are uneven, slippery, or sloped
18. Bending or twisting during manual handling	a. Force
a. Force	b. Repetition
b. Repetition	c. Awkward postures
c. Awkward postures	d. Static postures
d. Static posture	

Source: Courtesy of Occupational Safety and Health Administration.

ERGONOMIC RISK FACTORS

Ergonomic risk factors are the aspects of a job or task that impose a biomechanical stress on the worker. Ergonomic risk factors are the synergistic elements of MSD hazards. The following ergonomic risk factors are most likely to cause or contribute to an MSD:

- Force
- Vibration
- Repetition
- Contact stress
- Awkward postures
- Cold temperatures
- Static postures

These risk factors are described briefly in the following sections.

Force

Force refers to the amount of physical effort that is required to accomplish a task or motion. Tasks or motions that require application of higher force place higher mechanical loads on muscles, tendons, ligaments, and joints. Tasks involving high forces may cause muscles to fatigue more quickly. High forces may also lead to irritation, inflammation, strains, and tears of muscles, tendons, and other tissues.

The force required to complete a movement increases when other risk factors are also involved. For example, more physical effort may be needed to perform tasks when the speed or acceleration of motions increases, when vibration is present, or when the task also requires awkward postures. Force can be internal, such as when tension develops within the muscles, ligaments, and tendons during movement. Force can also be external, as when a force is applied to the body, either voluntarily or involuntarily. Forceful exertion is most often associated with the movement of heavy loads, such as lifting heavy objects on and off a conveyor, delivering heavy packages, pushing a heavy cart, or moving a pallet. Hand tools that involve pinch grips require more forceful exertions than those that allow other grips, such as power grips.

Repetition

Repetition refers to performing a task or series of motions over and over again with little variation. When motions are repeated frequently (e.g., every few seconds) for prolonged periods (e.g., several

Table 11.3   Repetition and Body Area

Body Area	Frequency Repetition per Minute	Level of Risk	Very High Risk if Modified by Either
Shoulder	More than 2.5	High	High external force, speed, high static load, extreme posture
Upper arm/elbow	More than 10	High	Lack of training, high output demands, lack of control
Forearm/wrist	More than 10	High	Long duration of repetitive work
Finger	More than 200	High	

Source: Courtesy of Occupational Safety and Health Administration.

hours, a work shift), fatigue and strain of the muscle and tendons can occur because there may be inadequate time for recovery. Repetition often involves the use of only a few muscles and body parts, which can become extremely fatigued while the rest of the body works very little. The following figure shows the frequency of repetition and length of task cycles that are associated with increased risk of injury in repetitive motion jobs (see Table 11.3).

Awkward Postures

Awkward postures refer to positions of the body (e.g., limbs, joints, back) that deviate significantly from the neutral position while job tasks are performed. For example, when a person's arm is hanging straight down (i.e., perpendicular to the ground) with the elbow close to the body, the shoulder is said to be in a neutral position. However, when employees are performing overhead work (e.g., installing or repairing equipment, grasping objects from a high shelf) their shoulders are far from the neutral position. Other examples include wrists bent while typing, bending over to grasp or lift an object, twisting the back and torso while moving heavy objects, and squatting. Awkward postures are often significant contributors to MSDs because they increase the work and the muscle force that is required.

Static Postures

Static postures (or static loading) refer to physical exertion in which the same posture or position is held throughout the exertion. These types of exertions put increased loads or forces on the muscles and tendons, which contribute to fatigue. This occurs because not moving impedes the flow of blood that is needed to bring nutrients to the muscles and to carry away the waste products of muscle metabolism. Examples of static postures include gripping tools that cannot be put down, holding the arms out or up to perform tasks, or standing in one place for prolonged periods.

Vibration

Vibration is the oscillatory motion of a physical body. Localized vibration, such as vibration of the hand and arm, occurs when a specific part of the body comes into contact with vibrating objects such as powered hand tools (e.g., chain saw, electric drill, chipping hammer) or equipment (e.g., wood planer, punch press, packaging machine). Whole-body vibration occurs when standing or sitting in vibrating environments (e.g., driving a truck over bumpy roads) or when using heavy vibrating equipment that requires whole-body involvement (e.g., jackhammers).

Contact Stress

Contact stress results from occasional, repeated, or continuous contact between sensitive body tissue and a hard or sharp object. Contact stress commonly affects the soft tissue on the fingers, palms, forearms, thighs, shins, and feet. This contact may create pressure over a small area of the body (e.g., wrist, forearm), which can inhibit blood flow, tendon and muscle movement, and nerve function. Examples of contact stress include resting wrists on the sharp edge of a desk or workstation while performing tasks, pressing of tool handles into the palms, especially when they cannot be put down, tasks that require hand hammering, and sitting without adequate space for the knees.

Cold Temperatures

Cold temperatures refer to exposure to excessive cold while performing work tasks. Cold temperatures can reduce the dexterity and sensitivity of the hand. Cold temperatures, for example, cause the worker to apply more grip force to hold hand tools and objects. Also, prolonged contact with cold surfaces (e.g., handling cold meat) can impair dexterity and induce numbness. Cold is a problem when it is present with other risk factors and is especially problematic when it is present with vibration exposure.

Of these risk factors, force (i.e., forceful exertions), repetition, and awkward postures, especially when occurring at high levels or in combination, are most often associated with the occurrence of MSDs. Exposure to one ergonomic risk factor may be enough to cause or contribute to a covered MSD. However, most often ergonomic risk factors act in combination to create a hazard. Jobs that have multiple risk factors have a greater likelihood of causing an MSD, depending on the duration, frequency, and magnitude of exposure to each. Thus, it is important that ergonomic risk factors be considered in light of their combined effect in causing or contributing to an MSD.

Examples of Risky Activities

Some practical examples of the aforementioned risk factors are

- Pulling meat off a bone on a meat cutting assembly line
- Pulling hard to tighten bolts or screws in assembly line work
- Squeezing hard on a pair of pliers
- Pulling hard on a long wrench to tighten or loosen a bolt
- Chuck boner job in a beef processing plant
- Shaking crab meat from Alaskan king crab legs
- Holding an extrusion nozzle while checking each hole (50 holes) to ensure it is the appropriate size
- Holding a jar in one hand while attempting to remove the lid with the other hand
- Working on a hot pack used in extruding plastic with heat-resistant gloves
- Holding a chicken leg while wearing cut-resistant gloves

Examples of awkward postures are

- Throwing 20 lb bundles of printed material to overhead conveyors
- Bolting or screwing a new part into an auto that is on a lift
- Attaching doors on the bathroom vanity assembly line
- Capping and cupping cookies on an assembly line
- Threading extruded fiber onto a spool that is 15 in. above the floor
- Activating palm switches that are 60 in. above the floor
- Cradling a phone on the shoulder
- Holding the arms on the top half of a steering wheel

- Working at a computer workstation where the operator must lean forward to see the screen
- Working in a chair on an uneven floor
- Holding the head of a cow on a slippery surface while attempting to remove meat
- Holding a small part while assembling it
- Drilling a hole in a part that the worker has to hold
- Using the hips or thighs to hold a part in place while working on the part
- Using the hands to wring out a mop
- Bending sideways using the shoulder to hold a door panel in place while fastening the hinges
- Holding a part in place overhead while inserting fasteners

Examples of force and extreme postures are

- Throwing items into an overhead container
- Reaching over the bagging area to place bags of groceries into shopping carts

Examples of reaching are

- Reaching above the head to activate a press or other machine
- Reaching frequently for small parts in a bin that is at or close to the limit of the arm's reach
- Reaching down and behind the back to pick up parts to feed to a press or place on a conveyor
- Reaching across a conveyor to pick up items
- Reaching to pick up items on the other side of the scanner on a grocery checkout conveyor

Examples of contact stress are

- Using the hand as a hammer (an example of force plus contact stress)
- Operating a carpet kicker with the knees
- Working at a computer placed on a folding table
- Holding an injection molded part at eye level by resting the elbows on the work surface
- Watching a computer monitor that is above eye level
- Holding a mouse that is located in front of the keyboard
- Working in a chair where the seat pan is too long
- Working in chair with armrests that are too close to the body
- Extensive use of shears or scissors
- Using a tool with a small, thin handle that digs into the palm
- Using tools with grooved handles that press against the side of fingers
- Leaning against a metal work bench with a square edge
- Using a keyboard on a standard table or desk with nonrounded edges
- Sitting in a bench or chair that does not have a padded seat
- Pounding on a two-part mold to get it to seat or come together properly
- Hitting a palm button to activate a machine
- Striking two parts to separate them
- Striking the handle of a vise to loosen it
- Holding a pane of glass while attaching hardware
- Using the knee to position a pump while making the electrical connection
- Holding onto a nut while turning the bolt
- Wearing latex gloves that are too tight
- Selecting cases in a frozen foods warehouse while wearing knit gloves under thermal gloves

Examples of static contraction are

- Doing extensive repair work when the automobile is overhead on a vehicle lift
- Holding out the arm to use a mouse that is on a surface more than 15 in. from the body because the keyboard tray is not big enough to hold the mouse

- Working on a vertical drafting table
- Sitting at a grinding bench where the grinding wheel is 24 in. above the floor

Examples of repetition are

- Packing bags of potato chips into shipping boxes
- Intensive keying of information into computer

Examples of forceful repetition are

- Filleting fish in a processing plant
- Constantly using screwdriver to drive screws into wood

Examples of repetition in awkward posture are

- Sorting parts or letters into bins of different heights and locations (e.g., behind the employee)
- Working with bent wrists to assemble small circuit breakers

Examples of cold temperatures are

- Trimming chicken or turkey breasts in a processing plant
- Working in an operating room of a hospital
- Working in the plant's cooler for several hours
- Standing to direct traffic on a busy road in the winter
- Using a knife to process catfish fillets
- Using a socket wrench to change out equipment on the roof in the winter

Examples of static contraction are

- Standing in one place for long periods
- Holding and gripping a knife for long periods of time
- Holding a pipe overhead while preparing a fitting
- Holding an uncooperative animal on the exam table

Examples of hand and power tools issues are

- Weight and size of tool
- Tool handles and grips
- Tool activation (repetitively, one finger)
- Tool kickback, vibration, and maintenance
- Using powered driver to run and tighten nuts on bolts and opposing force when the driver reaches the end of the tightening process
- Constantly pressing the trigger to activate a drill with the index finger
- Reaching over a barrier to operate a rivet gun
- Squatting to tighten 20 bolts on a pipe flange
- Constantly holding knife used to trim chicken breasts in poultry plant
- Holding a wire wrap gun
- Using a screwdriver with edges on the handle to tighten bolts on an assembly line
- Using small wire clippers (handles press into the palm) to remove component lead after wave solder
- Cutting trees with chain saw
- Using grinding tools to form dentures

Examples of vibration issues are

- Working near a 100 ton press
- Working near a vibratory bowl
- Operating a fork truck over rough dock plates or gravel

- Leaning against a grinding machine while it is operating
- Holding a wheel while operating a sewing machine
- Manually aligning sections of a newspaper using a vibrating table
- Driving a fork truck over rough surfaces in a frozen food warehouse
- Using vibrating etching tools in a clean room

Examples of manual handling are

- Lifting a resident, who has little ability to assist, from the toilet to a wheelchair
- Lifting a 150 lb package from a loading dock into a van
- Pushing a 300 lb pump away from the paper machine
- Pushing a heavy cart up a sloped ramp
- Carrying several 50 lb bags of feedstock material to the basement
- Carrying a resident of a nursing home to the bath tub
- Pushing a heavy box on a nonpowered conveyor
- Carrying a hot pack used in extruding plastic to the repair cart
- Carrying a carboy of nitric acid
- Picking up a 35 lb spool of yarn from a peg above shoulder height
- Picking a 40 lb item from a 60 in. high shelf in a grocery warehouse
- Lifting a 50 lb motor off a pallet
- Pushing a cart with the hands above midchest height
- Pulling a wooden pallet across the floor
- Carrying large, bulky boxes of machine parts where the worker is unable to carry the box with a horizontal hold
- Carrying a large piece of furniture down steps
- Pushing a cart of restaurant supplies from the delivery truck to the restaurant
- Pushing a patient on a gurney to physical therapy
- Carrying trash cans to the garbage truck
- Carrying water bottles to the cooler
- Moving 30 lb motors from a workstation to a conveyor perpendicular (90°) to the workstation
- Moving a patient from the bed to a wheelchair
- Loading luggage into the cargo hold of an airplane
- Lifting a 40 lb fuel pump out of a tank of mineral oil
- Lifting wet watermelons out of a box (which requires the worker to use excessive grip force)
- Lifting a patient with little ability to assist out of bed
- Pushing a large box of potatoes in a product warehouse
- Carrying a keg of beer
- Carrying machined parts to a degreaser
- Carrying a side of beef
- Shoveling grain
- Lifting bags of laundry from a wet floor
- Pushing a laundry hamper across a wet floor
- Pushing a file cabinet on a carpeted floor
- Pushing a wheelchair through gravel
- Pushing a cart on a cracked concrete floor
- Carrying boxes of metal scraps down steps
- Carrying boxes of paper up a ramp into the computer room

PHYSICAL WORK ACTIVITIES AND CONDITIONS

Physical work activities and conditions include

- Physical demands of work
- Workplace and workstation conditions and layout

- Characteristics of object(s) that are handled or used
- Environmental conditions

Employers should examine a job in which an MSD has occurred to identify the physical work activities and workplace conditions and then evaluate the risk factors to assess the work environment.

LIMITS OF EXPOSURE

To make a determination as to the real risk, you need to look at the duration, frequency, and magnitude (i.e., modifying factors) of the employee's exposure to the ergonomic risk factors. These risk factors do not always rise to the level that poses a significant risk of injury. This may be because the exposure does not last long enough, is not repeated frequently enough, or is not intensive enough to pose a risk.

Duration

Duration refers to the length of time an employee is continually exposed to risk factors. The duration of job tasks can have a substantial effect on the likelihood of both localized and general fatigue. Table 11.4 shows the physical work activities and workplace conditions that are associated with those physical aspects.

In general, the longer the period of continuous work (i.e., the longer the tasks require sustained muscle contraction), the longer the recovery or rest time required. Duration can be mitigated by changing the sequence of activities or recovery time and pattern of exposure. Breaks or a short pause in the work routine helps to reduce the effects of the duration of exposure.

Frequency

The response of the muscles and tendons to work is dependent on the number of times the tissue is required to respond and the recovery time between activities. The frequency can be viewed at the microlevel, such as grasps per minute or lifts per hour. However, often a macroview will be sufficient, such as time in a job per shift, or days per week in a job.

Magnitude

Magnitude (or intensity) is a measure of the strength of the risk factor, for example, how much force, how deviated the posture, how great the velocity or acceleration of motion, and how much pressure due to compression. Magnitude can be measured either in absolute terms or relative to an individual's capabilities. There are many qualitative and quantitative ways to determine the magnitude of exposure. Often all it takes is to ask employees to describe the most difficult part of the job, and the answer will indicate the magnitude of the risk factor. A common practice for assessing forceful exertion is to ask the employee to rate the force required to do the task. When magnitude is assessed qualitatively, the employer is making a relative rating; that is, the perceived magnitude of the risk factor relative to the capabilities of the worker. Relative ratings are very useful in understanding whether the job fits the employees currently doing the job.

As mentioned above, ergonomic risk factors are synergistic elements of MSD hazards. In other words, the total effect of these risk factors is greater than the sum of their parts. As such, employers need to be especially watchful for situations where risk factors occur simultaneously. Levels of risk factors that may pose little risk when found alone are much more likely to cause MSDs when they occur with other risk factors.

Table 11.4   Physical Aspects of Jobs and Workstations

Physical demands of work
- Exerting considerable physical effort to complete a motion
- Doing the same motion over and over again
- Performing motions constantly without short pauses or breaks in between
- Maintaining same position or posture while performing tasks
- Sitting for a long time
- Using hand as a hammer
- Using hands or body as a clamp to hold object while performing tasks
- Moving objects or people significant distances

Layout and condition of the workplace or workstation
- Performing tasks that involve long reaches
- Working surfaces too high or too low
- Vibrating working surfaces, machinery, or vehicles
- Workstation edges or objects press hard into muscles or tendons
- Horizontal reach is long
- Vertical reach is below knees or above the shoulders
- Floor surfaces are uneven, slippery, or sloped

Characteristics of the objects handled
- Using hand and power tools
- Gloves bulky, too large, or too small
- Objects or people moved are heavy
- Object is slippery or has no handles

Environmental conditions
- Cold temperatures
- Temperature extremes and humidity
- Vibration
- Noise
- Illumination
- Colors

Source: Courtesy of Occupational Safety and Health Administration.

ERGONOMIC CONTROLS

Controls that reduce a risk factor focus on reductions in the risk modifiers (frequency, duration, or magnitude). By limiting exposure to the modifiers, the risk of an injury is reduced. Thus, in any job, the combination of the task, environment, and the worker create a continuum of opportunity to reduce the risk by reducing the modifying factors. The closer the control approach comes to eliminating the frequency, duration, or magnitude, the more likely it is that the MSD hazard has been controlled. Conversely, if the control does little to change the frequency, duration, or magnitude, it is unlikely that the MSD hazard has been controlled.

In determining control, ask employees in the problem job for recommendations about eliminating or materially reducing the MSD hazards. Second, identify, assess, and implement feasible controls (interim and permanent) to eliminate or materially reduce the MSD hazards. This includes prioritizing the control of hazards, where necessary. Thirdly, track your progress in eliminating or materially reducing the MSD hazards. This includes consulting with employees in problem jobs about whether the implemented controls have eliminated or materially reduced the hazard, and last, identify and evaluate MSD hazards when you change, design, or purchase equipment or processes in problem jobs.

Identify Controls

There are different methods you can use and places you can go to identify controls. Many employers rely on their internal resources to identify possible controls. These in-house experts may include

- Employees who perform the job and their supervisors
- Engineering personnel
- Workplace safety and health personnel or committee
- Maintenance personnel
- On-site health-care professionals
- Procurement staff
- Human resource personnel

Possible controls can also be identified from sources outside the workplace, such as

- Equipment catalogs
- Vendors
- Trade associations or labor unions
- Conferences and trade shows
- Insurance companies
- OSHA consultation services
- Specialists

You can refer to Appendix F for information on possible controls for various risk factors that were discussed earlier in this chapter.

Assess Controls

The assessment of controls is an effort by you, with input from employees, to select controls that are reasonably anticipated to eliminate or materially reduce the MSD hazards. You may find that several controls would be reasonably likely to reduce the hazard. Multiple control alternatives are often available, especially when several risk factors contribute to the MSD hazard. You need to assess which of the possible controls should be tried. Clearly, a control that significantly reduces several risk factors is preferred over a control that only reduces one of the risk factors.

Selection of the risk factor(s) to control, and control measures to try, can be based on numerous criteria. An example of one method involves ranking all of the ergonomic risk factors and possible controls according to how well they meet these four criteria:

- Effectiveness—greatest reduction in exposure to the MSD hazards
- Acceptability—employees most likely to accept and use this control
- Timeliness—takes least amount of time to implement, train, and achieve material reduction in exposure to MSD hazards
- Cost—elimination or material reduction of exposure to MSD hazards at the lowest cost

Implement Controls

Because of the multifactorial nature of MSD hazards, it is not always clear whether the selected controls will achieve the intended reduction in exposure to the hazards. As a result, the control of MSD hazards often requires testing selected controls and modifying them appropriately before implementing them throughout the job. Testing controls verifies that the proposed solution actually works and indicates what additional changes or enhancements are needed.

TRACKING PROGRESS

First, evaluating the effectiveness of controls is top priority in an incremental abatement process. Unless they follow up on their control efforts, employers will not know whether the hazards have been adequately controlled or whether the abatement process needs to continue. That is, if the job is not controlled, the problem solving is not complete.

Second, the tracking of progress is also essential in cases where you need to prioritize the control of hazards. It tells you whether they are on schedule with their abatement plans.

Third, tracking the progress of control efforts is a good way of determining whether the elements of the program are functioning properly and quantifying their success. Some of the measures to use include

- Reductions in severity rates, especially at the very start of the program
- Reduction in incidence rates
- Reduction in total lost workdays and lost workdays per case
- Reduction in job turnover or absenteeism
- Reduction in workers' compensation costs and medical costs
- Increases in productivity or quality
- Reduction in reject rates
- Number of jobs analyzed and controlled
- Number of problems solved

PROACTIVE ERGONOMICS

Sometimes this concept is referred to as proactive ergonomics or safety through design. The concept encompasses facilities, hardware, equipment, tooling, materials, layout and configuration, energy controls, and environmental concerns and products. Designing or purchasing to eliminate or materially reduce MSD hazards in the design process helps to avoid costly retrofitting. It also results in easier and less costly implementation of ergonomic controls.

Ergonomists endorse the hierarchy of controls, which accords first place to engineering controls, because they believe that control technologies should be selected based on their reliability and efficacy in eliminating or reducing the workplace hazard (risk factors) giving rise to the MSD. Engineering controls are preferred because these controls and their effectiveness are

- Reliable
- Consistent
- Effective
- Measurable
- Not dependent on human behavior (that of managers, supervisors, or workers) for their effectiveness
- Do not introduce new hazards into the process

In contrast to administrative and work practice controls or personal protective equipment, which occupies the second and third tiers of the hierarchy, respectively, engineering controls fix the problem once and for all. However, because there is such variability in workplace conditions you will need to use any combination of engineering, work practice, or administrative controls as methods of control for MSD hazards.

EDUCATION AND TRAINING

Education and training can be used in a variety of ways. The foremost is to train all employees in ergonomic hazard awareness, your program and procedures, sign and symptom identification, and types of injuries and illnesses. Second, train some of the workforce in ergonomic assessment so you will have teams of both management and labor to evaluate ergonomic hazards and make recommendations for controlling the potential risk factors on the jobs in your workplace. With proper training you will have an educated workforce who can be an asset rather than a liability in solving MSD problems.

Ergonomic principles are most effectively applied to workstations and new designs on a preventive basis, before injuries or illnesses occur. Good design with ergonomics provides the greatest economic benefit for industry. Design strategies should emphasize fitting job demands to the capabilities and limitations of employees. To achieve this, decision makers must have appropriate information and knowledge about ergonomic risk factors and ways to control them. They need to know about the problems in jobs and the causes. Designers of in-house equipment, machines, and processes also need to have an understanding of ergonomic risk factors and know how to control them. For example, they may need anthropometric data to be able to design to the range of capabilities and limitations of employees.

It is also important that persons involved in procurement have basic knowledge about the causes of problems and ergonomic solutions. For example, they need to know that adjustable chairs can reduce awkward postures and that narrow tool handles can considerably increase the amount of force required to perform a task. In addition, to prevent the introduction of new hazards into the workplace, procurement personnel need information about equipment needs.

Ergonomics is a continuous improvement process. If you can show that you have made an organized effort to identify ergonomic stressors, to educate affected employees on ergonomic principles, to implement solutions, and to have a system to identify when a solution is not working and needs to be readdressed, you have taken giant steps toward mitigating your ergonomic problems.

CHAPTER 12

Addressing Illnesses: Industrial Hygiene

Industrial hygienist taking a noise reading.

INTRODUCTION TO OCCUPATIONAL ILLNESS PREVENTION

In this chapter you will find a discussion of how to prevent illnesses by carrying out an environmental assessment of the workplace. You will need to make the decision as to whether you have the expertise to make an effective and accurate evaluation of your worksite. In carrying out such an assessment, you may not have the expertise or even the equipment that you might need to do a viable and proper assessment to protect your health and that of your workforce. From the information in this chapter you should be able to determine when you have reached your limitations as well as understand your role in working with an industrial hygienist (IH).

Industrial hygiene has been defined as "that science or art devoted to the anticipation, recognition, evaluation, and control of those environmental factors or stresses, arising in or from the workplace, which may cause sickness, impaired health and well-being, or significant discomfort and inefficiency among workers or among the citizens of the community."

The IH, although trained in engineering, physics, chemistry, or biology, has acquired by undergraduate or postgraduate study and experience, the knowledge of the effects on health of chemical and physical agents under various levels of exposure. The IH is involved with the monitoring and analytical methods required to detect the extent of exposure and the engineering and other methods used for hazard control.

The Occupational Safety and Health Act (OSHAct) has brought a restructuring of programs and activities relating to safeguarding the health of workers. Uniform occupational health regulations now apply to all businesses engaged in commerce, regardless of their locations within the jurisdiction. Nearly every employer is required to implement some element of an industrial hygiene or occupational health or hazard communication program, to be responsive to Occupational Safety and Health Administration (OSHA) and the OSHAct and its health regulations.

INDUSTRIAL HYGIENIST

The IH is a very diversely trained individual. IHs has a strong background in chemistry. They must have a background in engineering, biological sciences, and behavioral/social sciences. The IH has specific training in environmental sampling and is prepared to make recommendations involving solutions and controls for environmental factors that can cause health effects in your workplace.

The IH can perform the following for your operation:

- Identify potential risk factors that can create health effects in your workforce.
- Evaluate the chemicals that you are using and make recommendations on controls.
- Select and conduct sampling methods for chemical and other environmental factors.
- Recommend the appropriate personal protective equipment. You should definitely have an IH select and recommend the type of respirator that is needed for your operation.
- If you need more ventilation in workplace, IHs are trained to assist and advise you.
- If you are faced with ergonomic issues, the IH has the type of background that can help you solve them.
- If biological agents exist in your workplace, the IH can help identify, evaluate, and develop controls for you.
- IH can address other diverse health hazards faced in the workplace such as radiation, temperature extremes, vibration, and noise issues, to name a few.

ENVIRONMENTAL FACTORS OR STRESSORS

You can begin to get the picture of the usefulness of an IH when faced with workplace environmental issues. The IH looks at specific environmental factors (stressors) or hazards. These factors are physical, biological, ergonomic, and chemical.

Physical Hazards

Physical hazards include excessive levels of nonionizing and ionizing radiation, noise, vibration, and extremes of temperature and pressure. Any of these have or can have serious adverse effects upon your workforce. You should identify any of these that exist in your work environment and that present a risk to your employees.

Physical hazards are defined as the types of hazards that can cause harm to a worker from an external source. Types of physical hazards are loud noise (equipment), temperature extremes (working in personal protective equipment), radiation (exposures to the infrared or gamma rays), chemical burn (acids or caustics), fire, and explosions. Other physical hazards include, but are not limited to, slips and falls, exposed machinery because of improper guarding, live electrical circuits or conductors, equipment moving about on site, confined spaces, and falling objects.

Noise is a serious hazard when it results in temporary or permanent hearing loss, physical or mental disturbance, any interference with voice communications, or the disruption of a job, rest, relaxation, or sleep. Noise is any undesired sound and is usually a sound with varying intensity that bears no information. It interferes with the perception of wanted sound and is likely to be harmful, cause annoyance, and interfere with speech.

The noise created by circular saws, planers, or high-speed grinders and similar power tools is narrow band noise. High-frequency noise is very damaging to the inner ear. Impulse type noise is generated by energy bursts occurring repetitively or one at a time. Noise from a jackhammer is an example of repetitive impulse noise. The firing of a gun is an example of a singular impulse noise. All types of noise can harm you if they are of high intensity or the exposure time is prolonged or repetitive.

A healthy young person can detect sounds in the 20 to 20,000 cycles per second range. As aging takes place, some hearing is lost. Higher frequencies cause the most damage to our ears and most people who have hearing loss have high-frequency losses first. Loudness or softness is determined by the intensity or sound pressure. The more powerful the driving force of the sound the higher the pressure involved. This is measured with an instrument called a sound level meter (SLM) in units called decibels (dB). Sounds that can just be heard by a person with very good hearing in an extremely quiet location are assigned the value of 0 dB. Ordinary speech is around 50–60 dB. At about 120 dB the threshold of pain is reached. This would be like hearing a jet engine about 50 ft. away.

Noise dose limits are now required for workplaces to minimize hearing loss from occupational exposure. Although louder noise is allowed for brief periods during the workday, the mandatory noise level limit (set by OSHA) is 90 dBA [(dBA = the A weighted on SLM) time-weighted average, TWA] over 8 h. An employer must make hearing protection available, provide training, and provide hearing tests when the noise level exceeds 85 dBA, TWA. As a basic rule, if you cannot hear the snap of your fingers at arm's length you should be using hearing protection. Over 90 dBA, employers must assure that protection is used.

Heat stress is a serious physical hazard that should always be considered on a construction job site, especially during the summer months. The chance of developing heat stress increases with increased humidity, hot environments, and the use of personal protective equipment. Sweating is the most effective means of losing excess heat, as long as adequate fluids are taken in to replace the sweat. When individuals are severely stressed by the heat, they may stop sweating with the most severe consequences of heat stress occurring. Adequate rest periods, availability of large amounts of replacement fluids, and frequent monitoring are essential to prevent the consequences of heat stress, which may occur without warning symptoms. The body maintains a normal temperature (98.6° F) in a hot environment by two methods:

- Sending more blood to the skin
- Sweating

Cold stress occurs when temperatures go down; the body maintains its temperature by reducing blood flow to the skin. This causes a marked decrease in skin temperature. The most severe effect is on the extremities (fingers, toes, earlobes, and nose). When hands and fingers become cold, they become numb and insensitive, and there is an increased possibility of accidents. If the restriction of blood flow to the skin is not adequate to maintain temperature, then shivering occurs. If this is not adequate to warm the body, then a marked decrease in temperature (hypothermia) may occur. Workers who may be at increased risk are

- Those doing hard labor who become fatigued or wet either from sweating or contact with water
- Those taking sedatives or drinking alcohol before or during work
- Workers with chronic diseases that affect the heart or blood vessels of the hands or feet
- Those not physically fit or who have not worked in a cold environment recently
- Those who use pavement breakers or other vibrating equipment

Radiation is divided into two major categories, based on its effect on living tissue: (1) ionizing and (2) nonionizing radiation. Ionizing radiation has the ability to change or destroy the atomic (chemical) structure of cells, whereas nonionizing radiation does not. Some types of nonionizing radiation that we are exposed to everyday include microwave energy used for cooking and radio waves used in broadcasting over radio and television. The types of ionizing radiation we are exposed to are cosmic rays from the sun and stars, terrestrial radiation from the earth, nuclear radiation from reactors, and medical radiation from x-rays.

Although nonionizing radiation is not as hazardous as ionizing, there are exposures that can cause severe injuries. Nonionizing radiation is generated by sun, lamps, welding arcs, lasers, plastic sealers, and radio or radar broadcast equipment. Since the eye is the primary organ at risk to all types of nonionizing radiation, eye protection is very important. Protective glasses should be selected based on the type of radiation exposure; for example, sunlight or welding flashes. Ionizing radiation is so named because it has enough energy to change (ionize) atoms and molecules, the building blocks of all matter. There are four natural types of ionizing radiation—alpha and beta particles, gamma rays, and neutrons. Vibration is a much more difficult physical factor to address since it is often difficult to connect the symptoms with the exposure. In addition, our ability to measure vibration and determine the measurements that will cause ill effects to workers is very limited.

Biological Hazards

Biological hazards include vermin, insects, molds, fungi, viruses, and bacterial contaminants. Sanitation and housekeeping activities such as potable water, removal of industrial waste and sewage, food handling, and personal cleanliness have the potential to exacerbate the potential risk of biological hazards.

Biological agents may be a part of the total environment or may be associated with certain occupations such as agriculture. Biological agents in the workplace include viruses, rickettsiae (organisms that cause diseases), bacteria, and parasites of various types. Diseases transmitted from animals to humans are common. Infections and parasitic diseases may also result from exposure to insects or by drinking contaminated water. Exposure to biohazards may seem obvious in occupations such as nursing, medical research, laboratory work, farming, and handling of animal products (slaughterhouses and meat packing operations). Sting of bees, which many workers are allergic to, is not an obvious biological hazard.

Biohazards may be transmitted to a person through inhalation, injection, ingestion, or physical contact. Many plants and animals produce irritating, toxic, or allergenic (causing allergic reactions) substances. Dusts may contain many kinds of allergenic materials, including insect scale, hairs, and fecal dust, sawdust, plant pollens, and fungal spores. Other hazards include bites or attacks by domestic and wild animals. Workers on hazardous waste sites may risk exposure to bites from venomous snakes or poisonous spiders.

Ergonomic Hazards

Ergonomic hazards include improperly designed tools or work areas. Improper lifting or reaching, poor visual conditions, or repeated motions in an awkward position can result in accidents or illnesses in the occupational environment. Designing the tools and the job to be done to fit the worker should be of prime importance.

When repetitive motion injuries occur, they often result from continuous use of a body part often in an unnatural posture, employing more force than is normal for the body part. This may result in irritation, fluid buildup, or thickening of the tendons and ligaments in the wrists, or damage to nerves or blood vessels. Severe pain may occur along with numbness, and loss of movement may occur. Weakness of the hand, arm, or other body part may occur, making it difficult to hold objects and perform grasping motions. The worker may drop objects, be unable to use keys, or count change because of these injuries. Surgical treatment may be necessary if the symptoms are severe and if other measures do not provide relief.

Other ergonomic hazards include manual handling of objects and materials where lifting and carrying are done. Lifting is so much a part of many everyday jobs that most of us do not think about it. However, it is often done wrong, with unfortunate results such as pulled muscles, disk injuries, or painful hernias. Intelligent application of engineering and biomechanical principles is required to eliminate hazards of this kind (see Chapter 11).

Chemical Hazards

Chemical hazards arise from excessive airborne concentrations of mists, vapors, gases, or solids that are in the form of dusts or fumes. In addition to the hazard of inhalation, many of these materials may act as skin irritants or may be toxic by absorption through the skin. There are thousands and thousands of potentially harmful chemicals found in the workplace. Workers face the possibility of exposure on a daily basis to these harmful chemicals.

The majority of occupational health hazards arise from inhaling chemical agents in the form of vapors, gases, dusts, fumes, and mists, or by skin contact with these materials. The degree of risk of handling a given substance depends on the magnitude and duration of exposure.

To recognize occupational factors or stresses, a health and safety professional must first know about the chemicals used as raw materials and the nature of the products and by-products manufactured. This sometimes requires great effort. The required information can be obtained from the material safety data sheet (MSDS), which must be supplied by the chemical manufacturer or importer to the purchaser for all hazardous materials under the hazard communication standard. The MSDS is a summary of the important health, safety, and toxicological information on the chemical or the mixture ingredients. Other stipulations of the hazard communication standard require that all containers of hazardous substances in the workplace be labeled with appropriate warning and identification labels. If the MSDS or the label does not give complete information but only trade names, it may be necessary to contact the manufacturer of the chemicals to obtain this information.

Many industrial materials such as resins and polymers are relatively inert and nontoxic under normal conditions of use, but when heated or machined, they may decompose to form highly toxic by-products. Information concerning these types of hazardous products and by-products must also be included in the company's hazard communication program. Breathing of some materials can irritate the upper respiratory tract or the terminal passages of the lungs and the air sacs, depending on the solubility of the material. Contact of irritants with the skin surface can produce various kinds of dermatitis.

The presence of excessive amounts of biologically inert gases can dilute the atmospheric oxygen below the level required to maintain the normal blood saturation value for oxygen and disturb cellular processes. Other gases and vapors can prevent the blood from carrying oxygen to the tissues or interfere with its transfer from the blood to the tissue, thus producing chemical asphyxia or suffocation. Carbon monoxide and hydrogen cyanide are examples of chemical asphyxiants.

Some substances may affect the central nervous system and brain to produce narcosis or anesthesia. In varying degrees, many solvents have these effects. Substances are often classified according to the major reaction that they produce, as asphyxiants, systemic toxins, pneumoconiosis-producing agents, carcinogens, irritant gases, or high dust levels.

MODES OF ENTRY FOR CONTAMINANTS

In order for a harmful agent to exert its toxic effect, it must come into contact with a body cell and must enter the body through inhalation, skin absorption, ingestion, or injection. Chemical compounds in the form of liquids, gases, mists, dusts, fumes, and vapors can cause problems by inhalation (breathing), absorption (through direct contact with the skin), or ingestion (eating or drinking).

Inhalation

Inhalation involves those airborne contaminants that can be inhaled directly into the lungs and can be physically classified as gases, vapors, and particulate matter that includes dusts, fumes, smokes, and mists. Inhalation, as a route of entry, is particularly important because of the rapidity with which a toxic material can be absorbed in the lungs, pass into the bloodstream, and reach the brain. Inhalation is the major route of entry for many hazardous chemicals in the work environment.

Absorption

Penetration through the skin can occur quite rapidly if the skin is cut or abraded. Intact skin, however, offers a reasonably good barrier to chemicals. Unfortunately, many compounds can be absorbed through intact skin. Some substances are absorbed by way of the openings for hair follicles and others dissolve in the fats and oils of the skin, such as organic lead compounds, many nitro compounds, and organic phosphate pesticides. Compounds that are good solvents for fats (such as toluene and xylene) also can cause problems by absorption through the skin.

Many organic compounds, such as cyanides, and most aromatic amines, amides, and phenols can produce systemic poisoning by direct contact with the skin. Absorption of toxic chemicals through the skin and eyes is the next most important route of entry after inhalation.

Ingestion

In the workplace, people may unknowingly eat or drink harmful chemicals if they do not wash themselves before eating or if they store drinking containers in the workplace. Toxic compounds are capable of being absorbed from the gastrointestinal tract into the bloodstream. Lead oxide can cause serious problems if people working with this material are allowed to eat or smoke in work areas. In this situation, careful and thorough washing is required both before eating and at the end of every shift.

Inhaled toxic dusts can also be ingested in amounts that may cause trouble. If the toxic dust swallowed with food or saliva is not soluble in digestive fluids, it is eliminated directly through the intestinal tract. Toxic materials that are readily soluble in digestive fluids can be absorbed into the blood from the digestive system.

Injection

It is possible using force such as compressed air or contaminated sharp objects to inject a hazard into the body. An example would be the injection of the AIDS virus into a hospital worker from a contaminated needle. It is important that an industrial hygienist studies all routes of entry when evaluating the work environment (candy bars or lunches in work area, solvents used to clean work clothing and hands, in addition to air contaminants in working areas).

TYPES OF AIR CONTAMINANTS

There are precise meanings for certain words commonly used in industrial hygiene. These must be used correctly to understand the requirements of OSHA's regulations; effectively communicate with other workers in the field of industrial hygiene; and intelligently prepare purchase orders to procure health services and personal protective equipment.

For example, a fume respirator is worthless as protection against gases or vapors. Too frequently, terms (such as gases, vapors, fumes, and mists) are used interchangeably. Each term has a definite meaning and describes a certain state of matter. Air contaminants are commonly classified as either particulate contaminants, or gas and vapor contaminants. The most common particulate contaminants include dusts, fumes, mists, and fibers.

Particulates

- Dusts are solid particles generated by handling, crushing, grinding, rapid impact, detonation, and decrepitation (breaking apart by heating) of organic or inorganic materials, such as rock, ore, metal, coal, wood, and grain. Dust is a term used in industry to describe airborne solid particles that range in size from 0.1 to 25 μm. One micrometer is a unit of length equal to one millionth of a meter. A micrometer is also referred to as a micron and is equal to 1/25,400 of an inch. Dust can enter the air from various sources, such as the handling of dusty materials or during processes such as grinding, crushing, blasting, and shaking. Most industrial dusts consist of particles that vary widely in size, with small particles greatly outnumbering the large ones. Consequently (with few exceptions), when dust is noticeable in the air near a dusty operation, probably more invisible dust particles than visible ones are present. A process that produces dust fine enough to remain in the air long enough to be breathed should be regarded as hazardous until proven otherwise. An airborne dust of a potentially toxic material will not cause pulmonary illness if its particle size is too large to gain access to the lungs. Particles 10 μm in diameter and larger are known as nonrespirable. These particles will be deposited in the respiratory system long before they reach the alveolar sacs—the most important area in the lungs. Particles less than 10 μm in diameter are known as respirable. Since these particles are likely to reach the alveoli in great quantities, they are potentially more harmful than larger particles. By using a size-selective device (such as a cyclone) ahead of a filter at a specific airflow sampling rate, it is possible to collect respirable-sized particles on the filter. This allows one to determine the dust concentration of respirable particles.
- Fumes are formed when the material from a volatilized solid condenses in cool air. The solid particles that are formed make up a fume that is extremely fine—usually less than 1.0 μm in diameter. In most cases, the hot vapor reacts with the air to form an oxide. Gases and vapors are not fumes, although the terms are often mistakenly used interchangeably. Welding, metalizing, and other operations involving vapors from molten metals may produce fumes; these may be harmful under certain conditions. Arc welding volatilizes metal vapor that condenses—as the metal or its oxide—in the air around the arc. In addition, the rod coating is partially volatilized. These fumes, because they are extremely small, are readily inhaled. Other toxic fumes—such as those formed when welding structures that have been painted with lead-based paints, or when welding galvanized metal—can produce severe symptoms of toxicity rather rapidly in the absence of good ventilation or proper respiratory protection.
- Mists are suspended liquid droplets generated by condensation of liquids from the vapor back to the liquid state or by breaking up a liquid into a dispersed state, such as by splashing or atomizing. The term *mist* is applied to a finely divided liquid suspended in the atmosphere. Examples include oil mist produced during cutting and grinding operations, acid mists from electroplating, acid or alkali mists from pickling operations, and spray mist from spray finishing operations.
- Fibers are solid particles that have a slender, elongated structure with length several times as great as their diameter. Examples include asbestos, fibrous talc, and fiberglass. Airborne fibers may be found in construction activities, mining, friction product manufacturing and fabrication, and demolition operations.

Gases and Vapors

- Gases are formless fluids that expand to occupy the space or enclosure in which they are confined. They are a state of matter in which the molecules are unrestricted by cohesive forces. Examples are arc-welding gases, internal-combustion engine exhaust gases, and air.
- Vapors are the volatile form of substances that are normally in a solid or liquid state at room temperature and pressure. Evaporation is the process by which a liquid is changed into the vapor state and

mixed with the surrounding atmosphere. Some of the most common exposures to vapors in industry occur from organic solvents. Solvents with low boiling points readily form vapors at room temperature. Solvent vapors enter the body mainly by inhalation, although some skin absorption can occur.

EXPOSURE MONITORING

The role of monitoring is to tell you what contaminants are present and at what levels. Yet the limitations of many instruments mean that you cannot be sure of the readings unless all parameters are taken into consideration or you already know what is in the air. This seems to be a contradiction. After all, how can you know what is present if the instruments cannot tell you? Often, determining contaminant levels is possible only after extensive diagnostic work with a variety of sampling strategies. Air sampling instruments can provide very important information to clarify the hazards at a construction site. Monitoring surveys can help answer questions like

- What types of air contaminants are present?
- What are the levels of these contaminants?
- How far does the contamination range?
- What type of protective gear is needed for the workers?

Effective monitoring can be difficult work. It is much more than pushing buttons on a high-tech gadget. As you will see, it is more like detective work. The issues fall into three major categories:

- What are the limitations of the instruments used?
- What strategy should be used to get useful information?
- How do you evaluate the results that you get?

There are two types of air monitoring methods: (1) direct reading and (2) laboratory sampling. Direct-reading instruments have built-in detectors to give on the spot results. However, there is a trade-off between sophistication and the weight of the unit. The instruments must be truly portable to be useful. Therefore, it is important to be aware that there are limits to any given instrument.

The emphasis in laboratory sampling is on collecting a sample in the field and conducting the actual analysis later back at the laboratory. The disadvantage is the delay in obtaining results. An advantage is that the instruments in the laboratory do not have to be portable. They can be large and more sophisticated for more precise analysis. For example, laboratories can use an instrument known as a gas chromatograph and analyze a mixture of five different chemicals and separate them so that each can be examined separately. Laboratory procedures can also use computers to compare analytical results to known chemical properties. These unique properties serve to allow identification similar to fingerprinting.

It is common to use both types of procedures to investigate exposures in the workplace. Direct-reading methods are ideal for quick checks especially when the contaminants are known or suspected. However, they are limited in accuracy. No instrument can read every contaminant. Two common instruments, organic vapor analyzer (OVA) and the photoionizer (HNU), can detect hundreds of compounds but cannot detect important toxic chemicals such as phosgene, cyanides, arsenic, or chlorine.

Another example of a direct-reading instrument is an air pump and detector tubes, which are simple but important direct-reading instruments. There is a wide range of detector tubes for gases but accuracy is only about ±25%. The detectable range for each type of tubes must be reviewed carefully in addition to the number of strokes or the amount of sample needed. It is important to be aware of monitoring limits of any instrument. Most direct-reading instruments respond to several chemicals. For example, benzene detector tubes give the same response for the related chemicals toluene, xylene, and ethyl benzene.

You can begin to see that while direct-reading instruments can give you numbers on the spot, it takes longer to determine the actual amount of a substance present and determine the hazard to workers. You have to go through several steps to identify the chemical and then additional steps can be taken to determine the actual level of contaminants.

Calibrating direct-reading instruments is also an important step in getting accurate measurements. Calibration is the term used to describe checking the instrument response against a known source. This check is critical to ensure accuracy. Instruments can drift because of low batteries, rough handling, and several other factors. An uncalibrated instrument is like a clock that is 20 min too slow. It still works; however, it is just not accurate. An instrument is generally calibrated to see if it reads zero with no contaminant present and the correct amount with a known level of gas. For example, an OVA is calibrated with 0 and 100 ppm of methane. Sometimes special calibration is needed. For example, an oxygen meter must be calibrated for air pressure, due to different readings at extremes such as sea level or high elevations. Sometimes instruments are calibrated with different chemicals to aid in determining the level of a given chemical.

Care must be taken in using monitoring data for decision making about personal protective equipment. Higher levels of protection are needed on the job site at the early stages when only general information about exposures is known. Only when contaminants are further identified and exposure levels are more precisely known can the level of protective equipment be confidently lowered and the job site classified into various hazard areas. More accurate monitoring usually requires samples to be collected for laboratory analysis. Even when monitoring seems to have validity, it is important to realize that there is no guarantee that exposures will stay the same day after day especially on a construction site.

Worker exposures are influenced by several factors:

1. Change in location—Contaminants are not evenly distributed at most worksites. One area may have more solvents and fewer metals than another area. Monitoring often must be done when work is initiated due to the rapid change of conditions on a construction site.
2. Change in operation—Exposures tend to vary with jobs. Bagging out asbestos material will have a different exposure potential than removing the asbestos from the ceiling.
3. Site and environmental conditions—Construction work outside will have exposures that are variable with the wind and immediate weather conditions. Inside work will be more consistent with ventilation systems and the type of enclosure. Temperature, season, and even rainfall can affect contaminant exposures.
4. Mishaps—Leaks and spills can have obvious effects on exposure levels.

Because of these factors, the worker-exposure monitoring job is never done. It must be done on a periodic basis over the course of the entire job. Such is the case with asbestos and lead abatement work where continuous monitoring is required. Two major categories of samples collected to draw an exposure profile are normally analyzed by laboratories and are called area and personal. In general, direct-reading instruments are used to obtain area or background samples, and personal samples are obtained with laboratory-based analysis methods.

- Area samples are obtained in a given location. For example, a confined space might be checked for contaminants or oxygen level. Area samples are collected to verify background levels such as asbestos outside a regulated area. Sometimes high background levels prevent achieving clearance levels for reuse of space. Area samples are also a valuable tool in locating contaminant movement and documenting worst-case scenarios.
- Personal samples are obtained to determine a worker's exposure level without regard to respiratory equipment. They give the most accurate profile of the worker's daily exposure level. An air monitoring pump, drawing the same amount of air as a normal breath, is typically worn for the work shift and the results compared with an 8 h TWA permissible exposure limit (PEL) established by OSHA. It is not necessary to monitor every worker to obtain a valid exposure profile. Each type of job in an exposure area should be monitored.

Instead, workers who are representative of a typical job are usually sampled. It is best to choose those who are expected to have the highest exposure. Since occupational exposures are affected the most by worker activity, this type of sampling is typically done after work begins. Personal monitoring samples are typically taken in the worker's breathing zone, which is the area directly outside the respirator face piece within 1 ft. of the nose.

Immediately dangerous to life or health (IDLH) sampling is done at the beginning of a hazardous job and at appropriate periods throughout the job. This sampling is conducted to answer the question, Are dangerous conditions present? Personnel performing this type of work must wear appropriate personal protective equipment. Sampling should include worst-case conditions and should be conducted on the actual approach to worst-case conditions to give the monitoring person a degree of warning. A good example of this is a confined space with an oxygen level under 19.5%.

Worker exposure monitoring produces numbers. These numbers must be evaluated to be useful in decision-making. The skill and judgment used by an IH are critical. Interpreting the numbers correctly directly affects the health of workers and the profitability of a project. Several organizations either recommend or enforce exposure limits such as the National Institute for Occupational Safety and Health, American Conference of Governmental Industrial Hygienists (ACGIHs), and OSHA.

UNITS OF CONCENTRATION

In addition to the definitions concerning states of matter that find daily usage in the vocabulary of the IH, other terms used to describe degree of exposure include the following:

- ppm: This means parts per million parts of contaminated air on a volumetric basis. It is used for expressing the concentration of a gas or vapor.
- mg/m^3: This means milligrams of a substance per cubic meter of air. The term is most commonly used for expressing concentrations of dusts, metal fumes, or other particles in the air.
- mppcf: This means millions of particles of a particulate per cubic foot of air. This term is not widely used today.
- f/cc: This means the number of fibers per cubic centimeter of air. This term is used for expressing the concentration of airborne asbestos fibers.

The health and safety professional recognizes that air contaminants may exist as gases, dust, fumes, mist, or vapors in the workplace air. In evaluating the degree of exposure, the measured concentration of the air contaminant is compared to limits or exposure guidelines.

EXPOSURE GUIDELINES

Threshold Limit Values

Threshold limit values (TLVs) have been established for airborne concentrations of many chemical compounds. It is important to understand something about TLVs and the terminology in which their concentrations are expressed. The ACGIH annually publishes a list of TLVs and biological exposure indices. The lists are reviewed annually and values are updated as relative data become available. The ACGIH is not an official government agency. Membership is limited to professional personnel in government agencies or educational institutions engaged in occupational safety and health programs.

The data for establishing TLVs come from animal studies, human studies, and industrial experience, and the limit may be selected for several reasons. It may be because a substance is very irritating to the majority of people exposed, or other substances may be asphyxiants. Other reasons for establishing a TLV include the fact that certain chemical compounds are anesthetic, fibrogenic, or can cause allergic reactions or malignancies. Some additional TLVs have been established because exposure above a certain airborne concentration is a nuisance.

The basic idea of TLVs is fairly simple. They refer to airborne concentrations of substances and represent conditions under which it is believed that nearly all workers may be repeatedly exposed, day after day, without adverse effect.

Because individual susceptibility varies widely, an occasional exposure of an individual at (or even below) the threshold limit may not prevent discomfort, aggravation of a preexisting condition, or occupational illness. In addition to the TLVs set for chemical compounds, there are limits for physical agents, such as noise, microwaves, and heat stress.

Several important points should be noted concerning TLVs. First, the term TLV is a copyrighted trademark of the ACGIH. It should not be used to refer to the values published in OSHA or other standards. OSHA's limits are known as PELs and will be discussed later. The ACGIH TLVs are not mandatory federal or state employee exposure standards. These limits are not fine lines between safe and dangerous concentrations nor are they a relative index of toxicity.

Three categories of TLVs are specified and are as follows:

- Time-weighted average (TLV–TWA) is the TWA concentration for a normal 8 h workday or 40 h workweek, to which nearly all workers may be repeatedly exposed, day after day, without adverse effect. TWAs permit excursions above the limit provided they are time compensated by equivalent excursions below the limit during the workday.
- Short-term exposure limit (TLV–STEL) is the maximal concentration to which workers can be exposed continuously for a short period of time without suffering from any of the following:
  1. Irritation
  2. Chronic or irreversible tissue change
  3. Narcosis of sufficient degree to increase accident proneness, impair self-rescue, or materially reduce work efficiency
- The STEL is a 15 min TWA exposure, which should not be exceeded at any time during a workday, even if the 8 h TWA is within the TLV–TWA. Exposures above the TLV–TWA up to the STEL should not be longer than 15 min and should not occur more than four times per day. There should be at least 60 min between successive exposures in this range. The STEL is not a separate independent exposure limit; rather it supplements the TWA limit where there are recognized acute effects from a substance whose toxic effects are primarily of a chronic nature. STELs are recommended only where toxic effects have been reported from high short-term exposures in either humans or animals.
- Ceiling (TLV–C) is the concentration that should not be exceeded even instantaneously. Although the TWA concentration provides the most satisfactory, practical way of monitoring airborne agents for compliance with the limits, it is inappropriate for certain substances. Substances that are predominantly fast acting and whose threshold limit is more appropriately based on this particular response are in the latter group. Substances with this type of response are best controlled by a ceiling C limit that should not be exceeded. For some substances, for example, irritant gases, only one category, the TLV–C, may be relevant. For other substances, either two or three categories may be relevant, depending upon their physiologic action. It is important to observe that if any one of these three TLVs is exceeded, a potential hazard from that substance is presumed to exist.

Skin Notation

Nearly one-quarter of the substances in the TLV list are followed by the designation skin. This refers to potential significant contribution to the overall exposure by the cutaneous route, including mucous membranes and the eyes, usually by direct contact with the substance. This designation is intended to suggest appropriate measures for the prevention of cutaneous absorption.

OSHA Exposure Limits

The first compilation of health and safety standards promulgated by the Department of Labor's OSHA in 1970 was derived from the then existing federal standards and national consensus standards. Thus, many of the 1968 TLVs established by the ACGIH became federal standards or PELs. In addition, certain workplace quality standards known as maximal acceptable concentrations of the American National Standards Institute (ANSI) were incorporated as federal health standards in 29 CFR 1910.1000 as national consensus standards. These PEL values for general industry were subsequently updated in 1989.

Unlike the TLVs, OSHA's PELs are enforceable by law. Employers must keep employee exposure levels below the PELs of regulated substances. As with TLVs, there are three types of PELs. The most common is the 8 h TWA. The others are the STEL and the ceiling limit (C).

Time-Weighted Average

In adopting the TLVs of the ACGIH, OSHA also adopted the concept of the TWA concentration for a workday. The 8 h TWA is the average concentration of a chemical in air over an 8 h exposure period.

In general

$$\text{TWA} = \frac{C_a T_a + C_b T_b + \cdots + C_n T_n}{8}$$

where
T_a is the time of the first exposure period
C_a is the concentration of the contaminant in period a
T_b is another time period during the shift
C_b is the concentration during period b
C_n is the concentration during the nth time period
T_n is the nth time period

To illustrate the formula described above, assume that a substance has an 8 h TWA PEL of 100 ppm. Assume that an employee is subject to the following exposure:

Two hours exposure at 150 ppm
Two hours exposure at 75 ppm
Four hours exposure at 50 ppm

Substituting this information in the formula, we have

$$\text{TWA} = \frac{(150)(2) + (75)(2) + (50)(4)}{8} = 81.25 \text{ ppm}$$

Since 81.25 ppm is less than 100 ppm, the 8 h TWA limit, the exposure is acceptable.

WHEN YOU NEED AN IH

You will need to have the knowledge and courage to realize when you need to call an IH to help you. IHs have very special and specific training related to workplace environmental evaluations and assessments as well as the ability to make recommendations on controlling workplace hazards.

An IH concerned about exposure hazards associated with your workplace must be familiar with the various activities and processes that you have. The classic approach of recognition, evaluation, and control strategies used by IH applies to all industries. Sometimes exposures can be attributed to the job. For example, for a worker using a solvent to clean a piece of mechanical equipment, the IH may need to investigate organic vapor exposure, correct personal protective equipment use, surrounding environment, and possibly personal hygiene conditions.

Hazards involving normal work activities can usually be predicted by a trained IH. It is, however, very unpredictable how much airborne exposure a worker is subjected to from a particular source. Many times the same type of work conducted at one site is much different from an exposure condition at another. Inside exposures will remain more constant than outside where wind and weather conditions play a major role. For example, asbestos abatement work that is conducted in a controlled atmosphere inside should remain fairly constant if work practices such as negative air filtration are used and surfaces are wetted properly. Conversely, work on an asbestos roof on the outside, even though there is a difference in the type of asbestos, will depend more on weather conditions. Work practices such as location of the worker in relationship to the wind (up- or down-stream) and how intact the shingles are as they are removed also play an important part in overall exposure. The more broken up they are the more likely an asbestos exposure will result. Although inside exposures sometimes can vary vastly with the size of an area and individual work practices, it is not usually expected to be that way.

If the airborne exposure is to be determined for a particular job, the IH must be prepared to monitor quickly. The next day may be too late. Concentrations usually need to be high to find TWAs that exceed OSHA PELs. More often than not the construction worker is not conducting the same job for an 8 h period. Many tasks are usually required to accomplish a day's work, which also makes it difficult to evaluate a particular hazard. A worker welding, cutting, and burning all day on an outside project such as a painted bridge may have no exposure or wind up in the hospital undergoing chelation therapy with a blood lead level in the hundreds. Many variables affect the potential and real exposure levels such as work habits, weather, and type of paint on the steel as well as personal protective equipment used.

It is most appropriate to consult an IH when selecting personal protective equipment for a specific use such as which gloves are best for use with certain chemicals and which respirator should be used for exposure to a specific chemical. The IH is the only one who has the training and experience to determine the risk for exposure, the environmental sampling that is needed, the sampling techniques to use, and the controls that should be in place to prevent further exposure.

CHAPTER 13

Taking Action: Intervention, Controls, and Prevention

For failure to use controls one died and one lived. (Courtesy of Mine Safety and Health Administration.)

Before you start reading this chapter, please take a moment to circle the number of the answer option that best fits the state of your safety and health initiative at this time.

HAZARD PREVENTION AND CONTROLS

Occupational Safety and Health Administration (OSHA) require employers to protect their employees from workplace hazards such as machines, work procedures, and hazardous substances that can cause injury or illnesses. It is known from past practices and situations that something must be done to mitigate or remove hazards from the workplace. Actions taken often create other hazards, which had not existed before attempting to address the existing hazard.

Many companies have suggestion programs where workers receive rewards for suggestions that are implemented. It is no secret to anyone that the person who often has the best ideas on how to decrease or remove a hazard is the one who faces that hazard as part of doing normal work. Involve those who are impacted most in decision-making processes that affect their work is a sound management process.

Many ways to control hazards have been used over the years but usually these can be broken down into five primary approaches. The preferred ways to do this are through engineering controls,

Hazard Prevention and Control

Topic	Circle Answer	Answer Options
Timely and effective hazard control	5	Hazard controls are fully in place, known to and supported by workforce, with concentration on engineering controls and safe work procedures
	4	Hazard controls are fully in place with priority to engineering controls, safe work procedures, administrative controls, and personal protective equipment (PPE) (in that order)
	3	Hazard controls are fully in place, but there is some reliance on PPE
	2	Hazard controls are generally in place, but there is heavy reliance on PPE
	1	Hazard control is not complete, effective, and appropriate
Facility and equipment maintenance	5	Operators are trained to recognize maintenance needs and perform and order maintenance on schedule
	4	An effective preventive maintenance schedule is in place and applicable to all equipment
	3	A preventive maintenance schedule is in place and is usually followed except for higher priorities
	2	A preventive maintenance schedule is in place but is often allowed to slide
	1	There is little or no attention paid to preventive maintenance; breakdown maintenance is the rule

awareness devices, predetermined safe work practices, and administrative controls. When these controls are not feasible or do not provide sufficient protection, an alternative or supplementary method of protection is to provide workers with PPE and the knowledge to use it properly.

ENGINEERING CONTROLS

When a hazard is identified in the workplace, every effort should be made to eliminate it so that employees are not harmed. Elimination may be accomplished by designing or redesigning a piece of equipment or process. This could be the installation of a guard on a piece of machinery, which prevents workers from contacting the hazard. The hazard can be engineered out of the operation. Another way to reduce or control the hazard is to isolate the process, such as in the manufacture of vinyl chloride used to make such items as plastic milk bottles, where the entire process becomes a closed circuit. The results are no one is exposed to vinyl chloride gas, which is known to cause cancer. Thus, any physical controls that are put in place are considered to be the best approach from an engineering perspective. Keep in mind that you are a consumer of products. Thus, at times you can leverage the manufacturer to implement safeguards or safety devices on products that you are considering for purchase. Let your vendor do the engineering for you or do not purchase their product. This may not always be a viable option. The following are the engineering controls that can be used:

- Substitution
- Elimination
- Ventilation
- Isolation (see Figure 13.1)
- Process or design change

Figure 13.1 Using an enclosure to isolate workers from the machinery.

Presence-Sensing Devices

Machine safeguarding, presence-sensing devices (PSD) control machine operation and protect personnel by detecting the presence of objects that could interfere with or be harmed by machinery. Activating a machine guard, a presence-sensing device sounds a warning, ends a process, initiates a different machine action, or stops the machine altogether. Several types of guarding systems exist such as light curtains, safety laser scanners, safety mats, safety edges, proximity guarding, and area guarding,

Machine safeguarding uses many different technologies for presence-sensing devices. Contact strips and bumpers contain pressure-sensitive edges that stop the machine when contact is sensed. By contrast, emergency stop (e-stop) switches are activated by pushing a button or flipping a switch. Cable pull switches are also available. Radio frequency (RF) presence sensors transmit a high-frequency signal that, when disturbed, stops the machine or process. Similarly, ultrasonic sensors use high-frequency sound to monitor processes or machines. Laser presence sensors are used to sweep large areas and detect foreign objects. Safety mats are suitable for guarding the floor near a machine or robot. Light curtains use a photoelectric transmitter to generate multiple beams of light. When an opaque object interrupts one or more beams, the light curtain controller sends a stop signal to the guarded machine. Safety gates are movable barriers that protect operators before the machine cycle can be started. Intrinsic safety (IS) barriers are devices that limit the current, voltage, or total energy delivered to a sensor in a hazardous area or flammable environment. Safety interlock switches are available in a variety of form factors and with both contact and noncontact activation.

Machine safeguarding using presence-sensing devices is available for many different types of equipment. Examples include augers, presses, balers, cutters, crushers, rollers, mixers, drive shafts, conveyors, and feed systems. Some machine safeguarding, presence-sensing devices meet machine safeguarding requirements from the U.S. OSHA. Other machine safeguarding, presence-sensing devices comply with the B11 series of safety standards from the American National Standard Institute (ANSI). For example, the ANSI B.11.1.-2001 machinery safety standard covers mechanical power presses.

Various types of presence-sensing devices can be used to safeguard many different kinds of machines as per OSHA 29 CFR 1910.212 and 1910.217. These devices are widely used on robots, power presses, textile manufacturing equipment, and many other industrial machines. Presence-sensing devices provide maximum visibility of a machine's danger zone. This is desirable in most point-of-operation machine guarding. However, accident records and safety analyses of RF presence-sensing devices show that improper application, use, selection, design, or mounting of these devices can result in serious accidents. It is recommended that RF presence-sensing devices not be used as the primary safeguards or only safeguards for workers exposed to serious hazards.

The basic function of an RF presence-sensing device is to interrupt a machine's operating cycle when an object is detected within the device's sensing field. For the device to be effective, it must

- Be installed and tuned so that effective sensing occurs at least at a minimum safe distance
- Be able to detect penetration of a minimum hand size
- Stop or prevent the operation of a machine as long as any part of an employee's body is within the machine's danger zone
- Be interlocked with other guarding devices or have fixed guards to prevent human access to a machine's danger zone except through the sensing field
- Protect exposed employees during all phases of machine operation including start-up, shutdown, and maintenance
- Generate no false safe-to-operate signals and ignore received false safe-to-operate signals due to component malfunction or environmental conditions such as unexpected electrical or magnetic fields, humidity, and moisture

An RF presence-sensing device may not be able to provide effective protection if

- It is not installed and tuned so that its sensing field is at a safe distance
- Penetrating object is too small to be detected in time by the sensing field
- A penetrating object is between the field and the danger zone
- Its source of power fails while the machine remains operative
- It is installed on equipment not designed to stop upon field interruption at any point of the machine's operating cycle, such as full revolution presses

Capacitance proximity sensing (CPS) devices (RF presence-sensing devices) consist of a control unit, a coupler, a sensor antenna, and a cable connecting the control unit to the sensor. The antenna should have a geometrical shape that makes the generated sensing field form an envelope around the space where human detection is desired. This sensor antenna is made of electrical conducting material and is isolated from the ground. An oscillating electrical signal in the RF range of 150–400 kHz is generated in the control unit and sent through the cable and coupler to the sensor. At the sensor antenna the signal creates an electric potential and generates an electro-magnetic field. The developed field can be much larger than the volume defined by the antenna. The shape of the field is affected by the location of objects within its boundaries and by shielding at the sensor and is not normally well defined. Physical changes in the space surrounding the antenna alter this electromagnetic field. An electronic bridge circuit incorporating an adjustable reference capacitance is used to detect changes in the field capacitance. At a tuned level, the bridge becomes unbalanced, thus triggering machine stoppage.

Studies have shown that well-grounded conductors (people) are more easily detected by CPS system because a grounded conductor causes a greater change in capacitance in a CPS detection bridge than a weakly grounded conductor. In fact, ungrounded conductors may not be detected and poorly grounded objects must penetrate the field more deeply for CPS devices to detect the presence of such objects. Attention should be paid to potential problem situations such as those where employees can move from ground to an insulated platform while remaining in an antenna's detection space. Detectable changes of the sensing field in the presence of an operator may diminish when the operator is isolated from the ground. Intermittent energizing of outside electric fields (such as those from electric motors or power supplies) and the movement of objects into the sensing field (such as maintenance gear) should also be checked to ensure that they do not change the capacitance of the CPS system. Changes in machine tool or die layout may also adversely affect the detection field.

CPS devices employ a sensitivity control as an operating adjustment to compensate for changes in the field due to added or subtracted shielding such as tooling and support hardware. This adjustment has the effect of moving an effective sensing point toward or away from a machine's danger zone. Unfortunately, this can permit effective sensing at less than a safe distance. In addition, holes in the sensing field can be created if the effective sensing point is collapsed to the antenna. In this case the system might not detect the presence of an operator's arm in the sensing field. The safety distance between the sensing plane and the danger point cannot be set upon initial installation, but must be reset for each application.

Specific standards have not been adopted for the installation, adjustment, and maintenance of RF presence-sensing devices. Employers using such devices should have had safety hazard analyses performed to determine potential failures or limitations of the devices in their applications. Employers should be aware that the performance of a CPS device may be affected as environmental and physical conditions are changed.

In 1979, the National Institute for Occupational Safety and Health (NIOSH) issued a warning on the use of RF presence-sensing devices. The warning emphasizes that the following procedures be followed:

- Care must be taken to set a device's sensitivity for the operator's grounded condition and allow for variations in the operator's working position during a workday.
- Supervisors and workers should avoid drastic grounding changes once sensitivity is set and a device is in use. This includes changes in what the operator is standing or seated on.
- Manufacturer's instructions should be followed carefully in the construction and placement of RF antennas.
- Manufacturer of a device should be consulted about specific problems, such as extreme sensitivity variations in particular machine applications.

Because of the uncertainties involved with the ability of RF detectors to measure intrusion accurately due to ground changes, field interference, and other physical and environmental conditions, CPS devices should not be used as the primary or only safeguard to protect workers who are frequently exposed to a serious hazard. When used for applications involving infrequent human intervention such as perimeter guarding, CPS devices may provide adequate protection. Changes in sensing distance may not be critical if there is no need for an operator to normally reach into the danger zone. A combination of a CPS device and an automatic feeding device or other safe guarding method provides better protection.

There are uses for PSDI as safeguarding devices, and they should be carefully evaluated to assure that they meet regulations for worker protection and are able to control the machine in a safe manner. This technology and new devices should be employed in safeguarding if possible.

AWARENESS DEVICES

Awareness devices are linked to the senses. They are warning devices that can be heard and seen. They act as alerts to workers, but create no type of physical barrier. They are found in most workplaces and carry with them a moderate degree of effectiveness. Such devices are

- Backup alarms
- Warning signals both audible and visual
- Warning signs

WORK PRACTICES

Work practices concern the ways in which a job task or activity is done. This may mean that you create a specific procedure for completing the task or job. It may also mean that you implement special training for a job or task. It also presupposes that you might require inspection of the equipment or machinery before beginning work or when a failure occurs. An inspection should be done before restarting the process or task. It may also require that a lockout/tagout procedure be used to create a zero potential energy release.

ADMINISTRATIVE CONTROLS

A second approach is to control the hazard through administrative directives. This may be accomplished by rotating workers, which allows you to limit their exposure, or having workers work only in areas where no hazards exist during that part of their shift. This applies particularly to chemical exposures and repetitive activities that could result in ergonomics-related incidents. Examples of administrative controls are

- Requiring specific training and education
- Scheduling off-shift work
- Worker rotation

Management Controls

Management controls are needed to express the company's view of hazards and their response to hazards that have been detected. The entire program must be directed and supported through the management controls. If management does not have a systematic and set procedure in place for addressing the control of hazards, the reporting/identifying of hazards is a waste of time and money. This goes back to the policies and directives and the holding of those responsible accountable by providing them with the resources (budget) for correcting and controlling hazards. Some aspects of management controls are

- Policies
- Directives
- Responsibilities (line and staff)
- Vigor and example
- Accountability
- Budget

Your attempt to identify your worksite hazards and address them should be an integral part of your management approach. If the hazards are not addressed in a timely fashion, they will not be identified or reported. If dollars become the main reason for not fixing or controlling hazards, you will lose the motivation of your workforce to identify or report them.

PERSONAL PROTECTIVE EQUIPMENT

PPE includes a variety of devices and garments to protect workers from injuries. You can find PPE designed to protect eyes, face, head, ears, feet, hands and arms, and the whole body. PPE includes such items as goggles, face shields, safety glasses, hard hats, safety shoes, gloves, vests, earplugs, earmuffs, and suits for full-body protection (see Figure 13.2).

In November 2007, OSHA announced a final rule regarding employer-paid PPE. Under the rule, all PPE, with a few exceptions, will be provided at no cost to the employee. OSHA anticipates that this

Figure 13.2 Different types of PPE for hearing protection.

rule will have substantial safety benefits, which will result in more than 21,000 fewer occupational injuries per year.

Employees exposed to safety and health hazards may need to wear PPE to be protected from injury, illness, and death caused by exposure to those hazards. This final rule will clarify who is responsible for paying for PPE, which OSHA anticipates will lead to greater compliance and potential avoidance of thousands of workplace injuries each year.

The final rule contains a few exceptions for ordinary safety-toed footwear, ordinary prescription safety eyewear, logging boots, and ordinary clothing and weather-related gear. The final rule also clarifies OSHA's requirements regarding payment for employee-owned PPE and replacement PPE.

The final rule does not create new requirements regarding what PPE employers must provide. It does not require payment for uniforms, items worn to keep clean, or other items that are not PPE. The final rule contains exceptions for certain ordinary protective equipment, such as safety-toed footwear, prescription safety eyewear, everyday clothing and weather-related gear, and logging boots.

The final rule also clarifies OSHA's intent regarding employee-owned PPE and replacement PPE:

- It provides that, if employees choose to use PPE they own, employers will not need to reimburse the employees for the PPE. The standard also makes clear that employers cannot require employees to provide their own PPE and the employee's use of PPE they already own must be completely voluntary. Even when an employee provides his or her own PPE, the employer must ensure that the equipment is adequate to protect the employee from hazards at the workplace.
- It also requires that the employer pay for replacement PPE used to comply with OSHA standards. However, when an employee has lost or intentionally damaged PPE, the employer is not required to pay for its replacement.

The final rule requires employers to pay for almost all PPE that is required by OSHA's general industry, construction, and maritime standards. Employers already pay for approximately 95% of these types of PPE.

Hazard Assessment

Recent regulatory requirements make hazard analysis/assessment part of the PPE selection process. Hazard analysis and assessment procedures shall be used to assess the workplace to determine if hazards are present, or are likely to be present, which may necessitate the use of PPE. As part of this assessment, employees' work environment is to be examined for potential hazards, both health and physical, that are likely to present a hazard to any part of their bodies. If it is not possible to eliminate workers' exposure or potential exposure to the hazard through the efforts of engineering controls, work practices, and administrative controls, then proper PPE will need to be selected, issued, and worn. The hazard assessment certification form found in Figure 13.3 may be of assistance in conducting a hazard analysis and assessment.

Hazard assessment certification form

Date:	Location:

Assessment conducted by:

Specific tasks performed at this location:

Hazard assessment and selection of personal protective equipment

I. Overhead hazards
Hazards to consider include:
- Suspended loads that could fall
- Overhead beams or loads that could be hit against
- Energized wires or equipment that could be hit against
- Employees work at elevated site who could drop objects on others below
- Sharp objects or corners at head level

Hazards Identified:

Head protection

Hard hat:	Yes	No

If yes, type:

☐ **Type A** (impact and penetration resistance, plus low-voltage electrical insulation)

☐ **Type B** (impact and penetration resistance, plus high-voltage electrical insulation)

☐ **Type C** (impact and penetration resistance)

II. Eye and face hazards
Hazards to consider include:
- Chemical splashes
- Dust
- Smoke and fumes
- Welding operations
- Lasers/optical radiation
- Projectiles

Figure 13.3 Hazard assessment certification form. (Courtesy of Occupational Safety and Health Administration.)

Hazards identified:

Eye protection

Safety glasses or goggles	Yes	No
Face shield	Yes	No

III. Hand hazards
Hazards to consider include:
- Chemicals
- Sharp edges, splinters, etc.
- Temperature extremes
- Biological agents
- Exposed electrical wires
- Sharp tools, machine parts, etc.
- Material handling

Hazards identified:

Hand protection

Gloves	Yes	No
☐ Chemical resistant ☐ Temperature resistant ☐ Abrasion resistant ☐ Other (explain) ☐		

IV. Foot hazards
Hazards to consider include:
- Heavy materials handled by employees
- Sharp edges or points (puncture risk)
- Exposed electrical wires
- Unusually slippery conditions
- Wet conditions
- Construction/demolition

Hazards identified:

Figure 13.3 (continued)

(continued)

Foot protection		
Safety shoes	Yes	No
Types: ☐ Toe protection ☐ Metatarsal protection ☐ Puncture resistant ☐ Electrical insulation ☐ Other (explain)		

V. Other identified safety and/or health hazards:

Hazard	Recommended protection

I certify that the above inspection was performed to the best of my knowledge and ability, based on the hazards present on _____ .

_____
(Signature)

Figure 13.3 (continued)

When employees must be present and engineering or administrative controls are not feasible, it will be essential to use PPE as an interim control and not as a final solution. For example, safety glasses may be required in the work area. Too often PPE usage is considered the last thing to do in the scheme of hazard control. PPE can provide added protection to the employee even when the hazard is being controlled by other means. The drawbacks to the use of PPE are

- Hazard still looms
- Protection dependent upon worker using PPE
- PPE may interfere with performing task and productivity
- Requires supervision
- Is an ongoing expense

Many forms of PPE need to be addressed and required when a hazard assessment determines that PPE is the only option left for protecting the workforce. PPE includes the following:

- Eye and face protection (29 CFR 1910.133)
- Respiratory protection (29 CFR 1910.134)
- Head protection (29 CFR 1910.135)
- Foot and leg protection (29 CFR 1910.136)
- Electrical protective equipment (29 CFR 1910.137)
- Hand protection (29 CFR 1910.138)
- Respiratory protection of tuberculosis (29 CFR 1910.139)

Any other types of specialized protective equipment needed would be identified as part of the hazard assessment. Such equipment might include body protection for hazardous materials, protective equipment for material handling, protection for welding activities, or protection from exposure to biological agents.

Establishing a PPE Program

A PPE program sets out procedures for selecting, providing, and using PPE as part of an organization's routine operation. A written PPE program, although not mandatory, is easier to establish and maintain than a company policy and easier to evaluate than an unwritten one. To develop a written program you should consider including the following elements or information:

1. Identify steps taken to assess potential hazards in every employee's workspace and in workplace operating procedures.
2. Identify appropriate PPE selection criteria.
3. Identify how you will train employees on the use of PPE, including:
   a. What PPE is necessary?
   b. When is PPE necessary?
   c. How to properly inspect PPE for wear and damage?
   d. How to properly put on and adjust the fit of PPE?
   e. How to properly take off PPE?
   f. The limitations of PPE.
   g. How to properly care for and store PPE?
4. Identify how you will assess employee understanding of PPE training.
5. Identify how you will enforce proper PPE use.
6. Identify how you will provide for any required medical examinations.
7. Identify how and when to evaluate the PPE program.

Finally, use PPE for potentially dangerous conditions. Use gloves, aprons, and goggles to avoid acid splashing. Wear earplugs for protection from high noise levels and wear respirators to protect against toxic chemicals. The use of PPE should be the last consideration in eliminating or reducing the hazards the employee is subjected to because PPE can be heavy, awkward, uncomfortable, and expensive to maintain. Therefore, try to engineer the identified hazards out of the job.

RANKING HAZARD CONTROLS

In determining which hazard control procedures have the best chance of being effective, it is useful to rank them along a continuum. The five hazard controls that were espoused in the earlier part of this chapter are ranked in Figure 13.4. This should assist you in determining which control, if you have a choice of more than one, would be most effective for your purposes. The ranking goes from most effective to least effective.

OTHER TOOLS THAT CAN BE USED FOR HAZARD CONTROL

Accident and Incident Reporting

Accident and incident reporting process allows for the identification of hazards as well as the development of controls for the removal or mitigation of hazards. All incidents and accidents resulting in injury or causing illness to employees, and all events or (or near-miss accidents) must be reported to

Most effective

Elimination or substitution
- Change the process to eliminate human interaction
- Elimination of pinch points (increase clearance)
- Automated material handling

Engineering controls (safeguarding technology)
- Mechanical hard stops
- Barriers
- Interlocks
- Presence sensing devices
- Two-hand controls

More effective

Awareness means
- Lights, beacons, and strobes
- Computer warnings
- Signs
- Painted markings on floors, etc.
  for restrictive areas or envelopes
- Beepers
- Alarms
- Horns
- Public address systems
- Labels

Less effective

Training and procedures (administrative controls)
- Training
- Job rotation
- Off shift scheduling of work
- Safe job procedures
- Safety equipment inspections/audits
- Lockout/tagout

Least effective

Personal protective equipment
- Safety eyewear (face shield, etc.)
- Hearing protection (ear plugs or muffs)
- Fire proof clothing
- Gloves
- Safety shoes
- Respirators
- Whole-body protection (Tyveks, etc.)

Figure 13.4   Hazard control ranking.

1. Identify the hazards that caused the event.
2. Establish a written record of factors that caused injuries and illnesses or that caused occurrences (near-misses), which might have resulted in injury or illness but did not do this for bodily, property, and vehicle damage, and occurrences.
3. Maintain a capability to promptly investigate incidents and events to initiate and support corrective or preventive action and implement hazard controls.
4. Provide statistical information for use in analyzing all phases of incidents and events.
5. Provide the means for complying with the reporting requirements for occupational injuries and illnesses.
6. Improve OSHA compliance by identifying and removing or controlling hazards.

Your incident reporting system requirements should apply to all incidents involving company employees, on-site vendors, contractor employees and visitors, which result in (or might have resulted in) personal injury, illness, or property and vehicle damage.

Injuries and illnesses that require reporting include those occurring on the job that result in any of the following: lost work time, restrictions in performing job duties, required medical treatment, permanent physical bodily damages, or death. Examples of reportable injuries and illnesses include, but are not limited to, heat exhaustion from working in hot environments, strained back muscles from moving equipment, acid burns on fingers, etc.

Other incidents requiring reporting for company records include those occurring on the job that result in any of the following: injury or illness, first-aid treatment, damage to a vehicle; fire and explosion; property damage valued at more than $100; and chemical releases requiring large-scale evacuation as well as evacuations in the immediate area where a chemical spill has occurred.

Examples of nonreportable injuries and illnesses include small paper cuts, common colds, and small bruises not resulting in work restrictions or requiring only first aid or no medical attention.

Events (near-misses) and other incidents (near-misses) that, strictly by chance, do not result in actual or observable injury, illness, death, or property damage are required to be reported. The information obtained from such reporting can be extremely useful in identifying and mitigating problems before they result in actual personal or property damage. Examples of near-miss incidences that require reporting include the falling of a compressed-gas cylinder; overexposures to chemical, biological, or physical agents (not resulting in an immediately observable manifestation of illness or injury); and slipping and falling on a wet surface without injury (see Chapter 7).

Hazard Audits

The use of audit (inspection) instruments can help determine whether controls are in place and are utilized. It will convey to some extent the overall effectiveness of your hazard control process. Audits will provide information on control processes and techniques that are not functioning correctly. The effectiveness of your hazard control procedures can only be accomplished if they are implemented. Therefore, periodic reviews and audits will confirm or disavow that employees are familiar and following the hazard control processes and procedure that the company has implemented. These audits will help identify hazards and issues that have transpired since the previous audit or inspection (see Chapters 7 and 14).

Safe Operating Procedure

The final step in evaluating a job is to develop a safe job procedure to eliminate or reduce the potential accidents or hazards. The following criteria should be considered:

- Find a less hazardous way to do the job by using an engineering revision to find an entirely new and safe way to do a job. Determine the work goal and analyze the various ways of reaching this goal to establish which way is safest. Consider work-saving tools and equipment. An example of this would be to install gauges on the loader that would visually indicate the fluid level and prevent contact with the fluids.
- Change the physical conditions that created the hazard. If a new, less hazardous way of doing the job cannot be found, try to change the physical conditions that are creating the hazards. Changes made in the use of tools, materials, equipment, or the environment could eliminate or reduce the identified hazards. However, when changes are found, study them carefully to determine the potential benefits. Consider if the changes possess latent, inherent hazards that may be equally hazardous as the original condition. In this case, assess both conditions to determine which will be less hazardous and refer the decision to the proper level of management for approval and acceptance. An example of this would be the development of a new loader operator cab with better visibility for safer operations.
- To eliminate the hazards that cannot be engineered out of the job, change the job procedure. Changes in job procedures, which are developed to help eliminate the hazards, must be studied carefully. If the job changes are too arduous, lengthy, or uncomfortable, the employee will take risky shortcuts

to circumvent these procedures. Caution must be exercised when changing job procedures to avoid creating additional hazards. To help determine the effectiveness of procedural changes, two questions which might be asked are

1. To eliminate this particular hazard or prevent this potential accident, what should be the action of the employee?
2. How should the employee accomplish this? Answers must be specific and concrete if new procedures are to be useful and effective. Answers should precisely state what to do and how to do it. This might mean changing the loader walk-around inspection to include fluid inspection at the same time, thus saving time and possibly decreasing the time of exposure.

• Try to reduce the necessity of doing the job, or at least reduce the frequency at which it must be performed. Often maintenance jobs require frequent service or repair of the equipment, which is hazardous. To reduce the necessity of such a repetitive job, ask "What can be done to eliminate the cause of the condition that makes excessive repairs or service necessary?" If the cause cannot be eliminated, then try to improve the condition.

Job Safety Assessment

This is a process used to check how effective your safe operating procedures are and if there is a need to make changes to control the hazards of the job. Before the start of any task or operation, the designated competent or company authorized person should evaluate the task or operation to identify potential hazards and to determine the necessary controls. This assessment should focus on actual worksite conditions or procedures, which differ from, were not anticipated, or were not related to other hazard analyses. In addition, the competent person shall ensure that each employee involved in the task or operation is aware of the hazards related to the task or operation and of the measures or procedures to use for protection. Note that the job safety assessment is not intended to be a formal, documented analysis, but instead is more of a quick check of actual site conditions and a review of planned procedures and precautions.

HAZARD CONTROL SUMMARY

All identified hazardous conditions should be eliminated or controlled immediately. Where this is not possible, interim control measures are to be implemented immediately to protect workers, warning signs must be posted at the location of the hazard, all affected employees need to be informed of the location of the hazard and of the required interim controls, and permanent control measures must be implemented as soon as possible.

Controls come in all forms, from engineering devices and administrative policy, to PPE. The best controls can be placed upon equipment before involving people and, thus, either preclude or guard the workforce from hazards. Administrative controls rely upon individuals following policies, guidelines, and procedures to control hazards and exposure to hazards. However, as we all know, this certainly provides no guarantee that the protective policies and procedures will be adhered to unless effective supervision and enforcement exist. Again, this relies on the company having a strong commitment to occupational safety and health. The use of PPE will not control hazards unless individuals who are exposed to the hazards are wearing the appropriate PPE. The use of PPE is usually considered the control of last resort since it has always been difficult for companies to be sure that exposed individuals are indeed wearing the required PPE.

Where a supervisor or foreman is not sure how to correct an identified hazard or is not sure if a specific condition presents a hazard, the supervisor or foreman shall seek technical assistance from the designated competent person, safety and health officer, or technical authority.

It is important that all hazards are identified and an assessment is made of the potential risk from the hazard. This allows for the determination of the real danger. If a high degree of risk and danger exists, then efforts must be undertaken to alleviate or mitigate the potential danger.

CHAPTER 14

Using the Tools: Accident Prevention Techniques

Proper guarding and safe work practice could have prevented this accident.

Many different tools are available to those who have responsibility for occupational safety and health to assist in structuring the safety and health program. Some are administrative techniques; others are programs, while some will need to be developed by you. You will find these different accident prevention techniques to be very useful once you learn to make use of them.

SAFETY AND HEALTH AUDITS

Workplace audits are inspections that are conducted to evaluate certain aspects of the work environment for occupational safety and health. The use of safety and health audits has been shown to have a positive effect on a company's loss control initiative. In fact, companies that perform safety and health audits have fewer accidents/incidents than companies who do not perform audits.

Safety and health audits (inspections), which are often conducted in workplaces, serve a number of evaluative purposes. Audits or inspections can be performed to

- Check compliance with company rules and regulations
- Check compliance with OSHA rules
- Determine the safety and health condition of the workplace
- Determine the safe condition of equipment and machinery
- Evaluate supervisors' safety and health performance
- Evaluate workers' safety and health performance

- Evaluate progress in safety and health issues and problems
- Determine the effectiveness of new processes or procedural changes

Need for an Audit

First, determine what needs to be audited. You might want to audit specific occupations (e.g., machinist), tasks (e.g., welding), topics (e.g., electrical), teams (e.g., rescue), operators (e.g., crane operator), part of the worksites (e.g., loading and unloading), compliance with an OSHA regulations (e.g., hazard communication standard), or the complete worksite. You may want to perform an audit if any of the previous lists or activities have unique identifiable hazards, new tasks involved, increased risk potential, changes in job procedures, areas with unique operations, or areas where comparison can be made regarding safety and health factors.

In the process of performing audits, you may discover hazards in a new process, hazards once the process has been instituted, a need to modify or change processes or procedures, or situational hazards that may not exist at all times. These audits may verify that job procedures are followed and identify work practices that are both positive and negative. They may also detect exposure factors both chemical and physical and determine monitoring and maintenance methods and needs.

At times audits are driven by the frequency of injury; potential for injury; the potential severity of injuries; new or altered equipment, processes, and operations; and excessive waste or damaged equipment. These audits may be continuous, ongoing, planned, periodic, intermittent, or dependent on specific needs. Audits may also determine employees' comprehension of procedures and rules, the effectiveness of workers' training to assess the work climate, or perceptions held by workers and others, and to evaluate the effectiveness of a supervisor in his or her commitment to safety and health.

At many active workplaces, daily site inspections are performed by the supervisor or foremen to detect hazardous conditions, equipment, materials, or unsafe work practices. At other times periodic site inspections are conducted by the site safety and health officer or their staff. The use of teams comprising management and workers to carry out inspection has been found to be effective.

The frequency of inspections is established in the workplace safety and health program. The supervisor, in conjunction with the safety and health officer, determines the required frequency of these inspections, based on the level and complexity of the anticipated activities and on the hazards associated with these activities. When addressing site hazards and protecting site workers, the inspections should additionally review worksite conditions and activities as well as include an evaluation of the effectiveness of the company's safety and health program. The safety and health officer should revise the company's safety and health program as necessary, to ensure the program's continued effectiveness.

Before the start of each shift or new activity, a workplace and equipment inspection should take place. This should be done by the workers, crews, supervisor, and other qualified employees. At a minimum, they should check the equipment and materials that they will be using during the operation or shift for damage or defects that could present a safety hazard. In addition, they should check the work area for new or changing site conditions or activities that could also present a safety hazard.

All employees should immediately report any identified hazards to their supervisors. All identified hazardous conditions should be eliminated or controlled immediately. When this is not possible

- Interim control measures should be implemented immediately to protect workers.
- Warning signs should be posted at the location of the hazard.
- All affected employees should be informed of the location of the hazard and the required interim controls.
- Permanent control measures should be implemented as soon as possible.

When a supervisor is not sure how to correct an identified hazard, or is not sure if a specific condition presents a hazard, he or she should seek technical assistance from a competent person, a site safety and health officer, or from other supervisors or managers.

Safety and health audits should be an integral part of your safety and health effort. Anyone conducting a safety and health audit must know the workplace, the procedures or processes audited, the previous accident history, and the company's policies and operations. This person should also be trained in hazard recognition and interventions for safety and health.

Safety and health audits and inspections can be done on or for the entire plant (e.g., manufacturing), a department (e.g., quality control), a specific worker unit (e.g., boiler repair), a job or task (e.g., diving), a certain work environment (e.g., confined spaces), a specific piece of equipment (e.g., forklift), a worker performing a task (e.g., power press operator), or prevention of an event (e.g., fire). It is important to ensure that inspections and audits are tailored to meet the needs of your company in auditing for safety and health concerns. An example of a safety and health audit instrument can be found in Appendix D and further information can be found in Chapter 7. You will need to develop safety and health audit instruments that meet your specific needs.

SAFETY TALKS AND MEETINGS

Safety talks, sometimes called toolbox talks, are an important training tool for the safety and health director and the supervisor. These safety talks can be used to cover a wide range of important safety topics in real time or immediately after mishaps or near misses have occurred. Safety talks have the benefit of incorporating specific company issues and concerns.

Safety talks cannot be done in a helter–skelter fashion. They need to be approached in an organized manner using a planned approach or they become nonfunctional. Safety talks need to have meaning and purpose. Thus, they do not become a gripe session or have the appearance of being a seat-of-the-pants presentation.

Safety talks are used as a practical tool to communicate, motivate, and train workers in specific hazards, unique or arising hazards, incidents that have occurred, or updated job safety and health issues in a somewhat informal manner. The safety talk accomplishes efficiency, safety, and productivity (ESP). The safety talk accomplishes ESP because it has many advantages, which are listed here:

- Saves time
- Creates a cooperative communications climate
- Permits worker or individual participation
- Provides the same exposure for everyone
- Fosters group spirit
- Builds the speaker's image

In developing a safety talk the preparer should follow these steps: prepare, pinpoint, personalize, picturize, and prescribe. Seldom is a deliverer of a safety talk successful without preparation. A safety talk is best when it pinpoints or addresses a single topic. By personalizing the talk to the audience, the message becomes more relevant and better received. Most individuals usually say "I see what you mean" and not "I hear what you say," which is the reason the visuals/pictures are an important component when planning a safety talk (the eyes have it). At the end of the talk give the audience a prescription, help them answer the so what question, tell them how they can help, and ask for action from them.

Safety talks, as with other training, should be documented. This documentation needs to include the date of the talk, presenter's name, topic, a list of those in attendance, and any materials used, which should be attached to the safety talk log form, which can be found in Chapter 21.

ACCIDENT INVESTIGATIONS

Millions of accidents and incidents occur throughout the United States every year. The inability of people, equipment, supplies, or surroundings to behave or react as expected causes most of the accidents and incidents. Accident and incident investigations determine how and why each incident occurs. Using the information gained through an accident and incident investigation, a similar or perhaps more disastrous accident may be prevented. Accident and incident investigations should be conducted with accident prevention in mind. The mission is one of fact-finding; investigations are not to find fault.

An accident, by definition, is any unplanned event that results in personal injury or in property damage. When the personal injury requires little or no treatment, or is minor, it is often called a first-aid case. If it results in a fatality or in a permanent total, permanent partial, or temporary total (lost-time) disability, it is serious. Likewise, if property damage results, the event may be minor or serious. All accidents should be investigated regardless of the extent of injury or damage.

Accidents are part of a broad group of events that adversely affect the completion of a task. Accidents fall under the category of an incident. With this said, the most commonly used term for accidents and incidents is accident, which will be used to refer to both accidents and incidents, since the basic precepts are applicable to both.

An important element of a safety and health program is accident investigation. Although it may seem to be too little or too late, accident investigations serve to correct the problems that contribute to an accident and will reveal accident causes that might otherwise remain uncorrected.

The main purpose of conducting an accident investigation is to prevent a recurrence of the same or similar event. It is important to investigate all accidents regardless of the extent of injury or damage. The kinds of accidents that should be investigated and reported are

- Disabling injury accidents
- Nondisabling injury accidents that require medical treatment
- Circumstances that have contributed to acute or chronic occupational illness
- Noninjury, property damage accidents that exceed a normally expected operating cost
- Near accidents (sometimes called near misses) with a potential for serious injury or property damage

In spite of their complexity, most accidents are preventable by eliminating one or more causes. Accident investigations determine not only what happened, but also how and why. The information gained from these investigations can prevent recurrence of similar or perhaps more disastrous accidents. Accident investigators are interested in each event as well as in the sequence of events that led to an accident. The accident type is also important to the investigator. The recurrence of accidents of a particular type or those with common causes shows areas needing special accident prevention emphasis.

It is important to have some mechanism in place to investigate accidents and incidents to determine the basis of cause and effect relationships. You may determine these types of relationships only when you actively investigate all accidents and incidents that result in injuries, illnesses, or damage to property, equipment, and machinery.

Accident investigation becomes more effective when all levels of management, particularly top management, take a personal interest in controlling accidents. Management adds a contribution when it actively supports accident investigations. It is normally the responsibility of line supervisors to investigate all accidents; and in cases where there is serious injury or equipment damage, other personnel such as department managers and an investigation team might become involved as well.

Once you have determined the types of accidents and incidents that are transpiring, you can undertake prevention and intervention activities to ensure that you will have no recurrences. Even if you are not experiencing large numbers of accidents and incidents, you still need to implement activities that actively search for, identify, and correct the risk from hazards on jobsites. Reasons to investigate accidents and incidents include

- To know and understand what happened
- To gather information and data for present and future use
- To determine cause and effect
- To provide answers for the effectiveness of intervention and prevention approaches
- To document the circumstances for legal and workers' compensation issues
- To become a vital component of your safety and health program

If you have only a few accidents and incidents, you might want to move down one step to examine near misses and first aid-related cases. It is only a matter of luck or timing that separates the near miss or first-aid event from becoming a serious, recordable, or reportable event. The truth is you probably have been lucky by seconds or inches. (A second later and a tool would have hit someone or an inch more and it would have cut off a finger.) Truly, it pays dividends to take time to investigate accidents and incidents occurring in the workplace.

Reporting Accidents

When accidents are not reported, their causes usually go uncorrected, allowing the chance for the same accident to result again. Every accident, if properly investigated, serves as a learning experience to the people involved. The investigation should avoid becoming a mechanical routine. It should strive to establish what happened, why it happened, and what must be done to prevent a recurrence. An accident investigation must be conducted to find out the facts and not to place blame.

There are sound reasons for reporting accidents such as

- You learn nothing from unreported accidents.
- Accident causes go uncorrected.
- Infection and injury aggravations can result.
- Failure to report injuries tends to spread and become an accepted practice.

The results-oriented supervisor recognizes that the real value of investigation can only be achieved when his or her workers report every problem, incident, or accident that they know of. To promote conscientious reporting, it may be helpful to know some of the reasons why workers fail to, or avoid, reporting accidents. There are usually reasons that workers espouse for not reporting accidents and are as follows:

1. Fear of discipline
2. Concern about the company's record
3. Concern for their reputation
4. Fear of medical treatment
5. Dislike of medical personnel
6. Desire to keep personal record clear
7. Avoidance of red tape
8. Desire to prevent work interruptions
9. Concern about attitudes of others
10. Poor understanding of importance

How can you combat these reporting problems?

1. React in a more positive way.
2. Indoctrinate workers on the importance of reporting all accidents.
3. Make sure everyone knows what kinds of accidents should be reported.
4. Give more attention to prevention and control.

5. Recognize individual performance.
6. Develop the value of reporting.
7. Show disapproval of injuries neglected and not reported.
8. Demonstrate belief by action.
9. Do not make mountains out of molehills.

Let the worker know that you appreciate his or her reporting promptly. Inquire about his or her knowledge of the accident. Do not interrogate or grill the person. Stress the value of knowing about problems while they are still small. Focus on accident prevention and loss control. Emphasize compliance with practices, rules, protective equipment, and promptly commend good performance. Pay attention to the positive things workers do and give sincere, meaningful recognition where and when it is deserved. Use compliments as often as you use warnings. Use group and personal meetings to point out and pass on knowledge gained from past accidents. Give an accident example an important part of every job instruction. Show that you believe what you say by taking corrective action promptly. You can always do something right at the moment, even if permanent correction requires time to develop new methods, buy new equipment, or modify the building.

The first step in an effective investigation is the prompt reporting of accidents. You cannot respond to accidents, evaluate their potential, and investigate them, if they are not reported when they happen. Prompt reporting is the key to effective accident investigations. Hiding small accidents does not help prevent the serious accidents that kill people, put the company out of business, and take away jobs. If workers do not report accidents to the supervisor, they are stealing part of the supervisor's authority to manage his or her job.

JOB HAZARD ANALYSIS

Much of the information within this section comes from the U.S. Department of Labor's Mine Safety and Health Administration material entitled *The Job Safety Analysis Process: A Practical Approach*. It states that fatalities, accidents, and injuries can be reduced if we all work together and share our safety knowledge. An accident prevention method that has proven effective in industry is the job hazard analysis (JHA) program.

JHA is a basic approach to developing improved accident prevention procedures by documenting the firsthand experience of workers and supervisors and, at the same time, it tends to instill acceptance through worker participation. JHA can be a central element in a safety program; the most effective safety programs are those that involve employees. Each worker, supervisor, and manager should be prepared to assist in the recognition, evaluation, and control of hazards. Worker participation is important to ESP. Through the process of JHA, these benefits are fully realized.

JHA, also called job safety analysis (JSA), is a process used to determine hazards of, and safe procedures for, each step of a job or process. A specific job, process, or work assignment can be separated into a series of relatively simple steps. The hazards associated with each step can be identified and solutions can be developed to control each hazard.

Four Basic Steps of a JHA/JSA

JSA/JHA involves four basic steps:

1. Select a job to be analyzed.
2. Separate the job into its basic steps.
3. Identify the hazards associated with each step.
4. Control each hazard.

Looking at these four steps in detail will help explain the process and value of this type of analysis.

Selecting a Job to Be Analyzed

The first step of a JHA is to select a job to be analyzed. The sequence in which jobs are analyzed should be established when starting a JHA program. Potential jobs for analysis should have sequential steps and a work goal when these steps are performed.

To use the JHA program effectively, a method must be established to select and prioritize the jobs to be analyzed. The jobs must be ranked in the order of greatest accident potential. Jobs with the highest risks should be analyzed first. You may or may not be involved with the ranking process, but if you are asked to rank or prioritize jobs to be analyzed, the following criteria should be used:

1. Accident frequency
2. Accident severity
3. Judgment and experience
4. New jobs, nonroutine jobs, or job changes

Prime candidates for JSA/JHA result from jobs that are done rarely or never at a particular worksite. The hazards of the job might not be fully known. By applying the JHA process to these jobs, the likelihood of an accident occurring is greatly reduced.

After a job has been selected and the JHA has been initiated, a worksheet is prepared listing the basic job steps, the corresponding hazards, and the safe procedures for each step. The basic form generally has three columns. In the left column, the sequence of basic job steps is listed in the order in which the steps occur. The middle column describes all potential hazards. The right column lists the recommended safe job procedures that should be followed to guard against these hazards to prevent potential accidents (see Figure 14.1).

Changing Job Procedures

If the sequence of job steps or the deviations from established job steps are critical to the safe performance of a job, this should be noted in the JHA. The next part of the JSA process is to develop the recommended safe job procedure to eliminate or reduce potential accidents or hazards that have been identified. The following four points should also be considered for each hazard identified for the job step:

1. Can a less hazardous way to do the job be found?
2. Can an engineering revision take place to make the job or work area safer?
3. Is there a better way to do the job? This requires determining the work goal and then analyzing various ways to reach the goal to see which way is safest.
4. Are there work-saving tools and equipment available that can make the job safer?

If a new, less hazardous way to do the job cannot be found, physical conditions should be studied carefully. The steps involved in a JHA process have been outlined in the previous pages. It should be especially clear that the main point of doing a JHA is to prevent accidents by anticipating and eliminating hazards. JHA is a procedure for determining the sequence of basic job steps, identifying potential accidents or hazards, and developing recommended safe job procedures.

JHA is an accident prevention technique used in many successful safety programs. The JHA process is not difficult if it is taken with a commonsense approach on a step-by-step basis. JHAs should be reviewed often and updated with inputs from both supervisors and workers who do the

Job safety analysis worksheet

Title of job/operation   Date   No.

Position/title of person(s) who does job   Name of employee observed

Department   Analysis made by

Section   Analysis approved by

Sequence of basic job steps	Potential accidents or hazards	Recommended safe job procedures

1. Struck-by (SB)   5. Caught-on (CO)   9. Fall-to-below (FB)
2. Struck-against (SA)   6. Caught-in (CI)   10. Overexertion (OE)
3. Contacted-by (CB)   7. Caught-between (CBT)   11. Exposure (E)
4. Contact-with (CW)   8. Fall-same-level (FS)

Figure 14.1 Job safety/hazard analysis (JSA/JHA) worksheet. (Courtesy of Mine Safety and Health Administration.)

job everyday. The implementation of the JHA process will mean continuous safety improvements at your workplace with the ultimate goal of zero accidents. JHA takes a little extra effort, but the results are positive and helpful for everybody.

There are many advantages in using JHA. JHA provides training to new employees on safety rules and specific instructions on how the rules are to be applied to their work. This training is provided before the new employees perform the job tasks. JHAs also instruct new employees in safe work procedures. With JHAs, experienced employees can maintain safety awareness behavior and receive clear instructions for job changes or new jobs. Benefits also include updating current safety procedures and instructions for infrequently performed jobs.

It is important to involve workers in the JHA process. Workers are familiar with the jobs and can combine their experience to develop the JHA. This results in a more thorough analysis of the job. A complete JHA program is a continuing effort to analyze one hazardous job after another until all jobs with sequential steps have a written JHA. Once established, the standard procedures should be followed by all employees.

SAFE OPERATING PROCEDURES

Safe Operating Procedures (SOPs) or standard operating procedures should include safety as a part of the standard operating practices, which are delineated within it. Workers may not automatically understand a task just because they have experience or training. Thus, many jobs, tasks, and operations are best supported by an SOP. The SOP walks the worker through the steps of how to do a task or procedure in a safe manner and calls attention to the potential hazards at each step.

You might ask why an SOP is needed if the worker has already been trained to do the job or task. As you may remember from the previous section, a JHA usually focuses on those particular jobs that pose the greatest risk of injury or death. These are high-risk types of work activities and definitely merit the development and use of an SOP. You can use the JHA to develop an SOP. There are times when an SOP, or step-by-step checklist, is useful. Some of these times are when a

- New worker is performing a job or task for the first time
- Experienced worker is performing a job or task for the first time
- Experienced worker is performing a job that he or she has not done recently
- Mistakes could cause damage to equipment or property
- Job is done on an intermittent or infrequent basis
- New piece of equipment or different model of equipment is obtained
- Supervisors need to understand the safe operation to be able to evaluate performance
- Procedure or action within an organization is repetitive and is carried out in the same way each time
- Procedure is critically important, no matter how it is performed and must be carried out exactly according to detailed, stepwise instructions
- Need to standardize the way a procedure is carried out for ensuring quality control or system compatibility

SOPs are organizational tools that provide a foundation for training new employees, for refreshing the memories of management and experienced employees, and for ensuring that important procedures are carried out in a standard specified way. The principal function of an SOP is to provide detailed, step-by-step guidance to employees who are required to carry out a certain procedure. In this instance, it serves not only as a training aid, but also as a means of helping to ensure that the procedure is carried out in a standard, approved manner.

Another important function of an SOP is to keep management informed about the way functions are performed in areas under their supervision. A complete file of well-written, up-to-date SOPs is an indication of good management and provides management with instant access to information on functional details of the organization for which they are responsible. This is of enormous benefit during inspections and management reviews, to say nothing of providing timely answers to unanticipated questions from superiors.

The overridingly important feature of a good SOP is that it communicates what is to be done in a clear, concise, and stepwise manner. The most important person to whom it must communicate is typically the new employee who may have little or no experience with the procedure in question. Therefore, it is imperative that the writer of an SOP figuratively place him- or herself in the position of a new, inexperienced employee to appreciate what must be communicated and how to communicate it.

The content of an SOP should be comprehensive in terms of how to get the procedure accomplished, but should not encompass matters not directly relevant because digression does not directly address the issue of how to get the procedure accomplished and exactly who is to do it. In Table 14.1, an example of an SOP is provided. Some actual SOPs may be more complex than this example.

Perhaps the best advice concerning the content of an SOP is this. Ask yourself the questions— Who? What? Where? When? How? If the SOP answers all of these questions, it is complete. If not, revise it until it does in as clear and logical an order as possible.

Table 14.1   Safe Operating Procedure: Loading and Unloading a Glock 36.45 Cal. Pistol

What to Do	How to Do It	Key Points
Inspect the pistol for a loaded condition and unload	1. Always keep the pistol pointed in a safe direction. Press the magazine release and remove the magazine from the pistol (if needed).	1. Always keep the pistol pointed in a safe direction. Keep your finger off the trigger.
	2. Pull the barrel slide to the rear and lock into the rear position by pushing up on the slide release.	2. Do this over a table or flat surface to catch any ammunition that may eject from the pistol. Grip the pistol firmly and pull the slide directly back in line with the barrel.
	3. Inspect the barrel and chamber for ammunition or shell casings.	3. With the pistol pointed away from you look directly into the chamber to check for ammunition or shell casings.
	4. Remove any ammunition found.	4. Shake the pistol to dislodge any rounds. If necessary recycle the barrel slide to remove any rounds in the chamber. If ammunition does not eject from the chamber after several attempts seek qualified assistance for disassembly and cleaning. Keep all ammunition in a safe location. Maintain accountability of all ammunition.
Load the pistol	1. Insert a loaded magazine into the magazine well and push up until it locks into place.	1. Always keep the pistol pointed in a safe direction. Only use quality ammunition. Ensure that the proper number of rounds is in the magazine. Ensure the top round is fully seated in the magazine. You should hear a click when the magazine locks into place.
	2. Pull on magazine to ensure that it is locked into the magazine well.	2. Press up and lock into place if any movement is found.
	3. Push down on the slide release to release the barrel slide.	3. Keep fingers and hands clear of the slide when releasing. Interference may cause jams and misfed rounds. If the pistol does not chamber a round after 2–3 attempts cleaning or maintenance may be needed. Certain types of ammunition will not feed properly. Use only factory recommended ammunition. This pistol is loaded and in the ready to fire position, USE EXTREME CARE!
Unload the pistol	1. Always keep the pistol pointed in a safe direction. Press the magazine release and remove the magazine from the pistol.	1. Always keep the pistol pointed in a safe direction. Keep your finger off the trigger.
	2. Pull the barrel slide to the rear and lock into the rear position by pushing up on the slide release.	2. Pull the barrel slide to the rear and lock into the rear position by pushing up on the slide release.
	3. Inspect the barrel and chamber for ammunition.	3. With the pistol pointed away from you look directly into the chamber to check for ammunition.
	4. Remove any ammunition found.	4. Shake the pistol to dislodge any rounds. If necessary recycle the barrel slide to remove any rounds in the chamber. If ammunition does not eject from the chamber after several attempts seek qualified assistance for disassembly and cleaning.

JOB SAFETY OBSERVATION

This material on job safety observation was taken from the U.S. Department of Labor, Mine Safety and Health Administration's publication *Safety Observation (MSHA IG 84)*. Job safety observations are one of the accident prevention techniques that can be used to assess safe work performance.

There are many categories of accident causes and many terms used to describe these causes. To precisely determine the causes for each category, the terms person causes and environmental causes are often used. The actual and potential causes of accidents are generally accepted as the key factors in a successful loss prevention effort. Actual causes—direct and indirect—can only be considered after an accident has occurred. They can be found by asking the question, "What caused the accident?" Potential causes may be avoided before an accident actually occurs by asking the question, "What unsafe conditions (environmental causes) or unsafe procedures (person causes) could cause an accident?" Working with actual causes is similar to firefighting, with after-the-fact analysis and hindsight. The process of understanding, determining, and correcting potential causes is comparable to fire prevention or foresight.

All categories of accident causes must be considered and used in any complete loss prevention program. Safety observations and inspections are necessary phases in the overall safety effort. Making a safety observation is the process of watching how a person performs a specific job to detect unsafe behavior (person causes). Making a safety inspection is the process of visually examining the work area and work equipment to detect unsafe conditions (environmental causes). Detecting and eliminating potential causes of accidents may best be accomplished when supervisors understand safety observations and when safety inspections become separate phases of the loss prevention work. This section deals primarily with making safety observations.

The safety observation phase is initiated when a written set of procedures is prepared by management and safety personnel. The procedures should include prepared JSA ready for use, step-by-step safe job procedures, and the training of all supervisors in observation procedures. Objectives must be established for each step of the program. The establishment of definite goals at all levels of management will give direction to the safety effort.

Management should outline the purpose and types of job safety observations, including how to select a job or task for planned safety observations, how to prepare for a planned safety observation, how to use a checklist of activities to observe unsafe procedures, what the employees' role in the observation process is, what occurs after the observation, and how to deal with unsafe behavior and performance.

The basic idea of job safety observation is simple. It is a special effort to see how employees do their jobs. Planned safety observation involves more effort than an occasional or incidental observation of job procedures. Job safety observation is a way of determining unsafe practices and violations of safety rules. This accident prevention method emphasizes the importance of a proper supervisor/employee relationship. Becoming more interested in the employee through observations will lead to greater cooperation in the safety program. You will find a job safety observation form in Figure 14.2, which you can use to conduct your own job safety observation.

FLEET SAFETY PROGRAM

Fleet safety is often viewed as operator safety, which is definitely a key component of a company's attempt to protect its large investment in vehicles and mobile equipment. It is understood that many of the accidents that occur are a direct result of driver error. However, driver error is not the fault of the individual. It is the fault of management's failure to institute a fleet safety program that provides organization, direction, and accountability for the fleet of vehicles that the company owns.

Job safety observation form

Observation completed by: **Title:**   **Date:**

1. Job being observed:
2. Worker observed (Name):
3. Experience of worker at job or task: years   months
4. Is worker dressed appropriately for the job?   ☐ yes ☐ no
   Comments:
5. Is worker wearing all required personal protective equipment?   ☐ yes ☐ no
   Comments:

Steps in performing the job or task should be marked (S) Satisfactory, (R) Reobserve, or (U) Unsatisfactory.

Steps (describe)	Hazard involved	Worker performance
1.		
2.		
3.		
4.		
5.		

6. Did the worker perform the job according to the safety operating procedure?
   Comments:
7. Did the worker follow the safety and health rules?
   Comments:
8. Did the worker perform the job or task safely?
   Comments:
9. Did the worker have a good safety attitude?
   Comments:
10. Does the worker need training?
11. Should the worker be removed from the job or task?
12. Was the worker told about any deficiency?
13. Your final recommendations, if any

Figure 14.2 Job safety observation form. (From Reese, C.D. and Eidson, J.V., *Hand book of OSHA Construction Safety and Health*, 2nd edn., Boca Raton, FL: CRC/Lewis Publishers, 1999. Courtesy of Mine Safety and Health Administration.)

Commitment to a fleet safety program communicates the value that the company places on their property and employees. The care given to both vehicles/equipment and employees conveys the company's true view of the value of accident prevention.

A fleet safety program should consist of the following:

- Written fleet safety program
- Vehicle/equipment maintenance procedure
- Record keeping process
- Operator selection process
- Operator training requirements
- Operator performance requirements

The company should clearly state its policy regarding fleet safety and delineate what is expected to transpire as a result of its program. The overall intent of this program must be stated clearly. In turn, programs should incorporate the many facets of any good accident prevention effort. A program on fleet safety should provide the framework for safe management of the company's vehicles and equipment and employees. Program goals must be communicated to drivers and supervisory personnel.

There must be a person designated with responsibility for both job safety and compliance with regulations. This person must assume responsibility for compliance with existing regulations and for the implementation and enforcement of company rules and policies. He or she must oversee the qualification of operators/drivers and the care, safety, and maintenance of the company's fleet.

It is management's responsibility to recruit and screen new drivers, monitor driver qualifications and safety infractions, and provide training to upgrade drivers' skills and knowledge. Management should provide a formal mechanism for investigating and reviewing accidents and for monitoring maintenance and equipment safety. Management should also implement safe driving incentives and offer recognition to drivers who meet the required standard of performance. Each company must constantly monitor the effectiveness of its fleet safety program.

The cost of a fleet of vehicles is a staggering investment and a major cash outlay for companies. To reap the full benefits of your investment, start with a thorough purchasing process. You want quality and dependability for your money. This will entail some research on your part to ensure that you are getting the most for your money. Once you have your fleet in place, then you will want to get the most mileage out of your purchase. This can only be accomplished by a preventive maintenance program (PMP), which includes regularly scheduled maintenance, follow-up to operator complaints, and daily preshift inspections of vehicles and equipment. You will need a record keeping system for your maintenance program, which includes the following:

- Operator's inspection record—a checklist of things to be checked daily by operators and any corrections needed to ensure the safety of the vehicle. (This should go to the maintenance shop.)
- Schedule maintenance record—the maintenance shop record of routine or periodic service for each vehicle.
- Service record—to show all findings and results of the inspections, routine service, and repairs made along with the date of each such maintenance procedure.
- Vehicle history record—a complete history of the vehicle including, but not limited to, any accidents in which it was involved, any catastrophic failure or repairs (i.e., engine change), and the dates when tires were replaced.

The selection of operators is a key factor in your fleet safety program. Fleet safety may be viewed as vehicle safety or mechanical safety, but it depends on both maintenance and operators. An operator's job must include preoperation inspections and, upon completing his or her use of the vehicle, the reporting of any defects. This should be normal operating procedure for any PMP. Nevertheless, what you can expect is dependent on the quality of your operators. Operators are the center of your fleet safety program.

Thus, in a fleet safety program, it is important to select the best operators for the job. The operator is vital to the prevention of accidents, incidents, vehicle damage, and injuries. Careful selection of the operator is paramount to an effective fleet safety program. The selection process should involve access to the operator's past employment history, driving record (including accidents), accommodations, or awards, as well as previous experience, if any, on your type of equipment.

As a condition of employment and based on the criteria in a written job description, all potential operators should be able to pass a physical and mental examination and an alcohol/drug test. You will also need a well-written job description, and you may want your legal counsel and others to review it before its use.

To improve fleet safety, adequately qualified drivers must be recruited and their performance monitored. The great majority of preventable accidents can be shown to be directly related to the performance of the driver. It is, therefore, extremely productive to any fleet safety program to have careful new driver selection and adequate monitoring procedures for existing drivers.

An established formal procedure for interviewing, testing, and screening applicants needs to be in place. A defined standard of skill and knowledge should be met by successful applicants. Appropriate methods should be in place to check out previous employment histories and references of all potential operators. You must access and check the prior driving records of the applicants. Each applicant should undergo a physical examination that includes testing for drugs and alcohol.

Once an operator is hired, there should be a formal program for monitoring his or her performance. A periodic review of the driving record and that of the driver's health should be conducted. Operators should be monitored occasionally for drug and alcohol abuse. (Note: A company should have an established policy on drug and alcohol abuse.) A means should be in place for identifying deficiencies in an operator's skills and knowledge, and a procedure should be in place for remedial training. It is well worth the effort to establish a procedure for terminating unqualified operators.

All operators should undergo training related to company and government policies and procedures. This training should include record keeping, accident and incident reporting, driving requirements, and defensive driving. After classroom training, each operator should be required to take a supervised driving test, or hands-on supervised operational drive, to determine his or her competence. This should be done before the operator is released for work-related driving assignments. Even after the operator is released from training status, he or she may have a supervisor accompany him or her on work assignments.

Operators should be observed and evaluated on a periodic basis, retrained, if necessary, and supervised more closely. If you are sure that your vehicles are in proper and safe operating condition, your operators then become the key to your fleet safety program. Good, conscientious operators can prevent accidents from occurring; they are the focus of your fleet safety program.

Preoperation Inspection

Before placing a vehicle in service, an operator should conduct a preoperational inspection. The operator should evaluate at least the following items:

- Parking brake
- Fuel tank and cap
- Side marker lights
- Reflectors
- Tires and wheels (lugs)
- Mirrors
- Steering wheel (excess play)
- Trailer brakes
- All lights, including four-way flashers
- Fire extinguisher and warning devices
- Headlights
- Clearance lights
- Identification lights
- Stop lights
- Turn signals and four-way flashers
- Rear end protection (bumper)
- Cargo tie-downs/doors
- Safety chains
- Hoses and couplers
- Electrical connectors
- Couplings (fifth wheel, chains, lock devices)
- Engine

- Oil pressure (light or gauge)
- Air pressure or vacuum (gauge)
- Low air or vacuum warning device
- Instrument panel (lights or buzzers)
- Horn
- Backup alarm
- Windshield wiper and washer
- Heater and defroster

You should require that your operators perform a pretravel check before hitting the road using a standard company form similar to the one in Figure 14.3.

Preshift equipment checklist

Check any of the following defects before operating your equipment and vehicle and report those defects to your supervisor or maintenance department.

Type of equipment:_____ Identification number:_____

Date:_____

1. Walk around
_____ Tires
_____ Broken lights
_____ Oil leaks
_____ Hydraulic leaks
_____ Mirrors
_____ Tracks
_____ Damaged hose
_____ Bad connections fittings
_____ Cracks in windshields or other glass
_____ Damaged support structures
_____ Damage to body structures
_____ Fluid levels
_____ Oil
_____ Hydraulic
_____ Brake

2. Operation
_____ Engine starts
_____ Air pressure or vacuum gauge
_____ Oil pressure
_____ Brakes
_____ Parking brakes
_____ Horn
_____ Front lights
_____ Back lights
_____ Directional lights
_____ Warning lights
_____ Back-up alarm
_____ Noises or malfunctioning

　　　_____ Engine
　　　_____ Clutch
　　　_____ Transmission
　　　_____ Axles

2.Operation
_____ Seat belts
_____ Steering wheel play or alignment

3. While underway
_____ Engine, knocks, misses, overheats
_____ Brakes operate properly
_____ Steering loose, shimmy hard, etc.
_____ Transmission noisy, hard shifting, jumps out of gear, etc.
_____ Speedometer
_____ Speed control

4. Emergency equipment
_____ First aid kit
_____ Fire extinguisher
_____ Flags, flares, warning devices
_____ Reflectors
_____ Tire chains if needed

5. Cargo-related equipment
_____ Tie downs
_____ Cargo nets
_____ Tarps
_____ Spare parts

6. Other items
_____ Hand tools
_____ Spare parts
_____
_____
_____

Figure 14.3   Preshift equipment checklist.

```
_____ Fuel level
_____ Instrument panel
_____ Windshield wipers or washers
_____ Heater or defroster
_____ Mirrors

No defects noted: Operator signature _____ Date _____ Time _____

Describe any defects noted_____
_____
_____
_____

Defects corrected:    Defect correction unnecessary:   Defects corrected by  Date:
    (initials)                (initials)                    (Signature)

Defects corrected:              Operator's signature:              Date:
_____Yes_____ No
```

Figure 14.3 (continued)

Preventing Accidents

A preventable accident is one that occurs because the driver fails to act in a reasonably expected manner to prevent it. In judging whether the driver's actions were reasonable, one seeks to determine whether the driver drove defensively and demonstrated an acceptable level of skill and knowledge. The judgment of what is reasonable can be based on a company-adopted definition, thus establishing a goal for its safety management program.

The concept of a preventable accident is a fleet safety management tool, which achieves the following goals:

- It helps to establish a safe driving standard for the driver.
- It provides a criterion for evaluating individual drivers.
- It provides an objective for accident investigations and evaluations.
- It provides a means for evaluating the safety performance of individual drivers and the fleet as a whole.
- It provides a means for monitoring the effectiveness of fleet safety programs.
- It assists in dealing with driver safety infractions.
- It assists in the implementation of safe driving recognition programs.

Fleet safety driving performance is dependent on management's commitment to the implementation of a formal fleet safety program. An effective safety program will interact with most aspects of fleet operations and challenge the skills and knowledge of its supervisors and drivers.

PREVENTIVE MAINTENANCE PROGRAM

PMP depends heavily on an inspection form or checklist to ensure that a vehicle or equipment inspection procedure has been fully accomplished and its completion documented. It has long been noted that a PMP has benefits that extend beyond caring for equipment. Of course, equipment is expensive and, if cared for properly and regularly, will last a lot longer, cost less to operate, operate more efficiently, and have fewer catastrophic failures.

Remember, properly maintained equipment is also safer and there is a decreased risk of accidents occurring. The degree of pride in having safe operating equipment will transfer to the workers in the form of better morale and respect for the equipment. Well-maintained equipment sends a strong message regarding safe operation of equipment.

If you allow operators to use equipment, machinery, or vehicles that are unsafe or in poor operating condition, you send a negative message, which says, "I don't value my equipment, machinery, or vehicles, and I don't value my workforce either." A structured PMP will definitely foster a much more positive approach regarding property and the workforce. Reasons for establishing a PMP are as follows:

1. Improve operating efficiency of equipment, machinery, or vehicles.
2. Improve attitudes toward safety by maintaining good and safe operating equipment.
3. Foster involvement of not only maintenance personnel but also supervisors and operators, which forces everyone to have a degree of ownership.
4. Decrease risk for incidents or mishaps.

The first aspect of a PMP is to have a schedule for regular maintenance of all your equipment. Second, it motivates supervisors to make certain that all operators conduct daily inspections. Third, it assures the company that all defects are reported immediately. Last, it documents that repairs are made before operating vehicles or equipment. If this is impossible, the equipment should be tagged and removed from service.

You will need the following to have an effectively functional PMP:

1. Maintenance department, which carries out a regular and preventive maintenance schedule
2. Supervisors and operators who are accountable and responsible
3. Preshift checklist for each type of equipment, machinery, or vehicle (An example of a vehicle checklist can be found in Figure 14.4.)
4. Effective response system when defects or hazards are discovered
5. Commitment by management that your PMP is important and will be achieved

Preventive maintenance's ultimate purpose is to prevent accidents caused by vehicle deficiencies, machinery defects, or equipment hazards. Worn, failed, or incorrectly adjusted components can cause or contribute to accidents. Preventive maintenance and periodic inspection procedures help to prevent failures from occurring while the vehicle/machine/equipment is operated. Such procedures also reduce reliance on the operator, who may have limited skill and knowledge for detecting these deficiencies.

A preventive maintenance and inspection program should recognize wear of consumable components, which must be periodically replaced or serviced. It should take into account indicators of deterioration, which can be monitored at the driver inspection level. The driver should be trained in trouble shooting. Special attention must be paid to the condition of components that cannot be easily observed by the driver. Maintenance supervisors and mechanics should inspect components in which problems can occur but are not easily discernible.

To ensure that vehicles are in a safe operating condition while driven, the driver should check the whole vehicle carefully, pretravel and posttrip. These pretravel and posttrip inspection reports are an important part of a PMP. If something seems to be wrong with the vehicle, stop and check it out. Do not continue with the trip until you are satisfied it is safe to do so.

The driver should be the one held ultimately responsible to make sure that the vehicle that is driven is in a safe operating condition. Appropriate inspection procedures and reports assist in ensuring this. The driver is also in a position to detect vehicle deficiencies and refer them to maintenance for repairs. The driver should not operate a faulty vehicle. Federal and state laws require that the driver should not operate a vehicle unless fully satisfied that it is in a safe operating condition.

SPECIAL EMPHASIS PROGRAM

Special emphasis programs have been mentioned previously but should be reinforced as an effective accident prevention technique. Any time you institute a special program that targets a unique safety and health issue, you have developed an organized approach in prevention. The benefits of instituting a special program include the fact that the potential hazard is kept on everybody's mind, management receives feedback, and workers receive reinforcement for the desired performance. You can develop a program in any area where you feel the need. Some areas of focus could be ladder safety, back injuries, vehicle or equipment safety, power tool incidents, etc. For success, the program may contain goals to attain, rewards to receive, or even consequences for enforcement if the rules of the program are not followed. By setting up a program you are at least taking action to target accidents and prevent their occurrence. An example of a special emphasis program is a ladder safety program that will be used as an illustration of how this might be done.

USING SAFETY AND HEALTH CONSULTANTS

From time to time, each of us is faced with problems or issues that surpass our own training, experience, or expertise. An educated person is one who recognizes that he or she needs help. There are always plenty of individuals ready and willing to provide advice and help. However, make sure to get competent help that can truly and meaningfully provide the assistance that you desperately need.

When you know you do not have the knowledge to do what is required, it is good to be confident enough in yourself to understand your limitations. Find the appropriate individual to help you. This may take as much effort as you would use if you could have solved your own problem, but you cannot! Thus, you will want to make sure that you are getting what you paid for.

The consultant you need may be a specialist in a particular area of safety and health (e.g., ergonomist) or an engineer who can help with redesigning issues. No matter who the person you need, proceed in an organized fashion in selecting that individual and finally obtaining a solution to your problem.

A consultant will likely draw on a wide and diverse experience base in helping to address problems and issues. The consultant is not influenced by politics and allegiance and is, therefore, in a better position to make objective decisions. The professional consultant strives to give you cost-effective solutions since his or her repeat business is based on performance and professional reputation. In addition, consultants are temporary employee and the usual personnel issues are not applicable to them. Consultants work at the times of the day, week, or month when you have a need and not at their convenience. Most consultants have become qualified to perform these services through either education or experience.

In determining the need for a consultant, you will want to consider whether using a consultant will be cost-effective, faster, or more productive. You may also find the necessity for a consultant when you feel you need outside advice, access to special instrumentation, an unbiased opinion or solution, or an assessment that supports your initial solution.

A consultant can address many of your safety and health issues. He or she can also act as an expert witness during legal actions. However, primarily, the consultant is hired to solve a problem. Thus, the consultant must be able to identify and define the existing problem and then provide appropriate solutions to your problem. Consultants usually have a wide array of resources and professional contacts, which he or she can access to assist in solving the problem.

To get names and recommendations of potential consultants, you can contact professional organizations, colleagues, and insurance companies who often employ loss control or safety and health personnel. Furthermore, do not overlook your local colleges and universities; many times they also provide consultant services.

The consultant is not your friend. You have hired him or her based on your evaluation that he or she is the best person to help solve the problem that you cannot solve, do not have the time to solve, or do not have the resources to solve. The consultant should define the problem, analyze the problem, and make recommendations for a solution. You can make the consultant a part of your team by providing full information and support. This will assure that you reap the greatest value from your investment.

To summarize, the primary steps for employing a consultant are

1. You must know what you want from the beginning.
2. You should put the scope of work in writing.
3. Use a bid process for selecting the consultant.
4. Delineate your timeline for completion of work.
5. Decide whom the consultant will be reporting to.
6. Know your budget limitations (develop a formal budget).
7. Make sure that a written contract exists.
8. Have the consultant give his protocol for completing the work.
9. Discuss the evaluative approach used by the consultant before the contract is executed.
10. Have mechanisms in place to hold the consultant accountable for not completing the agreed work.
11. Establish penalties for not meeting preset schedules.
12. Have a tracking system to evaluate the work's progress. Be consistent in your expectations, for each time you change the scope of work or delay progress it will cost you extra.
13. Follow good business practices and maintain a professional relationship.

Before you select a consultant you should do your homework. Make sure that the consultant can service your geographic area or multiple site locations depending upon your needs. If special equipment is needed for sampling, for example, you may want to visit the consultant's office to ensure that he or she has the necessary resources to accomplish what you want him or her to do. You should also check the background of the consultant, which should include the consultant's education, industrial experience, membership in professional organizations, and any certification. Remember that you have to work with the consultant. There needs to be a rapport between you and the consultant. Consulting services are an important function in providing assistance relevant to your safety and health issues.

Most consultants are skilled safety professionals who have years of experience and unique expertise. The service that they offer is well worth the price. Using a planned approach will ensure that you get the provider you need for the job. However, be cautious; there are some individuals who are less scrupulous. They will deceive you, take your money, waste your time, and leave you with an inferior product or no product. As Linda F. Johnson in Occupational Safety and Health, *Choosing a Safety Consultant* advises, "As in all business decisions do your homework."

CHAPTER 15

Who Knows What: Safety and Health Training

Training is appropriate where safe work performance skills are lacking.

Before beginning to read this chapter take time to assess where your safety and health effort is at present with reference to the training of your workforce on safety and health. For each of the three topics, you are to circle the number of the option that best depicts your training effort at present.

Safety and Health Training

Topic	Circle Answer	Answer Options
Employees learn about employee hazards (how to protect themselves and others)	5	Facility is committed to high-quality hazard training, ensures that all participate, and provides regular updates; in addition, employees can demonstrate proficiency in, and support of, all areas covered by training
	4	Facility is committed to high-quality employee hazard training, ensures all participate, and provides regular updates
	3	Facility provides legally required training and makes effort to include all employees

(continued)

Safety and Health Training (continued)

Topic	Circle Answer	Answer Options
	2	Training is provided when the need is apparent; experienced employees are assumed to know the material
	1	Facility depends on experience and informal peer training to meet needs
Supervisors learn responsibilities and underlying reasons	5	All supervisors assist in worksite hazard analysis, ensure physical protection, reinforce training, enforce discipline, and can improve work procedures based on the training provided to them
	4	Most supervisors assist in worksite hazard analysis, ensure physical protection, reinforce training, enforce discipline, and can work procedures based on the training provided to them
	3	Supervisors have received basic training, appear to understand and demonstrate the importance of worksite hazard analysis, physical protection, training reinforcement, discipline, and knowledge of work procedures
	2	Supervisors make responsible efforts to meet safety and health responsibilities, but have limited training
	1	There is no formal effort to train supervisors in safety and health responsibilities
Managers learn safety and health program management	5	All managers have received formal training in safety and health management responsibilities
	4	All managers follow, and can explain, their roles in safety and health program management
	3	Managers generally show a good understanding of their safety and health roles and usually model them
	2	Managers are generally able to describe their safety and health roles, but often have trouble modeling them
	1	Managers generally show little understanding of their safety and health management responsibilities

If your workforce is not trained to perform their work in a safe and healthy manner, you are bound to have an ineffective safety and health effort. Seldom have managers been trained in safety and health principles and often the supervisors have had no training in how to manage safety and health or in how to perform the safety and health requirements of their job. Many times supervisors are selected for their job knowledge and communication skills without regard to their knowledge of safety and health. Just because you have an experienced workforce does not mean that they are attuned to hazards of their job and your expectations of how they should handle those hazards. These experienced workers should receive safety and health training regarding your initiatives and their role in the safety and health effort. Without fail any worker you hire should have safety and health training.

Many standards promulgated by Occupational Safety and Health Administration (OSHA) specifically require the employer to train employees in the safety and health aspects of their jobs. Other OSHA standards make it the employer's responsibility to limit certain job assignments to employees who are certified, competent, or qualified—meaning that employees have had special, previous training, in or out of the workplace. OSHA regulations imply that an employer has assured that a worker has been trained before he or she is designated as the individual to perform a certain task. In Appendix G you can find a listing of the OSHA regulations, which require that the employer must train his or her employees.

To make a complete determination of the OSHA requirement for training, one would have to go directly to the regulation that applies to the specific type of activity. The regulation may mandate hazard training, task training, and length of the training, as well as specifics to be covered by the training.

It is always a good idea for the employer, as well as the worker, to keep records of training. These records may be used by a compliance inspector during an inspection, after an accident resulting in injury or illness, and as a proof of good intentions, by the employer, to comply with training requirements for workers including new workers and those assigned new tasks.

TRAINING AND EDUCATION

Training and education is one of the most important elements of any safety and health program. Each training item should describe methods for introducing and communicating new ideas into the workplace, reinforcing existing ideas and procedures, and implementing your safety and health program into action. The training needs may range between manager and supervisor training, worker task training, employee updates, and new worker orientation. The content of new worker or new site training should include at least the following topics:

- Company safety and health program and policy
- Employee and supervisory responsibilities
- Hazard communication training
- Emergency and evacuation procedures
- Location of first-aid stations, fire extinguishers, and emergency telephone numbers
- Site-specific hazards
- Procedures for reporting injuries
- Use of personal protective equipment (PPE)
- Hazard identification and reporting procedures
- Review of each safety and health rule applicable to the job
- Site tour or map where appropriate

It is a good idea to have follow-up for all training, which may include working with a more experienced worker, supervisor coaching, job observations, and reinforced good/safe work practices.

Supervisors/forepersons are responsible for the prevention of accidents for tasks under their direction and for thorough accident prevention and safety training for the employees they supervise. Therefore, all supervisors/forepersons will receive training so that they have a sound theoretical and practical understanding of the site-specific safety program, OSHA construction regulations, and the company's specific safety and health rules. They should also receive training on the OSHA hazard communication standard, site emergency response plans, first aid and cardiopulmonary resuscitation (CPR), accident and injury reporting and investigation, and procedures for safety communications, such as toolbox safety talks. Beyond these training requirements, described previously, additional training might cover the implementation and monitoring of your safety and health program, personnel selection techniques, OSHA record keeping requirements, and motivating individuals and groups.

Safety and health training should be a major component of your written safety and health program. There are times to train and there are times when training will not make a difference. If the worker has the job skills and can perform the job but is not doing the job in a safe manner when he or she knows the safe way to perform the job, then you have a behavioral problem, which cannot be cured by more training. If someone does not have the skills needed to do the job or has not done the job in a long time, then training is a viable option. Training is not just for workers but should be designed for managers and supervisors since they are charged with safety and health initiatives and they need to have the same knowledge base. Thus, training is a critical component of safety and health.

Safety and health training is not the answer to accident prevention. Training cannot be the sum total answer to all accidents in the workplace. In fact, training is only applicable when a worker has not been trained previously, is new to a job or task, or when safe job skills need to be upgraded.

If a worker has the skills to do his or her job safely, then training will not address unsafe job performance. The problem is not the lack of skill to do the job safely, but the worker's unsafe behavior. Do not construe these statements to suggest that safety and health training does not have an important function as part of an accident prevention program. Without a safety and health training program, a vital element of workplace safety and health is missing.

Safety and health training should include not only workers but supervisors and management. Without training for managers and supervisors, it cannot be expected or assumed that they are cognizant of the safety and health practices of your company. Without this knowledge, they will not know safe from unsafe, how to implement the loss control program, or even how to reinforce, recognize, or enforce safe work procedures unless proper training on occupational safety and health has transpired.

Safety and health training is critical to achieving accident prevention. Companies who do not provide new-hire training, supervisory training, and worker safety and health training have an appreciably greater number of injuries and illnesses than companies who carry out safety and health training. You may offer all types of programs and use many of the recognized accident prevention techniques but without workers who are trained in their jobs and work in a safe manner, your efforts to reduce and prevent accidents and injuries will result in marginal success. If a worker has not been trained to do a job in a productive and safe manner, a very real problem exists. Do not assume that a worker knows how to do his or her job and will do it safely unless he or she has been trained to do so. Even with training, some may resist safety procedures and then you have a department or behavioral problem and not a training issue.

It is always a good practice to train newly hired workers and experienced workers who have been transferred to a new job. It is also important that any time a new procedure, new equipment, or extensive changes in job activities occur, workers receive training. Well-trained workers are more productive, more efficient, and safer.

Training for the sake of documentation is a waste of time and money. Training should be purposeful and goal or objective driven. An organized approach to on-the-job safety and health will yield the proper ammunition to determine your real training needs. These needs should be based on accidents/incidents, identified hazards, hazard/accident prevention initiatives, and input from your workforce. You may then tailor training to meet the company's needs and those of the workers.

Look for results from your training. Evaluate those results by looking at the reduced number of accidents/incidents, improved production, and good safety practices performed by your workforce. Evaluate the results by using job safety observations and safety and health audits, as well as statistical information on the numbers of accidents and incidents.

Many OSHA regulations have specific requirements on training for fall protection, hazard communication, hazardous waste, asbestos and lead abatement, scaffolding, etc. It seems relatively safe to say that OSHA expects workers to have training on general safety and health provisions, hazard recognition, as well as task-specific training. Training workers regarding safety and health is one of the most effective accident prevention techniques.

WHEN TO TRAIN

There are appropriate times when safety and health training should be provided. They are when

- Worker lacks the safety skills.
- New employee is hired.
- Employee is transferred to another job or task.
- Changes have been made in the normal operating procedures.
- Worker has not performed a task for some period of time.

OSHA TRAINING MODEL

The OSHA training model is designed to be one that even the owner of a business with very few employees can use without hiring a professional trainer or purchasing expensive training materials. Using this model, employers or supervisors can develop and administer safety and health training programs that address problems specific to their own business, fulfill the learning needs of their employees, and strengthen the overall safety and health program of the workplace.

Determining if Training Is Needed

The first step in any training process is to determine whether a problem can be solved by training. Whenever employees are not performing their jobs properly, it is often assumed that training will bring them up to standard procedure. However, it is possible that other actions (such as hazard abatement or the implementation of engineering controls) can enable employees to perform their jobs properly. Ideally, safety and health training should be provided before problems or accidents occur. This training should cover both general safety and health rules and specific work procedures; training should be repeated if an accident or near-miss incident occurs.

Problems that can be addressed effectively by training include those that arise from lack of knowledge of a work process, unfamiliarity with equipment, or incorrect execution of a task. Training is less effective (but still can be used) for problems arising from an employee's lack of motivation or lack of attention to the job. Whatever its purpose, training is most effective when designed in relation to the goals of the employer's total safety and health program. An example of a training needs assessment instrument that could be used with supervisors or workers can be found in Figure 15.1.

Identifying Training Needs

If the problem is one that can be solved, in whole or in part, by training, then the next step is to determine the type of training needed. For this, it is necessary to identify what the employee is expected to do and in what ways, if any, the employee's performance is deficient. This information can be obtained by conducting a job analysis, which pinpoints what an employee needs to know to perform a job.

When designing a new training program, or preparing to instruct an employee in an unfamiliar procedure or system, a job analysis can be developed by examining engineering data on new equipment or the safety data sheets on unfamiliar substances. The content of the specific federal or state OSHA standard, applicable to a business, can also provide direction in developing training content. Another option is to conduct a job hazard analysis. This is a procedure for studying and recording each step of a job, identifying existing or potential hazards, and determining the best way to

---

Safety and health training needs assessment

In trying to determine the safety and health training needs for supervisors and works at XYZ company, we are asking you to give us assistance by completing the following questionnaire. Please place on the back of this sheet under other comments any guidance or concerns which you have.

I. In order to provide you with an opportunity to have input on XYZ's safety and health training, select only 12 of the following topics that you deem most important for such a training program.

☐ Safety and health management	☐ Tracking safety performance
☐ Ergonomics	☐ Ladder safety
☐ Welding and cutting safety	☐ Fall protection
☐ Electrical safety	☐ Crane safety
☐ Lifting safety	☐ Hazardous chemicals
☐ Confined spaces	☐ Mobile equipment
☐ Hazard communications	☐ Powered industrial trucks
☐ Mobile work platforms and lifts	☐ Working around water
☐ Haulage equipment	☐ Material handling
☐ Fire prevention and protection	☐ Industrial hygiene
☐ Environmental management	☐ Accident investigation
☐ Promoting safety and health	☐ Walking and working surfaces
☐ Rigging safety	☐ Personal protective equipment
☐ Safety and health training	☐ Equipment and machine guarding
☐ Scaffolding safety	☐ Motivating safety and health
☐ Communicating safety and health	☐ Lockout/tagout
☐ Hand and power tool safety	☐ Accountability and responsibility for safety and health

II. Are there any other specific topics that you think should be covered that were not part of the previous list?

_____
_____

III. List and rank the five most important topics to you (1 would be the most important and 5 the least important):

_____
_____

IV. How should the training be conducted to make it most effective?

_____
_____

V. How long should each training module last? _____

VI. Other comments: What other recommendations or guidance do you have for the training?

_____
_____

Figure 15.1   Safety and health training needs assessment.

perform the job to reduce or eliminate risks. Information obtained from a job hazard analysis can be used as the content for training activity.

If an employer's training needs can be met by revising an existing training program rather than developing a new one, or if the employer already has some knowledge of the process or system to be used, appropriate training content can be developed through such means as

- Using company accident and injury records to identify how accidents occur and what can be done to prevent them from recurring.
- Requesting employees to provide, in writing and in their own words, descriptions of their jobs. These should include the tasks performed and the tools, materials, and equipment used.
- Observing employees at the worksite as they perform tasks, asking about the work, and recording their answers.
- Examining similar training programs offered by other companies in the same industry, or obtaining suggestions from such organizations as the National Safety Council, the Bureau of Labor Statistics, OSHA-approved state programs, OSHA full service area offices, OSHA-funded State consultation programs, or the OSHA Office of Training and Education.

Employees themselves can provide valuable information on the training they need. Safety and health hazards can be identified through the employees' responses to such questions as whether anything about their jobs frightens them, have they had any near-miss incidents, do they feel they are taking risks, or do they believe that their jobs involve hazardous operations or substances.

Identifying Goals and Objectives

Once the need for training has been determined, it is equally important to determine what kind of training is needed. Employees should be made aware of all the steps involved in a task or procedure, but training should focus on steps in which improved performance is needed. This avoids unnecessary training and tailors the training to meet the needs of the specific employees and processes. Once the employees' training needs have been identified, employers can then prepare objectives for the training. Instructional objectives, if clearly stated, will tell employers what they want their employees to do, to do better, or to stop doing.

Learning objectives do not necessarily have to be written, but in order for the training to be as successful as possible, clear and measurable objectives should be thought out before the training begins. For an objective to be effective it should identify as precisely as possible what the individuals will do to demonstrate what they have learned or that the objective has been reached. They should also describe the important conditions under which the individual will demonstrate competence and define what constitutes acceptable performance.

Using specific, action-oriented language, the instructional objectives should describe the preferred practice or skill and its observable behavior. For example, rather than using the statement: "The employee will understand how to use a respirator" as an instructional objective, it would be better to say, "The employee will be able to describe how a respirator works and when it should be used." Objectives are most effective when worded in sufficient detail, which other qualified persons can recognize when the desired behavior is exhibited.

Developing Learning Activities

Once employers have stated precisely what the objectives for the training program are, then learning activities in the form of training methods and materials can be identified and described. Methods and materials for the learning activity can be as varied as the employer's imagination and available resources will allow. The learning activities should enable employees to demonstrate that they have acquired the desired skills and knowledge.

Start by establishing objectives. To ensure that employees transfer the skills or knowledge from the learning activity to the job, the learning situation should simulate the actual job as closely as possible. Thus, employers may want to arrange the objectives and activities in a sequence that corresponds to the order in which the tasks are to be performed on the job, if a specific process is to be learned. For instance, if an employee must learn the start-up processes of using a machine, the sequence might be (1) check that the power source is connected, (2) ensure that the safety devices are in place and are operative, and (3) know when and how to throw the off switch.

Next, a few factors will help to determine the training methods to be incorporated into the training. One factor concerns the training resources available to the employer. For example, can a group training program using an outside trainer and a film be organized, or should the employer have individuals train employees on a one-to-one basis? Another factor concerns the kind of skills or knowledge to be learned. Is the learning oriented toward physical skills (such as the use of special tools) or toward mental processes and attitudes? Such factors will influence the type of learning activity designed by employers. For example, the training activity can be group-oriented, with lectures, role-playing, and demonstrations; or it can be designed for the individual with self-paced instruction.

Finally, decide on materials; the employer may want to use materials such as charts, diagrams, manuals, slides, films, viewgraphs (overhead transparencies), videotapes, audiotapes, a simple blackboard, or any combination of these and other instructional aids. Whatever the method of instruction, learning activities should be developed in such a way that the employees can clearly demonstrate that they have acquired the desired skills or knowledge.

Conducting the Training

After organizing the learning activity, the employer is ready to begin conducting the training. To the extent possible, training should be presented so that its organization and meaning are clear to employees. This will help to reinforce the message and the long-term recall. Employers or supervisors should (1)provide overviews of the material to be learned; (2) relate, wherever possible, the new information or skills to the employees' goals, interests, or experience; and (3) reinforce what the employees have learned by summarizing both the program's objectives and the key points of information covered. These steps will assist employers in presenting the training in a clear, unambiguous manner.

In addition to organizing content, employers must also develop the structure and format of the training. The content developed for the program should closely relate to the nature of the workplace, or other training site, and the resources available for training. Having planned this, employers will be able to determine for themselves the frequency of training activities, the length of the sessions, the instructional techniques, and the individuals best qualified to present the information.

To motivate employees to pay attention and learn from the training activities provided by the employer or supervisor, each employee must be convinced of the importance and relevance of the material. Some ways to develop motivation are

- Explain the goals and objectives of instruction.
- Relate the training to the interests, skills, and experiences of the employees.
- Outline the main points to be presented during the training sessions.
- Point out the benefits of training (e.g., the employee will be better informed, more skilled, and thus, more valuable both on the job and on the labor market; or the employee will, if he or she applies the skills and knowledge learned, be able to work at reduced risk).

An effective training program allows employees to participate in the training process and practice their skills or knowledge. This will help to ensure that they are learning the required knowledge or skills, and it permits correction if necessary. Employees can become involved in the training process by participating in discussions, asking questions, contributing their knowledge and expertise, learning through hands-on experiences, and through role-playing exercises.

Do not overlook within company resources and expertise. It has been found by many companies that using their own employees when they are qualified is often the most prudent answer to instructors and trainers. These trainers/instructors know the business, policies, and procedures used by the company and therefore are able to provide more relevant training and training that is more credible to the trainees. Outside instructors should be used for technical, specialty, or training beyond the expertise of in-house employees.

Evaluating Program Effectiveness

To make sure that the training program is accomplishing its goals, an evaluation of the training can be valuable. Training should have, as one of its critical components, a method of measuring its effectiveness. A plan for evaluating the training sessions, either written or thought-out by the employer, should be developed when the course objectives and content are developed. It should not be delayed until the training has been completed. Evaluation will help employers and supervisors determine the amount of learning achieved and whether an employee's performance has improved on the job. Methods of evaluating training include:

- Student opinion questionnaires—informal discussions with employees can help employers determine the relevance and appropriateness of the training program.
- Supervisors' observations—supervisors are in a good position to observe an employee's performance both before and after the training and note improvements or changes.
- Workplace improvements—the ultimate success of a training program may be changes throughout the workplace that result in reduced injury or accident rates.

However it is conducted, an evaluation of training can give employers the information necessary to decide whether employees achieved the desired results and whether the training session should be offered again at some future date.

Improving the Program

After evaluation, if it is clear that the training did not give the employees the level of knowledge and skill that was expected, then it may be necessary to revise the training program or provide periodic retraining. At this point, asking questions of employees and of those who conducted the training may be of some help. Among the questions that could be asked are

- Were parts of the content already known and, therefore, unnecessary?
- What material was confusing or distracting?
- Was anything missing from the program?
- What did the employees learn, and what did they fail to learn?

It may be necessary to repeat steps in the training process, that is, to return to the first steps and retrace one's way through the training process. As the program is evaluated, the employer should ask:

- When a job analysis was conducted, was it accurate?
- Was any critical feature of the job overlooked?
- Were the important gaps in knowledge and skill included?
- Was material already known by the employees intentionally omitted?
- Were the instructional objectives presented clearly and concretely?
- Did the objectives state the level of acceptable performance that was expected of employees?
- Did the learning activities simulate the actual job?
- Were the learning activities appropriate for the kinds of knowledge and skills required on the job?
- When the training was presented, was the organization of the material and its meaning made clear?
- Were the employees motivated to learn?
- Were the employees allowed to participate actively in the training process?
- Was the employer's evaluation of the program thorough?

A critical examination of the steps in the training process will help employers determine where course revisions are necessary.

EDUCATION/TRAINING AND TECHNOLOGY

Occupational safety and health has the opportunity to make use of technology for training purposes. Use of the computer and the Internet provides access to the most up-to-date information. Computer training modules or online training although expensive to generate provide a source of training for individuals or groups and can be utilized at the worksite or at home, which makes for convenience.

In addition, due to the hazards involved in the workplace from high-energy sources, hazardous chemicals/waste, critical processes, or jobs that require special skills Training is needed to perform them safely. Thus, the use of computer simulations, mockups, or virtual reality allows for practice in performing high-risk operations, handling toxic chemicals, using explosive materials, and demonstrating skills needed to safely operate high-risk or hazardous equipment or operations.

The use of technology can prevent injuries and illnesses that could transpire while learning to perform or handle dangerous materials. The technology used in training reduces the potential exposure for such individuals as firefighters or airplane pilots. By developing skills before actual performance of high-risk or hazardous jobs such as nuclear plant operations, moving hazardous waste, or operating larger pieces of equipment, potential disasters or release of poisonous materials possibly resulting in exposure for large numbers of individuals or workers can be minimized. Although these types of training approaches may be expensive, the cost is well worth the prevention that is provided.

TRAINING NEW HIRES

History has shown that individuals new to the workplace suffer more injuries and deaths than experienced workers. These usually occur within days or the first month on the job. It is imperative that new hires receive initial training regarding safety and health practices at the worksite. Some potential topics, which should be covered during the training of a worker new to your workplace, are

- Accident reporting procedures
- Basic hazard identification and reporting
- Chemical safety
- Company's basic philosophy on safety and health
- Company's safety and health rules
- Confined space entry
- Electrical safety
- Emergency response procedures (fire, spills, etc.)
- Eyewash and shower locations
- Fall protection
- Fire prevention and protection
- First aid and /CPR
- Hand tool safety
- Hazard communications
- Housekeeping
- Injury reporting procedures
- Ladder safety
- Lockout/tagout procedures
- Machine guarding

- Machine safety
- Material handling
- Mobile equipment
- Medical facility location
- Personal responsibility for safety
- Rules regarding dress code, conduct, and expectations
- Unsafe acts and conditions reporting procedures
- Use of PPE

TRAINING SUPERVISORS

The supervisor is the key person in the accomplishment of safety and health in the workplace. The supervisor will often be overlooked when it comes to safety and health training. Thus, the supervisor is frequently ill equipped to be the lead person for the company's safety and health initiative. A list of suggested training topics for supervisors includes

- Accident causes and basic remedies
- Building attitudes favorable to safety
- Communicating safe work practices
- Cost of accidents and their effect on production
- Determining accident causes
- First-aid training
- Giving job instruction
- Job instruction for safety
- Knowledge of federal and state laws
- Making the workplace safe
- Mechanical safeguarding
- Motivating safe work practices
- Number and kinds of accidents
- Organization and operation of a safety program
- Safe handling of materials
- Supervising safe performance on the job
- Supervisor's place in accident prevention
- Investigation and methods of reporting accidents to the company and government agencies
- The number and kinds of accidents
- The cost of accidents and their effect on production

TRAINING EMPLOYEES

The training of all workers in safety and health has been demonstrated to reduce costs and increase the bottom line. Workers must be trained in safety and health just as you would conduct job skill training in the form of on-the-job training (OJT) or job instruction training (JIT). Most employers and safety professionals realize that a skilled worker is a safe worker. The following are some recommendations that should be addressed to ensure that workers are trained or that training has been upgraded.

- Any new hazards or subjects of importance
- Basic skills training
- Explanation of policies and responsibilities
- Federal and state laws
- First-aid training
- Importance of first-aid treatment
- Methods of reporting accidents

- New safe operating procedures
- New safety rules and practices
- New skill training for new equipment, etc.
- Technical instruction and job descriptions
- Where to get first aid
- Where to get information and assistance
- Any other subjects of importance

TRAINING

Experience shows that a good safety program is based on a well-planned, ongoing training program. The training program should include both safety and the skills involved in performing the tasks to be accomplished. The training program should include a minimum of classroom-type experiences and continual reinforcement of training concepts in OJT situations. The length of the training sessions is not as important as their quality. A well-planned program will save time and increase the effectiveness of the training. The basic general objectives of the training program should include the following:

- An understanding of the company's basic philosophy and genuine concern for safety on the job:

Remember, behavior of the supervisor will say more than any training program. Safe behavior of the supervisor is a must for any safety program.

- Basic skills training. A skilled worker is a safe worker.
- Thorough knowledge of company safety policies and practices.
- Indoctrination of new employees.
- Annual training required, as necessary, or any required by law.

Subject areas that can sharpen supervision skills also enhance safety effectiveness, especially in the areas of

- Giving job instruction
- Supervising employees on the job
- Determining accident causes
- Building safety attitudes*

Training programs may be obtained from a variety of sources.[†] Much of the training is provided without charge by

- Government agencies, both federal and state
- Professional and industrial associations
- Equipment and product representatives
- Insurance companies
- Films and audiovisual materials (on loan)
- Qualified or experienced people in your own company
- Safety and health consultants

Attempt to keep safety subjects pertinent. Think of upcoming jobs and anticipate possible safety concerns. An excellent safety meeting may be nothing more than relating some precautions to a

---

* A side benefit of such training is a more effective supervisor for production and cost-effectiveness.

[†] Use variety to keep your safety training interesting. Outside speakers, audiovisual aids, incentives, and awards are all proven techniques.

```
┌─────────────────────────────────────────────────────────────────────────┐
│                     Individual worker's training form*                    │
│                                                                           │
│   Worker's name_____ Soc. Sec. # _____    │
│                                                                           │
│   Clock number _____  │
│                                                                           │
│   Subject              Date        Length of training   Instructor   Worker's signature │
│                                                                           │
│   New hire orientation                                                    │
│                                                                           │
│   Hazard communications                                                   │
│                                                                           │
│   Hazardous waste                                                         │
│                                                                           │
│   *Keep this form in the employee's personnel file.                       │
└─────────────────────────────────────────────────────────────────────────┘
```

Figure 15.2   Individual worker's training form.

crew when giving work instructions. Plan your training program in advance so that the time spent will be worthwhile.

Documenting Safety and Health Training

Each worker's safety and health training should be documented and placed in his or her permanent personnel file. The documentation should include the name of the worker, job title, social security number, clock number, date, topic or topics covered, duration of training, and the trainer's name. This should be signed and dated by the trainer. This is the documentation of training, which an OSHA inspector, legal counsel, or other interested parties would desire to verify. An example of the individual worker's training form used to document training can be seen in Figure 15.2. This can be used to document training at your company and to provide the worker a copy of training they attended.

In Figure 15.3, you can find a sample written statement for inclusion in your written safety and health program describing your company's commitment to safety and health training.

Figure 15.4 presents a more detailed company training form that the worker can sign and that can be kept as a part of individual personnel records for future reference and training tracking purposes.

```
┌─────────────────────────────────────────────────────────────────────────┐
│                                                                           │
│   All employees, from supervisors to workers, will receive safety training on all phases of work performed │
│   by (Name of Company). The following safety education and training practices will be implemented and │
│   enforced at all company facilities/jobsites.                            │
│                                                                           │
│       New employees and current employees who are transferred from another facility or jobsite must attend │
│   a facility or jobsite-specific new-hire safety orientation. This program provides each employee the basic │
│   information about the facility/jobsite-specific safety and health plan, federal and state OSHA standards, │
│   and other applicable safety rules and regulations. Attendance is mandatory before working for all company │
│   employees. The facility/jobsite supervisor is to document attendance using the company training form │
│   (see example, Figure 15.4) and all training records will be maintained by the company and placed in each │
│   worker's personnel file.                                                │
│                                                                           │
└─────────────────────────────────────────────────────────────────────────┘
```

Figure 15.3   Sample policy statement on training for your written safety and health program.

Company training form

Worker's name _____ Soc. Sec. #_____

Date(s) of training_____

Length of training _____ Hours

Subjects covered (check all that apply)

_____ New hire orientation	_____ Fall protection
_____ Company rules and policies	_____ Electrical safety
_____ Hazard communications	_____ Use and care of hand tools
_____ Fire safety and fire prevention	_____ Ladder safety
_____ Scaffolding safety	_____ Vehicle safety
_____ Machine and equipment guarding	_____ Trench/excavation safety
_____ Steel erection	_____ Rigging
_____ Material handling	_____ Explosive/blasting
_____ Personal protective equipment	_____ Confined space entry
_____ Respirator use	_____ Asbestos abatement
_____ Lead abatement	_____ Hazardous waste
Other	remediation

_____

_____

_____

Worker's signature _____ Date _____

Figure 15.4   Company training form.

CHAPTER 16

Guiding Light: OSHA Compliance

OSHA inspectors are charged with enforcing workplace laws and regulations.

As an employer it is critical that you understand how Occupational Safety and Health Administration (OSHA) works, achieves its mission, and strives to protect the American workforce. OSHA can be an ally or a thorn in your side, depending on your approach to job safety and health.

Workers should expect to go to work each day and return home uninjured and in good health. There is no logical reason that a worker should be part of workplace carnage. Workers do not have to become one of the yearly workplace statistics. Employers who enforce the occupational safety and health rules and safe work procedures are less likely to have themselves or their workers become one of the 6,500 occupational trauma deaths, one of 90,000 occupational illness deaths, or even one of the 6.8 million nonfatal occupational injuries and illnesses that occur each year in the United States.

OSHA and its regulations should not be the driving force that ensures workplace safety and health. Since OSHA has limited resources and inspectors, enforcement is usually based on serious complaints, catastrophic events, and workplace fatalities. The essence of workplace safety and the strongest driving catalyst should first be the protection of the workforce, followed by economic incentives for the employer. Employers who have a good safety and health program and record will reap the benefits: a better opportunity to win more customers; lower insurance premiums for workers' compensation; decreased liability; and, increased employee morale and efficiency. Usually safety and health are linked to the bottom line (company's income), which is seldom perceived as humanitarian.

This chapter will provide answers to many of the questions that are asked regarding OSHA and workplace safety and health and will suggest how employers and their workforce can work together to provide a safe and healthy workplace. This information is a guide to understanding OSHA, OSHA compliance, and ensuring safer and healthier worksites.

During the many years preceding OSHA, it became apparent that employers needed guidance and incentives to ensure safety and health on the jobsite. The employer needed to realize that workers had a reasonable right to expect a safe and healthy workplace. This guidance and the guarantee of a safe and healthy workplace came to fruition with the enactment of the Occupational Safety and Health Act (OSHAct) of 1970. OSHA was created by the Act to

- Encourage employers and employees to reduce workplace hazards and to improve existing safety and health programs or implement new programs
- Provide for research in occupational safety and health to develop innovative ways of dealing with occupational safety and health problems
- Establish separate-but-dependent responsibilities and rights for employers and employees for the achievement of better safety and health conditions
- Maintain a reporting and record keeping system to monitor job-related injuries and illnesses
- Establish training programs to increase the numbers and competence of occupational safety and health personnel
- Develop mandatory job safety and health standards and enforce them effectively
- Provide for the development, analysis, evaluation, and approval of state occupational safety and health programs

Thus, the purpose of OSHA is to ensure, as much as possible, a healthy and safe workplace free of hazardous conditions for workers in the United States.

OSHA STANDARDS

OSHA standards, found in the Code of Federal Regulations (CFR), include the standards for the following industry groups: construction; maritime; agriculture; the general industry, which includes manufacturing; transportation and public utilities; wholesale and retail trades; finance; insurance; and services. Some of the specific areas covered by regulations are found in Table 16.1.

OSHA standards and regulations for occupational safety and health are found in Title 29 of the CFR and can be obtained through the Government Printing Office (GPO). The standards for specific industries are found in Title 29 of the CFR (Table 16.2).

An employer can seek relief (variance) from an OSHA standard. The reasons for variances approved by OSHA are

- Employer may not be able to comply with the standard by its effective date.
- Employer may not be able to obtain the materials, equipment, or professional or technical assistance needed to comply.

Table 16.1   Standard Topics

Lockout/tagout	Electrical safety
Housekeeping	Training requirements
Noise exposure	Fire prevention
Hazard communication	Confined spaces
PPE	Ventilation requirements
Sanitation	Medical and first aid
Fall protection	Working with hazardous substances
Emergency planning	Guarding
Record keeping	Machine and equipment safety
Use of hand tools	Radiation
Ladders and scaffolds safety	Bloodborne pathogens
Explosives/blasting	Walking/working surfaces
Compressed gases	Hazardous chemicals
Diving (commercial)	Welding and cutting
Carcinogens	

- Employer already has processes or methods in place that provide protection to workers and are at least as effective as the standard's requirements.

A temporary variance that meets the criteria listed above may be issued until compliance is achieved or for 1 year, whichever is shorter. It can also be extended or renewed for 6 months (twice). Employers may obtain a permanent variance if the employer can document with a preponderance of evidence that existing or proposed methods, conditions, processes, procedures, or practices provide workers with protection equivalent to or better than the OSHA standard. Employers are required to post a copy of the variance in a visible area in the workplace and make workers aware of a request for a variance.

PROTECTIONS UNDER THE OSHAcт

Usually all employers and their employees are considered to be protected under the OSHAcт, with the exception of

- Self-employed persons
- Farms where only immediate family members are employed
- Workplaces already protected under federal statutes by other federal agencies such as the Department of Energy and Mine Safety and Health Administration
- State and local employees

Table 16.2   CFRs for Industry-Specific Regulations

General industry—29 CFR PART 1910

Shipyard employment—29 CFR PART 1915

Marine terminals—29 CFR PART 1917

Longshoring—29 CFR PART 1918

Gear certification—29 CFR PART 1919

Construction—29 CFR PART 1926

Agriculture—29 CFR PART 1928

Federal agencies—29 CFR 1960

NATIONAL INSTITUTE FOR OCCUPATIONAL SAFETY AND HEALTH (NIOSH)

Although the formation of NIOSH was a requirement of the OSHAct of 1970, NIOSH is not part of OSHA. NIOSH is one of the Centers for Disease Control and Prevention, headquartered in Atlanta, Georgia. NIOSH reports to the Department of Health and Human Services (DHHS) and not to the Department of Labor (DOL) as OSHA does. Its functions are to

- Recommend new safety and health standards to OSHA
- Conduct research on various safety and health problems
- Conduct health hazard evaluations (HHEs) of the workplace when called upon
- Publish an annual listing of all known toxic substances and recommend exposure limits (RELs)
- Conduct training that will provide qualified personnel under the OSHAct

An employer, worker's representative, or worker can request a HHE from NIOSH to have a potential health problem investigated. It is best to use the NIOSH standard form. It can be obtained by calling 1-800-35-NIOSH. The HHE request should include the following information:

- Description of the problem.
- Symptoms exhibited by the workers.
- Name of the suspected substance (trade or chemical name).
- Process in which the problem is occurring.
- Hazard warning from the label or material safety data sheet (MSDS) of the substance.
- Length of time worker is exposed to it.
- When the symptoms were first noticed.
- Is this a new or old process or material that is used?
- Has this problem occurred previously?
- Has the complaint been registered with OSHA or another government agency?

OCCUPATIONAL SAFETY AND HEALTH REVIEW COMMISSION (OSHRC)

OSHRC was established, under the OSHAct, to conduct hearings when OSHA citations and penalties are contested by employers or by their employees. As with NIOSH, OSHRC formation was a requirement of the OSHAct but it is a separate entity apart from OSHA.

EMPLOYER RESPONSIBILITIES UNDER THE OSHAct

The employer is held accountable and responsible under the OSHAct. The General Duty Clause, Section 5(a)(1) of the OSHAct states that employers are obligated to provide a workplace free of recognized hazards that are likely to cause death or serious physical harm to employees. Employers must

- Abide and comply with OSHA standards.
- Maintain records of all occupational injuries and illnesses.
- Maintain records of workers' exposure to toxic materials and harmful physical agents.
- Make workers aware of their rights under the OSHAct.
- Provide, at a convenient location and at no cost, medical examinations to workers when OSHA standards require them.
- Report within 8 h to the nearest OSHA office all occupational fatalities or catastrophes where three or more employees are hospitalized.
- Abate cited violations of the OSHA standard within the prescribed time period.

- Provide training on hazardous materials and make MSDSs available to workers on request.
- Assure workers are adequately trained under the regulations.
- Post information required by OSHA such as citations, hazard warnings, and injury/illness records.

WORKERS' RIGHTS AND RESPONSIBILITIES UNDER THE OSHAct

Workers have many rights under the OSHAct. These rights include the rights to

- Review copies of appropriate standards, rules, regulations, and requirements that the employer should have available at the workplace
- Request information from the employer on safety and health hazards in the workplace, precautions that may be taken, and procedures to be followed if an employee is involved in an accident or is exposed to toxic substances
- Access relevant worker exposure and medical records
- Be provided with personal protective equipment (PPE)
- File a complaint with OSHA regarding unsafe or unhealthy workplace conditions and request an inspection
- Not be identified to the employer as the source of the complaint
- Not be discharged or discriminated against in any manner for exercising rights under the OSHAct related to safety and health
- Have an authorized employee representative accompany the OSHA inspector and point out hazards
- Observe the monitoring and measuring of hazardous materials and see the results of the sampling, as specified under the OSHAct and as required by OSHA standards
- Review the occupational injury and illness records (OSHA No. 200 or equivalent) at a reasonable time and in a reasonable manner
- Have safety and health standards established and enforced by law
- Submit to NIOSH a request for a HHE of the workplace
- Be advised of OSHA actions regarding a complaint and request an informal review of any decision not to inspect or issue a citation
- Participate in the development of standards
- Speak with the OSHA inspector regarding hazards and violations during the inspection
- File a complaint and receive a copy of any citations issued and the time allotted for abatement
- Be notified by the employer if the employer applies for a variance from an OSHA standard, testify at a variance hearing, and appeal the final decision
- Be notified if the employer intends to contest a citation, abatement period, or penalty
- File a notice of contest with OSHA if the time period granted to the company for correcting the violation is unreasonable, within 15 working days of the employer's notice
- Participate at any hearing before the OSHA review commission or at any informal meeting with OSHA when the employer or a worker has contested an abatement date
- Appeal the OSHRC's decisions in the U.S. Court of Appeals
- Obtain a copy of the OSHA file on a facility or workplace

Along with rights go responsibilities, and workers should be expected to conform to these responsibilities. Workers are expected to

- Comply with the OSHA regulations and standards
- Not remove, displace, or interfere with the use of any safeguards
- Comply with the employer's safety and health rules and regulations
- Report any hazardous conditions to the supervisor or employer
- Report any job-related injuries and illnesses to the supervisor or employer
- Cooperate with the OSHA inspector during inspections when requested to do so

One point that should be kept in mind is that it is the employer's responsibility to ensure that employees comply with OSHA regulations and safety and health rules. Workers are not held financially accountable by OSHA for violations of OSHA regulations. It is entirely up to the employer to hold employees accountable. With the accountability and responsibility falling upon the employer, he or she must take control and direct the safety and health effort at the workplace.

DISCRIMINATION AGAINST WORKERS

Workers have the right to expect safety and health on the job without fear of punishment. This is spelled out in Section 11(c) of the OSHACT and under 49 U.S.C. 31105 (formerly Section 405) for the trucking industry. The law states that employers shall not punish or discriminate against workers for exercising rights such as

- Complaining to an employer, union, or OSHA (or other government agency) about job safety and health
- Filing a safety and health grievance
- Participating in OSHA inspections, conferences, hearings, or OSHA-related safety and health activities

If workers believe that they are discriminated against for exercising their safety and health rights, they should contact the nearest OSHA office within 30 days of the time they sense that discriminatory activity has started. To file a formal complaint a worker should visit, call, or write the nearest OSHA office or state OSHA office, if a state program exists there. If a worker calls or visits, then a written follow-up letter should be sent. This may be the only documentation of a complaint. Complaints should be filed only when the following is occurring:

- Discrimination has been continuing.
- Employer has been devious, misleading, or has been concealing information regarding the grounds for the worker's discriminatory treatment.
- Worker has attempted to use the grievance or arbitration procedures under the collective bargaining agreement during the 30 days.

When OSHA receives a Worker's Discrimination Complaint, OSHA will review the facts of the complaint and decide whether to conduct an investigation. If an investigation ensues, the worker and the employer will be notified of the results within 90 days.

If the investigation indicates the worker's case has merit to process the case through the courts, OSHA or the state agency will attempt to negotiate with the employer. The settlement might include reinstatement of the worker's job, full back pay, and purging of the worker's personnel records. The employer might also be required to post a notice on the jobsite warning about any further workplace safety and health discrimination.

At times employers may decide not to settle. In this instance, OSHA or the state agency will submit the case to the U.S. District Court. The court can order the employer to reinstate the employee, pay lost wages, purge the worker's personnel records, and protect him or her from further discrimination.

If the investigation determines that the worker does not have a case, the worker may feel the decision was an error and may appeal the decision of OSHA or the state agency. The worker will need to provide a detailed explanation, as well as documentation, for contesting the prior decision. Workers can file a discrimination complaint with federal OSHA if the worker's state program and its courts do not offer to protect them from discrimination.

RIGHT TO INFORMATION

Workers have a right-to-know. This means that the employer must establish a written, comprehensive hazard communication program, which includes provisions for container labeling, MSDSs, and an employee training program. The program must include

- List of the hazardous chemicals in the workplace
- Means the employer uses to inform employees of the hazards of nonroutine tasks
- Way the employer will inform other employers of the hazards to which their employees may be exposed

Workers have the right to information regarding the hazards to which they are or will be exposed. They have the right to review plans such as the hazard communication plan. They have a right to see a copy of an MSDS during their shift and receive a copy of an MSDS when requested. In addition, information on hazards that may be brought to the workplace by another employer should be available to workers. Other forms of information such as exposure records, medical records, etc., are to be made available to workers upon request.

ASSURING A SAFE AND HEALTHY WORKPLACE

Workers have the right to refuse hazardous work. This right is not free of stipulations. Workers must ensure that three criteria are met:

1. Workers have a reasonable belief, based on what they know at the time, that there is a real likelihood they could be killed or suffer serious injury (imminent danger).
2. When the employer or supervisor has been asked to eliminate the danger and does not take action, the worker should ask for another assignment while the previous one is made safe. The worker should not return to the jobsite unless ordered to do so.
3. If the employer does not respond, workers should call the nearest OSHA office to explain the circumstances. If OSHA cannot respond in a timely manner, because of time constraints, the worker has no other alternative but to refuse the work.

Workers have the right to receive the results of any OSHA test for vapors, noise, dusts, mists, fumes, radiation, etc. This includes observation of any measurement of hazardous materials in the workplace. If the hazards of the workplace are such that PPE is required for workers, workers are to be provided, at no cost, proper and well-maintained PPE appropriate for the job.

WORKERS' COMPLAINTS

Workers have the right to complain to OSHA regarding workplace safety and health concerns. Workers will need to contact the nearest federal OSHA office or state OSHA office if the state has a state plan. A listing of the OSHA offices for each state can be found in Appendix H or they may use the National OSHA hotline (after hours) 1-800-321-OSHA. An OSHA complaint should include the following information:

- Description of the problem, work process, or job
- Type of inspection requested
- Location of the hazard
- Identification of the problem as a health or safety hazard

- Number of workers endangered or exposed
- Identification of the intensity of the hazard and if there is an immediate danger to life and health
- Identification of the standard violated
- Record of a previous violation for this hazard
- Whether the work has been shut down by the employer or by a federal or state agency
- Time when the hazard was first noticed
- Notification of employer
- Request for confidentiality
- Request for an employee representative to accompany the inspector
- Request for a closing conference
- Notification of any worker who has been reprimanded or discriminated against for complaining about the hazard
- Notification if the complaint is going through any internal grievance procedure
- Proof of any written documentation regarding the hazard

OSHA INSPECTIONS

OSHA has the right to conduct workplace inspections as part of its enforcement mandate. OSHA can routinely initiate an unannounced inspection of a business. Other inspections occur due to fatalities and catastrophes, routine program inspections, or by referrals and complaints. These occur during normal working hours.

Workers have the right to request an inspection. The request should be in writing (either by letter or by using the OSHA complaint form to identify the employer and the alleged violations). Send the letter or form to the area director or state OSHA director. If workers receive no response, they should contact the OSHA regional administrator. It is beneficial to call the OSHA office to verify its normal operating procedures. If workers allege an imminent danger, they should call the nearest OSHA office.

These inspections include checking company records, reviewing compliance with the hazard communication standard, fire protection, PPE, and review of the company's health and safety plan. This inspection will include conditions, structures, equipment, machinery, materials, chemicals, procedures, and processes. OSHA's priorities for scheduling an inspection are rank-ordered as follows:

- Situations involving imminent danger
- Catastrophes or fatal accidents
- Complaints by workers or their representatives
- Referral from other state and federal agencies or media
- Regular inspections targeted at high-hazard industries
- Follow-up inspections

Usually no advance notice is given to an employer before an inspector appears at a jobsite. However, there are times when advance notice is an acceptable practice. They are

- In case of an imminent danger
- When it would be effective to conduct an inspection after normal working hours
- When it is necessary to ensure the presence of the employer or a specific employer or employee representatives
- When the area director determines that an advance notice would enhance the probability of a more thorough and effective inspection

No inspection will occur during a strike, work stoppage, or picketing action unless the area director approves such action. Usually this type of inspection would be due to extenuating circumstances such as an occupational death inside the facility. The steps of an OSHA inspection encompass the following:

- Inspector becoming familiar with the operation including previous citations, accident history, business demographics and gaining entry to the operation. OSHA is forbidden from making a warrantless inspection without the employer's consent. Thus, the inspector may have to obtain a search warrant if reasonable grounds for an inspection exist, and entry has been denied by the employer.
- Inspector holding an opening conference with the employer or a representative of the company. It is required that a representative of the company be with the inspector during the walk around and a representative of the workers be given the opportunity to accompany the inspector.
- Inspection tour could take hours or possibly days, depending on the size of the operation. The inspector usually covers every area within the operation while assuring compliance with OSHA regulations.
- Closing conference is conducted, which gives the employer an opportunity to review the inspector's findings. The inspector will request from the employer an abatement time for the violations to be corrected. An employee representative (union) will also be afforded an opportunity to have separate opening and closing conferences.
- Area director will issue, to the employer, the written citations with proposed penalties and abatement dates. This document is called notification of proposed penalty.

WORKERS' COMPLAINTS AND REQUESTS FOR INSPECTIONS

Requesting an OSHA inspection is a right that should be used in a prudent and responsible manner and only after all other options have been exhausted. Workers' complaints are the most frequent reason for OSHA inspections.

Requests for these types of inspections should be in writing, using the OSHA complaint form or letter. The complaint should include information on the ongoing work process, the number of workers affected, the nature of the problem, a safety or health hazard, and an indication that the worker has tried to get the employer to fix the problem or remove the hazard. A written complaint guarantees a written record. This means that OSHA has to keep the worker informed of the results and it will protect a worker against employer discrimination.

When OSHA receives a complaint, OSHA gathers information concerning the complaint and decides whether or not the complaint warrants sending a compliance officer (inspector) to the site. OSHA will not inspect if the complaint does not indicate adequate cause or if the complaint was aimed at harassing the employer. In the case of a noninspection, the complainant will be notified and a copy of the complaint sent to the employer. Workers requesting an inspection have the right to know of any actions OSHA takes concerning their request and may have an informal review when OSHA decides not to inspect. If OSHA decides to conduct an inspection, workers should do the following during the inspection:

- Cooperate with the OSHA compliance inspector.
- Have a worker representative accompany the inspector.

After a formal complaint is made, OSHA's normal time constraints for conducting an inspection are based on the seriousness of the complaint. The usual times are

- Within 24 h if the complaint alleges an imminent danger
- Within 3 days if the complaint is serious
- Within 20 days for all other complaints

Upon completion of the inspection, the workers' representative can request the inspector to conduct a closing conference for labor. There will be a separate closing conference for the employer. The employer must post all citations issued. During the closing conference workers should

- Ask the inspector to describe all hazards discovered and standard violations found
- Make sure the inspector has all the information, as well as information on other complaints
- Keep a written record and specific notes of the closing conference
- Ask about the procedures and results that will ensue from the inspection

CITATIONS, PENALTIES, AND OTHER ENFORCEMENT MEASURES

If violations of OSHA standards are detected during an inspection, the citations will include the following information:

- Violation
- Workplace affected by the violation
- Specific control measures to be taken
- Abatement period or time allotted to correct the hazard

Upon receipt of the penalty notification, the employer has 15 working days to submit a notice of contest, which must be given to the workers' authorized representative or, if no representative exists, it must be posted in a prominent location in the workplace. During the 15 days, it is recommended that the employer first request an informal conference with the area director. During the informal conference, the issues concerning the citations and penalties can be discussed. If the employer is not satisfied, a notice of contest can be filed. An employer who has filed a notice of contest may withdraw it before the hearing date by

- Showing that the alleged violation has been abated or will be abated
- Informing the affected employees or their designated representative of the withdrawal of the contest
- Paying the assessed fine for the violation

Workers can contest the length of the time period for abatement of a citation. They may also contest the employer's petition for an extension of time for correcting the hazard. Workers must do this within 10 working days of posting. Workers cannot contest the

- Employer's citations
- Employer's amendments to citations
- Penalties for the employer's citations
- Lack of penalties

Copies of the citation should be posted near the violation's location for at least 3 days or until the violation is abated, whichever is longer. Violations are categorized in the following manner (Table 16.3):

In describing these violations, the de minimis is the least serious and carries no penalty since it violates a standard that has no direct or immediate relationship to safety and health. An other-than-serious violation would probably not cause death or serious harm, but could have a direct effect on the safety or health of employees. Serious violations are those where a substantial probability of death or serious physical harm could result. Willful violations are violations where an employer has deliberately, voluntarily, or intentionally violated a standard. In addition, repeat violations are ones that occur within 3 years of an original citation. The values or penalties applied to citations are based on four criteria:

- Seriousness or gravity of the alleged violation
- Size of the business
- Employer's good faith in genuinely and effectively trying to comply with the OSHAcт before the inspection and, during and after the inspection, making a genuine effort to abate and comply
- Employer's history of previous violations

Table 16.3   OSHA Violations and Penalties

De minimis	No penalty
Other than serious	Up to $7,000 per violation
Serious	$1,500–$7,000 per violation
Willful, no death	Up to $70,000 per violation minimum of $5,000
Willful, repeat violations	Same as willful, no death
Willful, death results	Up to $250,000 or $500,000 for a corporation, and 6 months in jail
Willful, death results, second violation	$250,000 and 1 year in jail
Failure to correct a cited violation	$7,000/day till abated
Failure to post official documents	$7,000 per poster
Falsification of documents	$10,000 and 6 months in jail
Assaulting a compliance officer	Not more than $5,000 and not more than 3 years of imprisonment

Employers can contest either the citation or the penalty by requesting an informal hearing with the area director to discuss these issues and the area director can enter into a settlement agreement if the situation merits it. However, if a settlement cannot be reached, the employer must notify the area director, in writing, of a notice of contest of the citation, penalties, or abatement period within 15 days of receipt of the citation. Workers can challenge an OSHA decision on employer appeals, but have limited challenge rights regarding OSHA's decision. They are

- Time element in the citation for abatement of the hazard.
- Employer's petition for modification of abatement (PMA). Workers have 10 days to contest the PMA.

COMMON VIOLATIONS FOUND BY OSHA

No matter what type of work your company is involved in, you should be aware of the types of violations that OSHA has found during inspection. In addition, you need to know which violations are cited the most so that you can place emphasis on ensuring that these violations do not exist in your workplace. In Table 16.4, you can find the 100 most cited violations by OSHA, while in Appendix I you can find the 50 most frequent citations for the major industrial sectors. If you want to find more specific citation information and you know the more detailed standard industrial classification (SIC) for your particular type of operation, you can go to the OSHA Web site at http://www.osha.gov/oshstats/std1.html and conduct a detailed search for only those violations found in your specific industry.

100 MOST FREQUENT VIOLATIONS CITED BY OSHA FOR ALL INDUSTRIES

Table 16.4 shows the standards that were cited by federal OSHA for the specified SIC during the period October 2006 through September 2007. Remember that 1910 is for general industry and 1926 is for construction.

STATE OSHA PLANS

Most state plans provide for the state to take over the enforcement of workplace safety and health rather than to have federal OSHA perform this service within the state. A listing of the federal and state office addresses and telephone numbers can be found in Appendix H. Many states have

Table 16.4   100 Most Frequently Cited Violations

Standard	No. Cited	Description
1926.451	10,538	General requirements for all types of scaffolding
1926.501	6,901	Duty to have fall protection
1910.1200	6,607	Hazard communication
1910.147	4,003	The control of hazardous energy, lockout/tagout
1910.134	3,885	Respiratory protection
1910.178	3,552	Powered industrial trucks
1910.305	3,300	Electrical, wiring methods, components and equipment
1926.1053	2,967	Ladders
1910.212	2,745	Machines, general requirements
1910.303	2,576	Electrical systems design, general requirements
1926.453	2,021	Aerial lifts
1926.503	2,020	Fall protection training requirements
1926.20	1,978	General safety and health provisions
1926.651	1,922	Excavations, general requirements
1910.132	1,919	PPE, general requirements
1926.100	1,876	Head protection
1910.219	1,815	Mechanical power-transmission apparatus
1910.23	1,624	Guarding floor and wall openings and holes
1910.1030	1,556	Bloodborne pathogens
1910.215	1,484	Abrasive wheel machinery
1926.454	1,446	Training requirements, scaffolds
5A1	1,355	General duty clause (section of OSHA Act)
1910.157	1,345	Portable fire extinguishers
1910.37	1,336	Maintenance, safeguards, and operational features for exit routes
1926.652	1,334	Excavations, requirements for protective systems
1926.404	1,326	Electrical, wiring design and protection
1926.405	1,265	Electrical wiring methods, components and equipment general use
1904.29	1,096	Forms
1926.21	1,092	Safety training and education
1926.1052	1,069	Stairways
1910.22	1,061	Walking/working surfaces, general requirements
1910.213	1,028	Woodworking machinery requirements
1926.102	1,027	Eye and face protection
1926.502	999	Fall protection systems criteria and practices
1910.95	967	Occupational noise exposure
1910.146	891	Permit-required confined spaces
1926.452	860	Additional requirements applicable to specific types of scaffolds
1910.151	846	Medical services and first aid
1910.107	776	Spray finishing with flammable/combustible materials
1910.304	707	Electrical, wiring design and protection
1926.403	701	General requirements, electrical
1910.106	685	Flammable and combustible liquids
1910.217	671	Mechanical power presses
1910.266	667	Logging operations
1910.253	635	Oxygen-fuel gas welding and cutting
1926.1101	576	Asbestos

Table 16.4 (continued)

Standard	No. Cited	Description
1910.133	567	Eye and face protection
1926.62	541	Lead
1910.179	528	Overhead and gantry cranes
1904.32	521	Annual summary
1910.242	506	Hand and portable powered tools and equipment, general
1926.1060	503	Stairways and ladders, training requirements
1910.119	485	Process safety management, highly hazardous chemicals
1910.36	475	Design and construction requirements for exit routes
1926.95	441	Criteria for PPE
1926.416	424	General requirements, electrical
1926.760	398	Fall protection
1910.1025	396	Lead
1910.184	386	Slings
1910.141	380	Sanitation
1910.176	347	Materials handling, general
1926.701	332	Concrete/masonry, general requirements
1926.25	330	Construction, housekeeping
1910.1000	326	Air contaminants
1926.602	325	Material handling equipment
1910.24	290	Fixed industrial stairs
1926.304	288	Woodworking tools
1926.150	286	Fire protection
1926.1051	284	Stairways and ladders, general requirements
1926.550	278	Cranes and derricks
1926.350	271	Gas welding and cutting
1910.333	259	Electrical, selection and use of work practices
1910.334	247	Electrical, use of equipment
1910.67	239	Vehicle-mounted elevating and rotating work platforms
1910.252	237	General requirements, welding, cutting, and brazing
1910.138	233	Hand protection
1910.335	233	Safeguards for personnel protection
1910.38	222	Emergency action plans
1910.26	221	Portable metal ladders
1910.1052	220	Methylene chloride
1926.300	218	General requirements, tools, hand and power
1904.41	216	Annual OSHA injury and illness survey of 10 or more employers
1910.110	206	Storage and handling of liquefied petroleum gases
1910.101	186	Compressed gases (general requirements)
1910.243	181	Guarding portable power tools
1926.251	169	Rigging equipment for material handling
1910.120	166	Hazardous waste operations and emergency response
1910.332	158	Electrical, training
1926.152	153	Flammable and combustible liquids
1904.2	145	Partial exemption for establishments in certain industries
1926.50	138	Medical services and first aid
1910.1001	135	Asbestos tremolite, anthophyllite, and actinolite

(continued)

Table 16.4 (continued)

Standard	No. Cited	Description
1926.28	135	PPE
1903.19	134	Abatement verification
1926.850	132	Preparatory operations, demolition
1926.200	127	Accident prevention signs and tags
1910.1048	126	Formaldehyde
1904.40	110	Providing records to government representatives
1904.39	107	Reporting fatalities and multiple hospitalizations to OSHA
1910.27	104	Fixed ladders

opted to take on this responsibility (Table 16.5). The states of Connecticut, New Jersey, Virgin Islands, and New York have unique plans in that they cover only state and local employees (public sector), while federal OSHA covers the general and construction industries. If a state has a federally approved plan or program, the following conditions must exist:

- State must create an agency to carry out the plan.
- State's plan must include safety and health standards and regulations. The enforcement of these standards must be at least as effective as the federal plan.
- State plan must include provisions for right of entry and inspection of the workplace, including a prohibition on advance notice of inspections.
- State's plan must also cover state and local government employees.

If a state has a plan, are there state-specific standards and regulations? The answer is yes, and they must be at least as stringent as the federal standards and regulations. Some states have standards and regulations that go beyond the requirements of the existing federal standards and regulations, while others simply adopt the federal standards and regulations verbatim.

Anyone who feels their state program has not responded to requests for inspections, complaints of discrimination, or appeals on citations or variances can file a complaint with federal OSHA. Federal OSHA is responsible to monitor state programs and make evaluations on their effectiveness. A written Complaint Against State Program Administration (CASPA) should contain the following:

- Description of the attempts to get action from the state and the justification
- State's response(s) or action(s) that demonstrated poor administration of the state OSHA program
- Date the incident occurred
- Exact location where the incident occurred
- Name of the employer
- Name and occupations of those involved in the incident
- Notification to the state agency that a CASPA has been filed
- Statement requesting confidentiality during the investigation

A CASPA should be filed when the state plan agency has not

- Conducted an inspection in a timely and effective way
- Enforced state OSHA standards and regulations
- Responded to a request for an inspection
- Protected workers' rights against discrimination
- Issued citations for violations discovered
- Complied with proper procedures for granting variances

Table 16.5   State or Federal OSHA Jurisdiction

State Plan States	Federal Jurisdiction
Alaska	*Region 1*
Arizona	Connecticut
California	Massachusetts
Connecticut[a]	Maine
Hawaii	New Hampshire
Indiana	Rhode Island
Iowa	*Region 2*
Kentucky	New York
Maryland	New Jersey
Michigan	Virgin Islands
Minnesota	*Region 3*
Nevada	Delaware
New Jersey[a]	District of Columbia
New Mexico	Pennsylvania
New York[a]	West Virginia
North Carolina	*Region 4*
Oregon	Alabama
Puerto Rico	Florida
South Carolina	Georgia
Tennessee	Mississippi
Utah	*Region 5*
Vermont	Illinois
Virginia	Ohio
Virgin Islands	Wisconsin
Washington	*Region 6*
Wyoming	Arkansas
	Louisiana
	Oklahoma
	Texas
	Region 7
	Kansas
	Missouri
	Nebraska
	Region 8
	Colorado
	Montana
	North Dakota
	South Dakota
	Region 9
	America Samoa
	Guam
	Trust Territory of the Pacific Islands
	Region 10
	Idaho

[a] Public sector only.

Federal OSHA will evaluate the complaint and then notify the worker in writing of its decision. If the filer of a CASPA is not happy with OSHA's response, a written request for a reevaluation should be sent to the nearest OSHA area office.

WORKER TRAINING

Many standards promulgated by OSHA specifically require the employer to train employees in the safety and health aspects of their jobs. Other OSHA standards make it the employer's responsibility to limit certain job assignments to employees who are certified, competent, or qualified—meaning employees who have had special previous training, in or out of the workplace. OSHA regulations imply that an employer has assured that a worker has been trained before he or she is designated as the individual to perform a certain task.

To make a complete determination of the OSHA requirement for training, one would have to go directly to the regulation that applies to the specific type of activity. The regulation may mandate hazard training, task training, and length of the training, as well as specifics to be covered by the training.

It is always a good idea for the employer, as well as the worker, to keep records of training. These records may be used by a compliance inspector during an inspection, after an accident resulting in injury or illness, as proof of good intentions by the employer or to comply with training requirements for workers, including new workers and those assigned new tasks.

OCCUPATIONAL INJURIES AND ILLNESSES

The recording and reporting of occupational injuries and illness requirements can be found in 29 CFR 1904—recording and reporting occupational injuries and illnesses. This regulation has been revised and came into effect as of January 2002. These requirements are summarized in the following paragraphs.

Any illness that has been caused by exposure to environmental factors such as inhalation, absorption, ingestion, or direct contact with toxic substances or harmful agents and has resulted in an abnormal condition or disorder that is acute or chronic is classified as an occupational disease. Repetitive motion injuries are also included in this category. All illnesses are recordable, regardless of severity. Injuries are recordable when (see Figure 16.1)

- On-the-job death occurs regardless of length of time between injury and death
- One or more lost workdays occur
- Restriction of work or motion transpires
- Loss of consciousness occurs
- Worker is transferred to another job
- Worker receives medical treatment beyond first aid

Employers with more than 10 employees are required to complete and maintain occupational injury and illness records. The OSHA 301 (see Figure 16.2) injury and illness incident report, or equivalent, must be completed within 7 days of the occurrence of an injury at the worksite and the OSHA 301 must be retained for 5 years. In addition, the OSHA 300 log of work-related injuries and illnesses (see Figure 16.3) is to be completed within 7 days when a recordable injury or illness occurs and maintained for 5 years. The OSHA 300A summary of work-related injuries and illnesses (see Figure 16.4) must be posted yearly from February 1 to April 30. OSHA forms can now be maintained on the computer until they are needed.

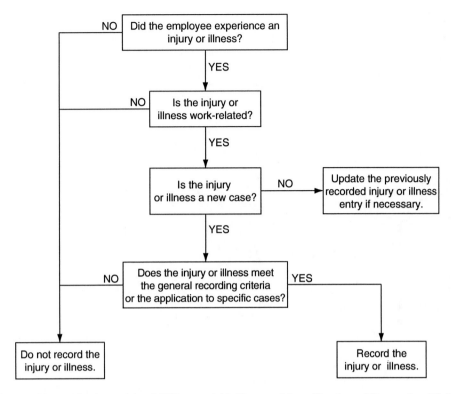

Figure 16.1 Diagram for determining OSHA-recordable illness or injury. (Courtesy of Occupational Safety and Health Administration.)

The following are lists of industries that are required to maintain occupational injury and illness records (Tables 16.6 and 16.7).

All employers with no more than 10 full- or part-time employees at any one time in the previous calendar year are not required to keep OSHA records. Employers in the retail trades of finance, insurance, real estate, and services industries are not required to keep OSHA records. Some employers are normally not required or are exempt from keeping OSHA records (Table 16.8).

Some employers and individuals who are not required to keep OSHA records are

- Self-employed individuals
- Partners with no employees
- Employers of domestics
- Employers engaged in religious activities

Some changes have occurred relevant to what is considered to be first aid versus medical treatment. The areas where changes have occurred are found in Table 16.9.

MEDICAL AND EXPOSURE RECORDS

Medical examinations are required by OSHA regulations for workers before they can perform certain types of work. This work includes at the present

OSHA's Form 301
Injury and Illness Incident Report

U.S. Department of Labor
Occupational Safety and Health Administration

Form approved OMB no. 121k-0176

Attention: This form contains information relating to employee health and must be used in a manner that protects the confidentiality of employees to the extent possible while the information is being used for occupational safety and health purposes.

This *Injury and Illness Incident Report* is one of the first forms you must fill out when a recordable work-related injury or illness has occurred. Together with the *Log of Work-Related Injuries and Illnesses* and the accompanying *Summary*, these forms help the employer and OSHA develop a picture of the extent and severity of work-related incidents.

Within 7 calendar days after you receive information that a recordable work-related injury or illness has occurred, you must fill out this form or an equivalent. Some state workers' compensation, insurance, or other reports may be acceptable substitutes. To be considered an equivalent form, any substitute must contain all the information asked for on this form.

According to Public Law 91-596 and 29 CFR 1904, OSHA's recordkeeping rule, you must keep this form on file for 5 years following the year to which it pertains.

If you need additional copies of this form, you may photocopy and use as many as you need.

Information about the employee

1) Full name _____

2) Street _____
   City _____ State ____ ZIP _____

3) Date of birth ___/___/___

4) Date hired ___/___/___

5) ☐ Male
   ☐ Female

Information about the physician or other health care professional

6) Name of physician or other health care professional _____

7) If treatment was given away from the worksite, where was it given?
   Facility _____
   Street _____
   City _____ State ____ ZIP _____

8) Was employee treated in an emergency room?
   ☐ Yes
   ☐ No

9) Was employee hospitalized overnight as an in-patient?
   ☐ Yes
   ☐ No

Information about the case

10) Case number from the Log _____ (Transfer the case number from the Log after you record the case.)

11) Date of injury or illness ___/___/___

12) Time employee began work _____ AM / PM

13) Time of event _____ AM / PM ☐ Check if time cannot be determined

14) *What was the employee doing just before the incident occurred?* Describe the activity, as well as the tools, equipment, or material the employee was using. Be specific. *Examples:* "climbing a ladder while carrying roofing materials"; "spraying chlorine from hand sprayer"; "daily computer key-entry."

15) *What happened?* Tell us how the injury occurred. *Examples:* "When ladder slipped on wet floor, worker fell 20 feet"; "Worker was sprayed with chlorine when gasket broke during replacement"; "Worker developed soreness in wrist over time."

16) *What was the injury or illness?* Tell us the part of the body that was affected and how it was affected; be more specific than "hurt," "pain," or "sore." *Examples:* "strained back"; "chemical burn, hand"; "carpal tunnel syndrome."

17) *What object or substance directly harmed the employee? Examples:* "concrete floor"; "chlorine"; "radial arm saw." *If this question does not apply to the incident, leave it blank.*

18) *If the employee died, when did death occur?* Date of death ___/___/___

Completed by _____

Title _____

Phone (___) ___ - ___   Date ___/___/___

Public reporting burden for this collection of information is estimated to average 22 minutes per response, including time for reviewing instructions, searching existing data sources, gathering and maintaining the data needed, and completing and reviewing the collection of information. Persons are not required to respond to the collection of information unless it displays a current valid OMB control number. If you have any comments about this estimate or any other aspects of this data collection, including suggestions for reducing this burden, contact: US Department of Labor, OSHA Office of Statistics, Room N-3644, 200 Constitution Avenue, NW, Washington, DC 20210. Do not send the completed forms to this office.

Figure 16.2   OSHA 301 report of incidents. (Courtesy of Occupational Safety and Health Administration.)

OSHA's Form 300

Log of Work-Related Injuries and Illnesses

Year 20____

U.S. Department of Labor
Occupational Safety and Health Administration

Attention: This form contains information relating to employee health and must be used in a manner that protects the confidentiality of employees to the extent possible while the information is being used for occupational safety and health purposes.

Form approved OMB no. 1218-0176

You must record information about every work-related death and about every work-related injury or illness that involves loss of consciousness, restricted work activity or job transfer, days away from work, or medical treatment beyond first aid. You must also record significant work-related injuries and illnesses that are diagnosed by a physician or licensed health care professional. You must also record work-related injuries and illnesses that meet any of the specific recording criteria listed in 29 CFR Part 1904.8 through 1904.12. Feel free to use two lines for a single case if you need to. You must complete an injury and illness Incident Report (OSHA Form 301) or equivalent form for each injury or illness recorded on this form. If you're not sure whether a case is recordable, call your local OSHA office for help.

Establishment name _____

City _____ State _____

Identify the person			**Describe the case**			**Classify the case**											

Figure 16.3 OSHA 300 log. (Courtesy of Occupational Safety and Health Administration.)

Public reporting burden for this collection of information is estimated to average 14 minutes per response, including time to review the instructions, search and gather the data needed, and complete and review the collection of information. Persons are not required to respond to the collection of information unless it displays a currently valid OMB control number. If you have any comments about these estimates or any other aspects of this data collection, contact: US Department of Labor, OSHA Office of Statistics, Room N-3644, 200 Constitution Avenue, NW, Washington, DC 20210. Do not send the completed forms to this office.

OSHA's Form 300A

Summary of Work-Related Injuries and Illnesses

Year 20___

U.S. Department of Labor
Occupational Safety and Health Administration

Form approved OMB no. 1218-0176

All establishments covered by Part 1904 must complete this Summary page, even if no work-related injuries or illnesses occurred during the year. Remember to review the Log to verify that the entries are complete and accurate before completing this summary.

Using the Log, count the individual entries you made for each category. Then write the totals below, making sure you've added the entries from every page of the Log. If you had no cases, write "0."

Employees, former employees, and their representatives have the right to review the OSHA Form 300 in its entirety. They also have limited access to the OSHA Form 301 or its equivalent. See 29 CFR Part 1904.35, in OSHA's recordkeeping rule, for further details on the access provisions for these forms.

Number of Cases

Total number of deaths	Total number of cases with days away from work	Total number of cases with job transfer or restriction	Total number of other recordable cases
(G)	(H)	(I)	(J)

Number of Days

Total number of days of job transfer or restriction	Total number of days away from work
(K)	(L)

Injury and Illness Types

Total number of . . .
(M)

(1) Injuries	____	(4) Poisonings	____
(2) Skin disorders	____	(5) All other illnesses	____
(3) Respiratory conditions	____		

Post this Summary page from February 1 to April 30 of the year following the year covered by the form.

Public reporting burden for this collection of information is estimated to average 50 minutes per response, including time to review the instructions, search and gather the data needed, and complete and review the collection of information. Persons are not required to respond to the collection of information unless it displays a currently valid OMB control number. If you have any comments about these estimates or any other aspects of this data collection, contact: US Department of Labor, OSHA Office of Statistics, Room N-3644, 200 Constitution Avenue, NW, Washington, DC 20210. Do not send the completed forms to this office.

Establishment information

Your establishment name _____

Street _____

City _____ State _____ ZIP _____

Industry description (e.g., *Manufacture of motor truck trailers*) _____

Standard Industrial Classification (SIC), if known (e.g., *SIC 3715*) _____

Employment information *(If you don't have these figures, see the Worksheet on the back of this page to estimate.)*

Annual average number of employees _____

Total hours worked by all employees last year _____

Sign here

Knowingly falsifying this document may result in a fine.

I certify that I have examined this document and that to the best of my knowledge the entries are true, accurate, and complete.

_____
Company executive   Title

( ) _____
Phone   Date

Figure 16.4 OSHA 300A summary. (Courtesy of Occupational Safety and Health Administration.)

Table 16.6   Industries Required to Record Injuries and Illnesses

Agriculture, forestry, and fishing

Oil and gas extraction

Construction

Manufacturing

Transportation

Wholesale trades

Utilities

- Asbestos abatement
- Lead abatement
- Hazardous waste remediation
- When workers are required to wear respirators for 30 days during a year

Exposure records (monitoring records) are to be maintained by the employer for 30 years. These records include personal sampling, air sampling, and other industrial hygiene sampling records. Medical records are to be maintained by the employer for the length of employment plus 30 years.

A worker must make a written request to obtain a copy of his or her medical records or to make them available to a representative or physician. (See a sample of a medical record request letter in Figure 16.5.) A worker's medical record is considered confidential and a request, in writing, from the worker to the physician is required for the records to be released.

If an employer goes out of business or sells the business, the medical records of employees can be transferred to the new owner or if no successor exists then the previous employer is to contact the affected workers to make their records available to them before disposal. In addition, the previous

Table 16.7   1904.2—Newly Covered Industries

SIC Code	Industry
553	Auto and home supply stores
555	Boat dealers
556	Recreational vehicle dealers
559	Automotive dealers not elsewhere classified
571	Home furniture and furnishing stores
572	Household appliance stores
593	Used merchandise stores
596	Nonstore retailers
598	Fuel dealers
651	Real estate operators and lessors
655	Land subdividers and developers
721	Laundry, cleaning, and garment services
734	Services to dwellings and other buildings
735	Miscellaneous equipment rental and leasing
736	Personnel supply services
833	Job training and vocational rehabilitation services
836	Residential care
842	Arboreta and botanical or zoological gardens

Table 16.8 Nonmandatory Appendix A to Subpart B—Partially Exempt Industries

SIC Code	Industry Description	SIC Code	Industry Description
525	Hardware stores	725	Shoe repair and shoeshine parlors
542	Meat and fish markets	726	Funeral service and crematories
544	Candy, nut, and confectionery stores	729	Miscellaneous personal services
545	Dairy products stores	731	Advertising services
546	Retail bakeries	732	Credit reporting and collection services
549	Miscellaneous food stores	733	Mailing, reproduction, and stenographic services
551	New and used car dealers	737	Computer and data processing services
552	Used car dealers	738	Miscellaneous business services
554	Gasoline service stations	764	Reupholstery and furniture repair
557	Motorcycle dealers	78	Motion picture
56	Apparel and accessory stores	791	Dance studios, schools, and halls
573	Radio, television, and computer stores	792	Producers, orchestras, entertainers
58	Eating and drinking places	793	Bowling centers
591	Drug stores and proprietary stores	801	Offices and clinics of medical doctors
592	Liquor stores	802	Offices and clinics of dentists
594	Miscellaneous shopping goods stores	803	Offices of osteopathic physicians
599	Retail stores, not elsewhere classified	804	Offices of other health practitioners
60	Depository institutions (banks and savings institutions)	807	Medical and dental laboratories
61	Nondepository institutions (credit institutions)	809	Health and allied services, not elsewhere classified
62	Security and commodity brokers	81	Legal services
63	Insurance carriers	82	Educational services (schools, colleges, universities, and libraries)
64	Insurance agents, brokers, and services	832	Individual and family services
653	Real estate agents and managers	835	Child day care services
654	Title abstract offices	839	Social services, not elsewhere classified
67	Holding and other investment offices	841	Museums and art galleries
722	Photographic studios, portrait	86	Membership organizations
723	Beauty shops	87	Engineering, accounting, research, management, and related services
724	Barber shops	899	Services, not elsewhere classified

Employers are not required to keep OSHA injury and illness records for any establishment classified in the following SIC codes, unless they are asked in writing to do so by OSHA, the Bureau of Labor Statistics (BLS), or a state agency operating under the authority of OSHA or the BLS. All employers, including those partially exempted by reason of company size or industry classification, must report to OSHA any workplace incident that results in a fatality or the hospitalization of three or more employees (see 1904.39).

Table 16.9 First-Aid List

1904.7 (b)(5)(ii) What is first aid?

For the purposes of Part 1904, first aid means the following:

(A) Using a nonprescription medication at nonprescription strength (for medications available in both prescription and nonprescription form, a recommendation by a physician or other licensed health-care professional to use a nonprescription medication at prescription strength is considered medical treatment for record keeping purposes)

(B) Administering tetanus immunizations (other immunizations, such as hepatitis B vaccine or rabies vaccine, are considered medical treatment)

(C) Cleaning, flushing, or soaking wounds on the surface of the skin

(D) Using wound coverings such as bandages, Band-Aids™, gauze pads, etc.; or using butterfly bandages or Steri-Strips™ (other wound closing devices such as sutures, staples, etc. are considered medical treatment)

(E) Using hot or cold therapy

(F) Using any nonrigid means of support, such as elastic bandages, wraps, nonrigid back belts, etc. (devices with rigid stays or other systems designed to immobilize parts of the body are considered medical treatment for record keeping purposes)

(G) Using temporary immobilization devices while transporting an accident victim (e.g., splints, slings, neck collars, back boards, etc.)

(H) Drilling of a fingernail or toenail to relieve pressure, or draining fluid from a blister

 (I) Using eye patches

 (J) Removing foreign bodies from the eye using only irrigation or a cotton swab

(K) Removing splinters or foreign material from areas other than the eye by irrigation, tweezers, cotton swabs, or other simple means

 (L) Using finger guards

(M) Using massages (physical therapy or chiropractic treatment are considered medical treatment for record keeping purposes)

(N) Drinking fluids for relief of heat stress

(iii) Are any other procedures included in first aid?

No, this is a complete list of all treatments considered first aid for Part 1904 purposes.

employer must notify the director of NIOSH 3 months in advance to determine if NIOSH will accept the records. If NIOSH does not respond, then the previous employer may dispose of the medical records.

POSTING

Employers are required to post in a prominent location the following:

- New job safety and health (OSHA 3165) poster developed by OSHA to be posted in the workplace (see Figure 16.6).
- Copies of any OSHA citations of violations of the OSHA standard are to be posted at or near the location of the violation for at least 3 days or until the violation is abated, whichever is longer.
- Copies of summaries of petitions for variances from any standard, including record keeping procedures.
- Summary portion of the log and summary of occupational injuries and illnesses (OSHA 300A summary) is to be posted annually from February 1 to April 30.

Industries required to record injuries and illnesses

- Agriculture, forestry, and fishing
- Oil and gas extraction
- Construction
- Manufacturing
- Transportation
- Wholesale trades
- Utilities

I _____ , hereby authorize _____
(full name of worker/patient)   (individual or organization holding

_____ to release to _____
    the medical records)   (individual or organization authorized

_____ , the following medical information from my personal
    to receive the medical information)

medical records: _____
                              (describe generally the information desired to be released)

I give permission for this medical information to be used for the following purpose

_____ ,

but I do not give permission for any other use or redisclosure of this information.

_____
(Full name of employee or legal representative)

_____
(Signature of employee or legal representative)

_____
(Date of signature)

Figure 16.5 Sample authorization letter for the release of employee medical information to a designated representative.

WHAT TO DO WHEN OSHA COMES KNOCKING

When an inspector from OSHA or a corporate or insurance company's safety and health professional shows up at a project or jobsite, there is nothing to worry about if a safety and health program has been implemented and its mandates are being enforced. To start, the following items should be in place:

1. Job safety and health protection poster (OSHA 3165) posted on a bulletin board that should be visible to all workers
2. Summaries of any petitions for variances
3. Copies of any new or unabated citations
4. Summary of the OSHA 300A summary, posted during the month of February through April
5. Following should be available for the workers' and inspector's examination:
   a. Any exposure records for hazardous materials
   b. Results of medical surveillance
   c. All NIOSH research records for exposure to potentially harmful substances
6. Verification that workers have been told:
   a. If exposures have exceeded the levels set by the standard and if corrective measures are taken
   b. If there are hazardous chemicals in their work area
7. Training records are available at the time of inspection

Figure 16.6 Job safety and health poster (OSHA 3165).

Inspection Process

The inspection process should be handled in a professional manner and mutual respect between the inspector and the employer or representative needs to be developed in a short period of time. It is appropriate to

1. Check the compliance officer's credentials and secure security clearance, if required.
2. Discuss company's safety and health program and its implementation.
3. Delineate activities and initiatives taken to improve safety and health on the job, as well as worker protection.
4. Ask for recommendations and advice that will improve what is being done.
5. Discuss any consultation programs or voluntary participation programs and pursue any inspection exemptions.
6. Ask the purpose, scope, and applicable standards for the inspection and obtain a copy of the employee complaint, if that is what triggered the inspection.
7. Make sure the employer's representative who accompanies the inspector is knowledgeable.
8. Include, if possible, an employees' representative.
9. Employer's representative must be familiar with the project and should try to choose an appropriate route for the inspection. However, the inspector's route cannot be dictated; he or she can choose the route for the inspection, if desired.
10. Make sure all observations, conversations, photographs, readings, and records examined are duplicated. Take good notes and ask appropriate questions.
11. Have records available for the inspector such as the OSHA 300 log, OSHA 301, exposure records, and training records.
12. Pay close attention to unsafe or unhealthy conditions that are observed. Discuss how to correct them with the inspector and take corrective actions immediately, if possible.
13. Never at any time interfere with employee interviews with the inspector.

Mitigating the Damage

There is no turning back; the inspection will occur. It is imperative that the inspection's outcome results in as little damage as possible. This can be accomplished in many ways during the inspection process. Some actions may seem redundant, but they need to be reinforced:

1. Ask for an OSHA consultation service or pursue an exemption if the inspector cannot tell you how to abate or correct a violation.
2. Know the jobsite and be familiar with all the processes and equipment.
3. Try to select the inspector's route, if possible. Save the known or suspect problem areas for last.
4. Take good notes and document the inspection process completely. Photograph anything that the inspector does.
5. Keep records; many benefits are gained from good record keeping.
6. Correct apparent violations immediately, if possible.
7. Maintain updated copies of any required written programs.

Closing Conference

During the closing conference, when the culmination of the inspection process occurs, adhere to the following items to maintain the overall continuity of the process:

1. Listen actively and carefully to the discussion of unsafe or unhealthy conditions and apparent violations.
2. Ask questions for clarification so as to avoid confrontation. Confrontation will accomplish nothing.
3. Make sure the inspector discusses the appeal rights, informal conference procedure rights, and procedures for contesting a citation.

4. Produce documentation to support the company's compliance efforts or special emphasis programs.
5. Provide information that will guide the inspector in setting the times for abatement of citations.

After the Inspector Leaves

Citations and notices will arrive by certified mail and will need to be posted at or near the area where the violation occurred for at least 3 days or until abated, whichever is longer. Any notice of contest or objection must be received by the OSHA area director in writing within 15 days of the receipt of any citations. The area director will forward the notice of contest to the OSHRC. It is also a good idea to request an informal meeting with the area director during the 15 day period.

The notice of contest will be assigned to an administrative law judge by the OSHRC. Once the judge rules on the contest notice, further review by OSHRC may be requested. If necessary, the OSHRC ruling can be appealed to the U.S. Court of Appeals.

Remember that all citations or violations must be corrected or abated by the prescribed date unless the citation or abatement date is formally contested. If the response to a citation or violation cannot be abated in the time allotted, because of factors that are beyond reasonable control, a petition to modify the time for abatement must be filed with the area director to extend the date.

Make the Inspection a Positive Experience

A proactive safety and health preparation can make for a quality safety program. This is a safeguard for property, equipment, profits, and liability, as well as for workers. Working with workers to correct deficiencies can foster better safety attitudes. A safer workplace is also a more productive workplace. This also safeguards a very important asset, the worker. Remember, OSHA has a great deal of expertise within its ranks. Use OSHA as a resource to improve your safety and health program. Before the knock by OSHA, it is necessary to implement a safety and health program. This includes, as stated earlier, the following:

1. Formal written program or safety manual
2. Standard operating procedures (SOPs) that incorporate OSHA standards.
3. Worker and supervisor training
4. Standard record keeping procedures
5. Workplace inspections, audits, job observations, and job safety analyses
6. Safety and health committees, if possible
7. Accident or incident investigation procedures
8. Hazard recognition or reporting procedures
9. First aid and medical facility availability
10. Employee medical surveillance or examinations
11. Consultation services available

With these as a prerequisite, an OSHA inspection quickly becomes a positive learning experience from which many benefits will be reaped. It can produce higher morale, better production, a safer workplace, and a better bottom line since many negatives have been avoided by good preactivity and planning.

MULTIEMPLOYER WORKSITES

On multiemployer worksites, citations are normally issued to the employer whose employees are exposed to workplace hazards (the exposing employer). In addition, the following employers normally shall be cited, whether or not their own employees are exposed:

- Employer who actually creates the hazard (the creating employer).
- Employer who is responsible, by contract or through actual practice, for safety and health conditions on the worksite; that is, the employer who has the authority for ensuring that the hazardous condition is corrected (the controlling employer).
- Employer who has the responsibility for actually correcting the hazard (the correcting employer).

Before issuing citations to an exposing employer, it must first be determined whether the available facts indicate that the employer has a legitimate defense to the citation. This is accomplished by answering the following questions:

- Did the employer create the hazard?
- Did the employer have the responsibility or authority to have the hazard corrected?
- Did the employer have the ability to correct or remove the hazard?
- Did the employer demonstrate that the creating, controlling, or correcting employers, as appropriate, have been specifically notified of the hazard to which their employees are exposed?
- Did the employer instruct employees to recognize the hazard?

Where feasible, an exposing employer must have taken appropriate, alternative means of protecting employees from the hazard; and when extreme circumstances justify it, to avoid a citation the exposing employer shall remove employees from the job. If an exposing employer has met all of the previous criteria, then the employer shall not be cited.

If all employers on a worksite who have employees exposed to a hazard meet the previous criteria, the citation shall be issued only to the employers who are responsible for creating the hazard or are in the best position to correct or ensure correction of the hazard. In such circumstances, the controlling employer or the hazard-creating employer shall be cited even though none of their employees are exposed to the condition that resulted in the violation. Penalties for such citations shall be appropriately calculated by using the exposed employees of all employers as the number of employees for probability assessment.

SUMMARY

It is envisioned that this chapter will be an asset to employers. It has given you a brief overview of what you can expect from OSHA. Knowledge has been shown to fix accountability, as well as responsibility, upon those who claim ignorance of it. The workplace is where both labor and management spend the bulk of their waking hours. Keeping this in mind, the safety and health of those in the workplace should be everyone's concern and responsibility.

Employers and safety and health professionals need to know how OSHA provides for worker safety and health on worksites. This will also assist in ensuring that workers' rights are protected and give them the knowledge to help mitigate health and safety issues and problems that may arise. This type of knowledge should ensure a safer and more productive worksite. Respect for the efficient, effective, and proper use of health and safety rules will have a positive effect upon those in the workplace.

Although it is the ultimate responsibility of the employer to provide for workplace safety and health, adherence to OSHA occupational safety and health rules is the foundation upon which a good safety and health program can be built. The program should hold everyone responsible for the well-being of those in the workplace, including the employer, managers, supervisors, and workers. All should abide by the safety and health rules and the OSHA standards. Together, and through cooperation, all parties can assure a safe and healthy workplace. A safe and healthy home away from home is, and should be, the ultimate goal.

CHAPTER 17

Golden Rules: OSHA Regulations

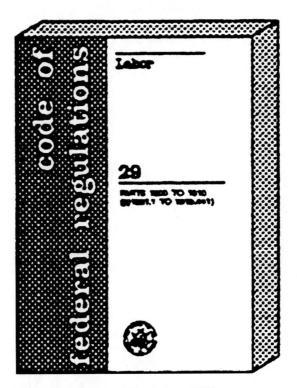

Regulations are found in the U.S. Code of Federal Regulations (CFR).

Occupational Safety and Health Administration (OSHA) was formed by the Occupational Safety and Health Act (OSHAct) of 1970. Congress gave OSHA a mandate to develop regulations and standards to protect the American worker. Not only was OSHA to develop these regulations but it was to implement (promulgate) them and enforce them to protect a valuable entity (the American worker). OSHA has no choice but to follow the mandate provided by Congress. Over the years OSHA has gone from a strict enforcement agency to an agency bent on trying to help employers comply with its regulations.

Few, if any, regulations are developed, implemented, and enforced unless there have been deaths, injuries, or illnesses that can be attributable to activities within the workplace. A great number of deaths and a great deal of bleeding and carnage attributed to a certain hazard precede the development of any regulation. Regulations are not developed without much justification. In fact, the regulatory process is long and laborious. The development of a new regulation usually takes years. One of the fastest regulations to be developed and implemented was the bloodborne pathogen standard and that was only because everyone was frightened by the possibility of contracting AIDS or hepatitis B.

To have a good safety and health effort in your workplace, you need to be familiar with what you need to know to be able to come into compliance with the regulations that affect your operation. Not all the regulations found in the CFR will apply to your operation. For instance, you will not be particularly interested in the commercial diving regulation if you do not conduct diving operations nor have divers employed. This chapter is meant to help you learn about regulations and how to use them to ensure that your safety and health effort is proceeding in the right direction and is lawful.

FEDERAL LAWS

Congress establishes federal laws (legislation or acts) and the president signs them into law. These laws often require that regulations (standards) be developed by the federal agencies that are responsible for the intent of the law.

OSHAcт

The OSHAcт of 1970 is such a law and is also called the Williams–Steiger Act. It was signed by President Richard Nixon on December 29, 1970 and became effective April 29, 1971. (The OSHAcт was not amended until November 5, 1990 by Public Law 101-552.) The OSHAcт assigned the responsibility of implementing and enforcing the law to a newly created agency, the OSHA, located in the Department of Labor (DOL).

Most such federal laws (acts) contain the following elements:

- Reason for the law
- Statement of the national policy related to the law
- Objectives/goals/outcomes expected of the law
- Authorization of the agency responsible for implementation
- Requirements and structure of the regulations to be developed
- Time frames for regulation, implementation, or deadlines
- Enforcement guidelines to be followed
- Fines or assessments available to the enforcing agency
- Specific actions required by the law

CONTENT OF THE OSHAcт

Before the OSHAcт, there were some state laws, a few pieces of federal regulations, and a small number of voluntary programs by employers. Most of the state programs were limited in scope and the federal laws only partially covered workers.

Another important reason for the OSHAcт was the increasing number of injuries and illnesses within the workplace. Thus, the OSHAcт was passed with the express purpose of assuring that every working man and woman in the nation would be provided safe and healthful work conditions while preserving this national human resource, the American worker. The OSHAcт is divided into sections each with a specific purpose. The full text of the OSHAcт, all 31 pages, can be obtained from your local OSHA office or on the OSHA Web site. As a quick reference to the OSHAcт, the following paragraphs summarize what each section includes.

The OSHAcт starts in Section 2 and contains congressional findings. Due to excessive injuries and illnesses, employers now have specific responsibilities regarding occupational safety and health (OS&H). It is the responsibility of the Secretary of Labor to institute OSHA. He or she will oversee the development and implementation of workplace health and safety standards, including any research and training required, as well as ensure the enforcement of OSHA standards, entice states to become involved, develop reporting requirements for injuries and illnesses, and foster joint labor/management efforts for OS&H.

Section 3 of the OSHAcт defines the employer as a person whose company is engaged in a business that affects commerce. This definition does not include the United States or other government entities. In addition, the definition of employees is those employed by an employer who affects commerce. There are also other definitions in this section that are pertinent to the OSHAcт.

Section 4 explains the applicability of the Act. In this section, the OSHAcт is described as not applicable to other federal agencies that exercise their own authority over OS&H. The Act supersedes other existing federal laws and regulations related to OS&H and will not have a similar effect on any workers' compensation laws that already exist.

Section 5 includes the General Duty Clause Section (5)(a)(1), which states that each employer shall furnish employment free from recognized hazards. This allows OSHA inspectors to cite an employer even if no OSHA regulation exists for an observed/known workplace hazard. In addition, it requires employers to comply with OSHA standards and employees to comply with rules and regulations.

Section 6 provides OSHA the authority to promulgate start-up standards without following a formal rulemaking procedure. This section addresses rulemaking procedures, emergency temporary standards, variances from standards, the use of the *Federal Register* for publishing the required public notices during the standard development process, as well as the final standard. Many other issues are also addressed: medical examinations, toxic materials, personal protective equipment (PPE), labels, etc. The main intent of this section is the promulgation of OS&H standards.

Section 7 delineates the responsibility of the Secretary of Labor to establish an advisory committee on OS&H and provide the resources for the mission and intent of the advisory committee. The procedures and resources available to the committee are explained. This section authorizes OSHA to make use of the services and personnel of state and federal agencies and to provide OS&H consultative services.

Section 8 deals with inspections, investigations, and record keeping. It gives the OSHA representative the authority to enter workplaces without delay, at reasonable times, and inspect during regular working hours. During the inspection, the OSHA inspector may be accompanied by an employer representative and an employee representative, if they so desire. The OSHA inspector has the authority to question, privately, employers and employees. [Note: The Marshall v. Barlow decision (1978) requires a warrant if denied entry]. Section 8 also provides OSHA subpoena power. Employers are required to maintain and post injury and illness records as well as exposure records. Workers can file a complaint with OSHA if they believe that their workplace is subject to physical hazards or imminent danger. OSHA will determine whether the complaint merits a formal inspection.

Section 9 states that employers who have violated Section 5 of the Act or any standard, regulation, rule, or order related to Section 6 of the Act shall be issued a citation. The citation will be in writing, describing the particular violation and reference the location of the standard, rule, regulation, or order in the Act. These citations are to be posted by the employer. Citations must be issued within 6 months following a detected violation.

Section 10 sets forth the enforcement procedures. The employer has the right to contest any citation, procedure, and time for abatement, and to receive information concerning how the contested citation will be handled. The employees' rights are limited to contesting the abatement time for a hazard only.

Section 11 provides for the appeal and review of any orders issued by the Occupational Safety and Health Review Commission (OSHRC). This section also addresses discrimination by the employer against workers who decide to exercise their right to complain formally or informally regarding safety and health issues.

Section 12 mandates the formation of the OSHRC, which is composed of three members appointed by the president for a 6 year term. The commission conducts hearings, when necessary, relevant to the OSHACT or reviews processes, violations, and concerns.

Section 13 requires the Secretary of Labor to take action to protect workers from imminent danger. The secretary can be held liable for arbitrary or capricious disregard of an imminent danger that is brought to his or her attention.

Section 14 provides for the Solicitor of Labor to represent the secretary during litigation.

Section 15 protects the trade secrets of a company by requiring that any information gathered during performance of an inspection, by either the secretary or his or her representative, be confidential.

Section 16 provides the secretary with the power to make variations, tolerances, and exemptions from any or all provisions of the OSHACT when the impairment of the national defense is threatened. This can take place for a period of 6 months without notifying employees or holding a hearing.

Section 17 deals with the issuance of citations and their accompanying penalties. The types of violations and the amounts of the penalties, as well as the reason for such penalties are discussed.

Section 18 allows for states to assume responsibility from the federal authorities for the safety and health program, but federal OSHA must approve the plan. If no federal standards are in effect, the states may issue their own standards. Federal OSHA will monitor, support, and evaluate the approved states' plans.

Executive Order 12196, Section 19, states the responsibilities of federal agencies regarding safety and health and requires these agencies to have effective OS&H programs.

Section 20 mandates that the Department of Health and Human Services (DHHS) be responsible for the research functions under the Act, and that the National Institute for Occupational Safety and Health (NIOSH) carry out most of these functions.

Section 21 requires DHHS to carry out training and employee education by using grants, contracts, and short-term training.

Section 22 mandates the establishment of NIOSH to conduct research and training relevant to OS&H.

Section 23 authorizes the DOL to make grants available to the states to assist them in the operation of their OS&H programs.

Section 24 provides for the collection and analysis of statistics concerning occupational fatalities, injuries, and illnesses. These data are to be collected and compiled by the Bureau of Labor Statistics (BLS).

Section 25 requires the recipients of grants to maintain records. It also gives authority to the secretaries of DHHS and DOL to conduct audits when deemed appropriate and necessary.

Section 26 requires the secretaries of DOL and DHHS to provide an annual report within 120 days of the convening of each regular session of Congress. This section also dictates the required content of these reports.

Section 27 establishes a National Committee on State Workers' Compensation Laws to study and evaluate the fairness and adequacy of the present laws.

Section 28 amends the Small Business Act and allows for loans to be given to small businesses in order for them to comply with the OSHACT.

Section 29 adds an Assistant Secretary of Labor for Occupation Safety and Health.

Section 30 allows for an additional 25 DOL and 10 DHHS administrative positions to aid in the implementation of the Act.

Section 31 amends the Federal Aviation Act of 1958 to require fixed-wing powered aircraft that are used in air commerce to have an emergency locator beacon.

Section 32 states that if any provision or application of the Act is invalid for any person, then the remainder of the Act or its application or provisions are held invalid for that person.

Section 33 gives authorization to OSHA to receive funding to carry out the mandate of the Act. This is based on Congress' approval of necessary funding levels.

Section 34 specifies that the effective date of this Act was 120 days after the date of its enactment.

REGULATORY PROCESS

OSHA was mandated to develop, implement, and enforce regulations relevant to workplace safety and health and the protection of workers. Time constraints prevented the newly formed OSHA from developing brand new regulations. Therefore, OSHA adopted previously existing regulations from other government regulations, consensus standards, proprietary standards, professional groups' standards, and accepted industry standards. This is the reason that today the hazardous chemical exposure levels, with a few exceptions, are the same as the existing threshold limit values (TLVs) published by the American Congress of Government Industrial Hygienists in 1968. Once these TLVs were adopted, it became very difficult to revise them. Even though research and knowledge in the past 30 years have fostered newer and safer TLVs, they have not been adopted by OSHA.

As stated previously, the original OSHA standards and regulations have come from three main sources: consensus standards, proprietary standards, and federal laws that existed when the OSHACT became law.

Consensus standards are developed by industry-wide standard-developing organizations and are discussed and substantially agreed upon through industry consensus. OSHA has incorporated into its standards the standards of two primary groups: the American National Standards Institute (ANSI) and the National Fire Protection Association (NFPA). As an example, ANSI A10.33, Safety and Health Program Requirements for Multi-Employer Projects, covers minimum elements and activities of a program. It also defines the duties and responsibilities of the individual construction employers who will be working on a construction project.

Another example comes from the NFPA standards. NFPA No. 30-1969, Flammable and Combustible Liquids Code, was the source standard for CFR Part 1910, Section 106. It covers the storage and use of flammable and combustible liquids that have flash points below 200°F.

Proprietary standards are prepared by professional experts within specific industries, professional societies, and associations. They are determined by a straight membership vote, not by consensus. An example of these standards can be found in the Compressed Gas Association, Pamphlet P-1, *Safe Handling of Compressed Gases*. This proprietary standard covers requirements for safe handling, storage, and use of compressed-gas cylinders.

Some of the preexisting federal laws that are enforced by OSHA include: the Federal Supply Contracts Act (Walsh–Healey), the Federal Service Contracts Act (McNamara–O'Hara), the Contract Work Hours and Safety Standard Act (Construction Safety Act), and the National Foundation on the Arts and Humanities Act. Standards issued under these Acts are now enforced in all industries where they apply.

When OSHA needs to develop a new regulation or even revise an existing one, it becomes a lengthy and arduous process. This is why it took so long to get the following regulations passed:

- Process chemical safety standard—7 years
- Hazard communications standard—10 years
- Lockout/tagout standard—12 years (still does not apply to construction)
- Confined spaces—17 years (still does not apply to construction)

However, it only took 3 years to get a new regulation passed covering lift-slab construction after the collapse of L'Ambience Plaza in Bridgeport, CT, where 28 workers died. In addition, only a short period of time lapsed in getting the bloodborne pathogen standard passed when people were scared to death of HIV (AIDS) and hepatitis B virus (HBV).

Standards are sometimes referred to as either horizontal or vertical in their application. Most standards are horizontal or general. This means they apply to any employer in any industry. Fire protection, working surfaces, and first-aid standards are examples of horizontal standards.

Some standards are only relevant to a particular industry and are called vertical or particular standards. Examples of these standards applying to the construction industry, the longshoring industry, and special industries are covered in Subpart R of 29 CFR 1910.

Through newspapers and conversations, we get the impression that OSHA is producing new standards each day, which will impact the workplace. This is not true. The regulatory process is very slow. Why in some cases is the time so long and others so short? Are not the same steps followed for each regulation? The answer is yes, the process is the same, but at each step the time and the stumbling blocks may not be the same. The steps are as follows:

1. Agency (OSHA) opens a regulatory development docket for a new or revised regulation.
2. This indicates that OSHA believes a need for a regulation exists.
3. Advanced notice of proposed rulemaking (ANPRM) is published in the *Federal Register* and written comments are requested to be submitted within 30–60 days.
4. Comments are analyzed.
5. Notice of proposed rulemaking (NPRM) is published in the *Federal Register* with a copy of the proposed regulation.
6. Another public comment period transpires, usually for 30–60 days.
7. If no additional major issues are raised by the comments, the process continues to step 10.
8. If someone raises some serious issues, the process goes back to step 4 for review and possible revision of the NPRM.
9. Once the concerns have been addressed, it continues forward to steps 5 and 6 again.
10. If no major issues are raised, a final rule (FR) will be published in the *Federal Register*, along with the date when the regulation will be effective (usually 30 to 120 days).
11. There can still be a petition of reconsideration of the FR. There are times when an individual or industry may take legal action to bar the regulation's promulgation.
12. If the agency does not follow the correct procedures or acts arbitrarily or capriciously, the court may void the regulation and the whole process will need to be repeated.

If you desire to comment on a regulation during the development process, feel free to do so; your comments are important. You should comment on the areas where you agree or disagree. This is your opportunity to speak up. If no one comments, it is assumed that nobody cares one way or the other. You must be specific. Give examples, be precise, give alternatives, and provide any data or specific information that can back up your opinion. Federal agencies always welcome good data that substantiate your case. Cost/benefit data are always important in the regulatory process and any valid cost data that you are able to provide may be very beneficial. However, make sure that your comments are based on what is published in the *Federal Register* and not based on hearsay. Remember that the agency proposing the regulation may be working under specific restraints. Make sure you understand these constraints. Due to restrictions the agency may not have the power to do what you think ought to be done.

Sometimes the agency feels that there is no need for the proposed regulation, but it has been mandated to develop it. Your comments could be useful in stopping the development of this regulation. Just be sure your comments are polite, not demeaning or combative. Remember that an individual has worked on this proposed regulation and is looking for constructive and helpful comments. Even if you are against this regulation, do not let your comments degenerate to a personal level. Focus on the regulation, not individuals.

FEDERAL REGISTER

The *Federal Register* is the official publication of the U.S. government. If you are involved in regulatory compliance, you should obtain a subscription to the *Federal Register*. The reasons for obtaining this publication are clear. It is official, comprehensive, and not a summary prepared by someone else. It is published daily and provides immediate accurate information. The *Federal Register* provides early notices of forthcoming regulations, informs you of comment periods, and gives the preamble and responses to questions raised about a final regulation. It provides notices of meetings, gives information on obtaining guidance documents, supplies guidance on findings and on cross-references, and gives the yearly regulatory development agenda. It is the Bible for regulatory development and is recognizable by its brown paper and newsprint quality printing (see Figure 17.1).

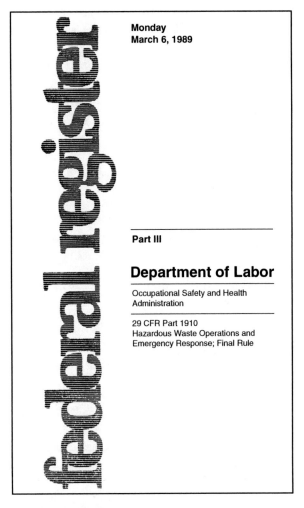

Figure 17.1 Sample cover for the *Federal Register*.

CODE OF FEDERAL REGULATIONS

Probably one of the most common complaints from people who use the *U.S.* CFR is "How do you wade through hundreds of pages of standards and make sense out of them?" From time to time you may have experienced this frustration and been tempted to throw the standards in the round file.

The CFR is a codification of the general and permanent rules published in the *Federal Register* by the executive departments and agencies of the federal government. The code is divided into 50 titles, which represent broad areas that are subject to federal regulations. Each title is divided into chapters that usually bear the name of the issuing agency. Each chapter is further subdivided into parts covering specific regulatory areas. Based on this breakdown, OSHA has designated Title 29—Labor, Chapter XVII (OSHA) and Part 1926 for the construction industry sector. The CFR related to OS&H for specific industries is shown in Table 17.1.

Each volume of the CFR is revised at least once each calendar year and issued on a quarterly basis. OSHA issues regulations at the beginning of the fourth quarter, or July 1 of each year (the approximate revision date is printed on the cover of each volume). An example of what the CFR looks like can be found in Figure 17.2.

The CFR is kept up-to-date by individual revisions issued in the *Federal Register*. These two publications (The CFR and the *Federal Register*) must be used together to determine the latest version of any given rule.

To determine whether there have been any amendments since the revision date of the U.S. Code volume in which you are interested, the following two lists must be consulted: The "Cumulative List of CFR Sections Affected," issued monthly; and the "Cumulative List of Parts Affected," appearing daily in the *Federal Register*. These two lists refer you to the *Federal Register* page where you may find the latest amendment of any given rule. The pages of the *Federal Register* are numbered sequentially from January 1 to January 1 of the next year.

As stated previously, Title 29, Chapter XVII, has been set aside for OSHA. Chapter XVII is broken down into parts and is further broken down into subparts, sections, and paragraphs.

REGULATION PARAGRAPH NUMBERING SYSTEM

To use the CFR, you need an understanding of the hierarchy of the paragraph numbering system. The numbering system is a combination of letters and numbers. Before 1979, italicized small case letters and small case roman numerals were used. A change was made after 1979.

Table 17.1   CFRs for Industry-Specific Regulations

General industry—29 CFR Part 1910

Shipyard employments—29 CFR 1915

Marine terminals—29 CFR 1917

Longshoring—29 CFR Part 1918

Gear certification—29 CFR Part 1919

Construction—29 CFR Part 1926

Agriculture—29 CFR Part 1928

Federal agencies—29 CFR 1960

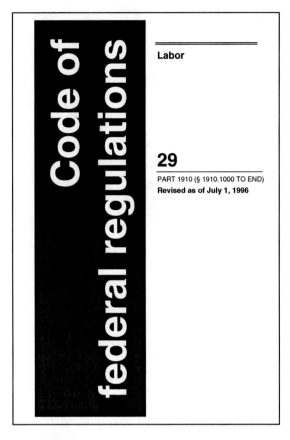

Figure 17.2   Sample cover for the CFR.

CFR Numbering Hierarchy	
<1979	**1980**
(a)	(a)
(1)	(1)
(i)	(i)
Italicized (a)	(A)
Italicized (1)	{1}
Italicized (i)	(i)

When trying to make use of the regulations, knowledge of the regulatory numbering system will help remove a lot of the headaches. This should make them easier to comprehend and more user-friendly. The following illustrates and explains the numbering system using an example from Subpart H of 29 CFR 1910.110—Storage and Handling of Liquefied Petroleum Gas.

29 CFR 1910.110 (b)(13)(ii)(b)(7)(iii)

Portable containers shall not be taken into buildings except as provided in paragraph (b)(6)(i) of this section.

Title	Code of Fed. Reg.	Part	Subpart	Section	Paragraph
29	CFR	1910	D	.110	

As can be seen from this example, the first number (29) stands for the title. Next comes CFR, which stands for the CFR, followed by 1910, which is the part 1910. Finally, there is a period followed by an Arabic number. This will always be the section number. In this case section .110 is the handling and storage of liquefied petroleum gas regulation. If the number had been .146, the section would pertain to permit-required confined spaces.

29 CFR 1910.110 (b)(13)(ii)(b)(7)(iii)
Portable containers shall not be taken into buildings except as provided in paragraph (b)(6)(i) of this section.

Title	Code of Fed. Reg.	Part	Subpart	Section	Paragraph
29	CFR	1910	D	.110	(b)

This means that the next breakdown of paragraphs will be sequenced using small case letters in parentheses (a), (b), (c), etc. If you had three major paragraphs of information under a section, they would be lettered .110(a), .110(b), and .110(c).

29 CFR 1910.110 (b)(13)(ii)(b)(7)(iii)
Portable containers shall not be taken into buildings except as provided in paragraph (b)(6)(i) of this section.

Title	Code of Fed. Reg.	Part	Subpart	Section	Paragraph
29	CFR	1910	D	.110	(b)(13)

The next level of sequencing involves the use of Arabic numbers. As illustrated, if there were three paragraphs of information between subheadings (a) and (b), they would be numbered (a)(1), (a)(2), and (a)(3).

29 CFR 1910.110 (b)(13)(*ii*)(*b*)(7)(*iii*)
Portable containers shall not be taken into buildings except as provided in paragraph (b)(6)(i) of this section.

Title	Code of Fed. Reg.	Part	Subpart	Section	Paragraph	
29	CFR	1910	D	.110	(b)(13)(ii)	(*b*)(7)(*iii*) Italicized

The next level uses the lower case roman numerals. An example would be between paragraphs (2) and (3). If there were five paragraphs of information pertaining to Arabic numeral (2) they would be numbered (2)(i), (2)(ii), (2)(iii), (2)(iv), and (2)(v).

If there are subparagraphs to the lower case roman numerals and the regulation was developed and implemented before 1979 as is the case, then an italicized small case letter is used such as (*a*), (*b*),..., (*c*) as in this example. Any other subparagraph falling under the italicized small case letter will be an italicized number such as (*1*), (*5*),..., (*8*) and subparagraphs to the italicized number are italicized small case roman numerals such as (*i*), (*ii*),..., (*iii*).

After 1979, the subparagraphs under the small case roman numerals became upper case letters such as (A), (B),..., (C). Any other subparagraph falling under an upper case letter is numbered using brackets, for example, {1}, {5},..., {23}, and any subparagraph to the bracketed numbers would be denoted by an italicized roman numeral as follows: (*i*), (*iv*),..., (*ix*).

If you are not using the OSHA Web site to access a copy of the OSHA regulations, you may have a copy of the CFR, which has a poor table of contents and a fair index, to help you to find information in a quick fashion. I usually have my students place a labeled tab at the beginning of each subpart (A–Z) and then I have them use a highlighter marking each section, major paragraphs, and subparagraphs. This will make using and finding information in your CFR easier.

Now let us see what can be done to simplify the use of the CFR. It is suggested that you color code your CFR book. Although there are many ways to do this, it is suggested to use the method shown in Figure 17.3.

It is suggested that you highlight every section heading in pink. All of the paragraphs, (a), (b), (c), etc., should be colored yellow by a horizontal line. At this point using a yellow dot for each subparagraph denoted by an Arabic character (1) becomes important since it is nearly impossible to tell the difference from an alpha character (l). This is also the case for the lower case roman alpha or numerals. The next step would be to yellow dot all the Arabic numbers. You can easily find the beginning of each paragraph by looking for the horizontal lines. The Arabic number subparagraphs are easily located by the yellow dots. Usually, color-coding two levels below the section heading is sufficient. At times you may want to select other colors for more detailed color coding.

Figure 17.3 Color-coding your CFR.

The general industry and construction standards, 29 CFR 1910 and 1926, are divided into 26 subparts lettered A through Z. In Appendix J, each subpart is highlighted for these regulations with an overview paragraph, a listing of all the sections in it, and a short checklist to assist you in deciding which subparts of these regulations apply to your type of work. If you check an entry on a subpart then you will need to comply with part or all of that subpart.

This chapter should help you understand OSHA regulations and how they are set up and how they might apply to your operation. If you have questions related to what a regulation means or if it applies to you, you should call OSHA and ask for clarification. Do not be afraid to call OSHA. They do not take names and numbers. An OSHA representative would prefer to answer your questions than to have to visit your facility or conduct an inspection of your facility. OSHA representatives will advise you regarding the application of a regulation to your workplace. However, remember, the final responsibility to comply with OSHA regulations rests with you.

CHAPTER 18

All Around: Workplace Environmental Issues

Industrial pollution of air, water, and soil is not acceptable today. (Courtesy of the U.S. Environmental Protection Agency.)

Occupational safety and health has undergone a change in that safety and health have been expanded to include the workplace environment. Today, most of you will hear the term Environmental Health and Safety (EHS), which is in use by industry and companies and their safety and health departments.

The reason for this is that many industrial processes have the potential to contaminate or pollute the worksite and its surroundings. If chemicals are able to escape from a process, they may result in the occurrence of a spill or the release of vapors, fumes, or particles into the atmosphere. These releases may result in creating an atmosphere that could be toxic (poisonous) to workers and the public. In addition, the escaped or spilled material may result in contamination of water or soil.

The U.S. Department of Environmental Protection Agency (USEPA) or your state Department of Environmental Protection may be alerted to these incidents and find you in violation of existing environmental laws. Whichever agency has jurisdiction, it may assess a penalty of $25,000 per day per violation until the violation has been abated or taken care of. These types of fines can result in the accumulation of large sums of money owed to the federal government or state agency. These fines usually surpass by many dollars the paltry fines levied by Occupational Safety and Health Administration.

Environmental law also addresses the disposal and storage of hazardous materials. Failure to properly store or dispose of hazardous material in a responsible manner will result in enforcement action by the federal or state environmental agency. In fact, any violation of an environmental law can result in enforcement action. In the past, businesses and companies have been polluters of the environment; with the formation of the USEPA and the laws under which it operates, this type of exploitation of the environment is no longer tolerated.

INDUSTRY TODAY

The responsible industry of today has the goal of producing a manufacturing process that will have minimal impact upon the environment. In the new environmentally friendly approach and designs that industry employs today, the emphasis is on minimizing raw material use, energy consumption, and waste production. In these newly designed processes and approaches, it is expected that there is to be a minimum impact upon the environment. These processes should be designed so that raw material, water, and waste are recycled as much as possible. The new process should use state-of-the-art technologies to decrease the effects on air, water, and solid waste emissions. Some of the ways technology can be applied to minimize environmental impact are as follows:

- Increase efficiency by optimization of processing operations
- Increase efficiency and maximize fuel use by using waste heat recovery systems to achieve full energy use
- Decrease pollution by using materials that minimize pollution
- Treat waste products efficiently by using advanced technologies
- Decrease waste production by using precision processing and machining systems (e.g., lasers)
- Maintain maximum energy efficiency, maximum utilization of raw materials, and minimum generation of pollutants by employing computerized control systems
- Achieve maximum recycling and minimum waste production by using processes that foster these outcomes

INTRODUCTION TO LAWS AND REGULATIONS

Laws and regulations are a major tool in protecting the environment. Congress passes laws that govern the United States and these are implemented in the form of regulations or standards by the enforcing agencies. To put those laws into effect, Congress authorizes certain government agencies, including the EPA, to create and enforce regulations. Here, you will find a basic description of how laws and regulations come to be, what they are, and where to find them, with an emphasis on environmental laws and regulations.

Creating a Law

Step 1: A member of Congress proposes a bill. A bill is a document that, if approved, will become law. To see the text of bills Congress is considering, or has considered, look on the Library of Congress' Thomas Web server.

Step 2: If both houses of Congress approve a bill, it goes to the President who has the option to either approve or veto it. If approved, the new law is called an act, and the text of the act is known as a public statute. Some of the better-known laws related to the environment are the Clean Air Act (CAA), the Clean Water Act (CWA), and the Safe Drinking Water Act (SDWA).

Step 3: Once an act is passed, the House of Representatives standardizes the text of the law and publishes it in the U.S. Code of Federal Regulations (CFR). The CFR is the official record of all federal laws.

Putting the Law to Work

Now that the law is official, how is it put into practice? Laws often do not include all the details. The CFR would not tell you, for example, what the speed limit is in front of your house. To make the laws work on a day-to-day level, Congress authorizes certain government agencies—including EPA—to create regulations.

Regulations set specific rules about what is legal and what is not. For example, a regulation issued by EPA to implement the CAA might state what levels of a pollutant—such as sulfur dioxide—are safe. It would tell industries how much sulfur dioxide they can legally emit into the air and what the penalty will be if they emit too much. Once the regulation is in effect, EPA then works to help Americans comply with the law and to enforce it.

Creating a Regulation

First, an authorized agency—such as EPA—decides that a regulation may be needed. The agency researches it and, if necessary, proposes a regulation. The proposal is listed in the *Federal Register* so that members of the public can consider it and send their comments to the agency. The agency considers all the comments, revises the regulation accordingly, and issues a final rule. At each stage in the process, the agency publishes a notice in the *Federal Register*. These notices include the original proposal, requests for public comment, notices about meetings where the proposal will be discussed (open to the public), and the text of the final regulation. (The *Federal Register* includes other types of notices, too.) A complete record of *Federal Register* notices issued by the entire federal government is available from the U.S. Government Printing Office.

Twice a year, each agency publishes a comprehensive report that describes all the regulations it is working on or has recently finished. These are published in the *Federal Register*, usually in April and October, as the Unified Agenda of Federal and Regulatory and Deregulatory Actions.

Once a regulation is completed and has been printed in the *Federal Register* as a final rule, it is codified by being published in the CFR. The CFR is the official record of all regulations created by the federal government. It is divided into 50 volumes, called titles, each of which focuses on a particular area.

Almost all environmental regulations appear in Title 40. The CFR is revised yearly with one fourth of the volumes updated every 3 months. Title 40 is revised every July 1.

The full text of CFR Title 40: Protection of Environment is retrievable by chapters, subchapters, and parts in portable document format (pdf) at the EPA Web site. A searchable database containing the complete CFR is available from the U.S. Government Printing Office.

Listing of the Laws

Among the major environmental laws enacted by Congress through which EPA carries out its efforts are the following:

- 1938 Federal Food, Drug, and Cosmetic Act
- 1947 Federal Insecticide, Fungicide, and Rodenticide Act (FIFRA)

- 1948 Federal Water Pollution Control Act (also known as the CWA)
- 1955 Clean Air Act (CAA)
- 1965 Shoreline Erosion Protection Act
- 1965 Solid Waste Disposal Act (SWDA)
- 1970 National Environmental Policy Act (NEPA)
- 1970 Pollution Prevention Packaging Act
- 1970 Resource Recovery Act
- 1971 Lead-Based Paint Poisoning Prevention Act
- 1972 Coastal Zone Management Act
- 1972 Marine Protection, Research, and Sanctuaries Act (MPRSA)
- 1972 Ocean Dumping Act
- 1973 Endangered Species Act
- 1974 Safe Drinking Water Act (SDWA)
- 1974 Shoreline Erosion Control Demonstration Act
- 1975 Hazardous Materials Transportation Act
- 1976 Resource Conservation and Recovery Act (RCRA)
- 1976 Toxic Substances Control Act (TSCA)
- 1977 Surface Mining Control and Reclamation Act
- 1978 Uranium Mill-Tailings Radiation Control Act
- 1980 Asbestos School Hazard Detection and Control Act
- 1980 Comprehensive Environmental Response, Compensation, and Liability Act (CERCLA)
- 1982 Nuclear Waste Policy Act
- 1984 Asbestos School Hazard Abatement Act
- 1986 Asbestos Hazard Emergency Response Act
- 1986 Emergency Planning and Community Right-to-Know Act (EPCRA)
- 1988 Indoor Radon Abatement Act
- 1988 Lead Contamination Control Act
- 1988 Medical Waste Tracking Act
- 1988 Ocean Dumping Ban Act
- 1988 Shore Protection Act
- 1990 National Environmental Education Act

INDUSTRY'S DILEMMA

With over 33 laws on the books and their companion regulations, it is an almost impossible task to know which laws and their regulations apply to your business. Some of this can be achieved by a process of elimination. If you do not have any medical waste, nuclear waste, or uranium mill tailings then the 1988 Medical Waste Tracking Act, the 1982 Nuclear Waste Policy Act, nor the 1978 Uranium Mill-Tailing Radiation Control Act would apply to your workplace. Many other laws do not apply to the majority of businesses in this country. However, many of the USEPA laws do apply or could affect your business.

To make this topic manageable for the average business or industry, the emphasis here will be on three major areas of the environment. They are air, water, and soil, each of which is critical for the preservation of life on the earth. The laws that are applicable to business and industry will be discussed as they apply to air, water, and soil.

MAJOR ENVIRONMENTAL LAWS

If you are interested in becoming active in environmental, health, and safety issues, you will need to understand many of the following federal laws. These laws, and others enacted by states, have various requirements and are enforced by various agencies. A brief description of the intent of each law follows. For more details, you should obtain a copy of the law from your local library, state library, or the relevant federal or state agency. Federal and state officials will help you gain a working knowledge of these laws.

National Environmental Policy Act

The National Environmental Policy Act (NEPA) was one of the first laws ever written that establishes the broad national framework for protecting our environment. NEPA's basic policy is to ensure that all branches of government give proper consideration to the environment before undertaking any major federal action that significantly affects the environment. NEPA requirements are invoked when airports, buildings, military complexes, highways, parkland purchases, and other such federal activities are proposed. Environmental assessments (EAs) and environmental impact statements (EISs), which are assessments of the likelihood of impacts from alternative courses of action, are required from all federal agencies and are the most visible NEPA requirements.

The basic purposes of NEPA are spelled out as a declaration of a national policy to encourage productive and enjoyable harmony between humans and the environment, to promote efforts that will prevent or eliminate damage to the environment and biosphere and stimulate the health and welfare of humans, to enrich the understanding of the ecological systems and natural resources important to the nation, and to establish a council on environmental quality (CEQ).

These purposes are followed by a Declaration of National Environmental Policy, which commits the federal government to work with other levels of government and other groups to improve environmental conditions, and also creates the CEQ in the Executive Office of the President.

NEPAs unique requirement for preparation of EISs has had a dramatic influence on federal agency decision-making, as numerous court rulings enforced strict compliance with the EA procedures for major programs and projects.

How Does This Act Affect You?

This Act has little direct impact upon you or your business. It is a policy directed at federal agencies to determine the impact upon the environment when undertaking any major federal action that might impact negatively on the environment.

Clean Air Act

The CAA is the comprehensive federal law that regulates air emissions from area, stationary, and mobile sources. This law authorizes the USEPA to establish National Ambient (Outdoor) Air Quality Standards (NAAQS) to protect public health and the environment. The goal of the Act was to set and achieve NAAQS in every state by 1975. This setting of maximum pollutant standards was coupled with directing the states to develop state implementation plans (SIPs) applicable to appropriate industrial sources in the state. The Act was amended in 1977 primarily to set new goals (dates) for achieving attainment of NAAQS since many areas of the country had failed to meet the deadlines. The 1990 amendments to the CAA in large part were intended to meet unaddressed or insufficiently addressed problems such as acid rain, ground level ozone, stratospheric ozone depletion, and air toxics. The Act establishes federal standards for mobile sources of air pollution, for sources of 188 hazardous air pollutants, and for the emissions that cause acid rain (see Appendix K). It establishes a comprehensive permit system for all major sources of air pollution. It also addresses the prevention of pollution in areas with clean air and protection of the stratospheric ozone layer.

The 1970 amendments established procedures under which EPA sets national standards for air quality, required a 90% reduction in emissions from new automobiles by 1975, established a program to require the best available control technology at major new sources of air pollution, established a program to regulate air toxics, and greatly strengthened federal enforcement authority.

The 1977 amendments extended deadlines and added the Prevention of Significant Deterioration program to protect air that is cleaner than national standards.

Changes to the Act in 1990 included provisions to

1. Classify nonattainment areas according to the extent to which they exceed the standard, tailoring deadlines, planning, and controls to each area's status.
2. Tighten auto emission standards and require reformulated and alternative fuels in the most polluted areas.
3. Revise the air toxics section, establishing a new program of technology-based standards and addressing the problem of sudden, catastrophic releases of toxics.
4. Establish an acid rain control program, with a marketable allowance scheme to provide flexibility in implementation.
5. Require a state-run permit program for the operation of major sources of air pollutants.
6. Implement the Montreal Protocol to phase out most ozone-depleting chemicals.
7. Update the enforcement provisions so that they parallel those in other pollution control acts, including authority for EPA to assess administrative penalties.

The Act requires EPA to establish NAAQS for several types of air pollutants. The NAAQS must be designed to protect public health and welfare with an adequate margin of safety. Using this authority, EPA has promulgated NAAQS for six air pollutants: sulfur dioxide (SO_2), particulate matter (2.5 and 10 ppm), nitrogen dioxide (NO_2), carbon monoxide (CO), ozone, and lead. The Act requires EPA to review the scientific data upon which the standards are based and revise the standards, if necessary, every 5 years. More often than not, however, EPA has taken more than 5 years in reviewing and revising the standards.

Industries must reduce emissions from existing facilities by 10% more than the emissions of any new facility opened in the area and impose reasonably available control technology (RACT) on all major sources emitting more than 100 tons per year for the nine industrial categories where EPA has already issued control technique guidelines that describe RACT before 1990.

How Does This Act Affect You?

This Act applies to you if you are emitting any material from your process including particles, gases, vapors, or mist that is visible or detectable by smell. If you are processing chemicals, conducting chemical reactions, doing grinding operations, producing a combustion reaction, or emissions from transportation vehicles or aircraft, you may be emitting air pollutants. Citizens or others are apt to report your company to the EPA. A visit by the EPA will result in air sampling from your operation followed by citation and monetary penalties if you are in violation with the NAAQS or regulations involving ambient air. You should not take this lightly since once on the list of perpetrators you will be a target of enforcement activities for some time to come. If you think that you have the possibility of emitting pollutants, you should have your own air sampling performed as an act of good faith and put your mind at rest regarding your violation of the CAA.

Clean Water Act

The CWA is a 1977 amendment to the Federal Water Pollution Control Act of 1972, which set the basic structure for regulating discharges of pollutants to waters of the United States. This law gave EPA the authority to set effluent standards on an industry-by-industry basis (technology-based) and continued the requirements to set water quality standards for all contaminants in surface waters. The CWA makes it unlawful for any person to discharge any pollutant from a point source into navigable waters unless a permit National Pollutant Discharge Elimination System (NPDES) is obtained under the Act. The 1977 amendments focused on toxic pollutants. In 1987, the CWA was reauthorized and again focused on toxic substances, authorized citizen suit provisions, and funded sewage treatment plants public owned treatment works (POTWs) under the Construction Grants Program.

The CWA provides for the delegation of many permitting, administrative, and enforcement aspects of the law by EPA to state governments. In states with the authority to implement CWA programs, EPA still retains oversight responsibilities.

The legislation declared as its objective the restoration and maintenance of the chemical, physical, and biological integrity of the nation's waters. Two goals were also established: zero discharge of pollutants by 1985 and, as an interim goal and where possible, water quality that is both fishable and swimmable by mid-1983. While those dates have passed, the goals remain, and efforts to attain the goals continue.

The CWA today consists of two major parts, one being the Title II and Title VI provisions, which authorize federal financial assistance for municipal sewage treatment plant construction. The other is the regulatory requirements, found throughout the Act, that apply to industrial and municipal dischargers.

The Act has been termed a technology-forcing statute because of the rigorous demands placed on those who are regulated by it to achieve higher and higher levels of pollution abatement. Industries were given until July 1, 1977, to install best practicable control technology (BPT) to clean up waste discharges. Municipal wastewater treatment plants were required to meet an equivalent goal, termed secondary treatment, by that date. (Municipalities unable to achieve secondary treatment by that date were allowed to apply for case-by-case extensions up to July 1, 1988. According to EPA, 86% of all cities met the 1988 deadline; the remainder was put under judicial or administrative schedules requiring compliance as soon as possible. However, many cities, especially smaller ones, continue to make investments in building or upgrading facilities needed to achieve secondary treatment.) Cities that discharge wastes into marine waters were eligible for case-by-case waivers of the secondary treatment requirement, where sufficient showing could be made that natural factors provide significant elimination of traditional forms of pollution and that both balanced populations of fish, shellfish, and wildlife and water quality standards would be protected.

Before the 1987 amendments, programs in the CWA were primarily directed at point source pollution and wastes discharged from discrete and identifiable sources, such as pipes and other outfalls. In contrast, except for general planning activities, little attention had been given to nonpoint source pollution (storm water runoff from agricultural lands, forests, construction sites, and urban areas), despite estimates that it represents more than 50% of the nation's remaining water pollution problems. As it travels across land surface toward rivers and streams, rainfall and snowmelt runoff picks up pollutants, including sediments, toxic materials, and conventional wastes (e.g., nutrients) that can degrade water quality. The 1987 amendments authorized measures to address such pollution by directing states to develop and implement nonpoint pollution management programs. States were encouraged to pursue groundwater protection activities as part of their overall nonpoint pollution control efforts. Federal financial assistance was authorized to support demonstration projects and actual control activities. These grants may cover up to 60% of program implementation costs.

The emphasis is on controlling toxic pollutants—heavy metals, pesticides, and other organic chemicals. In addition to these limitations applicable to categories of industry, EPA has issued water quality criteria for more than 115 pollutants, including 65 named classes or categories of toxic chemicals, or priority pollutants. These criteria recommend ambient, or overall, concentration levels for the pollutants and provide guidance to states for establishing water quality standards that will achieve the goals of the Act.

The NPDES permit, containing effluent limitations on what may be discharged by a source, is the Act's principal enforcement tool. EPA may issue a compliance order or bring a civil suit in U.S. district courts against persons who violate the terms of a permit. The penalty for such a violation can be as much as $25,000 per day. Stiffer penalties are authorized for criminal violations of the Act—for negligent or knowing violations—of as much as $50,000 per day, 3 years' imprisonment, or both. A fine of as much as $250,000, 15 years in prison, or both is authorized for knowing endangerment—violations that knowingly place another person in imminent danger of death or serious bodily injury.

How Does the CWA Apply to Your Business?

If you use water in your operation or processing, then you will need to develop control technologies to ensure that you control the discharge of chemical pollutants, sediment, or heated water back into the environment. You will need to ensure that your use of water is in compliance with the CWA or this could be costly to your company. As you can see, the dumping of contaminated water into oceans, rivers, streams, on the ground, into storm sewers, or in sanitary sewer systems of pollutants is prohibited.

Safe Drinking Water Act

The SDWA was established to protect the quality of drinking water in the United States. This law focuses on all waters actually or potentially designated for drinking use, whether from above- or underground sources. The Act authorized EPA to establish safe standards of purity and required all owners or operators of public water systems to comply with primary (health-related) standards. State governments, which assume this power from EPA, also encourage attainment of secondary standards (nuisance-related).

The SDWA, Title XIV of the Public Health Service Act, is the key federal law for protecting public water systems from harmful contaminants. First enacted in 1974 and substantively amended in 1986 and 1996, the Act is administered through regulatory programs that establish standards and treatment requirements for drinking water, control underground injection of wastes that might contaminate water supplies, and protect groundwater. The 1974 law established the current federal-state arrangement in which states may be delegated primary implementation and enforcement authority for the drinking water program; the 1986 amendments sought to accelerate contaminant regulation. The state-administered Public Water Supply Supervision (PWSS) program remains the basic program for regulating the nation's public water systems.

Further amendments required EPA to (1) issue regulations for 83 specified contaminants by June 1989 and for 25 more contaminants every 3 years thereafter, (2) promulgate requirements for disinfection and filtration of public water supplies, (3) ban the use of lead pipes and lead solder in new drinking water systems, (4) establish an elective wellhead protection program around public wells, (5) establish a demonstration grant program for state and local authorities that have designated sole-source aquifers to develop groundwater protection programs, and (6) issue rules for monitoring injection wells that inject wastes below a drinking water source. SDWA regulations must ensure that the standard or treatment techniques must minimize the overall health risk.

Toxic Substances Control Act

The TSCA of 1976 was enacted by Congress to test, regulate, and screen all chemicals produced or imported into the United States. Many thousands of chemicals and their compounds are developed each year with unknown toxic or dangerous characteristics. To prevent tragic consequences, TSCA requires that any chemical that reaches the consumer market place be tested for possible toxic effects before commercial manufacture.

Any existing chemical that poses health and environmental hazards is tracked and reported under TSCA. Procedures are also authorized for corrective action under TSCA in cases of cleanup of toxic materials contamination. TSCA supplements other federal statutes, including the CAA and the Toxic Release Inventory (TRI) under EPCRA.

The TSCA authorizes EPA to screen existing and new chemicals used in manufacturing and commerce to identify potentially dangerous products or uses that should be subject to federal control. EPA may require manufacturers and processors of chemicals to conduct tests and report the results to determine the effects of potentially dangerous chemicals on living things. Based on test results and other information, EPA may regulate the manufacture, importation, processing, distribution, use, and disposal of any chemical that presents an unreasonable risk of injury to human health or the environment. A variety of regulatory tools are available to EPA under TSCA ranging in severity from a total ban on production, import, and use to a requirement that a product bears a warning label at the point of sale.

Federal legislation to control toxic substances was originally proposed in 1971 by the President's CEQ. There existed a need to identify and control chemicals whose manufacture, processing, distribution, use, and disposal were potentially dangerous and not adequately regulated under other environmental statutes. Episodes of environmental contamination—including contamination of the Hudson River and other waterways by polychlorinated biphenyls (PCBs), the threat of stratospheric ozone depletion from chlorofluorocarbon (CFC) emissions, and contamination of agricultural produce by polybrominated biphenyls (PBBs) in the state of Michigan—increased the support for TSCA.

How Does This Affect You?

For starters, if you manufacture or import chemicals then the registration process defined under TSCA must be completed. If you use chemicals categorized as toxic then your process should not endanger the environment or your workforce. You are responsible if these chemicals are inadvertently released into the environment in any way or misused by you. In addition, you are responsible for the final disposal of the chemical and any waste products. This is to say that you must plan ahead on how to safely use the chemical, properly clean up spills or contamination, and properly dispose of the chemical. Some people call this a cradle-to-grave procedure or process.

Federal Insecticide, Fungicide, and Rodenticide Act

The primary focus of FIFRA was to provide federal control of pesticide distribution, sale, and use. EPA was given authority under FIFRA not only to study the consequences of pesticide usage but also to require users (farmers, utility companies, and others) to register when purchasing pesticides. Through later amendments to the law, users also must take exams for certification as applicators of pesticides. All pesticides used in the United States must be registered (licensed) by EPA. Registration assures that pesticides will be properly labeled and that, if used in accordance with specifications, they will not cause unreasonable harm to the environment.

FIFRA requires EPA to regulate the sale and use of pesticides in the United States through registration and labeling of the estimated 21,000 pesticide products currently in use. The Act directs EPA to restrict the use of pesticides as necessary to prevent unreasonable adverse effects on people and the environment, taking into account the costs and benefits of various pesticide uses. FIFRA prohibits sale of any pesticide in the United States unless it is registered and labeled indicating approved uses and restrictions. It is a violation of the law to use a pesticide in a manner that is inconsistent with the label instructions. EPA registers each pesticide for each approved use, for example, to control boll weevils on cotton. In addition, FIFRA requires EPA to reregister older pesticides based on new data that meet current regulatory and scientific standards. Establishments that manufacture or sell pesticide products must register with EPA. Facility managers are required to keep certain records and to allow inspections by EPA or state regulatory representatives.

Pesticides are broadly defined in FIFRA Section 2(u) as chemicals and other products used to kill, repel, or control pests. Familiar examples include pesticides used to kill insects and weeds that can reduce the yield and sometimes harm the quality of agricultural commodities, ornamental plantings, forests, wooden structures, and pastures. However, the broad definition of pesticide in FIFRA also applies to products with less familiar pesticidal uses. For example, substances used to control mold, mildew, algae, and other nuisance growths on equipment, in surface water, or on stored grains are pesticides. The term also applies to disinfectants and sterilants, insect repellents and fumigants, rat poison, mothballs, and many other substances.

Is There an Impact upon You?

Yes, there could be if you have untrained workers applying pesticide and they become ill, which will be reported by the medical community to their public health department and then to EPA. If EPA comes to your place of employment and finds that you are not using the pesticide in compliance with the label directions (the label is a legal document), they will undertake enforcement actions against you in the form of citations and financial penalties. Pesticides are poisons and can have severe effects on the health of humans. Thus, if you need to use pesticides, you must take appropriate precautions to protect individuals as well as the environment.

Comprehensive Environmental Response, Compensation, and Liability Act

CERCLA (pronounced SERK-la) provides a federal Superfund to clean up uncontrolled or abandoned hazardous waste sites as well as accidents, spills, and other emergency releases of pollutants and contaminants into the environment. Through the Act, EPA was given power to seek out the parties responsible for any release and assure their cooperation in the cleanup. EPA cleans up orphan sites when potentially responsible parties (PRPs) cannot be identified or located, or when they fail to act. Through various enforcement tools, EPA obtains private party cleanup through orders, consent decrees, and other small party settlements. EPA also recovers costs from financially viable individuals and companies once a response action has been completed. EPA is authorized to implement the Act in all 50 states and U.S. territories. Superfund site identification, monitoring, and response activities in states are coordinated through the state environmental protection or waste management agencies.

CERCLA authorizes the federal government to respond to spills and other releases (or threatened releases) of hazardous substances, as well as to leaking hazardous waste dumps. Hazardous substances are identified under the SWDA, the CWA, the CAA, and the TSCA, or are designated by the EPA. Response is also authorized for releases of pollutants or contaminants, which are broadly defined to include virtually anything that can threaten the health of any organism. Most nuclear materials and petroleum are excluded, except for those petroleum products that are specifically designated as hazardous substances under the jurisdiction of one of the laws mentioned previously. The fund is not to be used for responding to: (1) releases of naturally occurring unaltered substances; (2) releases from products that are part of the structure of residential buildings, businesses, or community structures (such as asbestos); or (3) releases into drinking water supplies due to ordinary deterioration of the water system. An exception to these three limitations is made, however, in cases of public health or environmental emergencies when no other person has the authority and capability to respond in a timely manner. EPA is to give priority to releases that threaten public health or drinking water supplies.

How Does This Affect You?

If you own a worksite and contaminate it during the course of your business, then you are liable for the cleanup (remediation) of that site. Usually this must occur before the sale of your worksite, which usually requires an EA. Even if you are able to sell your worksite and it is determined later that you had contaminated that site, the USEPA will hold you financially liable for the remediation of your previously owned worksite. Therefore, you should take pollution prevention as a serious matter around and within your workplace. The cost could bankrupt your company. This type of contamination could result in tainting of the groundwater and soil, which you would be financially responsible to clean up.

Superfund Amendments and Reauthorization Act

The Superfund Amendments and Reauthorization Act (SARA) of 1986 reauthorized CERCLA to continue cleanup activities around the country. Several site-specific amendments, definitions, clarifications, and technical requirements were added to the legislation, including additional enforcement authorities. Title III of SARA also authorized the EPCRA.

Resource Conservation and Recovery Act and Solid Waste Disposal Act

RCRA (pronounced rick-rah) gave EPA the authority to control hazardous waste from cradle to grave. This includes the generation, transportation, treatment, storage, and disposal (TSD) of hazardous waste. RCRA also sets forth a framework for the management of nonhazardous solid wastes.

The 1986 amendments to RCRA enabled EPA to address environmental problems that could result from underground tanks storing petroleum and other hazardous substances. RCRA focuses only on active and future facilities and does not address abandoned or historical sites.

Hazardous and Solid Waste Amendments (HSWA) (pronounced hiss-wa)—The federal HSWA are the 1984 amendments to RCRA, which required phasing out land disposal of hazardous waste. Some of the other mandates of this strict law include increased enforcement authority for EPA, more stringent hazardous waste management standards, and a comprehensive underground storage tank (UST) program.

The RCRC of 1976 established the federal program regulating solid and hazardous waste management. RCRA actually amended earlier legislation (the SWDA of 1965), but the amendments were so comprehensive that the Act is commonly called RCRA rather than its official title.

The Act defines solid and hazardous waste, authorizes EPA to set standards for facilities that generate or manage hazardous waste, and establishes a permit program for hazardous waste TSD facilities. RCRA was last reauthorized by the HSWA of 1984. The amendments set deadlines for permit issuance, prohibited the land disposal of many types of hazardous waste without prior treatment, required the use of specific technologies at land disposal facilities, and established a new program regulating USTs.

Federal solid waste law has gone through four major phases. The SWDA (passed in 1965 as Title II of the CAA of 1965) focused on research, demonstrations, and training. It provided for sharing with the states the costs of making surveys of waste disposal practices and problems and of developing waste management plans. The Resource Recovery Act of 1970 changed the whole tone of the legislation from efficiency of disposal to concern with the reclamation of energy and materials from solid waste. It authorized grants for demonstrating new resource recovery technology and required annual reports from the EPA on means of promoting recycling and reducing the generation of waste. In a third phase, the federal government embarked on a more active, regulatory role, embodied in the RCRA of 1976. RCRA instituted the first federal permit program for hazardous waste and prohibited open dumps. In a fourth phase, embodied in the HSWA of 1984, the federal government attempted to prevent future cleanup problems by prohibiting land disposal of untreated hazardous wastes, setting liner and leachate collection requirements for land disposal facilities, setting deadlines for closure of facilities not meeting standards, and establishing a corrective action program.

A waste is hazardous if it is ignitable, corrosive, reactive, or toxic, or appears on a list of about 100 industrial process waste streams and more than 500 discarded commercial products and chemicals. The 1976 law expanded the definition of solid waste, of which hazardous waste is a subset, to include sludge and other discarded material, including solid, liquid, semisolid, or contained gaseous material. The broadened definition is particularly important with respect to hazardous wastes, at least 95% of which are liquids or sludges. Some wastes are specifically excluded, however, including irrigation return flows, industrial point source discharges (regulated under the CWA), and nuclear material covered by the Atomic Energy Act.

Under RCRA, hazardous waste generators must comply with regulations concerning record keeping and reporting; the labeling of wastes; the use of appropriate containers; the provision of information on the wastes' general chemical composition to transporters, treaters, and disposers; and the use of a manifest system. Facilities generating less than 1000 kg of waste per month were initially exempt from the regulations; the 1984 amendments to RCRA lowered that exemption to 100 kg per month, beginning in 1986.

Transporters of hazardous waste must also meet certain standards. These regulations were coordinated by EPA with existing regulations of the Department of Transportation. A manifest system, effective since 1980, is used to track wastes from their point of generation, along their transportation routes, to the place of final TSD.

TSD facilities are required to have permits, to comply with operating standards, to meet financial requirements in case of accidents, and to close their facilities in accordance with EPA regulations. The 1984 amendments imposed a number of new requirements on TSD facilities with the intent of minimizing land disposal. Bulk or noncontainerized hazardous liquid wastes are prohibited from disposal in any landfill, and severe restrictions are placed on the disposal of containerized hazardous liquids, as well as on the disposal of nonhazardous liquids in hazardous waste landfills. The land disposal of specified highly hazardous wastes was phased out over the period from 1986 to 1990. EPA was directed to review all wastes that it has defined as hazardous and to make a determination as to the appropriateness of land disposal for them. Minimum technological standards were set for new landfills and surface impoundments requiring, in general, double liners, a leachate collection system, and groundwater monitoring.

The major (nonhazardous) solid waste provision in RCRA is the prohibition of open dumps. This prohibition is implemented by the states, using EPA criteria to determine which facilities qualify as sanitary landfills and may remain open. EPA was required to revise the sanitary landfill criteria for facilities that receive small quantities of generator hazardous waste or hazardous household waste. In general, the new criteria require liners, leachate collection, groundwater monitoring, and corrective action at municipal landfills.

To address a nationwide problem of leaking USTs, Congress established a leak prevention, detection, and cleanup program through the 1984 RCRA amendments and the 1986 SARA. The 1984 RCRA amendments created a federal program to regulate USTs containing petroleum and hazardous chemicals to limit corrosion and structural defects and thus minimize future tank leaks. The law directed EPA to set operating requirements and technical standards for tank design and installation, leak detection, spill and overfill control, corrective action, and tank closure. The UST program is to be administered primarily by states. It requires registration of most underground tanks, bans the installation of unprotected tanks, sets federal technical standards for all tanks, coordinates federal and state regulatory efforts, and provides for federal inspection and enforcement. EPA lacked explicit authority to clean up contamination from leaking underground petroleum tanks as Congress had specifically excluded petroleum products (although not petrochemicals) from the Superfund law. The new provisions authorized the federal government to respond to petroleum spills and leaks and created a Leaking UST Trust Fund to fund cleanup of leaks from petroleum USTs in cases where the UST owner or operator does not clean up a site.

What Does This Mean to You?

If you generate hazardous or solid waste, you are responsible for ensuring that it is transported, treated, and disposed of in compliance with EPA regulations. Each act has an enforcement clause in it, which makes use of substantial financial penalties for failure to comply. This act is not an exception. If you have an UST that has the potential to leak, then you should take steps to replace it, hopefully before it starts to leak. Once it has a leak then you will be responsible for any contamination of the soil and groundwater and you might be cited by EPA as well. You can understand why I am emphasizing your responsibility to assure that you follow EPA laws and regulations. If you do not, you will reap a large financial burden possibly beyond your means to absorb.

Emergency Planning and Community Right-to-Know Act

Also known as Title III of SARA, EPCRA was enacted by Congress as the national legislation on community safety. This law was designed to help protect local communities, public health, safety, and the environment from chemical hazards. To implement EPCRA, Congress required each state to appoint a State Emergency Response Commission (SERC). The SERCs were required to divide their states into Emergency Planning Districts and to name a Local Emergency Planning Committee (LEPC) for each district. Broad representation by firefighters, health officials, government and media representatives, community groups, industrial facilities, and emergency managers ensures that all necessary elements of the planning process are represented.

EPCRA established state commissions and local committees to develop and implement procedures for coping with releases of hazardous chemicals and mandated annual reporting to government officials on environmental releases of such chemicals by the facilities that manufacture or use them in significant amounts. EPA facilitates planning, enforces compliance when necessary, and provides public access to information about environmental releases of toxic chemicals.

EPCRA established a national framework for EPA to mobilize local government officials, businesses, and other citizens to plan ahead for chemical accidents in their communities. Subtitle A requires local planning to respond to sudden releases of chemicals that might occur in the event of a spill, explosion, or fire. It ensures that responsible officials know what hazardous chemicals are used or stored by local businesses and will be notified quickly in the event of an accident.

There are various reporting requirements for facilities. The information collected may be used to develop and implement emergency plans as well as to provide the public with general information about chemicals to which they may be exposed.

The Occupational Health and Safety Act (OSHAct) of 1970 requires most employers to provide employees with access to a material safety data sheet (MSDS) for any hazardous chemical. This right-to-know law for workers aims to ensure that people potentially exposed to such chemicals have access to information about the potential health effects of exposure and know how to avoid them.

EPCRA, Section 311 requires facilities covered by OSHAct to submit an MSDS for each hazardous chemical or a list of such chemicals to the LEPC, the SERC, and the local fire department. EPA has authority to establish categories of health and physical hazards and to require facilities to list hazardous chemicals grouped by such categories in their reports. An MSDS need only be submitted once, unless there is a significant change in the information it contains. An MSDS must be provided in response to a request by an LEPC or a member of the public. Hazardous chemicals are defined by the CFR, Title 29, at Section 1910.1200(c)(3).

EPCRA, Section 312, requires the same employers to submit annually an emergency and hazardous chemical inventory form to the LEPC, SERC, and local fire department. These forms must provide estimates of the maximum amount of the chemicals present at the facility at any time during the preceding year; estimates of the average daily amount of chemicals present; and the general location of the chemicals in the facility. Information must be provided to the public in response to a written request. EPA is authorized to establish threshold quantities for chemicals, below which facilities are not required to report.

Section 313 mandates development of the TRI, a computerized EPA database of toxic chemical releases to the environment by manufacturing facilities. It requires facilities that manufacture, use, or process toxic chemicals to report annually to EPA on the amounts of each chemical released to each environmental medium (air, land, or water) or transferred off-site. EPA makes TRI data available in raw or summarized form to the public. The public may obtain specific information (e.g., about a particular manufacturing facility) by submitting a request in writing to EPA. EPA distributes written and electronic, nationwide, and state-by-state summaries of annual data.

EPCRA Section 313 requires a report to EPA and the state from each manufacturer with 10 or more employees who either uses 10,000 lb or manufactures or processes 25,000 lb of any toxic chemical during the reporting year. EPCRA enumerates the following data reporting requirements for each covered chemical present at each facility whether it is manufactured, processed, or otherwise used, and the general category of use, the maximum amount present at each location during the previous year, treatment or disposal methods used, and amount released to the environment or transferred off-site for treatment or disposal.

EPCRA requires reporting by manufacturers, which the law defines as facilities in Standard Industrial Classification codes 20 through 39. The law authorized EPA to expand reporting requirements to additional industries. From August 3, 1993, President Clinton required reporting by all federal facilities (Executive Order 12856). The president announced extension of TRI requirements to federal contractors on August 8, 1995. On November 30, 1994, EPA exempted from standard reporting requirements facilities that manufacture, process, or otherwise use up to 1 million pounds of a toxic chemical per year, if they have less than 500 lb of reportable quantities of chemical per year. EPA promulgated a rule on May 1, 1997, requiring reports on toxic releases from additional industrial categories, including some metal mining, coal mining, commercial electric utilities, petroleum bulk terminals, chemical wholesalers, and solvent recovery facilities.

How Does This Affect You?

Almost any industry that has workers and uses chemicals will have chemicals that are deemed toxic or hazardous. This Act has the potential to affect your company the most since it requires reporting and planning by your company. I would suggest that you review in detail these requirements and evaluate your chemicals, amounts, hazard potential, and need for compliance.

Endangered Species Act

The Endangered Species Act provides a program for the conservation of threatened and endangered plants and animals and the habitats in which they are found. The U.S. Fish and Wildlife Service (FWS) of the Department of Interior maintains the list of 1046 endangered species (598 are plants) and 305 threatened species (146 are plants). Species include birds, insects, fish, amphibians, reptiles, mammals, crustaceans, flowers, grasses, and trees. Anyone can petition FWS to include a species on this list or to prevent some activity, such as logging, mining, or dam building. The law prohibits any action, administrative or real, that results in a taking of a listed species or adversely affects habitat. Likewise, import, export, interstate, and foreign commerce of listed species are all prohibited.

EPAs decision to register a pesticide is based in part on the risk of adverse effects on endangered species as well as environmental fate (how a pesticide will affect habitat). Under FIFRA, EPA can issue emergency suspensions of certain pesticides to cancel or restrict their use if an endangered species will be adversely affected. Under a new program, EPA, FWS, and USDA are distributing hundreds of county bulletins that include habitat maps, pesticide use limitations, and other actions required to protect listed species.

In addition, we are enforcing regulations under various treaties, including the Convention on International Trade in Endangered Species (CITES) of wild fauna and flora. The United States and 70 other nations have established procedures to regulate the import and export of imperiled species and their habitat. The FWS works with U.S. Customs agents to stop the illegal trade of species, including the black rhino, African elephants, tropical birds and fish, orchids, and various corals.

How Does This Affect You?

There really should not be an impact to your business unless you deal with the sale of plants, animals, or animal products. It usually will not affect your worksite unless you happen to be unlucky enough to have on your property a habitat for one of these endangered species and decide to bulldoze it and build upon it. Mining, logging, and dam building have larger impacts upon the environments and are more likely to face the endangered species problem. Any time that you may want to build or alter the landscape, you may come face to face with the Endangered Species Act.

Pollution Prevention Act

The PPA of 1990 focused industry, government, and public attention on reducing the amount of pollution produced through cost-effective changes in production, operation, and raw materials use. Opportunities for source reduction are often not realized because existing regulations, and the industrial resources required for compliance, focus on treatment and disposal. Source reduction is fundamentally different from and more desirable than waste management or pollution control. Pollution prevention also includes other practices that increase efficiency in the use of energy, water, or other natural resources and protect our resource base through conservation. Practices include recycling, source reduction, and sustainable agriculture.

The PPA of 1990 requires the EPA to establish an Office of Pollution Prevention, develop and coordinate a pollution prevention strategy, and develop source reduction models. In addition to authorizing data collection on pollution prevention, the Act requires owners and operators of manufacturing facilities to report annually on source reduction and recycling activities.

Enactment of the PPA of 1990 marked a turning point in the direction of U.S. environmental protection policy. From an earlier focus on the need to reduce or repair environmental damage by controlling pollutants at the point where they are released to the environment (i.e., at the end of the pipe or smokestack, at the boundary of a polluter's private property, in transit over public highways and waterways, or after disposal), Congress turned to pollution prevention through reduced generation of pollutants at their point of origin. Broad support for this policy change was based on the notion that traditional approaches to pollution control had achieved progress but should be supplemented in the future with new approaches that might better address cross-media pollution transfers, the need for cost-effective alternatives, and methods of controlling pollution from dispersed or nonpoint sources of pollution. Pollution prevention, also referred to as source reduction, is viewed as the first step in a hierarchy of options to reduce risks to human health and the environment. Where prevention is not possible or may not be cost-effective, other options would include recycling, followed next by waste treatment according to environmental standards, and as a last resort, safe disposal of waste residues.

The PPA states that it is the policy of the United States that "pollution should be prevented or reduced at the source whenever feasible; pollution that cannot be prevented should be recycled in an environmentally safe manner, whenever feasible; pollution that cannot be prevented or recycled should be treated in an environmentally safe manner whenever feasible; and disposal or other release into the environment should be employed only as a last resort and should be conducted in an environmentally safe manner."

Source reduction is defined as "any practice that

- Reduces the amount of any hazardous substance, pollutant, or contaminant entering any waste stream or otherwise released into the environment (including fugitive emissions) prior to recycling, treatment, or disposal
- Reduces the hazards to public health and the environment associated with the release of such substances, pollutants, or contaminants"

Source reduction is the preferred strategy for environmental protection because it is often cost-effective; offers industry substantial savings in reduced raw materials, pollution control costs, and liability costs; reduces risks to workers; and reduces risk to the environment and public health. The Act was meant to increase interest in source reduction and encourage adoption of cost-effective source reduction practices.

EPA was directed to develop and implement a detailed and coordinated strategy to promote source reduction, to consider the effect on source reduction of all EPA programs and regulations, and to identify and make recommendations to Congress to eliminate barriers to source reduction. EPA must also conduct workshops and produce and disseminate guidance documents as part of a training program on source reduction opportunities for state and federal enforcement officers of environmental regulations. EPAs strategy, issued in 1991, identifies goals, tasks, target dates, resources required, organizational responsibilities, and criteria to evaluate program progress. In addition, the Act requires EPA to promote source reduction practices in other federal agencies and to identify opportunities to use federal procurement to encourage source reduction.

To facilitate source reduction by industry, EPA is required to develop, test, and disseminate model source reduction auditing procedures to highlight opportunities; to promote research and development of source reduction techniques and processes with broad applicability; to disseminate information about source reduction techniques through a clearinghouse; to establish a program of state matching grants for programs to provide technical assistance to business; and to establish an annual award program to recognize innovative programs.

PPA requires these reports as part of the TRI to include information about the facility's efforts in source reduction and recycling. Specifically, reports must include

- Quantity of the toxic chemical entering any waste stream (or released to the environment) before recycling, treatment, or disposal
- Quantity of toxic substance recycled (on- or off-site)
- Source reduction practices used
- Quantities of toxic chemicals expected to enter waste streams and to be recycled in the 2 years following the year for which the report is prepared
- Ratio of production in the reporting year to production in the previous year
- Techniques used to identify opportunities for source reduction
- Amount of toxic chemicals released in a catastrophic event, remedial action, or other one-time event
- Amount of toxic chemicals treated on- or off-site

How Does This Affect You?

This is primarily an EPA effort to reduce pollution by coming up with better data, identifying state-of-the-art reduction technologies and processes, and attempting to elicit participation by industry and others in the effort.

Oil Pollution Act

The Oil Pollution Act (OPA) of 1990 streamlined and strengthened EPAs ability to prevent and respond to catastrophic oil spills. A trust fund financed by a tax on oil is available to clean up spills when the responsible party is incapable or unwilling to do so. The OPA requires oil storage facilities and vessels to submit to the federal government response plans detailing how they will respond to large discharges. EPA has published regulations for aboveground storage facilities; the Coast Guard has done so for oil tankers. The OPA also requires the development of Area Contingency Plans to prepare and plan for oil spill response on a regional scale.

How Does This Affect You?

This Act should not affect you or your company unless you ship or store larger quantities of oil or petroleum products.

Ocean Dumping Act

The Ocean Dumping Act has two basic aims: to regulate intentional ocean disposal of materials and to authorize related research. Title I of the MPRSA of 1972 (P.L. 92-532), which is often referred to just as the Ocean Dumping Act, contains permit and enforcement provisions for ocean dumping. Research provisions are contained in Title II, concerning general and ocean disposal research; Title IV established a regional marine research program; and Title V addresses coastal water quality monitoring. The third title of the MPRSA, not addressed here, authorizes the establishment of marine sanctuaries.

The nature of marine pollution requires that it be regulated internationally, since once a pollutant enters marine waters, it knows no boundary. Thus, a series of regional treaties and conventions pertaining to local marine pollution problems and more comprehensive international conventions providing uniform standards to control worldwide marine pollution have evolved over the last 25 years.

U.S. marine waters had been used extensively as a convenient alternative to land-based sites for the disposal of various wastes such as sewage sludge, industrial wastes, and pipeline discharges and runoff. Four federal agencies have responsibilities under the Ocean Dumping Act: EPA, the U.S. Army Corps of Engineers, the National Oceanic and Atmospheric Administration (NOAA), and the Coast Guard. EPA has primary authority for regulating ocean disposal of all substances except dredged spoils, which are under the authority of the Corps of Engineers. NOAA is responsible for long-range research on the effects of human-induced changes to the marine environment, while EPA is authorized to carry out research and demonstration activities related to phasing out sewage sludge, radioactive materials, and industrial waste dumping. The Coast Guard is charged with maintaining surveillance of ocean dumping.

What Does This Mean to You?

Ocean dumping should not be an issue with you. It is something that you should not be doing, so for all practical purposes this Act is not applicable to you.

FURTHER PLANNING AND ACTION

If you still have questions after reading about the environmental laws, I would suggest that you follow two modes of action. Call either your federal or state EPA for assistance. They are more interested in helping you comply than in enforcing the laws or you may decide to visit USEPAs Web site at http://www.epa.gov/, where you will find many types of material or publications, which help to simplify the requirements of the laws.

If you still have difficulties with understanding your role and the requirements of your workplace, you may want to hire an environmental consultant to assist you in assessing your needs and taking environmental samples where needed. Use the guidelines in Chapter 14 for hiring a consultant. Also, make sure that he or she is available to help you, should you receive a visit from the EPA.

You should have the proper personal protective equipment if your workforce has the potential to be exposed to airborne contaminants or spilled chemicals. You should tell your workforce what you expect them to do in case of an emergency. If they have been trained, you might want them to help in mitigating the problem, cleanup, or containment. If you expect them to evacuate, then you need to tell them so.

If you are depending upon your local fire department to be your hazmat responders, you should have informed them of the potential contaminants that they would face, which would include MSDSs for all chemicals on your facility. If you hire a hazmat company to respond to emergencies, they also need to be indoctrinated on your unique situation and the hazards that they would face.

You should have policies and procedures in place for releases, spills, storage, transportation, and disposal of any hazardous materials. Great care should be taken not to contaminate the soil, which could lead to long-term damage of the groundwater. You must also be very careful if you are using water in your processes or procedure and ensure that you have ways of preventing its release until such time as it has been treated and deemed fit for release or until you can dispose of it appropriately without contamination of other water or water sources.

Environmental problems are far more costly to deal with than worker safety and health issues and it is often financially draining to install the technology for pollution prevention. Therefore, it is wise to take preservation of the environment as a serious business rather than proceed irresponsibly and cause your business infinitely more problems and difficulties than if you had addressed the environmental issues upfront.

CHAPTER 19

Keep Me Safe: Workplace Security and Violence

Workplace security should also address workplace violence.

In recent years it has become apparent that workplace security is a major emphasis of occupational safety and health. Workers should feel that they can come to work and work at their jobs without the threat that they may come to harm in some way from violence during their work shift.

Thus, with the escalation of workplace violence in the past two decades violence in the workplace has reared its ugly head as a workplace issue, with homicide being the third leading cause of occupational death among all workers in the United States from 1980 to 1988 and the leading cause of fatal occupational injuries among women from 1980 to 1985. Higher rates of occupational homicides were found in the retail and service industries, especially among sales workers. This increased risk may be explained by contact with the public and the handling of money.

Research into the causes of the increasing incidence of death and serious injury to health care workers has led to the theory that exposure to the public may be an important risk. The risk is increased particularly in emotionally charged situations with mentally disturbed persons or when workers appear to be unprotected.

It is the employers' responsibility to provide a workplace free from hazards that could cause death or serious physical harm and this includes workplace violence. Thus, the employer of today must take into consideration the security of his or her workplace to endsure that employees can perform their work without the interference of outside sources of danger.

WORKPLACE VIOLENCE STATISTICS

In 1995, the Workplace Violence Institute reported that the cost of workplace violence was 35.5 billion dollars. The fatalities in the workplace in 2005 were 5702 with 564 of these being homicides. Homicide is the second leading cause of workplace deaths or one in every six fatal occupational injuries. Eight percent was caused by firearms and 20% were from bombings, stabbings, or beatings.

The U.S. Department of Justice indicates that during 1992–1996 2 million individuals were victims of violent crime or threatened with violent criminal acts in the workplace. Simple assault accounted for 1.5 million of these incidents.

In many cases (37%) the victims of workplace violence knew their offenders, but only 1% was victimized by a current or former spouse, boyfriend, or girlfriend. In cases where a spouse was the perpetrator of the crime it occurred to women 21% of the time and only 2% were men.

Of those committing workplace violence 67% are usually males and 33% females. The victims were injured only 12% of the time. Of the injured victims only one-half required medical attention.

The estimated annual victimizations for the years of 1992 through 1996 for workplace crimes counted by the Bureaus of Justice Statistics' National Crime Victimization Survey and the Bureau of Labor Statistics indicated that

- Simple assaults = 1.5 million
- Aggravated assaults = 396,000
- Robberies = 84,000
- Rapes and sexual assaults = 51,000
- Homicides = 1,000

The rate of violence per 1000 workers during the 5 year period for selected occupations was as follows:

- Law enforcement officer = 306
- Prison or jail correction officer = 218
- Taxi driver = 184
- Private security guard = 117
- Bartender = 91
- Mental health professional = 80
- Gas station attendant = 79
- Convenience or liquor store clerk = 68
- Mental health custodial worker = 63
- Junior high or middle school teacher = 57
- Bus driver = 45
- Special education teacher = 41
- High school teacher = 29
- Elementary school teacher = 16
- College or university teacher = 3

Fewer than half of all the nonfatal workplace crimes are reported to police.

RISK FACTORS

Some of the common risk factors for workers who could be affected by workplace violence are

- Contact with the public
- Exchange of money
- Delivery of passengers, goods, or services

- Having a mobile workplace such as a taxi or police cruiser
- Working with unstable or volatile persons in health care, social service, or criminal justice settings
- Working alone or in small numbers
- Working late at night or during early morning hours
- Working in high-crime areas
- Guarding valuable property or possessions
- Working in community-based settings

Risk factors may be viewed from the standpoint of (1) the environment, (2) administrative controls, and (3) behavior strategies.

PREVENTION STRATEGIES

Usually there are three main areas that must be considered when looking at attempts to provide security and safety for your workforce due to violent occurrences within and without your workplace. These strategies are a good starting point.

Environmental Designs

Commonly implemented cash-handling policies in retail settings include procedures such as using locked drop safes, carrying small amounts of cash, and posting signs and printing notices that limited cash is available. It may also be useful to explore the feasibility of cashless transactions in taxicabs and retail settings through use of debit or credit cards, especially late at night. These approaches can be used in any setting where cash is currently exchanged between workers and customers.

Physical separation of workers from customers, clients, and the public through the use of bullet-resistant barriers or enclosures has been proposed for retail settings, such as gas stations and convenience stores, hospital emergency departments, and social service agency claims areas. The height and depth of the counters (with or without bullet-resistant barriers) are also important considerations in protecting workers, since they introduce physical distance between workers and potential attackers. Nonetheless, consideration must be given to the continued ease of conducting business: a safety device that increases frustration for workers, customers, clients, or patients may be self-defeating. Visibility and lighting are also important environmental design considerations. Making high-risk areas visible to more people and installing good external lighting should decrease the risk of workplace assaults.

Access to and egress from the workplace are also important areas to assess. The number of entrances and exits, the ease with which nonemployees can gain access to work areas because doors are unlocked, and the number of areas where potential attackers can hide are issues that should be addressed. These issues have implications for the design of buildings and parking areas, landscaping, and the placement of garbage areas, outdoor refrigeration areas, and other storage facilities that workers must use during a work shift.

Numerous security devices may reduce the risk for assaults against workers and facilitate the identification and apprehension of perpetrators. These include closed-circuit cameras, alarms, two-way mirrors, card-key access systems, panic-bar doors locked from the outside only, and trouble lights or geographic locating devices in taxicabs and other mobile workplaces. Personal protective equipment such as body armor has been used effectively by public safety personnel to mitigate the effects of workplace violence. For example, the lives of more than 1800 police officers have been saved by Kevlar vests.

Administrative Controls

Staffing plans and work practices (such as escorting customers and visitors and prohibiting unsupervised movement within and between work areas) are issues that need to be addressed regarding security. Increasing the number of staff on duty may also be appropriate in any number of services and retail settings. The use of security guards or receptionists to screen persons entering the workplace and controlling access to actual work areas has also been suggested by security experts.

Work practices and staffing patterns during the opening and closing of establishments and during money drops and pickups should be carefully reviewed for the increased risk of assault they pose to workers. These practices include having workers take out garbage, dispose of grease, store food or other items in external storage areas, and transport or store money.

Policies and procedures for assessing and reporting threats allow employers to track and assess threats and violent incidents in the workplace. Such policies clearly indicate a zero tolerance of workplace violence and provide mechanisms by which incidents can be reported and handled. In addition, such information allows employers to assess whether prevention strategies are appropriate and effective. These policies should also include guidance on recognizing the potential for violence, methods for defusing or deescalating potentially violent situations, and instruction about the use of security devices and protective equipment. Procedures for obtaining medical care and psychological support following violent incidents should also be addressed. Training and education efforts are clearly needed to accompany such policies.

Behavioral Strategies

Training employees in nonviolent response and conflict resolution has been suggested to reduce the risk that volatile situations will escalate to physical violence. Also critical is training that addresses hazards associated with specific tasks or worksites and relevant prevention strategies. Training should not be regarded as the sole prevention strategy but as a component in a comprehensive approach to reducing workplace violence. To increase vigilance and compliance with stated violence prevention policies, training should emphasize the appropriate use and maintenance of protective equipment, adherence to administrative controls, and increased knowledge and awareness of the risk of workplace violence.

Perpetrator and Victim Profile

Only a small percentage of violence is perpetrated by the mentally ill. Gang members, distraught relatives, drug users, social deviants, or threatened individuals are often aggressive or violent. A history of violent behavior is one of the best indicators of future violence by an individual. This information, however, may not be available, especially for new workers, patients, or clients. Even if this information were available, workers not directly involved with these individuals would not have access to it.

Workers who make home visits or do community work cannot control the conditions in the community and have little control over the individuals they may encounter in their work. The victims of assault are often untrained and unprepared to evaluate escalating behavior or to know and practice methods of defusing hostility or protecting themselves from violence. Training, when provided, is often not required as part of the job and may be offered infrequently. However, using training as the sole safety program element creates an impossible burden on the employee for safety and security for him- or herself, coworkers, or other clients. Personal protective measures may be needed and communication devices are often lacking.

COST OF VIOLENCE

Little has been done to study the cost to employers and employees of work-related injuries and illnesses, including assaults. A few studies have shown an increase in assaults over the past two decades. In one reported situation of 121 workers sustaining 134 injuries, 43% involved lost time from work with 13% of those injured missing more than 21 days from work. In this same investigation, an estimate of the costs of assault was that the 134 injuries from patient violence cost $766,000 and resulted in 4,291 days lost and 1,445 days of restricted duty.

Additional costs may result from security or response team time, employee assistance program or other counseling services, facility repairs, training and support services for the unit involved, modified duty, and reduction of effectiveness of work productivity in all staff due to a heightened awareness of the potential for violence. The cost of not developing and providing security at your workplace could be disastrous to your business. Hence, it is imperative that a part of your occupational safety and health effort be directed toward security and the prevention of workplace violence.

PREVENTION EFFORTS

Although it is difficult to pinpoint specific causes and solutions for the increase in violence in the workplace, recognition of the problem is a beginning. Some solutions to the overall reduction of violence in this country may be found in actions such as eliminating violence in television programs, implementing effective programs of gun control, and reducing drug and alcohol abuse. All companies should investigate programs recently instituted by several convenience store chains or robbery deterrence strategies such as increased lighting, closed-circuit TV monitors, and visible money handling locations. If sales are involved, consider limiting access and egress and providing security staff. You might want to construct a response plan. Although it may not help to prevent incidents, a response plan should be incorporated into an overall plan of prevention. Training employees in management of assaultive behavior (MAB) or professional assault response has been shown to reduce the incidence of assaults. Administrative controls and mechanical devices are being recommended and gradually implemented.

Some safety measures may seem expensive or difficult to implement but are needed to adequately protect the health and well-being of workers. It is also important to recognize that the belief that certain risks are part of the job contributes to the continuation of violence and possibly the shortage of trained workers.

The guidelines provided in this chapter, while not exhaustive, include philosophical approaches as well as practical methods to prevent and control assaults. The potential for violence may always exist for workers; the cooperation and commitment of employers are necessary, however, to translate these guidelines into an effective program for the occupational health and safety of the workforce.

PROGRAM DEVELOPMENT AND ESSENTIAL ELEMENTS

In order to be consistent with the earlier chapters in this book related to safety and health program development, the four critical elements in a safety and health program will be employed to demonstrate how to put safety and health program development to use by taking security as the subject.

Management Commitment and Employee Involvement

Commitment and involvement are essential elements in any safety and health program. Management provides the organizational resources and motivating forces necessary to deal effectively with safety and security hazards. Employee involvement, both individually and collectively, is achieved by encouraging participation in the worksite assessment, developing clear effective procedures, and identifying existing and potential hazards. Employee knowledge and skills should be incorporated into any plan to abate and prevent safety and security hazards.

Commitment by Top Management

The implementation of an effective safety and security program includes a commitment by the employer to provide the visible involvement of everyone, so that all employees, from managers to line workers, fully understand that management has a serious commitment to the program. An effective program should have a team approach with top management as the team leader and should include the following:

- Demonstration of management's concern for employee emotional and physical safety and health by placing a high priority on eliminating safety and security hazards.
- Policy that places employee safety and health on the same level of importance as customer, patient, and client safety. The responsible implementation of this policy requires management to integrate issues of employee safety and security to ensure that this protection is part of the daily functioning of the workplace.
- Employer commitment to security through the philosophical refusal to tolerate violence in the workplace and the assurance that every effort will be made to prevent its occurrence.
- Employer commitment to assign and communicate the responsibility for various aspects of safety and security to supervisors, forepersons, lead workers, and other employees involved so that they know what is expected of them. Also to ensure that record keeping is accomplished and utilized to aid in meeting program goals.
- Employer commitment to provide adequate authority and resources to all responsible parties so that assigned responsibilities can be met.
- Employer commitment to ensure that each manager, supervisor, professional, and employee is responsible for the security and safety program in the workplace and is accountable for carrying out those responsibilities.
- Employer develops and maintains a program of medical and emotional health care for employees who are assaulted or suffer abusive behavior.
- Development of a safety committee, which evaluates all reports and records of assaults, and incidents of aggression. When this committee makes recommendations for correction, the employer should report back to the committee in a timely manner on actions taken on the recommendation.

Employee Involvement

An effective program includes a commitment by the employer to provide for and encourage employee involvement in the safety and security program and in the decisions that affect worker safety and health, as well as client well-being. Involvement may include the following:

- Employee suggestion/complaint procedure, which allows workers to bring their concerns to management and receive feedback without fear of reprisal or criticism of ability.
- Employees follow a procedure that requires prompt and accurate reporting of incidents with or without injury. If injury has occurred, prompt first aid or medical aid must be sought and treatment provided or offered.
- Employees participate in a safety and health committee that receives information and reports on security problems, makes facility inspections, analyzes reports and data, and makes recommendations for corrections.
- Employees participate in incident review to identify problems, which may help employees to identify potentially violent behavior patterns and discuss safe methods of managing difficult situations.
- Employees participate in security response teams that are trained and possess required professional assault response skills.

- Employees participate in training and refresher courses in professional assault response training to learn techniques such as recognizing escalating agitation, deflecting or controlling the undesirable behavior and, if necessary, of controlling assaultive behavior, and protecting customers and staff members.
- Training in dealing with the hostile individual, or the police department program on personal safety, should be provided and required to be attended by all involved employees.

Effective implementation requires a written program for job safety, health, and security, which is endorsed and advocated by the highest level of management. This program should outline the employer's goals and objectives. The written program should be suitable for the size, type, and complexity of the workplace and its operations and should permit these guidelines to be applied to the specific hazardous situation of each operation.

The written program should be communicated to all personnel regardless of the number of workers or work shift. The program should establish clear goals and objectives that are understood by all members of the company or organization. The communication needs to be extended to all levels of the workforce.

Hazard Identification and Analysis

Worksite hazard identification and analysis identifies existing hazards and conditions, operations and situations that create or contribute to hazards, and areas where hazards may develop. This includes close scrutiny and tracking of injury or illness and incident records to identify patterns that may indicate causes of aggressive behavior and assaults. The objectives of worksite hazard identification and analyses are to recognize, identify, and plan to correct security hazards. Analysis uses existing records and worksite evaluations should include record review and identification of security hazards.

Record Review

Analyze medical, safety, and insurance records, including the Organizational Safety and Health Administration (OSHA) 300 log and information compiled for incidents or near incidents of assaultive behavior from workers or visitors. This process should ensure confidentiality of records of employees and others. This information should be used to identify incidence and severity and to establish a baseline for identifying change.

Identify and analyze any apparent trends in injuries relating to particular departments, units, job titles, unit activities or workstations, activities, or time of day. It may include identification of sentinel events such as threatening of workers or identification and classification of where aggressive behavior could be anticipated, and by whom.

Identification of Security Hazards

Worksite hazard identification and analysis should use a systematic method to identify the areas needing in-depth scrutiny of security hazards. This analysis should do the following:

- Identify work positions in which workers are at risk of assaultive behavior.
- Use a checklist for identifying high-risk factors that includes components, such as type of people contacts, physical risk factors of the building, isolated locations/job activities, lighting problems, high-risk activities or situations, problem workers, service and delivery personnel or customers, uncontrolled access, and areas of previously encountered security problems.
- Identify low-risk positions for light or relief duty or restricted activity work positions when injuries do occur.

- Determine if risk factors have been reduced or eliminated to the extent feasible. Identify existing programs in place, and analyze effectiveness of those programs, including engineering control measures and their effectiveness.
- Apply analysis to all newly planned and modified facilities or any public services program to ensure that hazards are reduced or eliminated before involving the public, customers, or employees.
- Conduct periodic surveys at least annually (or whenever there are operation changes) to identify new or previously unnoticed risks and deficiencies and to assess the effects of changes in the building designs, work processes, patient services, and security practices. Evaluation and analysis of information gathered, and incorporation of all this information into a plan of correction and ongoing surveillance, should be the result of the worksite analysis.

Hazard Prevention and Control

Select work settings to apply methods of reducing hazards. You will need to make use of general engineering concepts, specific engineering and administrative controls, work practice controls, and personal protective equipment as appropriate to control hazards.

General Building, Work Station, and Area Designs

Workplace designs are appropriate when they provide secure, well-lighted protected areas, which do not facilitate assaults or other uncontrolled activity.

- Design of facilities should ensure uncrowded conditions for workers and customers. Areas for privacy and protection are needed, although isolation should be avoided. For example, doors must be fitted with windows so that other workers can view any aberrant behavior.
- Work areas should be designed and furniture arranged to prevent entrapment of the workers and others.
- Reception areas should be protected by enclosures, which prevent molesting, throwing objects, reaching into the work area or otherwise creating a hazard or nuisance to the worker; such barriers should not restrict communication but should be protective.
- Lockable and secure bathroom facilities and other amenities must be provided for workers separate from customer or public restrooms.
- Public or customer access to workers' workstations and other facility areas must be controlled; that is, doors from waiting rooms must be locked and all outside doors locked from the outside to prevent unauthorized entry, but permit exit in cases of emergency or fire.
- Metal bars or protective decorative grating on outside ground level windows should be installed (in accordance with fire department codes) to prevent unauthorized entry.
- Bright and effective lighting systems must be provided for all indoor building areas as well as grounds around the facility or workplace, especially in the parking areas.
- Curved mirrors should be installed at intersections of halls or in areas where an individual may conceal his or her presence.
- All permanent and temporary employees who work in secured areas should be provided with keys or swipe cards to gain access to work areas whenever on duty.
- Metal detectors should be installed to screen visitors, customers, service personnel, and visitors in high-security areas. In other situations implement handheld metal detectors to use in identifying weapons.

Maintenance

Maintenance must be an integral part of any safety and security system. Prompt repair and replacement programs are needed to ensure the safety of workers and customers. Replacement of burned out lights, broken windows, etc., is essential to maintain the system in safe operating condition.

If an alarm system is to be effective, it must be used, tested, and maintained according to strict policy. Any personal alarm devices should be carried and tested as required by the manufacturer and facility policy. Maintenance on personal and other alarm systems must take place monthly. Batteries and operation of the alarm devices must be checked by a security office to ensure the function and safety of the system. Any mechanical device used for security and safety must be routinely tested for effectiveness and maintained on a scheduled basis.

Engineering Control

Alarm systems are imperative for use in psychiatric units, hospitals, mental health clinics, high-hazard areas, emergency rooms, or where drugs are stored. Whereas alarm systems are not necessarily preventive, they may reduce serious injury when a person is acting in an abusive manner or threatening with or without a weapon. Many other engineering controls can be used such as the following:

- Alarm systems that rely on the use of telephones, whistles, or screams are ineffective and dangerous. A proper system consists of an electronic device that activates an alert to a dangerous situation in two ways—visually and audibly. Such a system identifies the location of the room or action of the worker by means of an alarm sound and a lighted indicator, which visually identifies the location. In addition, the alarm should be sounded in a security (or other response team) area to summon aid. This type of alarm system typically utilizes a pen-like device, which is carried by the employee and can be triggered easily in an emergency. Back-up security personnel must be available to respond to the alarm.

  An emergency personal alarm system is of the highest priority. An alarm system may be of two types: the personal alarm device or the type that is triggered at a desk or counter. This desk system may be silent at the desk or counter, but audible in a central assistance area. It must clearly identify the location in which the problem is occurring.

  These alarm systems must be relayed to security, police, or locations where assistance is available 24h per day. A telephone link to the local police department should be established in addition to other systems.

- Panic buttons are needed at times when someone is confronted with an abusive person. Any such alarm system may incorporate a telephone paging system to direct others to the location of the disturbance, but alarm systems must not depend on the use of a telephone to summon assistance.
- Video screening of high-risk areas or activities may be of value and permits one security guard to visualize a number of high-risk areas, both inside and outside the building. Closed-circuit TV monitors may be used to survey concealed areas or areas where problems may occur.
- Metal detection systems, such as handheld devices or other systems to identify persons with hidden weapons, should be considered. These systems are in use in courts, boards of supervisors, some departments of public social service, schools, and emergency rooms. Although controversial, the fact remains that many people, including homeless and mentally ill persons, carry weapons for defense while living on the streets. Some system of identifying persons who are carrying guns, knives, ice picks, screw drivers, etc. may be useful and should be considered when situations merit them. Signs posted at the entrance will notify workers, customers, and visitors that screening will be performed.
- Reception areas should be designed so that receptionists and staff may be protected by safety glass and locked doors to their work areas.
- First-aid kits should be available.
- Materials and equipment to meet the requirements of the bloodborne pathogen standard should be available.

- Strictly enforced limited access to work areas is needed to eliminate entry by unwanted or danger-ous persons. Doors may be locked or key-coded.
- To provide some measure of safety and to keep the employee in contact with headquarters or another source of assistance, cellular car phones should be installed or provided for official use when work-ers are assigned to duties that take them into private homes and the community. These workers may include parking enforcers, union business agents, psychiatric evaluators, public social service work-ers, children's service workers, visiting nurses, and home health aides.
- Handheld alarm or noise devices or other effective alarm devices are highly recommended to be provided for all field personnel.
- Beepers or alarm systems should be investigated and provided to alert a central office of problems.
- Other protective devices such as pepper spray should be investigated and provided.

Administrative Controls and Work Practices

A sound overall security program includes administrative controls that reduce hazards from inadequate staffing, insufficient security measures, and poor work practices.

Employees are to be instructed not to enter any location where they feel threatened or unsafe. This decision must be the judgment of the employee. Procedures should be developed to assist the employee to evaluate the relative hazard in a given situation. In hazardous cases, the managers must facilitate and establish a buddy system. This buddy system should be required whenever an employee feels insecure regarding the time of activity, the location of work, the nature of the individuals in that location, and past history of aggressive or assaultive behavior by these individuals.

- Employers must provide a program of personal safety education for the field staff. This program should be at the minimum, one provided by local police departments or other agencies. It should include training on awareness, avoidance, and action to take to prevent mugging, robbery, rapes, and other assaults.
- Procedures should be established to assist employees to reduce the likelihood of assaults and rob-bery from those seeking drugs or money, as well as procedures to follow in the case of threatening behavior and provision for a fail-safe back-up in administration offices.
- Fail-safe back-up system is provided in the administrative office at all times of operation for employees in the field who may need assistance.
- All incidents of threats or other aggression must be reported and logged. Records must be main-tained and used to prevent future security and safety problems.
- Police assistance and escorts should be required in dangerous or hostile situations or at night. Procedures for evaluating and arranging for such police accompaniment must be developed and training provided.
- Security guards must be provided. These security guards should be assigned to areas where there may be problems such as emergency rooms or psychiatric services.
- To staff safely, a written guideline should be established that evaluates the level of staff or worker coverage needed. Provision of sufficient staff interaction and clinical activity is important because patients and clients need access to medical assistance from staff. Possibility of violence often threatens staff when the structure of the patient/nurse relationship is weak. Therefore, sufficient staff members are essential to allow formation of therapeutic relationships and a safe environment.
- It is necessary to establish on-call teams, reserve or emergency teams of staff who may provide services, such as responding to emergencies, transportation or escort services, dining room assistance, or many of the other activities where potential hazards exist.
- Methods should be developed to communicate to workers who come to work about any potential security breaches or violence potentials.
- Workers should be instructed to limit physical intervention in altercations whenever possible, unless there are adequate numbers to assist them or emergency response teams and security are

called. In a case where serious injury could occur, emergency alarm systems should always be activated. Administrators need to give clear messages to everyone that violence is not permitted and that legal charges will be pressed when violence or the threat of it occurs. Management should provide information to workers who may be in danger. Policies must be provided with regard to the safety and security of workers in confronting or querying unrecognized individuals in the workplace, key and door opening policy, open vs. locked seclusion policies, and evacuation policy in emergencies.

- Escort services by security should be arranged so that workers should not have to walk alone in parking lots or other parking areas in the evening or late hours.
- Visitors and maintenance persons or crews should be escorted and observed while in any locked or secured facility. Often they have tools or possessions, which could be inadvertently left unattended and thus could become weapons.
- Management needs to work with local police to establish liaison and response mechanisms for assistance when calls are made for help. They should also make clear policies on how they wish the workers to respond.
- It is not wise to allow workers to confront an aggressive or threatening individual, nor is it appropriate to allow aggressive behavior to go unchecked. Workers should respond according to the company's policy and procedures.
- It is a wise policy to require badging of all workers and require them to visibly wear their picture badges at all times. Anyone who does not have a badge in restricted work areas should be confronted and reported to the supervisor or security should be called.
- Security guards trained in principles of human behavior and aggression should be provided where there are large numbers of customers, clients, patients, or visitors. Guards should be provided where there may be psychologically stressed clients or persons who have taken hostile actions, such as in emergency facilities, hospitals where there are acute or dangerous patients, or areas where drug or other criminal activity is commonplace.
- No employee should be permitted to work or stay alone in an isolated area without protection from some source.
- Clothing and apparel that will not contribute to injury should be worn, such as low-heeled shoes, conservative earrings or jewelry, and clothing that is not provocative.
- Keys should be kept covered and worn in such a manner to avoid incidents, yet be available.
- All protective devices and procedures should be required to be used by all workers.
- After dark, all unnecessary doors should be locked and access into the workplace limited and patrolled by security.
- Emergency or hospital staff that has been assaulted should be permitted and assisted to request police assistance or file charges of assault against any customer, client, visitor, patient, or relative who injures, just as a private citizen has that right. Being at work does not reduce the right of pressing charges or damages.
- Visitors should sign in and out and have an issued pass that identifies them as a visitor and specifies the locations they are permitted to access in the workplace.

Training and Education

A major program element in an effective safety and security program is training and education. The purpose of training and education is to ensure that employees are sufficiently informed about the safety and security hazards to which they may be exposed and thus, are able to participate actively in their own and their coworkers' protection. All employees should be periodically trained in the employer's safety and security program.

Training and education are critical components of a safety and security program for employees who are potential victims of assaults. Training allows managers, supervisors, and employees to understand security and other hazards associated with a job or location within the facility, the prevention and control of these hazards, and the medical and psychological consequences of assault.

Participants in the Training Program

A training program should include all affected employees who could encounter or be subject to abuse or assaults. This means all employees, engineers, security officers, maintenance personnel, supervisors, managers, and workers at all levels. The following should be the elements to balance the program.

- The program should be designed and implemented by qualified persons. Appropriate special training should be provided for personnel responsible for administering the training program.
- Several types of programs are available and have been utilized, such as MAB, professional assault response training (PART), police department assault avoidance programs, or personal safety training. A combination of such training may be incorporated depending on the severity of the risk and assessed risk. These management programs must be provided and attendance required at least yearly. Updates may be provided monthly or quarterly.
- The program should be presented in the language and at a level of understanding appropriate for the individuals trained. It should provide an overview of the potential risk of illness and injuries from assault, the causes and early recognition of escalating behavior, or recognition of situations that may lead to assaults. The means of preventing or defusing volatile situations, safe methods of restraint or escape, or use of other corrective measures or safety devices that may be necessary to reduce injury and control behavior are critical areas of training. Methods of self-protection and protection of coworkers, the proper treatment of staff and patient procedures, record keeping, and employee rights need to be emphasized.
- The training program should also include a means for adequately evaluating its effectiveness. The adequacy of the frequency of training should be reviewed. The whole program evaluation may be achieved by using employee interviews, testing and observing, or reviewing reports of behavior of individuals in situations that are reported to be threatening in nature.
- Employees who are potentially exposed to safety and security hazards should be given formal instruction on the hazards associated with the unit or job and facility. This includes information on the types of injuries or problems identified in the facility, the policy and procedures contained in the overall safety program of the facility, those hazards unique to the unit or program, and the methods used by the facility to control the specific hazards. The information should discuss the risk factors that cause or contribute to assaults, etiology of violence and general characteristics of violent people, methods of controlling aberrant behavior, methods of protection, reporting procedures, and methods to obtain corrective action.

Training for affected employees should consist of both general and specific job training. Specific job training is discussed in the following section or may be found in administrative controls in the specific work location section.

Job-Specific Training

New employees and reassigned workers should receive an initial orientation and hands-on training before they are placed in a treatment unit or job. Each new employee should receive a demonstration of alarm systems, protective devices, and the required maintenance schedules and procedures. The training should also contain the use of administrative or work practice controls to reduce injury.

Initial Training Program

The initial training program should include

- Care, use, and maintenance of alarm tools and other protection devices
- Location and operation of alarm systems

- MAB, PART, or other training
- Communication systems and treatment plans
- Policies and procedures for reporting incidents and obtaining medical care and counseling
- Hazard communication program
- Bloodborne pathogen program, if applicable
- Rights of employees, treatment of injury, and counseling programs

On-the-job training should emphasize employee development and use of safe and efficient techniques, methods of deescalating aggressive behavior, self-protection techniques, methods of communicating information, which will help other staff to protect themselves, and discussions of rights of employees in the work setting.

Specific measures at each location, such as protective equipment, location, and use of alarm systems, determination of when to use the buddy system as needed for safety, must be part of the specific training. Training unit coworkers from the same unit and shift may facilitate team work in the work setting.

Training for Supervisors and Managers, Maintenance and Security Personnel

Supervisors and managers are responsible for ensuring that employees are not placed in assignments that compromise safety and that employees feel comfortable in reporting incidents. They must be trained in methods and procedures that will reduce the security hazards and train employees to behave compassionately with coworkers when an incident does occur. They need to ensure that employees have safe work practices and receive appropriate training to enable them to do this. Supervisors and managers, therefore, should undergo training comparable to that of the employee and such additional training as will enable them to recognize a potentially hazardous situation, make changes in the physical plant, patient care treatment program, staffing policy and procedures, or other such situations that contribute to hazardous conditions. They should be able to reinforce the employer's program of safety and security, assist security guards when needed, and train employees as the need arises.

Training for engineers and maintenance personnel should consist of an explanation or a discussion of the general hazards of violence, the prevention and correction of security problems, and personal protection devices and techniques. They need to be acutely aware of how to avoid creating hazards in the process of their work.

Security personnel need to be recruited and trained whenever possible for the specific job and facility. Security companies usually provide general training on guard or security issues. However, specific training should include psychological components of handling aggressive and abusive individuals, types of disorders, and the psychology of handling aggression and defusing hostile situations. If weapons are used by security staff, special training and procedures need to be developed to prevent inappropriate use of weapons and the creation of additional hazards (see the sample security program in Appendix L).

Medical Management

A medical program that provides knowledgeable medical and emotional treatment should be established. This program shall ensure that victimized employees are provided with the same concern that is shown to the victims. Violence is a major safety hazard in psychiatric and acute care facilities, emergency rooms, homeless shelters, and other health care settings. Medical and emotional evaluation and treatment are frequently needed, but often difficult to obtain.

The consequences to employees who are abused by others may include death and severe, life-threatening injuries, in addition to short- and long-term psychological trauma, posttraumatic stress, anger, anxiety, irritability, depression, shock, disbelief, self-blame, fear of returning to work, disturbed sleep patterns, headache, and change in relationships with coworkers and family. All have been reported by workers after assaults, particularly if the attack has come without warning. They may also fear criticism by managers, increase their use of alcohol and medication to cope with stress, suffer from feelings of professional incompetence, physical illness, powerlessness, increased absenteeism, and may experience performance difficulties.

Managers and supervisors have often ignored the needs of the physically or psychologically abused or assaulted staff, requiring them to continue working, obtain medical care from private medical doctors, or blame the individual for irresponsible behavior. Injured staff must have immediate physical evaluations, be removed from the unit, and treated for acute injuries. Referral should be made for appropriate evaluation, treatment, counseling, and assistance at the time of the incident and for any required follow-up treatment. Medical services include

- Provision of prompt medical evaluation and treatment whenever an assault takes place, regardless of severity. A system of immediate treatment is required regardless of the time of day or night. Injured employees should be removed from the unit until order has been restored. Transportation of the injured to medical care must be provided if it is not available onsite or in an employee health service. Follow-up treatment must also be provided at no cost to employees.
- Trauma-crisis counseling or critical incident debriefing program must be established and provided on an ongoing basis to staff who are victims of assaults. This counseling program may be developed and provided by in-house staff as part of an employee health service, by a trained psychologist, psychiatrist, or other clinical staff members such as a clinical nurse specialist, or a social worker; a referral may also be made to an outside specialist. In addition, peer counseling or support groups may be provided. Any counseling provided should be by well-trained psychosocial counselors whether through Employee Assistance Programs, in-house programs, or by other professionals away from the facility who must understand the issues of assault and its consequences.
- Reassignment of staff should be considered when assaults have taken place. At times it is very difficult for staff to return to the same unit to face the assailant. Assailants often repeat threats and aggressive behavior and actions need to be taken to prevent this from occurring. Staff development programs should be provided to teach staff and supervisors to be more sensitive to the feelings and trauma experienced by victims of assaults. Some professionals advocate joint counseling sessions including the assaultive client and staff member to attempt to identify the motive when it occurs in inpatient facilities and to defuse situations that may lead to continued problems.
- Other workers should also receive counseling to prevent blaming the victim syndrome and to assist them with any stress problems they may be experiencing as a result of the assault. Violence often leaves staff fearful and concerned. They need to have the opportunity to discuss these fears and to know that the administration is concerned and will take measures to correct deficiencies. This may be called a defusing or debriefing session, and unit staff members may need this activity immediately after an incident to enable them to continue working. First-aid kits or materials must be provided on each unit or facility.
- The replacement and transportation of the injured workers must be provided for at the earliest possible time. Do not leave the workplace understaffed in the event of an assault. The development of an employee health service, staffed by a trained occupational health specialist, may be an important addition to the hospital team. Such employee health staff can provide treatment, arrange for counseling, refer to a specialist, and should have procedures in place for all shifts. Employee health nurses should be trained in posttraumatic counseling and may be utilized for group counseling programs or other assistance programs.
- Legal advice regarding pressing charges should be available, as well as information regarding workers' compensation benefits and other employee rights must be provided regardless of apparent injury. If assignment to light duty is needed or disability is incurred, these services are to be provided without hesitation. Reporting to the appropriate local law enforcement agency and assistance in making this report is to be provided. Employees may not be discouraged or coerced when making reports or workers' compensation claims.

- All assaults must be investigated, reports made, and needed corrective action determined. However, methods of investigation must be such that the individual does not perceive blame or criticism for assaultive actions taken by the attacker. The circumstances of the incident or other information that will help to prevent further problems need to be identified, but not to blame the worker for incompetence and compound the psychological injury that is most commonly experienced.

Record Keeping

Within the major program elements, record keeping is the heart of the program, providing information for analysis, evaluation of methods of control, severity determinations, identifying training needs, and overall program evaluations.

Records shall be kept of the following:

- OSHA 300 log. OSHA regulations require entry on the injury and illness log of any injury that requires more than first aid, is a lost time injury, requires modified duty, or causes loss of consciousness. Assaults should be entered on the log. Doctors' reports of work injury and supervisors' reports shall be kept of each recorded assault.
- Incidents of abuse, verbal attacks, or aggressive behavior, which may be threatening to the worker, but not result in injury, such as pushing, shouting, or an act of aggression toward other clients requiring action by staff, should be recorded. This record may be an assaultive incident report documented in some manner, which can be evaluated on a monthly basis by the department safety committee.
- System of recording and communicating should be developed so that all workers who may provide care for escalating or potentially aggressive, abusive, or violent individuals will be aware of the status of those individuals and of any problems experienced in the past. This information regarding history of past violence should be noted on those individual's records, communicated in shift change report, and noted in an incident log.
- Information gathering system should be in place, which will enable incorporation of past history of violent behavior, incarceration, probation reports, or any other information that will assist health care staff to assess violence status. Employees are to be encouraged to seek and obtain information regarding history of violence whenever possible.
- Emergency room staff should be encouraged to obtain and record, from police and relatives, information regarding drug abuse, criminal activity, or other information to adequately assist in assessing a patient. This would enable them to appropriately house, treat, and refer potentially violent cases. They should document the frequency of admission of violent clients or hostile encounters with relatives and friends.
- Records need to be kept concerning assaults, including the type of activity, i.e., unprovoked sudden attack, patient-to-patient altercation, and MAB actions. Information needed includes who was assaulted and circumstances of the incident without focusing on any alleged wrongdoing of staff. These records also need to include a description of the environment, location, or any contributing factors, corrective measures identified, including building design, or other measures needed. Determination must be made of the nature of the injuries sustained, whether severe, minor, or the cause of long-term disability, and the potential or actual cost to the facility and employee. Records of any lost time or other factors that may result from the incident should be maintained.
- Minutes of safety meetings and inspections shall be kept. Corrective actions recommended as a result of reviewing reports or investigating accidents or inspections need to be documented with the management's response, and completion dates of those actions should be included in the minutes and records.
- Records of training program contents and sign-in sheets of all attendees should be kept. Attendance records of all training should be retained. Qualifications of trainers shall be maintained along with records of training.

Evaluation of the Program

Procedures and mechanisms should be developed to evaluate the implementation of the safety and security programs and to monitor progress and accomplishments. Top management and supervisors should review the program regularly. Semiannual reviews are recommended to evaluate success in meeting goals and objectives. Evaluation techniques include some of the following:

- Establishment of a uniform reporting system and regular review of reports.
- Review of reports and minutes of safety and security committees.
- Analyses of trends and rates in illness, injury, or incident reports.
- Surveys of employees.
- Before and after surveys/evaluations of job or worksite changes or new systems.
- Up-to-date records of job improvements or programs implemented.
- Evaluation of employee experiences with hostile situations and results of medical treatment programs provided. Follow-up should be repeated several weeks and several months after an incident.
- Results of management's review of the program should be a written progress report and program update, which should be shared with all responsible parties and communicated to employees. New or revised goals arising from the review identifying jobs, activities, procedures, and departments should be shared with all employees. Any deficiencies should be identified and corrective action taken. Safety of employees should not be given a lesser priority than client safety as they are often dependent on one another. If it is unsafe for employees, the same problem will be the source of risk to other clients or patients.
- Managers and supervisors should review the program frequently to reevaluate goals and objectives and discuss changes. Regular meetings with all involved including the safety committee, union representatives, and employee groups at risk should be held to discuss changes in the program.

If you are to provide a safe work environment, it must be evident from managers, supervisors, and peer groups that hazards from violence will be controlled. Employees in psychiatric facilities, drug treatment programs, social services, customer relations, human resource management, emergency rooms, law enforcement, service industries, convalescent homes, taxicab services, community clinics or community settings are to be provided with a safe and secured work environment and injury from assault is not to be accepted or tolerated and is no longer part of the job.

Procedures and mechanisms should be developed to evaluate the implementation of the security program and to monitor progress. This evaluation and record keeping program should be reviewed regularly by top management and the medical management team. At least semiannual reviews are recommended to evaluate success in meeting goals and objectives.

TYPES OF WORKPLACE VIOLENCE EVENTS

When one examines the circumstances associated with workplace violence, events can be divided into three major types. However, it is important to keep in mind that a particular occupation or workplace may be subject to more than one type. In all three types of workplace violence events, a human being, or hazardous agent, commits the assault.

In Type I, the agent has no legitimate business relationship to the workplace and usually enters the affected workplace to commit a robbery or other criminal acts. In Type II, the agent is either the recipient, or the object, of a service provided by the affected workplace or the victim, e.g., the assailant is a current or former client, patient, customer, passenger, criminal suspect, inmate, or prisoner. In Type III, the agent has some employment-related involvement with the affected workplace. Usually this involves an assault by a current or former employee, supervisor, or manager; by a current or former spouse or lover; a relative, friend, or some other person who has a dispute with an employee of the affected workplace.

In bullying, the agent may not be considered violent or assaultive to the target, but this type of behavior may lead to violent acts by either party depending on the interaction occurring. Bullying has not been defined as workplace violence, but needs to be discussed in some context, because of

Figure 19.1 There are degrees of violence.

its deleterious effect on the workplace. A more detailed discussion of workplace bullying is found in Chapter 20. The characteristics of the establishments affected, the profile and motive of the agent or assailant, and the preventive measures differ for each of the three major types of workplace violence events (see Figure 19.1).

Type I Events

The majority (60%) of workplace homicides involve a person entering a small late-night retail establishment, e.g., liquor store, gas station, or a convenience food store, to commit a robbery. During the commission of the robbery, an employee or, more likely, the proprietor is killed or injured.

Employees or proprietors who have face-to-face contact and exchange money with the public, work late at night and into the early morning hours, and work alone or in very small numbers are at greatest risk of a Type I event. While the assailant may feign to be a customer as a pretext to enter the establishment, he or she has no legitimate business relationship to the workplace.

Retail robberies resulting in workplace assaults usually occur between 11 in the evening and six in the early morning hours and are most often armed (gun or knife) robberies. In addition to employees who are classified as cashiers, many victims of late-night retail violence are supervisors or proprietors who are attacked while locking up their establishment for the night or janitors who are assaulted while cleaning the establishment after it is closed.

Other occupations or workplaces may be at risk of a Type I event. For instance, assaults on taxicab drivers also involve a pattern similar to retail robberies. The attack is likely to involve an assailant pretending to be a bona fide passenger during the late night or early morning hours who enters the taxicab to rob the driver of his or her fare receipts. Type I events also involve assaults on security guards. It has been known for sometime that security guards are at risk of assault when protecting valuable property, which may be the object of an armed robbery.

Prevention Strategies for Type I Events

To many people, Type I workplace violence appears to be part of society's crime problem, and not a workplace safety and health problem. Under this view, the workplace is an innocent bystander, and the solution to the problem involves societal changes, not occupational safety and health principles. The ultimate solution to Type I events may indeed involve societal changes, but until such changes occur it is still the employer's legal responsibility to provide a safe and healthful place of employment for their employees. Employers with employees who are known to be at risk for Type I events should be required to address workplace security hazards to satisfy the regulatory requirement of establishing, implementing, and maintaining an effective security program. The first step toward obtaining this goal is strong management commitment to violence prevention. Employers at risk for Type I (as well as Types II and III) events should include as a part of their establishment's security program

- A system for ensuring that employees comply with safe and healthy work practices, including ensuring that all employees, including supervisors and managers, comply with work practices designed to make the workplace more secure and do not engage in threats or physical actions that create a security hazard to other employees, supervisors, or managers in the workplace.
- A system for communicating with employees about workplace security hazards, including a means that employees can use to inform the employer of security hazards at the worksite without fear of reprisal.
- Procedures for identifying workplace security hazards including scheduled periodic inspections to identify unsafe conditions and work practices whenever the employer is made aware of a new or a previously unrecognized hazard.
- Procedures for investigating occupational injury or illness arising from a workplace assault or threat of assault.
- Procedures for correcting unsafe conditions, work practices, and work procedures, including workplace security hazards, and with attention to procedures for protecting employees from physical retaliation for reporting threats.
- Training and instruction about how to recognize workplace security hazards, measures to prevent workplace assaults and what to do when an assault occurs, including emergency action and postemergency procedures.

The cornerstone of an effective workplace security plan is appropriate training of all employees, supervisors, and managers. Employers with employees at risk of workplace violence must educate them about the risk factors associated with the various types of workplace violence and provide appropriate training in crime awareness, assault and rape prevention, and defusing hostile situations. In addition, employers must instruct their employees about the steps to be taken during an emergency incident.

Type II Events

A Type II workplace violence event involves an assault by someone who is either the recipient or the object of a service provided by the affected workplace or the victim. Even though Type I events represent the most common type of fatality, type II events involving victims who provide services to the public are also increasing. Type II events accounted for approximately 30% of workplace homicides. Further, when more occupation-specific data about nonfatal workplace violence becomes available, nonfatal Type II events involving assaults to service providers, especially to health care providers, may represent the most prevalent category of workplace violence resulting in physical injury. Type II events involve fatal or nonfatal injuries to individuals who provide services to the public. These events involve assaults on public safety and correctional personnel, municipal bus or railway drivers, health care and social service providers, teachers, sales personnel, and other public or private service sector employees who provide professional, public safety, administrative, or business services to the public.

Law enforcement personnel are at risk of assault from the object of public safety services (suspicious persons, detainees, or arrestees) when making arrests, conducting drug raids, responding to calls involving robberies or domestic disputes, serving warrants and eviction notices, and investigating suspicious vehicles. Similarly, correctional personnel are at risk of assault while guarding or transporting jail or prison inmates. Of increasing concern, though, are Type II events involving assaults to the following types of service providers:

- Medical care providers in acute care hospitals, long-term care facilities, outpatient clinics, and home health agencies
- Mental health and psychiatric care providers in inpatient facilities, outpatient clinics, residential sites, and home health agencies
- Alcohol and drug treatment providers
- Social welfare service providers in unemployment offices, welfare eligibility offices, homeless shelters, probation offices, and child welfare agencies
- Teaching, administrative, and support staff in schools where students have a history of violent behavior
- Other types of service providers, e.g., justice system personnel, customer service representatives, and delivery personnel

Unlike Type I events, which often represent irregular occurrences in the life of any particular at-risk establishment, Type II events occur on a daily basis in many service establishments and, therefore, represent a more pervasive risk for many service providers.

Prevention Strategies for Type II Events

An increasing number of fatal and nonfatal assaults and threats involve an employee who is providing a service to a client, patient, customer, passenger, or other types of service recipient. Employers who provide service to recipients or service to those known or suspected to have a history of violence must also integrate an effective workplace security component into their security program. An important component of a workplace security program for employers at risk for Type II events is supervisor and employee training in how to effectively defuse hostile situations involving their clients, patients, customers, passengers, and members of the public to whom they must provide services.

Employers concerned with Type II events need to be aware that the control of physical access through workplace design is also an important preventive measure. This can include controlling access into and out of the workplace and freedom of movement within the workplace, in addition to placing barriers between clients and service providers. Escape routes can also be a critical component of workplace design. In certain situations, the installation of alarm systems or panic buttons may be an appropriate back-up measure. Establishing a buddy system to be used in specified emergencies is often advisable as well. The presence of security personnel should also be considered where appropriate.

Type III Events

A Type III workplace violence event consists of an assault by an individual who has some employment-related involvement with the workplace. Generally, a Type III event involves a threat of violence, or a physical act of violence resulting in a fatal or nonfatal injury to an employee, supervisor, or manager of the affected workplace by the following types of individuals:

- Current or former employee, supervisor, or manager
- Some other person who has a dispute with an employee of the affected workplace, e.g., current or former spouse or lover, relative, friend, or acquaintance

Type III events account for a much smaller proportion of fatal workplace injuries. Type III events accounted for only 10% of workplace homicides. Nevertheless, Type III fatalities often attract significant media attention and are incorrectly characterized by many as representing the workplace violence problem. In fact, their media visibility makes them appear much more common than they actually are.

Most commonly, the primary target of a Type III event is a coemployee, a supervisor, or a manager of the assailant. In committing a Type III assault, an individual may be seeking revenge for what he or she perceives as unfair treatment by a coemployee, a supervisor, or a manager. Increasingly, Type III events involve domestic or romantic disputes in which an employee is threatened in the workplace by an individual with whom he or she has a personal relationship outside of work.

At first glance, a Type III assailant's actions may defy reasonable explanation. Often, his or her actions are motivated by perceived difficulties in his or her relationship with the victim, or with the affected workplace, and by psychosocial factors that are peculiar to the assailant.

Even though incomplete, existing data indicate that the number of Type III events resulting in nonfatal injury, or in no physical injury at all, greatly exceeds the number of fatal Type III events. Indeed, the most prevalent Type III event may involve threats and other types of verbal harassment.

Prevention Strategies for Type III Events

In a Type III event, the assailant has an employment-related involvement with the workplace. Usually, a Type III event involves a threat of violence, or a physical act of violence resulting in fatal or nonfatal injury, to an employee of the affected workplace by a current or former employee, supervisor or manager, or by some other person who has a dispute with an employee of the affected workplace, e.g., a current or former spouse or lover, relative, friend or, acquaintance.

Employers who have employees with a history of assaults or who have exhibited belligerent, intimidating, or threatening behavior in the workplace need to establish and implement procedures to respond to workplace security hazards when they are present and to provide training as necessary to their employees, supervisors, and managers to satisfy the regulatory requirement of establishing, implementing, and maintaining an effective injury and illness program (IIP).

Since Type III events are more closely tied to employer and employee relations than are Type I or II events, an employer's considerate and respectful management of his or her employees represents an effective strategy for preventing Type III events. Some workplace violence researchers have pointed out that employer actions that are perceived by an employee to place his or her continuing employment status in jeopardy can be triggering events for a workplace violence event, e.g., layoffs or reduction-in-force actions and disciplinary actions such as suspensions and terminations. Thus, where actions such as these are contemplated, they should be carried out in a manner that is designed to minimize the potential for related Type III events.

Some mental health professionals believe that belligerent, intimidating, or threatening behavior by an employee or supervisor is an early warning sign of an individual's propensity to commit a physical assault in the future, and that monitoring and appropriately responding to such behavior are necessary parts of effective prevention. To prevent these events a checklist or profiling can be used (see Figures 19.2 and 19.3) to validate security.

Many management consultants who advise employers about workplace violence stress that to effectively prevent Type III events from occurring, employers need to establish a clear anti-violence management policy, apply the policy consistently and fairly to all employees, including supervisors and managers, and provide appropriate supervisory and employee training in workplace violence prevention.

Lastly, an important subset of Type III workplace violence events affects women disproportionately. Domestic violence is now spilling over into the workplace and employers need to take appropriate precautions to protect at-risk employees. For instance, when an employee reports threats from an individual with whom he or she has (or had) a personal relationship, employers should take appropriate precautions to ensure the safety of the threatened employee, as well as other employees who are in the zone of danger and who may be harmed if a violent incident occurs in the workplace. One option is to seek a temporary restraining order (TRO) and an injunction on behalf of the affected employee.

Types I, II, and III violence events checklist

Preventive measures

Make your store unattractive to robbers by

- Removing clutter, obstructions, and signs from the windows so that an unobstructed view of the store counter and cash register exists.
- Keeping the store and parking lot as brightly lit as local law allows.
- Keeping an eye on what is going on outside the store and reporting any suspicious persons or activities to the police.
- Keeping yourself busy with other tasks away from the cash register, when there are no customers in the store.
- Posting emergency police and fire department numbers and the store's address by the phone.
- Mounting mirrors on the ceiling to help you keep an eye on hidden corners of the store. Consider surveillance cameras to record what goes on in the store and to act as a deterrent.
- Posting signs that inform customers that you have a limited amount of cash on hand. Make sure they are placed so that they are easy to spot from the outside of the store.
- Limiting accessible cash to a small amount and keeping only small bills in the cash register.
- Using a time access safe for larger bills and depositing them as they are received.
- Using only one register after dark and leaving unused registers open with empty cash drawers tilted up for all to see.

Event measures

- If you are robbed at gunpoint, stay calm and speak to the robber in a cooperative tone. Do not argue or fight with the robber and offer no resistance whatsoever. Hand over the money.
- Never pull a weapon during the event (it will only increase your chances of getting hurt).
- Always move slowly and explain each move to the robber before make it.

Postevent measures

- Make no attempt to follow or chase the robber.
- Stay where you are until you are certain the robber has left the immediate area, then lock the door of your store, and call the police immediately.
- Do not touch anything the robber has handled.
- Write down everything you remember about the robber and the robbery while you wait for the police to arrive.
- Do not open the door of the store until the police arrive.

Figure 19.2 Types I, II, and III violence events checklist.

Any employer may seek a TRO/injunction on behalf of an employee when he or she has suffered unlawful violence (assault, battery, or stalking).

SUMMARY

Effective security management to prevent all three types of workplace violence events also includes postevent measures such as emergency medical care and debriefing employees about the incident. After a workplace assault occurs, employers should provide postevent trauma counseling to those who desire such intervention to reduce the short- and long-term physical and emotional effects of the incident.

Workplace security profile for types I and II

Date: Inspection No.

Employer name: Address:

Nature of business: Hours of operation:

Describe the physical layout of the establishment. Indicate its location to other businesses or residences in the area and access to the street.

Number/gender of employees on-site between 10 p.m. and 5 a.m.:

Describe nature and frequency of client/customer/patient/passenger/other contact:

Are cash transactions conducted with the public during working hours? If yes, how much cash is kept in the cash register or in another place accessible to a robber?

Is there a safe or lockbox on the premises into which cash is deposited?

What is the security history of the establishment and environs?

What physical security measures are present?

What work practices has the employer implemented to increase security?

Has the employer provided security training to employees? If so, has the training been effective?

Figure 19.3 Workplace security profile for types I and II.

Workplace safety and health hazards affecting employees have traditionally been viewed as arising from unsafe work practices, hazardous industrial conditions, or exposures to harmful chemical, biological, or physical agents, not from violent acts committed by other human beings. Recently, though, employees, as well as supervisors and managers, have become, all too frequently, victims of assaults or other violent acts in the workplace that entail a substantial risk of physical or emotional harm. Many of these assaults result in fatal injury, but an even greater number result in nonfatal injury, or in the threat of injury, which can lead to medical treatment, missed work, lost wages, and decreased productivity.

A single explanation for the increase in workplace violence is not readily available. Some episodes of workplace violence, like robberies of small retail establishments, seem related to the larger societal problems of crime and substance abuse. Other episodes seem to arise more specifically from employment-related problems.

What can be done to prevent workplace violence? Any preventive measure must be based on a thorough understanding of the risk factors associated with the various types of workplace violence. Moreover, even though our understanding of the factors that lead to workplace violence is not perfect, sufficient information is available, which if utilized effectively, can reduce the risk of workplace violence. However, strong management commitment, and the day-to-day involvement of managers, supervisors, employees, and labor unions, is required to reduce the risk of workplace violence.

Workplace violence has become a serious occupational health problem requiring the combined efforts of employers, employees, labor unions, government, academic researchers, and security professionals. The problem cannot be solved by government alone.

CHAPTER 20

Mean Ones: Workplace Bullying

Bullying can be hard to identify or recognize.

BULLYING

Bullying has always existed but has not been addressed as an occupational safety and health (OSH) issue until recently. It has become more of a concern with the rise of workplace violence as an OSH issue. Workplace bullying is not always thought of as an aspect of safety and health. It is commonly viewed as more of a social problem that is interpersonal in nature, one that individuals should work out between themselves. However, it is not that simple.

Workplace bullying is a human phenomenon that is inherent in most settings where individuals must work together on a daily basis, either as part of a team, a department, an office, or in the public setting. Myriad relationships develop in any workplace. They can range from formal to informal, brief to long-term, implied to intentional, and social by nature. In those relationships, there are gender factors. Other defining aspects include age, race, ethnic background, and more. However, suspiciously, those aspects that are typically associated with discrimination are not necessarily the same ones that show up when bullying occurs.

One would think that bullying would be considered discrimination in the workplace. However, that is not the clear rule, nor do company policy, government regulation, and law define it as such in all cases. Its prevalence is more widely accepted than in the past, but the research and study of bullying have yet to result in the same level of importance as an issue to be prevented through government regulation and law. It is, I believe, the psychological aspect of this hazard that is difficult to pin down and in many cases there is a hesitance to reach a unanimous conclusion to treat it, prevent it, or invoke punishment for its results on the livelihood of workers. In some ways, many of us may wonder if we have been the victim or target of bullying in the past. We may also wonder if we had become the bully in our actions.

Data Regarding Bullying

According to the Workplace Bullying and Trauma Institute (WBTI) and the research led by Dr. Gary Namie, there are many specifics that can be tallied from well-recognized, national surveys taken in 2000 and 2003 from workplace respondents who were targets of this particular workplace hazard. The total number of respondents numbered 2335 individuals from all walks of life, the largest such survey of its kind done on this subject. The facts speak for themselves:

- Of the individuals targeted, in 2003, 80% were women and 20% were men, and in 2000, 77% were women and 23% were men.
- Average age of the target was 43 (mean).
- Eighty-four percent were college educated, 63% degreed, 17% with graduate degrees, and 4% were PhDs, doctors, or lawyers.
- Length of career was a mean of 21.4 years. A total of 6.7 years in the workplace where the bullying occurred.
- Type of employer was 36% corporate, 31% government, 12% nonprofit, and 11% small business.
- Of coworkers, 96% were aware of the target's abuse. The targets had told 87% of their coworkers.
- Average length of time that the target was exposed to bullying was 16.5 months, men reporting 18.38 months and women reporting 15.74 months.
- Majority (67%) reported that they had never been a victim before. They also were not victims of trauma in the past (62%).
- Bullies had targeted others 77% of the time on average and 88% of the time in government employment.

Of the targets, 37% were fired, 33% quit employment, and 17% transferred to another position that took them away from the bully. The effect on the individual target was dramatic. The Canadian Safety Council found that the targets of bullying wasted between 10% and 52% of their time at work consumed by the effects of their tormenter. They are demotivated, highly stressed, absent more often, fearful, anxious, angered, and 41% of the time they become depressed. It is a well-known statistic that depression affects about 10% of the workforce at any one time and the estimated cost of untreated depression in the workplace is $24 billion annually in lost productivity alone. Depression results in 200 million lost workdays and costs the U.S. economy $43.7 billion annually. According to the World Health Organization, depression may affect up to 20% of the world's workforce at a global cost of $240 billion. This means that the bully in the workplace has the potential to cause substantial economic damage to their own company through their actions.

The effects go on to include the family of the targets, damaging marriages and transferring feelings of guilt and shame onto their children. Physical aches and pain are common, including headache, nausea, stomachaches, skin rashes, sleep difficulties, and other stress-related illnesses. Stress can lead to heart disease, suicide, diabetes, and complicate a host of other illnesses.

The bullies have their own set of statistics. The data derived from the same surveys indicate

- Women were 58% of the bullies while men were 42% of the bullies.
- In only 25% of the cases was the target a member of a protected class and the bully was not. In 15% of the cases, the bully was the protected class.

- Women bullies chose women targets 87% of the time, men 13% of the time. Men who were bullies chose women targets 71% of the time and chose men 29% of the time.
- Woman on woman bullying represents 50% of all workplace incidents. Man on woman was 30%, man on man was 12%, and woman on man was 8%.
- Probability of women targets to be bullied by women is 63% and the probability for men to be the targets of men is 62%.
- Bullies had a higher rank than the target 71% of the time; men 76% and women were 67%.
- Only 17% of the time was the bully a coworker or peer of the target.
- Bullying from the lower ranks upward occurred 12% of the time with women being the target 36% and men 23%.
- Women were more likely to target more than one worker at 68%, while men would target more than one 63% of the time.
- Bullies did the targeting by themselves only 23% of the time, 77% of the time they enlisted the help of others, by bullying another target to help 32% of the time and using some help from others 45% of the time.
- Target's coworkers were the bully's help 48% of the time. Women enlisted help 53% and men 42%.
- When bullies worked in groups they were an average of 3.5 people, male groups were 3.9 and women 3.2.
- Bullies were overwhelmingly bosses who had the authority to terminate their targets at will.

Facts about Bullying

What is a bully and what are the typical characteristics? Here are some definitions from a wide sample of research that is attempting to address the problem. This is an excerpt of facts regarding bullying from the 2003 Report on Abusive Workplaces sponsored by the WBTI; principal investigator was Gary Namie, PhD.

- Bullying is the repeated, health endangering mistreatment of a person (the target) by a cruel perpetrator, the bully.
- It is best understood by the bully's behaviors—acts of commission and omission—which are all driven by the bully's need to control other people.
- Initially involves the bully deciding who is targeted, when, and where and psychological violence inflicted. Later, others may be coerced to participate in the assaults.
- Illegitimate behavior as opposed to tough behavior, it interferes with an employee's work production and the employer's business interests.
- It escalates from 1:1 harassment before being reported followed by a limited or nonexistent employer response, eventually engulfing an entire work unit in fear, paralyzing productivity.

The following are some of the characteristics that have been recognized by the same institute in a report titled; U.S. Hostile Workplace Survey, 2000. Dr. Gary Namie was the research director.

- Screaming, yelling, public attempts to humiliate, seeking to do battle when and where he or she chooses, needs to compete and win to feel good.
- Controls all resources (time, budget, support, and training) so as to prevent you from being successful at your job, undermining, set you up to fail.
- Constant, personal verbal assaults on your character, name calling, belittling, zealous attention to unimportant details, committed to systematic destruction of your confidence in your competence.
- Manipulates the impression others have of you, splits the work group into taking sides, and defames you with higher-up and at next job, killing your reputation.
- Bullies prefer public sites in front of witnesses for humiliating their targets. Occasionally the bullying was private and sometimes behind closed doors with the intent of being heard.

The following are the results of a study on bullying from the Mental Health Association of New South Wales, Australia, published by the Mental Health Information Service in conjunction with the

Australian Council of Trade Unions. Workplace bullying is a serious health and safety hazard. It is usually characterized by

- Unreasonable demands and impossible targets
- Restrictive and petty work rules
- Required to perform tasks without adequate training
- Forced to stay back to finish work or additional tasks
- Compulsory overtime, unfair work schedules, or allocation of work
- Constant intrusive surveillance or monitoring
- No say in how your job is done
- Interference with personal belongings or sabotage of work
- Shouting or using abusive language
- Open or implied threat of termination or demotion
- Oppressive, unhappy work environment
- People afraid to speak up about conditions, behaviors, or health and safety

Types of Bullies

There are two types of workplace bullies, a regular bully and a lethal bully. With a regular bully, the obvious sign is that he or she is intimidating. They evoke fear and are typically demanding. The real read on the bully is how you feel when talking to them. You will feel tense and anxious. The regular bully is easy to spot. They are someone who always looks good and makes everyone else look bad. They are not team players. They do not collaborate with others. They spell win–win as me–me. They are a very coercive presence. They may continue talking while you are trying to work. Regular bullies are about violating boundaries; they stand too close to you and they may come up behind you at your desk. A lethal bully is far more dangerous. There is usually a personality disorder. The abuse borders on violence using verbal abuse. Name-calling and vulgarity are common.

In the Society for Human Resource Professionals, *HR Magazine*, October 1999, vol. 44, no. 10, "Lurking in the Shadows," Rudy M. Yandrick suggests that some of the bully identifiers are as follows:

- Workplace bullies exhibit a variety of behaviors that fall under the rubric of low-level violence. Workplace bullying is commonly thought of as harassment, emotional abuse, and targeted aggression. Bullying follows a pattern of behavior and is not an isolated incident.
- Examples include yelling at or ridiculing a coworker or subordinate publicly for disagreeing, while exhibiting stony silence in private.
- Undermining an individual or group with vindictive or humiliating words or acts.
- Stalking and other forms of intimidation.
- As it applies to males more than females, an implied threat of physical attack, especially in blue-collar environments.

Effects of Bullying

The effects on the individual can range from minor and fleeting to the more common effect of changed behavior that damages the psyche, which can be long-lasting. The impact on the target or targets is recognizable only in that it is similar to the impact of other workplace hazards, accidents, and illness. All OSH hazards have the potential to damage lives of employees, just as they can damage a company's reputation, business goals, and financial welfare. Bullying has an insidious nature to it in that it may not be recognized immediately because social relationships have to be assessed from different points of view. The individuals involved are typically supervisory and subordinate, but not exclusively. In most cases, the management responsible for that assessment is not always close to the setting and people involved, so clear lines of cause and effect cannot be determined. The bully often disguises the behavior as an element of work, a responsibility, or the result of conditions created by others.

There are effects on the bully, but in the majority of cases researched the bully was not punished or given any counseling or training. Rather the bully simply got away with it, left to go on and do it again to someone else. In some cases, the bully was even supported by management or human resources and the target employee was labeled a complainer, whiner, or troublemaker. This usually results in voluntary or involuntary job termination when the unwanted behavior cannot be tolerated any longer. At this point a number of symptoms are observable including, poor concentration, poor morale (among other employees as well), feelings of anxiety, loss of control, trouble sleeping, daytime fatigue, gastrointestinal irritation, disturbed family life, and clinical depression, among others. The bully simply goes on to satisfy his or her insecurities or anger at a new victim.

Why Bullying Occurs

First, we need to look at the individual who is the bully. They share the same characteristics from person to person although the individual people are quite different in many ways. The bully is typically punished only 8% of the time, while the bully's boss, if he is aware of the problem will side with the bully 42% of the time. The bullying stops when the target goes away.

We know the basic characteristics as they mimic the general behaviors of disruptive employees, typically managers. They are rude, condescending with the intention to insult and demean their targets. Verbally they sound hostile and angry. They may shout, throw things, slam doors, and berate staff in front of others. They seem to have an inherent insensitivity toward others' lives. They are often disrespectful to their peers, their support staff, and anyone they consider less important than themselves. They usually do not take criticism well and will counterattack anyone who directs blame at them. Bullies generally dislike authority and believe they are above the rules and policies that apply to others.

Psychologically, they are insecure about themselves and their place in the world. They may have chronic conditions that complicate their negative feelings toward themselves and of others. There are cases of alcohol and substance abuse, family problems, broken marriages, and financial problems. These are all things that everyone has to deal with at some time in their lives but the bully has not learned to cope and work through the personal issues that life's challenges present. Instead, they look to release frustration by targeting those that they see as vulnerable or deserving of some of the misery, they themselves endure. As far as personalities, some are simply unpleasant people who have found that they can get what they want by imposing their wrath on others and do not care or understand the impact they leave behind.

There is another theory about the motivation of the bully, about the conditions that allow them to develop and cause havoc for others. The idea revolves around the concept of their workplace stress. Many company cultures are accepting of aggressive behavior and liken it to healthy competition. Competition is the name of the game in today's business. It is a global competition at that. Companies that reward competitive behavior with compensation and bonus packages that depend on unbridled competition create stressful relationships for management to work with and strong pressures to weed out those perceived as weak and noncompetitive. Stress leads to aggression more often than not. The basic nature of the bully is one that deals with stress by displaying outward aggression and this may be a formula for trouble from the start.

Combine this with some of the characteristics of a bully as noted in Namies' "Survey of Abusive Workplaces," and you begin to see a volatile mix. Namies found a trend for bullies to have less integrity than their targets, less ethicality, superiority in organizational politics, more committed to personal advancement, inferior emotion control and more of their identity defined by their job, their role at work, and their employment in general. The last characteristic is a key to open the door to the bully's mind at work. People who allow themselves to rely on their work to define who they are and where they fit into life, lack self-confidence and use what they perceive as strength to justify the majority of their actions. If they see their only strengths as derived from their job, then much of their emotions are influenced by the ups and downs of a career, as opposed to emotions built on family, faith, personal experience, or social relationships outside of work. When things go bad at work, they go bad through every aspect of the bully's life, whether at work, at home, or in social settings.

In the worst-case scenario, two potential disasters can occur. The bully can reach a point of frustration and emotional instability and their actions are played out as physical violence directed at a target. The consequences now multiply tenfold. There is injury or even death to the target, the bully, or even innocent bystanders. Workplace morale will suffer its worst blow, with employees suffering psychological and physical health problems, poor productivity in the workplace, and the potential for company liability for creating conditions conducive to violence, or worse not doing anything about obvious warning signs. The other potential disaster comes from the targeted employee. It happens where the emotions and psychological pressures brought on by bullying become too much to bear. The unstable person, who may be predisposed to a breakdown, suffering depression or other psychotic illness, becomes the time bomb. We have heard about it in the news; it has happened before. This suffering individual decides that he or she must place blame on someone other then themselves and show up at work with the intent to maim or kill their perceived tormenter and possibly others who may have played a part or been totally innocent and just bystanders.

Prevention of Bullying

The prevention of workplace bullying must be approached as any other hazard; through a workplace and hazard analysis, research and compilation of the data collected, consultation with all appropriate management personnel, the hiring or contracting of expert services that are suited to recommend a plan of action, the development of the plan and its incorporation into the company's overall safety and health plan, the assignment of roles and responsibilities that includes the necessary training of management, the rollout of the plan, including the implementation of training and awareness of all employees, and finally, a method of measurement of the impact of the plan that will prove positive results. In the process, it may be necessary to address existing problems, as they are uncovered, through counseling, discipline, or other means.

In establishing a plan and making it a part of corporate culture that is in line with business goals, you must be careful to use the appropriate resources to understand and prevent bullying from becoming a workplace hazard. There are specific experts in a variety of fields who can be tapped for guidance. Certainly, a company safety and health officer should take the lead in the organization and actions should be taken. A list of professionals who could be consulted would include industrial and occupational psychologists, occupational therapists, occupational physicians, occupational health nurses, risk management professionals, industrial security specialists, design or process engineers, organizational psychologists, and other organizational consultants.

Other specific tools can help analyze the workplace. Occupational Safety and Health Administration (OSHA) has voluntary and advisory guidelines relating to workplace violence, stress reduction, conflict resolution, risk assessment, and health issues related to this hazard. The National Institute for Occupational Safety and Health (NIOSH) has similar resources available. Many other private organizations will provide information and leads to professionals who practice the services needed. Of course, all of these organizations should be used throughout the process of developing and implementing the plan.

The tools for understanding the problem or potential for bullying, as a workplace hazard, are similar to those used to assess workplace violence, stress, and mental and physical health. They include the following: analysis of existing records, the monitoring of trends, incidents, and illness, employee and management surveys, focus groups, performance reviews, administrative and organizational procedure review, security and risk assessment, as well as outside observation of the business operation to lend an unbiased view of current activities.

Of course, the purpose of any aspect of a safety and health plan is to prevent the identified hazard from occurring and controlling it to a safe conclusion when it does occur. The people in an organization who are most likely to be involved include senior management, human resources, and the safety and health point person for the company. Section 5a1 of the federal Occupational Safety and Health Act of 1970 requires employers to provide a safe and healthy workplace free from hazards for all employees. Commonly called the General Duty clause, this standard could be used to hold an employer liable in a bullying incident that results in illness or injury and is provable by facts that can hold up in court. It would be a first step in any plan for the above-mentioned people to approach their task with this liability aspect in mind. Dependant on state laws, it is also possible that these individuals, as well as the bully, could be held personally liable for a lack of action to prevent or control bullying as a workplace hazard. Additional legislation might be needed on the bullying issue and it is likely that it will happen, as the economic and personal costs of bullying become a workplace focus. Will it happen? You bet it will. The United States is actually well behind other countries in addressing this workplace hazard. England and Canada have already enacted legislation and others are close to the same.

One of the areas that managers spend considerable time on is conflict resolution. It may be as much as 20% of their time. Effective training in conflict resolution can help managers not only deal with bullies and their targets but it can help them manage themselves and their own behavior. Additional training might include the general employee population on how to deal with conflict, change, and competition. Union representatives may also be able to provide training, advice, and mediation among employees they represent. Key points to bring into any of this training is the need for open communication, management commitment to address and prevent bullying, and a no retaliation policy for those who bring any hazard to the attention of management. Most of all convince people to bring bullying out into the open, expose bullies for what they are, and the damage that is done.

Management must take corrective action, uphold company policy, and administer disciplinary procedures up to and including termination. The rules and policies should already be in place to set and communicate expectations, measure performance against those expectations, provide feedback, manage poor performance with corrective actions, and document everything.

SUMMARY

Bullying is a form of organizational violence and a potential source of work-related stress. Bullying is primarily an employee relations issue, best dealt with by employers' internal and disciplinary procedures before it becomes a risk to an employee's health.

Bullying is usually viewed as occurring to schoolchildren but often happens in the work environment. As with children, adults feel threatened or ill at ease in reporting perceived incidents of workplace bullying. Probably many adult workers even fail to recognize that a behavior could be construed as bullying.

Bullying is a behavior that should not be tolerated by employers, but the employer may need to educate all employees regarding what constitutes bullying and what employees are to do if they feel that bullying is taking place. The company should have a policy and enforce it, as well as take bullying serious, which has not been the case in the past. The company or organization departments that are most likely to be involved in bullying include senior management, human resources, and the safety and health point person for the company. The companies with the organizational approach had developed a set of values and convictions that reflected on the treatment of antisocial behavior and management response. A distinct element in these organizations was that they each had a reason for developing these plans that was the result of factors outside the company such as new state regulation or prior litigation that had cost the company time and expense. The leaders in these companies recognized that in order to be effective, they had to grow trust and support from the inside of their companies which taught managers the importance of their roles in influencing the behavior of their employees.

One of the areas that managers spend considerable time on is conflict resolution. It may be as much as 20% of their time. Effective training in conflict resolution can help managers not only deal with bullies and their targets but it can help them manage themselves and their own behavior. Additional training might include the general employee population on how to deal with conflict, change, and competition. Union representatives may also be able to provide training, advice, and mediation among employees they represent. Key points to bring into any of this training is the need for open communication, management commitment to address and prevent bullying, and a no retaliation policy for those who bring any hazard to the attention of management. Most of all convince people to bring bullying out into the open, expose the bullies for what they are, and the damage that is done.

Management must take corrective action, uphold company policy, and administer disciplinary procedures up to and including termination. The rules and policies should already be in place to set and communicate expectations, measure performance against those expectations, provide feedback, manage poor performance with corrective actions, and document everything.

CHAPTER 21

Let Us Find a Way: Safety Communications

It is not always easy to communicate in the workplace.

It seems that most of us feel that we do a good job of communicating. I sense that we usually do a rather ineffective job in the workplace. Sometimes it is the way the message is communicated. The timing of messages is often too late and in many cases it is never delivered at all. No matter how hard you try, someone is going to misinterpret your message, read something into it other than what you meant, or only focus on the part they perceive to be most useful to them.

Finding ways to communicate your safety and health messages can be more of a task if your supervisors and workforce do not perceive its importance to you or them. In addition, their perception of the amount of real risk comes into play. If you have a rather safe operation, almost everyone feels that there is no need to emphasize safety since little risk exists. They may make statements such as "We have only had one accident in the last 2 years;" "It could have happened to anyone;" "It was just bad luck. But when John fell he broke his thighbone. The doctors put a pin and screws in it. He'll be as good as new." John did not return to work for 6 months.

What has not been communicated to the workforce is that John will never be as good as new. It took him 6 months to return to work and workers' compensation did not pay him his full and normal wages during that time. John's medical bills cost $50,000 and your insurance premiums increased because of this. You had to hire someone who was not as skilled as John to do his job, which could create a further hazard. John was a friend to everyone. He knew better than to jump off a piece of equipment. He was an experienced worker, but he disobeyed the safety rules.

This scenario is not unusual. Similar situations are played out everyday across the 6,000,000 workplaces in the United States. How do we do a better job of communicating our safety and health concerns? It definitely starts with management's commitment and attitude toward occupational safety and health. If you do not have the fire in your belly or drive to push safety and health in your workplace, then your communications will be viewed as nothing more than a noise from a hollow log. There is a saying that you can not give the measles to someone unless you have the measles. This same logic goes for communicating your message on safety and heath. You have to believe that it is important and you must truly believe in having a safe and healthy workplace.

Many safety problems arise because we assume that everyone knows the proper and safe way to do a job. In actual practice this is not so. It is imperative that management ensures that everyone on the property knows the safety policies of the company and the proper methods to use in performing their job. This comes about by effective training and good communications.

A large percentage of injuries occur when people are not aware of the policies, methods, or basic skills needed to perform a job safely. The responsibility for communicating these concepts rests with the management, but is often assigned to the supervisor who is already overloaded and does not see safety and health communications as an important part of his or her job. After all, there are no consequences to them if their communications are not effective.

A line of communication that constantly furnishes information to all employees must be established. Some methods of communicating the safety and health message are

- Management setting the example by abiding by their own rules
- Safety meetings
- Job training
- Joint/labor management committees
- Employee safety representatives
- Employee involvement
- Safety bulletin boards
- Use of computers
- Accident investigation reports or summaries
- New employee indoctrination
- Safe behavior on the part of the supervisors
- Job analysis for safety
- Job safety observation
- Toolbox/safety talks
- Motivational efforts
- Safe operating procedures
- Written and visual materials

Many of these techniques that help communicate the safety and health message have been covered in other chapters to an extent with which you should be able to perceive how they would fit your communication process.

COMMUNICATOR

You are the communicator. What this means is that you are the primary conveyor of safety and health messages whether you are the manager, supervisor, or safety and health professional. To paraphrase Ralph Waldo Emerson, how you act speaks so loudly, no one can hear what you say.

This means that your failure to visibly support safety and health, follow the company's safety and health rules, or wear the required personal protective equipment sends a stronger nonverbal message than your words can ever do, and will undermine the safety and health at your worksite.

You must be a role model for safety and health. Effective communication for you goes well beyond the written and verbal forms of communication. You must walk the walk before you can credibly talk the talk.

This does not suggest that you should not use written communications or vocally espouse your support for safety and health at your workplace. However, this is all for naught when you sidestep your safety and health rules and policies when it is inconvenient to follow them. When this occurs it is perceived that you are placing production above safety and health in your workplace. Is this what you want to communicate? If you look the other way when your supervisors do not follow your occupational safety and health policies and procedures, your actions are more effective than anything you have said or written regarding commitment to safety and health.

Before attempting to set up communications that will work in your situation, you need to make sure that you have a good understanding of the culture that exists and this is especially true of the safety culture. The safety culture needs to be assessed to make sure that your communications will be received in a manner that will positively influence your safety and health effort.

SAFETY CULTURE

Defining Safety Culture

Safety culture is a concept defined at group level or higher, which refers to the shared values among all the group or company, corporation, or organization members. Safety culture is concerned with formal safety issues in an organization and is closely related to, but not restricted to, the management and supervisory system. Safety culture emphasizes the contribution from everyone at every level of the organization. The safety culture of the business entity has an impact on all members of the workforce and their behavior at work. Safety culture is usually reflected in the relationship between the reward system and safety performance. A positive safety culture is indicative of an organization's willingness to develop, change, and learn from errors, incidents, and accidents. Safety culture is ingrained, enduring, stable, and very resistant to change.

In summary, safety culture is the enduring values and priority placed on workers and management by everyone within the organization. It refers to the extent to which individuals and groups commit to personal responsibility for safety; act to preserve, enhance, and communicate safety concerns; strive to actively learn, adapt, and modify (both individuals and the organization) behavior based on lessons learned from mistakes: and be rewarded in a manner consistent with these values.

Developing or Changing a Safety Culture

Without exception development or change starts with the management's commitment by actions and not words. Words have no substance, but action or behaviors have substance. During this process, actions and development must be maintained by a continuous, constant, and consistent effort. It must be worked at, a new plan must exist, a new approach must be fostered, and support must be gained from all involved.

During development and change, individuals are not merely encouraged to work toward change: they must take action when it is needed. Inaction in the face of safety problems must be unacceptable. This will result in pressure from all facets, both peers and leaders, to achieve a change in culture. There is no room in the culture of safety for those who uselessly point a finger and say "Safety is not my responsibility" and file a report and turn their back on helping to fix it.

These changes are contingent on leaders who are committed to change and open to sharing safety information. Open communications without fear of reprisal and knowledge that action will be taken when reported safety issues are tendered are vital to change. Senior management must support and commit to culture change by demonstrating commitment to safety and providing resources to achieve results. The message must be consistent and sustained since it takes a long time to develop or change a safety culture.

Describing a Safety Culture

Safety culture is the value that is placed on workplace safety and health by an organization, corporation, or company. The message communicated and espoused throughout the place of employment is the outward manifestation of the culture. If the safety culture is not a positive one, then all of the other communication attempts are only window dressing.

Safety culture is best described using an example of a worker wearing his or her protective safety eyewear without being told to do so. This behavior is an example of the worker's attitude to the wearing of personal protective equipment. This is an outward act that demonstrates the value held by the organization related to the true value the company places on safety and health.

Thus, the value the organization places on safety and health is demonstrated by the attitudes that not only the workers have, but also the attitudes that supervisors and managers hold related to safety and health issues. It is even more revealing of a positive safety and health culture when supervisors and managers in the same work environment are seen to be wearing protective eyewear. The behavior of management indicates that they do not view themselves above the values espoused by the company.

However, most of the time safety culture is not this clear-cut. For example, one company that was surveyed espoused the values of accident prevention outwardly. The management seemed to support all that was right in protecting the workforce. Actually, the real value held by the workers especially and the management by acceptance was that getting injured was just part of the job. Getting injured was a part and was acceptable and expected by those performing the work. Even though the company voiced a positive safety and health message, the reality was that the value held by everyone in that workplace was that injury was an acceptable part of doing business. This is an example of a communicated message that was truly not representative of the actual safety culture of the organization.

Workplace culture is not something that can be categorized into certain specific types of culture. Culture is what everyone in the workplace believes about the company, themselves, and safety. These opinions, assumptions, values, perceptions, stereotypes, rituals, leadership, and stories all mesh together to form the culture, which translates into policies, procedures, and accident/incidents. There are many factors invisible from the surface: the taboos, assumptions, and norms, which are never written down. These are the true forces behind outward safety behavior that reflect the real value of safety within the organization. No one espouses these deeply buried parts of the culture, but everyone knows what they are.

Many factors influence the culture of the workplace. They include

- Intelligence and job knowledge
- Emotions and emotional illness
- Individual motivation toward safe work
- Physical characteristics and handicaps
- Family situations
- Peer groups
- The company itself
- Existing society and its values
- Consequence of the work itself

Since everyone at the workplace knows what the culture is, it is necessary to be very observant and listen before trying to communicate. It does not take long to discover the culture. It is seen in the safety, productivity, quality, and discipline of the work and the workforce. The culture often stymies efforts to communicate even when the outcomes are good. It is the sovereign duty of all to maintain the old culture as it is. Thus, the process of changing the culture is time consuming and quite complex.

It is bad enough that the workforce has its own culture, but this is compounded by the company's culture, which affects, and at times is diametrically opposed to, the culture of the workforce. The goal would be to meld the two cultures together in a viable and productive relationship.

Industries and companies that have labor unions are seldom able to perfectly integrate safety into their culture even though safety should be in everyone's best interest. Unions often hold safety hostage as a bargaining chip, tool, or weapon against management. Thus, the culture does not see safety as a level playing field on which everyone can participate equally. However, in the same light, management often fights the battles related to safety and health regulations, which were instituted to provide safety and health for its workforce and is intent on fighting compliance.

Thus, the culture that exists is an amalgamation of the culture of labor and the company. At times there is reasonable compatibility and the two parts form a culture that strives for effective safety while at other times they are so opposite that no one benefits from their inability to merge the culture into a useful entity where effective and rational communications occur.

The only way the culture can be changed is when an urgent need to change is the motivator. First, this could be a number of occupational deaths, a catastrophe, or a series of occupational injuries or illnesses, which forge partnerships in troubled times. Second, the resources must be available and the ability to change needs to be present. Third, a road map or plan must be developed and agreed upon by all parties involved to transition to the new culture. Most of us are reticent to change. Thus, gaining consensus for change is going to be difficult. It will be the leadership's responsibility to decide that there is a need for change, communicating it to the organization, getting consensus of all parties, and directing the implementation.

If this discussion of culture seems ominous and overwhelming, then you have some feeling of how difficult it is to communicate if the culture is not receptive to your message. This makes it imperative to understand the culture that you are trying to communicate within. As for safety, if the culture is antisafety then your communication regarding safety will fall on deaf ears. Therefore, to communicate safety in your workplace you may first understand the culture and then begin to change it. This is a communicate-or-lose situation, which is not the best of all worlds. In most cases you will be able to accomplish some degree of safety communications based in a large part on the perception of how you really feel about safety and health at your workplace.

Positive Safety Culture

An organization, corporation, or company with a positive safety culture is one that gives appropriate priority to safety and realizes that occupational safety and health has to be managed like other areas of the business. Safety culture is more than just avoiding accidents or even reducing the number of accidents, although these are most likely to be the outcomes or measurement of a positive or successful safety culture.

A positive safety culture is doing the right thing at the right time in response to normal or emergency operation. The quality and effectiveness of training will play an important role in determining the attitudes and performance toward safe production or performance. These attitudes toward safety are in a large part a mirror image of the culture set up by the company or corporation.

The key to achieving a positive safety culture are

- Recognizing that all accidents are preventable by following correct procedures and established best practices
- Maintaining constant awareness and thinking about safety
- Trying to improve safety on a continuous basis

It is unusual for different types of accidents to occur within an organization, and most that continue to occur after the unsafe condition has been removed are the result of unsafe acts or behavior. The unsafe actions are usually the result of error, violations of best practices, or disregard for established rules and can be avoided. Most often those making such errors are aware of their mistakes. They take shortcuts even though they have received training, but fall prey to a culture that allows workers to take calculated risks. The goal is to have effectively trained workers who are motivated to self-regulate themselves and take personal responsibility to work safely using the safest practices, where these have become an integrated part of the workers' value system and company culture.

Assessing Safety Culture

You might ask yourself the following questions regarding the safety culture at your place of business or operation.

- Is health and safety a top priority in the organization?
- Is safe work performance reinforced, recognized, and rewarded?
- Are business decisions made with safety and health given major consideration?
- Is the requirement of safe and healthy attitudes and behaviors expected of all employees including management?
- Do employees feel comfortable to voice their concerns regarding any hazards or safety issues about the workplace to management?
- Does peer pressure act as positive or negative in support of a positive safety culture?
- Are production deadlines the cause for safe and healthy work practices to be overlooked?

If the safety culture does not coincide or integrate well with the business approach of the organization, corporation, or company, the indication is that a safety culture does not exist or only marginally exists.

Summary

A positive safety culture exhibits characteristics such as the following:

- Importance of committed leadership from top to bottom
- Clear set of explained expectations for front-line management
- Involvement of all employees
- Effective communication
- Commonly understood and agreed set of safety and health goals
- Learning organization that is responsive to change
- Zeal and attention to detail regarding safety and health
- Questioning attitude and problem-solving environment, not fault finding
- Trust permeating the organization
- Decision making reflecting safety first

The failure to consider and address safety culture can prove to be costly.

COMMUNICATION TOOLS

The following are some the communication techniques that can be used to make communications related to safety and health more effective.

- Written materials
- Bulletin boards

- Electronic signs
- Computers
- Posters
- Public address systems
- Safety and health talks

Written Materials

Written materials should be ergonomically sound so that they are easily readable, legible, and understandable by the workforce. They should also be attention getting. This can be accomplished by using print black on white or yellow paper, as this is more visually favorable than other combinations. Make the message as simple as possible. Do not make the page(s) too cluttered. Actually, double columns (such as those used by newspapers) are more readable than single ones.

It is a wasted communication tool if no one will read it. Your workforce may have poor reading skills or the culture of the workforce may be such that they do not like to read. If this be the case, you are wasting your money investing in reading types of materials.

Written materials are only useful for a short time period. It is a good idea to use them sparingly, maybe once a month as a stuffer in the payroll envelope or during a safety talk. If you post it on the bulletin board, it will be read within a week. You should then take it down since it has probably been read by all of those who will take the time to read it. At times, you may develop plasticized cards that contain new rules, inspection directions, operator guidelines, or changes in procedures, which you cannot expect a worker to memorize. These cards should fit into a shirt pocket and can act as reminders to workers if they have questions when no one is around. At times I recommend that safety operating procedures be plasticized and placed on machinery or equipment that is complicated to operate or not operated frequently. If you use written materials, test them out on the audience before mass distribution. Written materials have their place in fostering workplace safety and health communications.

Bulletin Boards

Well-designed and maintained bulletin boards can communicate safety and health messages. The most effective bulletin boards that I have seen are those placed in well-lighted areas convenient to the workforce. They are usually spacious and well designed to provide good spacing between items. The bulletin board should have a single focus on safety and health specifically to display information, for example, on accident statistics or confined spaces. It must be kept current especially if statistical data are displayed. In other words, a 6 month old date should not be displayed if monthly data are available. You should change the topic on a monthly basis to keep it fresh, i.e., switch from lifting to ladder safety.

Electronic Signs

A computer-controlled electronic message sign in the workplace, along highways, and outside of businesses is an important communication tool. The advantage is that the message can be changed and updated quickly. The message, alert, or notice is, or can be, a function of real time. It can be easily changed to convey the new hot topic. This also compels workers to pay attention to it so that they do not miss anything that is new or important to them. I saw one company sign thanking the workers for completing a special order without any injuries. These types of signs may be the best of the safety and health communication tools. If you have a large facility with multiple entrances, you will need to have one for each entrance. If these signs help keep current safety and health issues on the workers' minds, then they are probably well worth the cost.

Computers

The computer appears in almost all workplaces today and a large part of the workforce has a computer at their workstation or at least ready access to one. It behooves businesses and companies to make use of this very fast communication device. It can be used to alert workers to emergencies, timely reporting of potentially harmful hazards, short safety alerts, and safety messaging using instant message capabilities, or providing short or long training modules that can be accessed at the workers; workstation or at another time such as at home on their own computer. Instantaneous feedback can be provided on training modules.

It can target specific workers who may need special attention regarding safety and health issues. It can be used to post short safety messages to individual workers, work groups, or address family safety issues. It can be used to submit safety suggestions and respond to questions posed regarding safety and health issues. It can readily disperse time-sensitive information on safety and health organization wide at the push of a button. The only limitation is the creativity of the operator. The computer is probably the number one communications tools.

Safety and Health Posters

The posters seen in many workplaces that communicate a number of messages relevant to safety and health with color and graphics are eye-catching and provide a simplistic safety and health message. These posters usually have one message at a time. You must have a poster program to ensure that you are constantly rotating or replacing posters on at least a monthly basis. The posters become stale with time. They might impress visitors but they quickly lose effectiveness with the workforce if they are not current and ever changing (see Figure 21.1).

Figure 21.1 Example of a safety and health poster.

Public Address System

If your company has a public address system that reaches all areas of the workplace and loud noise is not a problem, then short safety and health messages or alerts can be communicated during the workday. These types of announcements should be written ahead of time and should not be announced so often that they become something the workers tune out. They could be a reminder of a hazard that exists, an explanation of an accident/incident, or the safety message of the day. These messages should be short and to the point.

Safety Talks

Safety talks are especially important to supervisors in the workplace and on worksites because they afford each supervisor the opportunity to convey, in a timely manner, important information to workers. Safety talks may not be as effective as one-on-one communications, but they still surpass a memorandum or written message. In the 5–10 min before the workday, during a shift, at a break, or as needed, this technique helps communicate time-sensitive information to a department, crew, or work team.

In these short succinct meetings, supervisors convey changes in work practices, short training modules, facts related to an accident or injury, specific job instructions, policies, and procedures, rules and regulation changes, or other forms of information that the supervisor feels are important to every worker under his or her supervision.

Although safety talks are short, they should not become just a routine part of the workday. Thus, to be effective, they must cover current concerns or information, be relevant to the job, and have value to the workers. Plan safety talks carefully to effectively convey a specific message and a real accident prevention technique. Select topics applicable to the existing work environment; plan the presentation and focus on one issue at a time. Use materials to reinforce the presentation and clarify the expected outcomes (see Figure 21.2). Some guidelines to follow are

1. Plan a safety talk training schedule in advance and post a notice.
2. Prepare supporting materials in advance.
3. Follow a procedure in the presentation: explain goals, try to answer questions, restate goals, and ask for action.
4. Make attendance mandatory.
5. Make each employee sign a log for each session.
6. Ask for feedback from employees on the topic at hand or other proposed topics.
7. Involve employees by reacting to suggestions or letting them make presentations when appropriate.
8. Reinforce the message throughout the work week.

No matter how effectively you communicate with your workforce, you still need to be sure that your workforce has the competence and basic skills to perform the tasks they have been assigned. The following is a policy statement on safety talks to use in your written safety and health program (see Figure 21.3).

Figure 21.4 provides a tool for documentation of safety talks or toolbox meetings, which the company can maintain to verify such efforts.

SUMMARY

Plan your lines of communication and keep them open. A safety and health program or effort may be good on paper, but unless it is communicated to the workers, it is useless. Even more important is your credibility. Do you mean what you say about commitment to safety and health?

Environment, Health, and Safety
Shop Talk

Safe Operation for Scissor Lift Table

Lifting tables are designed to provide day-to-day lifting, stacking, and handling of a wide variety of materials. These tools make it easier to perform our jobs as well as help us to avoid injuries like back, shoulder, arm, and groin strains and sprains. However, like all tools, it is important to understand their operation and safety procedures that should be followed to use them safely. The photograph below shows a typical lift table and a list of the Do's and Don'ts that should be followed for safe operation.

For repairs contact Electric Truck Repair ext. 5–7932

Push only sticker

Capacity tag must be visible and legible

Pendant control

Charging unit

Wheels

Foot brake

Warning information

Pinch points

Don'ts	Do's
• Work under a suspended load	• Protect the pendant control cord
• Stand on the charging unit	• Know the lifting capacity
• Overload the table	• Push the table from the handle side
• Use on soft or uneven surfaces	• Be aware of the pinch points
• Pull the table	• Read warning information
• Stand on table	• Set brake when loading material

THINK! BEFORE YOU LIFT

Figure 21.2 Example of handout for a safety talk.

To maintain awareness, update training, and convey important safety and health information, supervisors or forepersons will conduct at least weekly toolbox safety meetings, usually before the start of work. These toolbox meetings may be held more frequently, depending on the circumstances (i.e., fatality, injury, new operations, etc.). Each supervisor will complete the toolbox meeting form (see example, Figure 21.4), which includes the topic and attendees.

Figure 21.3 Sample of policy statement for your safety and health program.

Toolbox meeting form

Subject _____

Presenter _____

Date _____

Length of time_____

Worker's name	Signature	Soc. sec. #
_____	_____	_____
_____	_____	_____
_____	_____	_____
_____	_____	_____

Note: Staple any handouts or materials used during the toolbox meeting.

Figure 21.4 Toolbox meeting form.

Do you demonstrate that commitment? Are you always reinforcing the safety performance that you desire? Continue to work on communicating information on safety and health to your workforce.

CHAPTER 22

Everything Costs Dollars: Budgeting for Safety and Health

Incidents

Training

Equipment

Personnel

A budget is an organized way to disperse safety and health dollars.

The intent of this book is to portray occupational safety and health (OSH) as an element that should be managed the same as any other company function. As a unique and separate management undertaking, it must have a separate and distinct budget with dollars allocated for the various responsibilities, mandates, and requirements placed on the OSH initiative.

Without an extensive history of cost related to OSH, the development of a budget for safety and health can be a rather inexact undertaking. Without such a history of past budgets or spending on OSH, budgeting becomes somewhat of a guesstimate. This is not to suggest that a reasonable and logical budget cannot be formulated. However, the budget will require much more effort, research, and justification.

During my career I have witnessed few seminars or classes regarding budgeting for OSH and the only book that gives the topic a better than adequate coverage is *Safety and Health Management Planning* by James P. Kohn and Theodore S. Ferry. It seems safe to say that most safety and health professionals have had little or no training or experience related to budgeting for OSH unless they have had hands-on experience from having to develop budgets. This is a function of doing while you learn and probably results in some painful and time-consuming lessons.

One chapter in a book cannot provide a comprehensive coverage of this topic. This is especially true since a budget must be specific for the function of the company and the resources available. Budgets are usually considered to be a road map or planning document for the completion of the assigned tasks and responsibilities based on the resources allocated by the company and are never all that they should be or what a safety and health professional would want them to be.

BUDGET ITEMS

There are a host of items that may need to be budgeted for regarding OSH. The following list is not comprehensive but is a representation of potential budget items:

- Personnel cost

  a. Salaries
  b. Fringe benefits
  c. Social security
  d. Health insurance
  e. Workers' compensation
  f. Retirement
  g. Disability insurance
  h. Unemployment benefits
  i. Skill levels
  j. Training
  k. Professional meetings
  l. Travel

- Equipment—expendable
- Equipment—nonexpendable

  a. Purchase
  b. Rental
  c. Repair agreements
  d. Calibration or certification cost

- Travel
- Administrative cost

  a. Office supplies
  b. Computers
  c. Computer software
  d. Computer hardware
  e. Office equipment
  f. Repair of office equipment
  g. Ergonomic furniture
  h. Storage cabinets

    i. Other office furniture
    j. Telecommunication services (Internet, etc.)
    k. Cell phones
    l. Postage

- Compliance cost

    a. Program development
    b. Personal protective equipment
    c. Medical examinations
    d. Supplies (first aid, etc.)
    e. Professional services (e.g., industrial hygienist)
    f. Consultants (e.g., engineers for redesign of safeguards)

- Contracts (e.g., hazardous waste disposal)
- Facilities

    a. Utilities
    b. Repairs and upgrades
    c. Maintenance

- Liability insurance
- Budgeting for hiring new personnel
- Budgeting for long-term or multiyear projects
- Allowance for unforeseen emergencies
- Cost for Occupational Safety and Health Administration (OSHA) citations and violations

BUDGET APPROACH

While it would be great to be able to predict the future and plan for events and items as though they would certainly happen, it is unlikely that the use of the budget as a precise document to follow will occur. Many factors are not in the control of the person responsible for OSH.

Thus, the developers of an OSH budget must hitch their proverbial wagon to as many real to life safety and health issues as possible. This means that the best approach is to tie as much of the expenditures to compliance with regulatory requirement as humanly possible. Another approach is to show the use of dollars as intervention and prevention of potential cost (a cost avoidance strategy).

A budget that is developed identifies specific items that are to be completed for a specific cost with proper justification and provides the resources to complete the agreed upon safety and health task.

A budget should be broken down into several identifiable categories such as

- Workplace health issues
- Workplace safety issues
- Safety and health management issues
- Environmental safety issues
- Product safety issues

Health Budgeting

Health budgeting must address the existing health factors and the required controls and need for sampling or screening. This may require that an industrial hygienist be hired full-time, part-time, or on contract. The cost of purchasing sampling instrument and maintaining their calibration as well as the expense of having samples analyzed by a certified laboratory must be factored into the budget.

The best use of dollar resources will determine the best approach to the effective use of the money available. Programs that may require sampling are hazardous chemicals, radiation, and noise. If new chemicals are to be put to use, this may require more sampling than previously. If OSHA citations have been issued then more sampling may be necessary to maintain compliance. The presence of hazardous waste may require training in spill containment and remediation as well as contracts for disposal and spill cleanup or remediation. The health issues involved in the company's operation may result in special types of personal protective equipment being needed and purchases. Training will be needed on the use of the equipment, which is a cost factor in loss of production and time and must be accounted for.

Safety Budgeting

Safety budgeting is a function of energy since the sources of accident or incidents are usually the result of energy being released. The budget should include the cost of placing safeguards on machines and equipment, protection from exposure to electricity, protection and training of equipment operators with controls or personal protective equipment, the damage to equipment, implementation of controls, fire prevention and firefighting needs, the actual cost of injuries as well as disruption of production, and recertification of equipment (e.g., cranes). Any time a safety incident can be prevented, a cost saving is accomplished. The safety budget should emphasize the cost of safety prevention activities versus the cost of occurrences of safety-related incidents.

Management Budgeting

Safety and health management is more of an administrative function even though action items and their completion are an integral part of the safety and health initiative. The functions of time and cost for staff and workers involved in safety meetings, safety talks, and participatory programs are an indicator of the company's commitment to safety and health.

Cost and time to conduct audits and inspections to identify hazards and make recommendations for intervention and controls are management functions. The emphasis on time may seem irrelevant, but time is money or cost. Audits may result in the need for new equipment, revamping of processes and procedures, reengineering of equipment, revising programs, training or retraining, or changes in safe operating procedures.

As a safety and health profession in charge, you must make sure that you have the proper staff with the right skill levels to perform the action elements contained within the budget that you have developed. If you do not have the individuals with the skills needed, then you may have to contract or use temporary qualified staff. You must include this cost in your budget.

When a company implements a new program that is needed to effectively reduce the risk of accidents/incidents, it results in an investment in resources again of time and dollars.

When an accident, incident, or fatality occurs, a myriad time-consuming and costly factors accompany it. Not just the cost of the investigation, but the loss of production, equipment damage, loss of productive worker, litigation, OSHA enforcement activities, loss of supervisor's time, and the list goes on and costs mount.

There is always the cost of new-hire training, new job or task training, new process or program training, safety and health training for managers and supervisors, and refresher training for such areas as hazardous waste remediation, asbestos abatement, and lead abatement. At times first aid and cardiopulmonary resuscitation training or retraining is necessary. The components of managing safety and health must be planned for in the OSH budget.

Environmental Budgeting

In many companies environmental safety is addressed as a combined effort with OSH, entitled environmental or occupational safety and health (EOSH). There is a link between the two since environmental issues can cause both safety and health hazards to workers such as hazardous material exposure from spills or emergencies from the release of toxic gases. The development of a budget for environmental safety and health should follow a similar process as OSH because failure to comply with the U.S. Environmental Protection Agency's regulation can be much more costly than failure to comply with OSHA's regulations with regard to appreciable higher fine assessment for violations.

Product Safety Budgeting

Product safety is not an item that should be the responsibility of the OSH department. It should be the responsibility of the product development, quality control, or consumer safety personnel within the company since the applicable standards are found in a different regulatory arena than either the environment or OSH. This not to suggest that EOSH professionals should not lend their expertise to other departments, when they can contribute. Product safety should not be part of the EOSH budget as product safety does not deal with worker safety and health.

COMPLIANCE FACTOR

The OSHA database regarding the average cost of a citation can be used as a lever to get management's attention to the investment in safety and health requested in the budget. It is a real eye opener for management when safety and health professionals can say that the cost of compliance is $500 for bloodborne pathogens and the average violation produces a fine of $1000 per citation. In addition, the OSH professionals can research the most frequent violation for their particular industry sector. In many cases, the most frequent violation is the requirement for a program and training under the hazard communication standard.

Using compliance is by far one of the best ways to justify expenditures. Some of the regulations that require programs for compliance are

- Personal protective equipment
- Respiratory protection program
- HAZMAT program
- Employee emergency and fire prevention plans
- Bloodborne pathogen program
- Medical and first aid
- Ventilation systems

Some of the safety and health regulations require medical examinations to be in compliance such as

- Hearing examinations
- Hazardous materials and waste
- Respiratory clearance examinations

An OSHA violation or citation strikes fear in most employers and is an excellent lever to use to justify budget requests. Other such levers include

- Cost of accidents and incidents
- Medical costs

- Workers' compensation costs
- Real dollar savings
- Loss potential from not performing the action item

WRITTEN BUDGET

When it is time to put a budget in writing make sure that you are prepared to make good judgment decisions, which means that you have to use all the resources at your disposal such as historical-based data and that you have the resources needed to fulfill your requested budget. Ask yourself the question, am I getting the "best bang for the buck?" Is it really worth doing a certain task? Is there a gain in dollars or other benefits? Could it have long-range cost benefits? Is there going to be a return on the money and resources invested? Is the spending going to accomplish your goals? Answers to these questions will help you prioritize the items that should receive emphasis in your budget.

Whether you are developing an element or task budget, variable budget, or a percent of allocation by departments budget, it is important to identify operating costs and potential benefits such as direct benefits (reduced labor cost, lower accidents, reduced insurance cost, or productivity gains). Indirect benefits should also be considered in the light of quality improvements—reduced scrap; less rework; reduced product liability, exposure, or product recall expenses; improved corporate image, or increased market share. At times indirect benefits have improved employee morale, which reduces absenteeism and increases turnover, teamwork, and ownership. The potential reduction in the numbers of compliance penalties can be a benefit.

You might want to develop a form such as the one in Figure 22.1 to help in formulating and tracking your budget needs. Since budgets are unique to the particular type of business or operation, the type of budget dictated by the company, and the existing accounting system, you will want to develop tools that meet your needs in your unique situation.

CONTROLLING COST

The safety and health professional who develops a responsible budget that looks for ways to control cost and maintain or improve the company's bottom line will be viewed as a part of the team and not just a necessary evil.

The safety and health professional can look at ways to control cost by sharing resources, using his or her available resources to the fullest, working on projects with other departments, lending expertise when possible, using mail order to cut cost, using cost-bid service to obtain more for dollars spent, using in-house engineering expertise when possible, volunteering use of staff expertise when possible, using consultants or contractors for temporary or short-term projects for cost saving, performing as much as possible with own staff, being productive, and taking low-cost action by being a good shopper.

SUMMARY

There is always a degree of inaccuracy in any budget. Prevent cost overruns if possible since overruns almost always have detrimental effects. Make sure that you use the most qualified person to develop the OSH budget. Remember that there will always be budget expenditure and issues beyond your control since actual cost may not be the same as planned cost. Link budget elements to accomplishments, milestones, needs, and compliance. Provide as much substance as possible to your

**Budget worksheet
(budget tracking sheet)**

Budget action item: _____

Projected start date: _____

Projected completion date: _____

Projected total cost: _____

Budget items components	Proposed cost	Actual cost	Goal	#

Personnel
  Specific needs:
    1._____
    2._____

Equipment
  Specific needs:
    1._____
    2._____
    3._____

Materials
  Specific needs:
    1._____
    2._____
    3._____
    4._____

Others needs:
    1._____
    2._____
    3._____

Time line/% completion:
Start date: ¼   ½   ¾   Finished

Date/time

Justification (benefits): _____

Figure 22.1   Budget worksheet or tracking form.

budget. Quantified justifications have more substance than qualified opinions. Numbers are more understandable to senior management than reducing severity rates. Use all your experience and tools to develop a workable and reasonable budget.

CHAPTER 23

All's Well That Ends Well: Summary

Safety and health are everybody's job.

As you must know by now there are many common threads that run through the chapters within this book. First, management is ultimately accountable and responsible for its workforce's safety and health on the job. A key to an effective safety and health effort is the implementation of the many facets of management techniques for safety and health, the least of which is planning and an organized approach. It is imperative that you must identify your problem areas (hazards) and make an honest effort to address them or control them in some way.

It should be apparent to you that everyone has a role to play relevant to occupational safety and health. An atmosphere of genuine cooperation concerning safety and health must prevail. The welfare of those working for your company is a common theme, which everyone can support. You certainly want everyone to go to their homes after work with the same number of fingers and toes and not feeling as though they are ill or sick.

It is paramount that everyone be involved in ensuring that their workplace is safe, healthy, and free from hazards that might make them sick or cause them physical harm. This cannot be accomplished by management or supervisors alone. It must involve those who perform the work. After all, who knows the most about the work they have been assigned to accomplish especially the dangers of doing their jobs than the employees themselves?

You need to address all of the many facets of developing and implementing a functional safety and health effort. All safety and health management techniques need to be addressed as an integral part of the overall safety and health approach employed at your workplace. This includes the process, techniques, and the people side of managing safety and health.

It is imperative when addressing safety and health that you really mean what you say and support your efforts by every action, word, and deed. If you drop the ball on any part of your support for safety and health, the perception will exist that you have only given lip service to safety and health and really do not care about your workforce.

Employees and workers need to feel that they are valued by their employer. Many companies you know say, "Our employees are our most valuable asset." However, their actions do not convey that message. Certainly, it is a sacrifice both in personal effort and in resources to make others believe that you really do care. So many of us drop, trip, and fall on this one most important issue. In surveys done recently, employees in a large manufacturing facility stated that the most important aspect of their job was to be treated with respect and as a valued employee. These desires are evaluated as a statistically representative answer for this population. If any of us would venture a guess about what the results of such a survey would be at our own workplaces, surely they would be identical.

The tone or culture set by the company and reinforced by ethical practices go a long way toward ensuring that effective safety and health management is an integral part of a company's persona.

ETHICS OF THE SAFETY AND HEALTH PROFESSION

There is an old saying that goes like this, "Credibility is a lot like Virginity" in that once you have lost it, it is gone. Credibility once lost is nearly impossible to regain and in my estimation the same holds true for the ethics of safety and health conduct. If a company or the person responsible for safety and health (i.e., safety manager) breaches the ethics regarding safety and health of workers, the words ethical behavior may just as well be a foreign language. The trust or credibility that existed is gone and will take a great deal of effort to recover.

Safety and health is viewed as a profession that has at its core responsibility the health and welfare of people. This puts the health and safety professional on a similar pedestal as a medical professional. The safety and health professionals are expected to function under a code of ethics and exhibit behavior that is compatible with that code.

Ethics

It is my sense that the U.S. Military provides a glimpse of what is meant by ethics. Ethics are standards by which one should act based on values. Values are core beliefs such as duty, honor, and integrity that motivate attitudes and actions. Not all values are ethical values (integrity is; happiness is not). Ethical values relate to what is right and wrong and thus take precedence over nonethical values when making ethical decisions.

Primary ethical values include

- Honesty: Being truthful, straightforward, and candid are aspects of honesty. Truthfulness is required. Deceptions are usually easily uncovered. Lies erode credibility and undermine public confidence. Untruths told for seemingly altruistic reasons (to prevent hurt feelings, to promote good will, etc.) are nonetheless resented by the recipients.

  Straightforwardness adds frankness to truthfulness and is usually necessary to promote public confidence and to ensure effective, efficient conduct of operations. Truths presented in such a way as to lead recipients to confusion, misinterpretation, or inaccurate conclusions are not productive. Such indirect deceptions can promote ill will and erode openness, especially when there is an expectation of frankness.

Candor is the forthright offering of unrequested information. It is necessary according to the gravity of the situation and the nature of the relationships. Candor is required when a reasonable person would feel betrayed if the information were withheld. In some circumstances, silence is dishonest; yet in other circumstances, disclosing information would be wrong and perhaps unlawful.

- Integrity: Being faithful to one's convictions is part of integrity. Following principles, acting with honor, maintaining independent judgment, and performing duties with impartiality help to maintain integrity and avoid conflicts of interest and hypocrisy.
- Loyalty: Fidelity, faithfulness, allegiance, and devotion are all synonyms for loyalty. Loyalty is the bond that holds nations and companies together and the hedge against dissension and conflict. It is not blind obedience or unquestioning acceptance of the status quo. Loyalty requires careful balancing among various interests, values, and institutions in the interest of harmony and cohesion.
- Accountability: Professionals are required to accept responsibility for their decisions and the resulting consequences. This includes avoiding even the appearance of impropriety. Accountability promotes careful, well-thought-out decision making and limits thoughtless action.
- Fairness: Open-mindedness and impartiality are important aspects of fairness. Safety and health professionals must be committed to justice in the performance of their official duties. Decisions must not be arbitrary, capricious, or biased. Individuals must be treated equally and with tolerance.
- Caring: Compassion is an essential element when individuals are served. Courtesy and kindness, both to those we serve and to those we work with, help to ensure that individuals are not treated solely as a means to an end. Caring for others is the counterbalance against the temptation to pursue the expected outcomes at any cost.
- Respect: To treat people with dignity, to honor privacy, and to allow self-determination are critical in a corporate environment of diverse people. Lack of respect leads to a breakdown of loyalty and honesty within a company and brings chaos to functionality.
- Promisekeeping: No company can function for long if its commitments are not kept. Safety and health professionals are obligated to keep their promises to promote trust and cooperation. Because of the importance of promisekeeping, safety and health professionals must make commitments only within their authority.

Occupational Safety and Health Ethics

The major factors affecting ethics related to safety and health are character and culture. If safety and health are not values within the company culture then a safety and health professional with all the character in the world cannot be effective. Safety and health must have support within the company; then a safety and health professional can do what is right. This means that he or she must ask himself or herself the following trite questions:

- Would you be embarrassed if your friends and family found out what you did?
- What if this happened to you?
- Is it legal and does it comply with regulatory requirements?
- Does it feel right?

If you can answer these question positively, then do what is right. This means that safety and health professionals must

- Really care about the welfare of everyone; it will show
- Treat everyone and all aspects of safety and health equally

- Treat everyone with respect and dignity
- Take all concerns of employees seriously and respond in a timely and professional manner
- Try to avoid the gray areas concerning values and expectations
- Set up acceptable standards for behavior regarding right and wrong safety and health issues
- Lead by example
- Not accept unsafe acts or behavior
- Continue to build trust and commitment
- Know that reprisal is never an appropriate response
- Hold themselves and others accountable for safety and health practices in an evenhanded and fair manner

Ethics are the moral principles and values that govern the actions and decisions of an individual or group. Operating a company according to the law does not necessarily mean that the practice is ethical. Again, the cornerstone of ethics is character and culture.

PRINCIPLES OF MANAGEMENT TODAY

1. Management is ultimately responsible for occupational safety and health.
   Thus, the need for commitment, budgeting, and planning for safety falls upon management's shoulders.
2. Poor safety conditions and safety performance by the workforce result from management's failure to effectively manage workplace safety and health.
   Accidents and incidents result from management's inability to manage safety and health as they would any other company function.
3. Worker and supervisor involvement are critical to good workplace safety and health.
   A workforce that is not involved in safety and health has no ownership and thus feels no investment, responsibility, or accountability for it.
4. Workplace safety and health are not dynamic fast evolving components of the workplace since they should go hand-in-hand with your normally evolving business.
   In safety and health there is little that is new since we know the causes of occupational injuries and illnesses as well as how to intervene, mitigate, and prevent their occurrence. There should be no excuses for accidents and incidents since the philosophy should be that all are preventable.
5. You cannot have an effective safety and health program without specifically holding the first-line supervisors accountable for themselves and their employees' safety and health performance (as well as other management personnel).
   First-line supervisors are the key to the success or failure of a safety and health program. All your planning, budgeting, and goal setting are for naught if supervisors are not accountable and committed to safety and health.
6. Hazard identification and analysis are critical functions in assuring a safe and healthy work environment.
   If management and the workforce do not tolerate the existence of hazards and constantly ask the question, "How could this have happened?" they are better able to get to the basic causes of adverse workplace events.
7. Management's philosophy, actions, policies, and procedures regarding safety and health in the workplace put workers into situations where they must disregard good safety and health practices to perform their assigned task or work.
   Workers in most cases perform work in an unsafe or unhealthy manner when they have no choice or are forced to do so by existing conditions and expectations.

8. It is critical to obtain safe and healthy performance or behavior by effective communications and motivational procedures that are compatible with the culture of your workforce.

If you do not understand what it is that fulfills the needs of your workforce, you will not be able to communicate or motivate them regarding safety and health outcomes no matter how good your management approach.

TAKING ANOTHER LOOK

After you have made a real effort to rectify or strengthen your safety and health effort, take time to use the following assessment instrument to evaluate your progress toward an effective safety and health program. Circle the most applicable answer for each subheading under the major headings that best describes your safety and health effort. An individual score for each subheading and totals for the major heading will denote areas that you need to pay attention to. If you do not have an average score individually or averaged between 4 and 5, your safety and health effort needs work. Go ahead and see where you are now that you have made some changes.

Safety and Health Program Checkup

Management Leadership and Employee Involvement

Topic	Circle Answer	Answer Options
Clear worksite safety and health (S&H) policy	5	We have an S&H policy and all employees accept, can explain, and fully understand it
	4	We have an S&H policy and majority of employees can explain it
	3	We have an S&H policy and some employees can explain it
	2	We have a written (or oral, where appropriate) policy
	1	We have no policy
Clear goals and desired objectives are set and communicated	5	All employees are involved in developing goals and can explain results and how results are measured
	4	Majority of employees can explain results and measures for achieving them
	3	Some employees can explain results and measures for achieving them
	2	We have written (or oral, where appropriate) goals and objectives
	1	We have no safety and health goals and objectives
Management leadership	5	All employees can give examples of management's commitment to safety and health
	4	Majority of employees can give examples of management's active commitment to safety and health
	3	Some employees can give examples of management's commitment to safety and health
	2	Some evidence exists that top management is committed to safety and health
	1	Safety and health are not top management values or concerns
Management example	5	All employees recognize that managers in this company always follow the rules and address the safety behavior of others
	4	Managers follow the rules and always address the safety behavior of others

(continued)

Management Leadership and Employee Involvement (continued)

Topic	Circle Answer	Answer Options
	3	Managers follow the rules and usually address the safety behavior of others
	2	Managers generally follow basic safety and health rules
	1	Managers do not follow basic safety and health rules
Employee involvement	5	All employees have ownership of safety and health and can explain their roles
	4	Majority of employees feel that they have a positive impact on identifying and resolving safety and health issues
	3	Some employees feel that they have a positive impact on safety and health
	2	Employees generally feel that their safety and health input will be considered by supervisors
	1	Employee involvement in safety and health issues is neither encouraged nor rewarded
Assigned safety and health responsibilities	5	All employees can explain what performance is expected of them
	4	Majority of employees can explain what performance is expected of them
	3	Some employees can explain what performance is expected of them
	2	Performance expectations are generally spelled out for all employees
	1	Specific job responsibilities and performance expectations are generally unknown or hard to find
Authority and resources for safety and health	5	All employees believe they have the necessary authority and resources to meet their responsibilities
	4	Majority of employees believe they have the necessary authority and resources to meet their responsibilities
	3	Authority and resources are spelled out for all, but there is often a reluctance to use them
	2	Authority and resources exist, but most are controlled by supervisors
	1	All authority and resources come from supervision and are not delegated
Accountability	5	Employees are held accountable and all performance is addressed with appropriate consequences
	4	Accountability systems are in place, but consequences used tend to be for negative performance only
	3	Employees are generally held accountable, but consequences and rewards do not always follow performance
	2	There is some accountability, but it is generally hit or miss
	1	There is no effort toward accountability
Program review (quality assurance)	5	In addition to a comprehensive review, a process is used that drives continuous correction
	4	A comprehensive review is conducted at least annually and drives appropriate program modifications
	3	A program review is conducted, but it does not drive all necessary program changes
	2	Changes in programs are driven by events such as accidents or near-misses
	1	There is no program review process

Emergency and Medical Planning

Topic	Circle Answer	Answer Options
Emergency planning and preparation	5	There is an effective emergency response plan and employees know immediately how to respond as a result of effective planning, training, and drills
	4	There is an effective emergency response plan and employees have a good understanding of responsibilities as a result of planning, training, and drills
	3	There is an effective emergency response plan and team, but other employees may be uncertain of their responsibilities
	2	There is an effective emergency response plan, but training and drills are weak and roles may be unclear
	1	Little effort is made to prepare for emergencies
Emergency equipment	5	Facility is fully equipped for emergencies; all systems and equipment are in place and regularly tested; all personnel know how to use equipment and communicate during emergencies
	4	Facility is well-equipped for emergencies with appropriate emergency phones and directions; majority of personnel know how to use equipment and communicate during emergencies
	3	Emergency phones, directions, and equipment are in place, but emergency teams know what to do
	2	Emergency phones, directions, and equipment are in place, but employees show little awareness
	1	There is little or no effort made to provide emergency equipment and information
Medical program (health providers)	5	Occupational health providers are regularly on-site and fully involved
	4	Occupational health providers are involved in hazard assessment and training
	3	Occupational health providers are consulted about significant health concerns in addition to accidents
	2	Occupational health providers are available, but normally concentrate on employees who get hurt
	1	Occupational health assistance is rarely requested or provided
Medical program (emergency care)	5	Personnel fully trained in emergency medicine are always available on-site
	4	Personnel with basic first-aid skills are always available on-site, all shifts
	3	Either on-site or near-by community aid is always available on day shift
	2	Personnel with basic first-aid skills are usually available, with community assistance nearby
	1	Neither on-site nor community aid can be ensured at all times

Hazard Identification

Topic	Circle Answer	Answer Options
Hazard identification (expert survey)	5	Comprehensive expert surveys are conducted regularly and result in corrective action and updated hazard inventories
	4	Comprehensive expert surveys are conducted periodically and drive appropriate corrective action
	3	Comprehensive expert surveys are conducted, but corrective action sometimes lags
	2	Expert surveys in response to accidents, complaints, or compliance activity only
	1	No comprehensive surveys have been conducted
Hazard identification (inspection)	5	Employees and supervisors are trained, conduct routine joint inspections, and all items are corrected
	4	Inspections are conducted and all items are corrected; repeat hazards are seldom found
	3	Inspections are conducted and most items are corrected, but some hazards are still uncorrected
	2	An inspection program exists, but corrective action is not complete; hazards remain uncorrected
	1	There is no routine inspection program in place and many hazards can be found
Hazard reporting system	5	A system exists for hazard reporting, employees feel comfortable using it, and employees feel comfortable correcting hazards on their own initiative
	4	A system exists for hazard reporting and employees feel comfortable using it
	3	A system exists for hazard reporting and employees feel they can use it, but the system is slow to respond
	2	A system exists for hazard reporting but employees find it unresponsive or are unclear how to use it
	1	There is no hazard reporting system and employees are not comfortable reporting hazards
Accident/incident investigation	5	All loss-producing incidents and near-misses are investigated for root cause with effective prevention
	4	All Occupational Safety and Health Administration (OSHA) reportable incidents are investigated and effective prevention is implemented
	3	OSHA-reportable incidents are generally investigated; accident cause identification and correction may be inadequate
	2	Some investigation of incidents takes place, but root cause is seldom identified and correction is spotty
	1	Injuries are either not investigated or investigation is limited to report writing required for compliance

Hazard Analysis

Topic	Circle Answer	Answer Options
Hazard analysis	5	All workers and supervisors involved in assessing hazards and deriving solutions
	4	Only supervisors are involved in analyzing hazards and addressing interventions
	3	Only serious hazards are analyzed and controls recommended
	2	Hazards are analyzed after accidents/incidents have occurred
	1	No routine hazard analysis takes place
Root cause analysis	5	There is a system in place to evaluate the root cause of all accidents/incidents and even near-misses
	4	Supervisors and the safety professional determine the root cause of accidents/incidents
	3	Someone looks at the root cause of more serious accidents/incidents
	2	There are times when accidents/incidents are evaluated further than fixing blame on the injured worker
	1	There is no root cause analysis
Change analysis	5	Every planned or new facility, process, material, or equipment is fully reviewed by a competent team, along with affected workers
	4	Every planned or new facility, process, material, or equipment is fully reviewed by a competent team
	3	High-hazard planned or new facility, process, material, or equipment is reviewed
	2	Hazard reviews of planned or new facilities, processes, materials, or equipment are problem driven
	1	No system for hazard review of planned or new facilities exists
Hazard identification (job and process analysis)	5	A current hazard analysis exists for all jobs, processes, and material; it is understood by all employees; and employees have had input into the analysis for their jobs
	4	A current hazard analysis exists for all jobs, processes, and material and it is understood by all employees
	3	A current hazard analysis exists for all jobs, processes, or phases and it is understood by many employees
	2	A hazard analysis program exists, but few are aware of it
	1	There is no routine hazard analysis system in place
Injury and illness analysis	5	Data trends are fully analyzed and displayed, common causes are communicated, management ensures prevention; and employees are fully aware of trends, causes, and means of prevention
	4	Data trends are fully analyzed and displayed, common causes are communicated, and management ensures prevention
	3	Data are centrally collected and analyzed and common causes are communicated to supervisors for prevention
	2	Data are centrally collected and analyzed but not widely communicated for prevention
	1	Little or no effort is made to analyze data for trends, causes, and prevention

Hazard Prevention and Control

Topic	Circle Answer	Answer Options
Timely and effective hazard control	5	Hazard controls are fully in place, known to and supported by workforce, with concentration on engineering controls and safe work procedures
	4	Hazard controls are fully in place with priority to engineering controls, safe work procedures, administrative controls, and personal protective equipment (in that order)
	3	Hazard controls are fully in place, but there is some reliance on personal protective equipment
	2	Hazard controls are generally in place, but there is heavy reliance on personal protective equipment
	1	Hazard control is not complete, effective, and appropriate
Facility and equipment maintenance	5	Operators are trained to recognize maintenance needs and perform and order maintenance on schedule
	4	An effective preventive maintenance schedule is in place and applicable to all equipment
	3	A preventive maintenance schedule is in place and is usually followed except for higher priorities
	2	A preventive maintenance schedule is in place but is often allowed to slide
	1	There is little or no attention paid to preventive maintenance; break-down maintenance is the rule

Safety and Health Training

Topic	Circle Answer	Answer Options
Employees learn about hazards (how to protect themselves and others)	5	Facility is committed to high-quality employee hazard training, ensures that all participate, and provides regular updates; in addition, employees can demonstrate proficiency in, and support of, all areas covered by training
	4	Facility is committed to high-quality employee hazard training, ensures that all participate, and provides regular updates
	3	Facility provides legally required training and makes effort to include all employees
	2	Training is provided when the need is apparent; experienced employees are assumed to know the material
	1	Facility depends on experience and informal peer training to meet needs
Supervisors learn responsibilities and underlying reasons	5	All supervisors assist in worksite hazard analysis, ensure physical protections, reinforce training, enforce discipline, and can explain work procedures based on the training provided to them
	4	Most supervisors assist in worksite hazard analysis, ensure physical protections, reinforce training, enforce discipline, and can explain work procedures based on the training provided to them
	3	Supervisors have received basic training, appear to understand and demonstrate the importance of worksite hazard analysis, physical protections, training reinforcement, discipline, and knowledge of work procedures

Safety and Health Training (continued)

Topic	Circle Answer	Answer Options
	2	Supervisors make responsible efforts to meet safety and health responsibilities, but have limited training
	1	There is no formal effort to train supervisors in safety and health responsibilities
Managers learn safety and health program management	5	All managers have received formal training in safety and health management responsibilities
	4	All managers follow, and can explain, their roles in safety and health program management
	3	Managers generally show a good understanding of their safety and health roles and usually model them
	2	Managers are generally able to describe their safety and health roles, but often have trouble modeling them
	1	Managers generally show little understanding of their safety and health management responsibilities

To score the results add all of your responses together to obtain a total. If you have a cumulative score for your operation between 109 and 135, then you have an excellent program, 82–108 a very good program (needs some improvements), 55–81 is average (average is the best of the worst and worst of the best), 28–54 very poor (much effort is needed), and 1–27 a nonexistent program (this is totally unacceptable).

FIVE PRINCIPLES FOR SAFETY AND HEALTH

The principles of safety and health are those that have been identified and espoused in this book. They are as follows:

1. All accidents and incidents are preventable.
2. All levels of management are responsible for safety and health.
3. All employees have the responsibility to themselves, their coworkers, and their family to work in a safe and healthy manner.
4. To eliminate accidents and incidents, management must ensure that all employees are properly trained on how to perform every job task efficiently and in a safe and healthy manner.
5. Every employee must be involved in every area of the safety, health, and production process.

SUPERVISOR'S TEN COMMANDMENTS OF SAFETY AND HEALTH

Supervisors have a key role to play in the safety and health of the workplace. They must make an effort as a role model for safety and health to all others. Their commandments are as follows:

1. Care for your charges at work as you would care for your family at home. Be sure each one understands and accepts personal responsibility for safety and health.
2. Know the rules for safety and health that apply to the work that you supervise. Never let it be said that one of your charges was injured or became ill because you were not aware of the precautions required on their job.

3. Anticipate the risks that may arise from changes in equipment or methods. Make use of expert safety and health advice; this is available to help you guard against such new hazards.
4. Encourage your charges to discuss with you the hazards of their work. No job should proceed where a question concerning safety and health remains unanswered. When you are receptive to the ideas of the workforce, you tap a source of firsthand knowledge, which will help you prevent needless loss and suffering.
5. Instruct your charges to work safely, as you would guide and counsel your family—with persistence and patience.
6. Follow up your instructions consistently. Ensure that your charges make use of the safeguards provided to them. If necessary, enforce the safety and health rules by disciplinary action. Do not fail the company that sanctioned these rules—or your charges who need them.
7. Set a good example. Demonstrate safety and health in your own work habits and personal conduct. Do not appear a hypocrite in the eyes of your charges.
8. Investigate and analyze every accident and incident—however slight—that befalls any of your charges. Where minor injuries and illnesses go unheeded, crippling accidents and effects may strike later.
9. Cooperate fully with those in the organization who are actively concerned about workforce safety and health. Their dedicated purpose is to keep your charges fully able and on the job and to cut down the heavy toll of accidents and incidents.
10. Remember that accident and incident prevention not only reduces human suffering and loss, but, from a practical viewpoint, it is no more than good business. Safety and health, therefore, are one of a supervisor's primary obligations—to your company, to your fellow managers, and to your fellow workers.

TEN COMMANDMENTS OF SAFETY AND HEALTH FOR YOUR WORKFORCE

The following are 10 guidelines that apply to everyone regarding safety and health in the workplace.

1. Learn the safe and healthy way to do your job before you start.
2. Think safety and health, and act safely at all times.
3. Obey safety and health rules and regulations.
4. Wear proper clothing and protective equipment.
5. Conduct yourself properly at all times—horseplay is prohibited.
6. Operate only the equipment you are authorized to use.
7. Inspect tools and equipment for safe conditions before starting.
8. Advise your superior promptly of any unsafe conditions or practices.
9. Report any injury or illnesses immediately to your superior.
10. Support your safety and health program and take an active part in the safety and health meeting.

Yes, if you take your commitment to the safety and health of your workforce seriously, you will find that all's well that ends well. You will reap the positive benefits of a safe and healthy workforce. If the workforce is not worried about their degree of wellness then they will be much more productive. Ultimately, management has the ultimate and final say on whether you will have or be known as a company that cares about the safety and health of its workforce.

Emergency Action Plan—Fire Evacuation

1. An emergency escape route chart is posted on the facility bulletin board, indicating by department, a primary and secondary exit or escape route in the event of fire or other emergencies.
2. In the event of an emergency within the facility, such as fire, the supervisor in that department will immediately make repeated announcements over the paging system that an emergency exists and the areas to be evacuated, if known, otherwise he or she would evacuate the entire plant.
3. In event of an emergency notification from outside the plant, such as a bomb threat, the person receiving the call will immediately make repeated announcements over the paging system to evacuate the entire facility.
4. When an evacuation signal is given, each supervisor involved will assume station in the vicinity of the designated exit. The supervisor will ensure that all personnel are evacuated and will assist employees requiring the same.
5. Once evacuation of all employees has been completed, employees are to proceed to the previously designated accounting area for an additional head count by the supervisor or fire or emergency warden. The supervisor will report the department's status to the manager or individual in charge.
6. Should medical attention be required, first aid will be given by the supervisor and others trained in first aid.
7. Emergency telephone numbers are posted in the plant office, at the time clock, and bulletin board in all departments and are listed below. It is essential that the appropriate emergency service be called immediately.

EMERGENCY TELEPHONE NUMBERS

Fire _____
Police _____
Ambulance or rescue squad _____
Doctor _____
Hospital _____

Written Safety and Health Program*

MANAGEMENT'S COMMITMENT

Safety is a management function, which requires the management's participation in planning, setting objectives, organizing, directing, and controlling the program. Management's commitment to safety and health is evident in every decision the company makes and every action this company takes. Therefore, the management of Name of Company assumes total responsibility for implementing and ensuring the effectiveness of this safety and health program. The best evidence of our company's commitment to safety and health is this written program, which will be fully implemented on each company construction project.

ASSIGNING RESPONSIBILITY

The individual assigned with the overall responsibility and authority for implementing this safety and health program is Name of Individual, safety director. The management fully supports the safety director and will provide the necessary resources and leadership to ensure the effectiveness of this safety and health program.

The safety director will supplement this written safety and health program by

- Establishing workplace objectives and safety recognition programs
- Working with all government officials during accident investigations and safety inspections
- Maintaining safety and individual training records
- Encouraging reporting of unsafe conditions and promoting a safe workplace (some of these responsibilities will be delegated to supervisors for implementation)

SAFETY AND HEALTH POLICY STATEMENT

To all Employees:

Name of Company is committed to providing a safe and healthful workplace, which is free from recognized hazards. The safety and health of our employees is one of the highest priorities of the Name of Company. It is the policy of this company that accident prevention will be given primary

---

*Reese, C.D., Moran, J.B., and Lapping, K. *Model Construction Safety and Health Program.* Washington, DC: Laborers' Health and Safety Fund of North America, 1993.

importance in all phases of operation and administration. Therefore, the management has developed this safety and health program to reduce injuries and illnesses that are so prevalent in construction.

The effectiveness of this program depends on the cooperation and communication of management officials, supervisors, and employees. Everyone must be capable of recognizing hazards in the workplace and understanding their role. Each supervisor will make the safety and health of all employees an integral part of his or her regular management functions. In addition, each employee will adhere to established company safety rules and procedures. Participation of all employees is essential to ensure the effectiveness of this program.

The management will make every effort to provide adequate safety training to employees before allowing them to begin work. Employees in doubt about how to do a job or task safely are required to ask a qualified person for assistance. Employees must report all injuries and unsafe conditions to the management as soon as possible so that corrective measures can be taken to prevent future accidents.

Please read this safety and health program and follow the safe work procedures described. Safety is everyone's business and everyone (management officials, supervisors, employees) will be held accountable for participating in this program.
Please think safety and always work safely.

President

COMPANY SAFETY GOALS AND OBJECTIVES

On each Name of Company project jobsite, the site superintendent will be accountable to the management for the successful achievement of targeted company safety and health goals. Name of Company's safety and health goals are as follows:

- Zero fatalities or serious injuries
- Reduced injuries, lost workday accidents, and workers' compensation claims
- Prevention of damage or destruction to company property or equipment
- Increased productivity through reduction of injuries
- Reduced workers' compensation costs
- Enhanced company image by working safely
- Safety a paramount part of the workers' daily activities
- Recognition and reward of safe work practices
- Improved morale and productivity

SAFETY ENFORCEMENT POLICY

Whenever a violation of safety rules occurs, the following enforcement policy will be implemented:

- First offense—Employee receives verbal warning and proper instruction pertaining to the specific safety violation. (A notation of the violation may be made and placed in the employee's personnel file.)
- Second offense—Written warning with a copy placed in the employee's personnel file.
- Third offense—Receipt of two written reprimands in any 12 month period may result in suspension.
- Fourth offense—Employee is dismissed from employment.

Note: The company reserves the right to terminate immediately any employee who acts unsafely on Name of Company jobsites.

Responsibilities for safety and health include the establishment and maintenance of an effective communication system between management officials, supervisors, and workers. To this end, all personnel are responsible for assuring that their messages are received and understood by the intended receiver. Specific safety and health responsibilities for Name of Company personnel are as follows.

MANAGEMENT OFFICIALS

Active participation in, and support of, safety and health programs are essential. Therefore, all management officials of the Name of Company will display their interest in safety and health matters at every opportunity. At least one manager (as designated) will participate in safety and health meetings, accident investigations, and worksite inspections. Each manager will establish realistic goals for accident reduction in his or her area of responsibility and will establish the necessary implementing instructions for meeting the goals. Goals and implementing instructions must be within the framework established by this document. Incentives may be included as a part of implementing instructions.

SUPERVISORS

The safety and health of the employees are primary responsibilities of the supervisors. To accomplish this obligation, supervisors will

1. Assure that all safety and health rules, regulations, policies, and procedures are understood by conducting preworksite safety orientations with all workers and reviewing rules as the job or conditions change or when individual workers show a specific need
2. Require the proper care and use of all necessary personal protective equipment to protect workers from hazards
3. Identify and eliminate job hazards expeditiously through hazard analysis procedures
4. Receive and take initial action on employee suggestions, awards, or disciplinary measures
5. Conduct foreman/crew meetings the first 5 min of each work shift to discuss safety matters and work plans for the workday
6. Train employees (both new and experienced) in the safe and efficient methods to accomplish each job or task
7. Review accident trends and establish prevention measures
8. Attend safety meetings and actively participate in the proceedings
9. Participate in accident investigations and safety inspections
10. Promote employee participation in this safety and health program
11. Actively follow the progress of injured workers and display an interest in their rapid recovery and return to work

EMPLOYEES

Safety is a management responsibility; however, each employee is expected, as a condition of employment for which he or she is paid, to work in a manner that will not inflict self-injury or cause injury to fellow workers. Each employee must understand that responsibility for his or her own safety is an integral job requirement. Each employee of Name of Company will

- Observe and comply with all safety rules and regulations that apply to his or her trade
- Report all on-the-job accidents and injuries to his or her supervisor immediately

- Report all equipment damage to his or her supervisor immediately
- Follow instructions and ask questions of his or her supervisor when in doubt about any phase of his or her operation
- Report all unsafe conditions or situations that are potentially hazardous
- Operate only equipment or machinery that he or she is qualified to operate
- When in doubt, ask for directions
- Know what emergency telephone numbers to call in case of fire or personal injury
- Help to maintain a safe and clean work area
- Talk with management at any reasonable time concerning problems that affect his or her safety or work conditions

The most important part of making this program effective is the individual employee. Without your cooperation, the most stringent program can be ineffective. Protect yourself and your fellow workers by following the rules. Remember: Work safely so that you can return home each day the same way you left. Your family needs you and this company needs you! Do not Take Chances—Think Safety First!

COMPETENT/QUALIFIED PERSONS

Occupational Safety and Health Administration's (OSHA) construction standards (29 CFR 1926) require every employer to designate competent persons to conduct frequent and regular inspections of the jobsite, materials, and equipment.

To comply with OSHA competent/qualified person requirements, each project will have a project competent person capable of identifying existing and predictable hazards with authority to take prompt corrective measures to eliminate them. This individual may designate other competent persons to perform certain tasks, such as supervising scaffold erection.

Competent/qualified persons will be designated for each project and listed on the company's Safety and Health Competent Person Assignments Form. This form will be completed and displayed at all operations requiring the presence of a competent/qualified person. The form should be updated and replaced as necessary to reflect current designated competent/qualified persons and their area of expertise and responsibility.

The core of an effective safety and health program is hazard identification and control. Periodic inspections and procedures for correction and control provide methods of identifying existing or potential hazards in the workplace and eliminating or controlling them. The hazard control system provides a basis for developing safe work procedures and injury and illness prevention training. If hazards occur or recur, this reflects a breakdown in the hazard control system.

This written safety and health program establishes procedures and responsibilities for the identification and correction of workplace hazards. The following activities will be used by this company to identify and control workplace hazards:

- Jobsite inspections
- Accident investigation
- Safety and health committee

Jobsite Safety Inspections

Safety inspections of the jobsite will occur periodically every (insert frequency), when conditions change, or when a new process or procedure is implemented. These inspections should focus on the identification and correction of potential safety, health, and fire hazards. Individuals should use the site evaluation worksheet when conducting jobsite safety inspections. In addition, the safe work procedures should be reviewed by personnel conducting safety inspections of the jobsite.

As part of this safety and health program, the work procedures for each company worksite will

- Identify high-hazard areas of operation and determine inspection priorities
- Establish inspection responsibilities and schedules
- Develop an administrative system to review, analyze, and take corrective action on inspection findings

Accident Investigation

All accidents will be investigated to determine causal factors and prevent future recurrences of similar accidents. A written report of investigation findings will be prepared by each injured employee's supervisor and submitted to management for review. Written reports for accidents resulting in fatalities or serious injuries will also be submitted to company attorneys.

Whenever an accident is reported, the supervisor of the injured workers should respond to the scene of the accident as soon as possible and complete the supervisor's accident report. All witnesses should be interviewed privately as soon as possible after the accident. If possible, the supervisor should interview the workers at the scene of the accident so that events leading up to the accident can be reenacted.

Photographs should be taken as soon as possible after the accident and include the time and date taken.

Supervisors are required to submit accident investigation reports that answer the questions who, what, when, where and how:

- Who was involved? The investigation report should identify the injured workers' name and occupation.
- What happened? The investigation report should describe the accident, the injury sustained, eye-witnesses, the date, time, and location of the accident.
- Why did the accident occur? All the facts surrounding the accident should be noted here, including, but not limited to, the following:

  a. What caused the situation to occur?
  b. Were the workers qualified to perform the function involved in the accident?
  c. Were they properly trained?
  d. Were operating procedures established for the task involved?
  e. Were procedures followed, and if not, why?
  f. Where else might this or a similar situation exist, and how can it be avoided?
  g. What should be done? Methods for preventing future accidents of a similar nature should be identified.
  h. What has been done? A follow-up report will be completed by the site safety representative to determine if the suggested action was implemented, and if so, whether similar accidents were prevented as a result of such implementation.

Safety and Health Committee

Each Name of Company worksite will establish a safety and health committee to assist with implementation of this program and the control of identified hazards. The Safety and Health Committee will be composed of employees and management representatives. The committee should meet regularly, but not less than once a month. Written minutes from safety and health committee meetings will be available and posted on the project bulletin board for all employees to see.

The safety and health committee will participate in periodic inspections to review the effectiveness of the safety program and make recommendations for improvement of unsafe and unhealthy conditions. This committee will be responsible for monitoring the effectiveness of this program. The committee will review safety inspection and accident investigation reports, and where appropriate, submit suggestions for further action. The committee will also, upon request from OSHA, verify abatement action taken by Name of Company in response to safety and health citations.

Objectives of project labor/management safety and health committee:

1. Reduce accidents through a cooperative effort to identify and eliminate as many unsafe conditions and acts as possible.
2. Promote employee training in areas of recognition, avoidance, and prevention of workplace hazards.
3. Encourage employee participation in the company safety and health program.
4. Establish a line of communication for the worker to voice his or her concerns on existing or potential hazards and receive positive feedback.
5. Develop a mechanism that enables workers to provide suggestions on how to improve safety and health on the jobsite.
6. Provide a forum for joint labor–management cooperation on safety and health issues in the workplace.

Functions of project labor/management safety and health committee:

1. Involve workers in problem solving.
2. Examine accident and injury statistics and set safety objectives.
3. Communicate accident prevention information to the workforce.
4. Review reports of recent accidents.
5. Identify and correct hazardous conditions and practices.
6. Assist in identifying the causes of hazards.
7. Regularly review minutes of previous meetings to ensure that action has been taken.

MONTHLY PROJECT OR PRODUCTION SAFETY MEETING

A monthly safety meeting will be conducted on each worksite/jobsite to provide the affected parties with relevant information concerning existing or potential worksite hazards, corrective actions, and abatement. Minutes from these meetings should be recorded and a copy sent to the corporate safety office. The following parties should attend these monthly safety meetings:

- Company president/CEO or designated representative
- Middle manager
- Supervisor
- Foreman or lead person
- Safety and health representative

All employees from managers to workers will receive safety education and training through all phases of work performed by Name of Company. The following safety education and training practices will be implemented and enforced at all company projects/jobsites.

NEW-HIRE SAFETY ORIENTATION

New employees or current employees who are transferred from another project must attend a project-specific new-hire safety orientation. This program provides each employee the basic information about the Name of Company worksite safety and health rules, federal and state OSHA standards, and other applicable safety rules and regulations. Employee attendance is mandatory before working on the construction project. The site superintendent will record attendance using the New-Hire Safety Orientation Form and maintain a file documenting all workers who attend new-hire safety orientation.

The project/jobsite safety orientation program will introduce new employees to

- Company safety and health program and policy
- Project/jobsite and the employee's role within it
- Hazard communication requirements
- Emergency procedures
- Location of first-aid stations, fire extinguishers, telephone, lunchroom, washroom, and parking
- Site-specific hazards
- Safety and health responsibilities
- Reporting of injuries and hazardous conditions
- Use of personal protective equipment
- Tool handling and storage
- Review of each safety and health rule applicable to the job
- Introduction to safety and health representatives
- Introduction to supervisor
- Site tour or map where appropriate

Management understands that a new employee can absorb only so much information in the first few days. Therefore, each new employee will be paired with a veteran employee who can reinforce the new employee's training while, at the same time, raising the safety awareness of the experienced buddy.

SUPERVISOR TRAINING

The supervisor/foreman is responsible for the prevention of accidents for tasks under his or her direction, as well as thorough accident prevention and safety training for employees he or she supervises. Therefore, all supervisors/foremen will receive training so that they have a sound theoretical and practical understanding of the following:

- Site-specific safety program
- Occupational Safety and Health Act (OSHAct) and applicable regulations
- OSHA hazard communication standard
- Site emergency response plan
- First-aid and Cardiopulmonary resuscitation (CPR)
- Accident and injury reporting and investigation procedures
- Hazard assessment in their areas of expertise and topics appropriate for toolbox talks
- OSHA record keeping requirements
- Communication techniques

In addition to the training requirements described here, managers will receive additional training on, but not limited to, the following topics:

- Implementation and monitoring of the company's safety program
- Personnel selection techniques

- Jobsite planning
- Contractor supervision
- Worksite documents
- OSHA record keeping requirements

SAFETY BULLETIN BOARD

A safety bulletin board will be located on each worksite/jobsite where it will be visible to all employees. The bulletin board will contain information such as the following:

- Safety and health committee meeting minutes
- Safety promotions/awards
- Safety meeting dates and times
- OSHA 300A form (February 1 to April 30 of each year)
- Available safety training
- Safety inspection findings
- Emergency phone numbers
- Any violation cited by OSHA
- "It's the Law" OSHA job safety and health poster

Additional items may be posted with the management's approval.

SAFETY TALKS

Supervisors/foremen will conduct weekly work group sessions, also known as safety talks/ toolbox meetings each _____ immediately before start of work. These safety talks may be held more frequently depending on the circumstances (i.e., fatality, injury, new operations, etc.). The supervisor/foreman will provide appropriate materials (handouts, audio/visual aids, etc.) to discussion leaders in advance of each meeting. Discussion leaders will be selected for each meeting by the supervisor/foreman. These weekly meetings should not exceed 15 min. Active employee participation and a question-and-answer session are recommended during each meeting.

Meetings will be scheduled whenever new operations are introduced into the workplace to ensure that all employees are familiar with the safe job procedures and requirements for performing the job safely.

Employee attendance at a toolbox meeting must be recorded on the employee training record form. If discussion at the meeting identifies a suspected safety or health hazard, a copy of the form must be forwarded to the site superintendent.

Various types of reports are necessary to meet OSHA record keeping requirements, insurance carriers, and other government regulatory agencies. Additionally, some clients may require additional site record keeping requirements.

The Name of Company has established uniform record keeping procedures for all company worksites/jobsites to measure the overall safety and health performance of each project.

OSHA RECORDS

OSHA requires Name of Company to record and maintain injury and illness records. These records are used by management to evaluate the effectiveness of this safety and health program. The safety director shall be responsible for following the OSHA record keeping regulations listed here:

- Obtain a report on every injury or illness requiring medical treatment.
- Record each injury or illness on the OSHA form "Log of Work-Related Injuries and Illnesses" (Form 300).
- Prepare a supplementary record of the occupational injuries and illnesses within 7 days of the incident on an employer's "Injury or Illness Incident Report" (Supplementary Record, Form 301).
- Prepare the "Summary of Work-Related Injuries and Illnesses" OSHA form 300A, post it no later than February 1, and keep it posted where employees can see it until April 30 of each year as well as provide copies as required or requested.
- Maintain these records in company files for 5 years.

MEDICAL/EXPOSURE RECORDS

Medical/exposure records will be maintained for 30 years from the time of the end of an employee's employment unless a different retention period is specified by a specific standard. These records are confidential information and will remain in the custody of the safety director. Information from an employee's medical record will be disclosed only to the employee or his or her designated representative after written consent from the employee.

All employees will be informed by posted notice of the existence, location, and availability of medical/exposure records at the time of initial employment and at least annually thereafter. Name/Title of Individual is responsible for maintaining and providing access to these records.

TRAINING RECORDS

Training records will be maintained in each employee's personnel file and available for review upon request. Experience indicates that supervisors/foremen who receive basic first aid and CPR training are much more safety conscious and usually have better crew safety performance records. Therefore, all field supervisory personnel will be required to attend basic first aid and CPR training unless they possess a valid first aid and CPR card issued in their name.

Each Name of Company facility/worksite/jobsite will have adequate first-aid supplies and certified, trained personnel available for the treatment of personnel injured on the job. It is also imperative that all treatments be documented in the construction first-aid log. Prompt medical attention should be sought for any serious injury or if there is doubt of an employee's condition.

FIRST-AID SUPPLIES

First-aid supplies will be available and in serviceable condition at all company worksites. Items that must be kept sterile in the first-aid kit shall be contained in individual packaging. All first-aid kits will contain, but not be limited to, the following items:

- 1 pkg of adhesive bandages, 1 in. (16 per pkg)
- 1 pkg of bandage compress, 4 in. (1 per pkg)
- 1 pkg of scissors and tweezers (1 each per pkg)
- 1 pkg of triangular bandage, 40 in. (1 per pkg)
- 1 pkg of antiseptic soap or pads (3 per pkg)

MEDICAL SERVICES

Each Name of Company worksite will have medical services available either on the worksite or at a location nearby. Emergency phone numbers will be posted on the jobsite for employees to call in the event of an injury or accident on the worksite. Nurses will be available from _____ a.m. until _____ p.m. to respond to medical emergencies. First aid will be available from the Name of Fire Department at all other times.

JOBSITE FIRST-AID LOG

A first-aid log should be maintained in the Name of Company first-aid facility. This log should reflect the following information:

- Injured employee's name
- Immediate supervisor
- Date and time of injury
- Nature of the injury
- Injured employee's job or craft
- Treatment rendered and disposition of employee (returned to work or sent for medical attention)

EMERGENCY PROCEDURES

All employees will be provided with the locations of the first-aid stations on each worksite/jobsite. Instructions for using first-aid equipment are located in each station. In the event of an emergency, employees should contact any supervisor or individual who is trained in first aid. Supervisors and employees trained in first aid will be visible by a first-aid emblem on their hard hat or jacket.

Fire

Fire is one of the most hazardous situations encountered on a worksite/jobsite because of the potential for large losses. Prompt reaction to, and rapid suppression of, any fire is essential. Name of Company will develop a fire protection program for each worksite/jobsite. The program shall provide for effective firefighting equipment to be available without delay and designed to effectively meet all fire hazards as they occur. In addition, each fire protection program shall require that

- All firefighting equipment be conspicuously located and readily available at all times
- All firefighting equipment be inspected and maintained in operating condition
- All fire protection equipment be inspected no less than once monthly with documentation maintained for each piece of equipment inspected
- Discharged extinguishers or damaged equipment be immediately removed from service and replaced with operable equipment
- All supervisors and employees seek out potential fire hazards and coordinate their abatement as rapidly as possible
- Each individual assigned safety responsibilities receive the necessary training to properly recognize fire hazards, inspect and maintain fire extinguishers, and ensure their proper use
- A trained and equipped firefighting brigade be established, as warranted by the project, to assure adequate protection to life

Evacuation

Some emergencies may require company personnel to evacuate the worksite/jobsite. In the event of an emergency that requires evacuation from the workplace, all employees are required to go the area adjacent to the project that has been designated as the safe area. The safe area for this project is located: Description of location. Employers should have a written evacuation plan and it should be drilled or practiced.

SUMMARY

This is a sample written safety and health program and should be used as a template that would be added to or revised to meet the specific needs of a company. No one example such as this one would meet the needs of all companies.

Hazard Identification Tool

The entries in this checklist can be used to determine the types of hazards that exist in your workplace. Check the appropriate line if you identify a hazard.

ELECTRICAL

_____ Is there any electrical equipment used in this operation?
_____ Is there any equipment that is not listed in a nationally recognized testing laboratory?
_____ Is there any equipment new since October 1, 1999 or equipment that has been modified?
_____ Are there pieces of equipment that were purchased before October 1, 1999?
_____ Has this equipment been in use for less than 5 years?
_____ Has this equipment been in use continuously for more than 5 years?
_____ Have incidents occurred that resulted in injury or damaged equipment recently?
_____ Do these operations involve any electrical equipment that was built locally?
_____ Does this operation involve work on or near exposed conductors with voltage less than 50 V and power less than 1000 W with stored energy less than 10 J, or voltage greater than 50 V and current less than 5 mA with stored energy less than 10 J?
    _____ Are these conductors de-energized?
    _____ Are these conductors energized during the work (e.g., connecting test probes, signal tracing, and voltage measurements)?
_____ Does this operation involves work on or near exposed conductors with voltage less than 50 V and power greater than 1000 W with stored energy less than 10 J, or voltage from 50 to 250 V and current greater than 5 mA with stored energy less than 10 J, or voltage greater than 250 V and current less than 500 A with stored energy less than 10 J?
    _____ Are these conductors de-energized?
    _____ Are these conductors energized during diagnostics and testing (e.g., connecting test probes, signal tracing, and voltage measurements)?
    _____ Are these conductors energized during the work?
_____ Does this operation involve work on or near exposed conductors with voltage greater than 250 V and current greater than 500 A or stored energy greater than 10 J?
    _____ Are these conductors de-energized?
    _____ Are these conductors energized during diagnostics and testing (e.g., connecting test probes, signal tracing, and voltage measurements)?
    _____ Are these conductors energized during the work?

MACHINES AND EQUIPMENT

_____ Are there any mechanical hazards such as motors, pulleys, machinery or shop equipment, fork-lifts, hoists and cranes, or sources of kinetic or potential mechanical energy present in this operation?

_____ Does this operation include the use of typical shop equipment?

_____ Does this operation include the use of robotics?

_____ Does this operation include the use of a hoist, crane, or rigging?

_____ Does this operation use a crane?

_____ Does this operation use a hoist?

_____ Does this operation include the use of rigging?

_____ Does the rigging equipment experience a continuous static load?

_____ Does this operation include the use of hydraulic or pneumatic lifts?

_____ Does this operation include the use of a forklift?

HAND TOOLS

_____ Do these operations involve the use of electrically powered hand tools?

_____ Do these operations involve the use of pneumatically powered hand tools?

CONFINED SPACES

_____ Does this operation include any space that might meet the definition of a confined space?

_____ Will personnel be required to enter confined spaces?

ELEVATED WORK AREAS

_____ Could personnel be required to perform their duties from an elevated position (including ladders)?

WELDING AND CUTTING

_____ Will welding or cutting or spark and flame producing operations be conducted in association with this operation?

TOXIC OR HAZARDOUS CHEMICALS

_____ Are there any chemicals or toxic materials (including wastes) handled, generated, used, stored, or processed in this operation?

_____ Are any chemicals or chemical wastes used, stored, or generated in this operation either known or suspected human carcinogens?

_____ Are any category 1 chemicals used, stored, or generated in this operation?

_____ Are any of the chemicals used, stored, or generated in this operation pyrophoric?

_____ Does this operation include the use, storage, or generation of cyanide or hydrofluoric acid?

_____ Does this operation involve the use, storage, or generation of peroxide-forming chemicals, shock-sensitive chemicals, or picric acid?

_____ Does this operation involve the use, storage, or generation of toxic or highly toxic gases?

_____ Does this operation use, generate, or store flammable or combustible gases, liquids, or solids?

_____ Does this operation involve the use of hydrogen gas?

_____ Does this operation use, generate, or store chemical sensitizers?

_____ Do these operations have the potential for skin absorption of toxic chemicals and wastes?

_____ Are multiple chemicals (chemicals used or mixed together) or chemical procedures used in the chemical work area?

_____ Do you have any tasks where chemicals are mixed that will create an explosive mixture?

_____ Does your operation involve an exothermic chemical reaction? (Example: polymerization.)

_____ Does your operation involve an endothermic chemical reaction? (Example: pyrolysis.)

_____ Will this operation involve the transportation of chemicals?

_____ Does this operation involve the use, storage, or generation of caustic or corrosive chemicals or wastes?

_____ Does this operation use, generate, or store chemicals, which are reproductive hazards?

_____ Will this operation involve the use of cryogens?

_____ Will this operation involve the use of beryllium—other than articles made of beryllium or that contain beryllium?

RADIATION

_____ Do devices found in this workplace release or give off any type of radiation?

_____ Are radiation-generating devices (RGDs) capable of creating a high radiation area (>100 mrem/h at 30 cm)?

_____ Are RGDs capable of creating a radiation area?

_____ Does the RGD only produce radiation incidental to its primary function (such as electron microscopes, electron beam welders, and ion implantation equipment)?

_____ Does this operation use RGDs that are built locally or commercially available units that have been modified?

_____ Is the RGD an intentional x-ray generating device, which produces radiation as part of the primary function (i.e., x-ray diffractometers, x-ray machine).

_____ Is the device built locally or been modified or is it used outside design specifications?

RADIOACTIVE MATERIALS

_____ Are radioactive materials (including sealed sources and wastes) generated, handled, processed, used, or stored?

_____ Are radioactive materials part of the normal operation in this workplace?

_____ Does this operation involve radionuclides listed in the radionuclide threshold table in amounts that exceed the quantity listed?

_____ Are radioactive materials used in the processes in this workplace dispersible?

_____ Will any radioactive material or waste be transported as a result of this operation?

_____ Does this operation involve any accountable sources?

NONIONIZING RADIATION

_____ Do you work with any of the following nonionizing radiation (NIR) sources:

_____ Permanently installed radio frequency micro wave (RFMW) gear capable of radiating over 1 W into an open area at frequencies between 3 kHz and 300 GHz or of emitting over 100 W if the output is normally completely enclosed by coaxial cables, waveguides, or dummy or real loads;

_____ Satellite and permanently installed communications transmitters (not receivers).

_____ Portable walkie-talkie communications sets capable of radiating over 7 W at frequencies between 100 kHz and 450 MHz, and over 7 (450/f) W at frequencies between 450 MHz and 1.0 GHz (f in MHz).

_____ Induction heaters. (Microwave ovens used as a household appliance, cellular phones, video display terminals, and radar speed guns are exempt.)

_____ Any equipment that would expose personnel to high levels of subradio frequency electric or magnetic fields including static and DC magnetic fields.

_____ Any equipment that would expose personnel to high levels of visible light and near infrared (400–300 nm) (>1 cd/cm^2), infrared (770–3000 nm) (>10 w/cm^2), or ultraviolet (UV) radiation (180–400 nm) (see table for allowable limits).

_____ Any infrared heat lamp or any near-infrared source where a strong visual stimulus is absent (luminance of less than E-06 cd/cm^2).

LASERS

_____ Does this operation involve the use of lasers?

_____ Do personnel use or have the potential to be exposed to class 3b or 4 lasers?

_____ Does this operation involve class 2 or 3a lasers?

_____ Does this operation involve a class 1 laser with embedded 3b or 4 lasers?

_____ Have any of the lasers involved in this operation been built locally or have any commercially available lasers been modified?

_____ Are laser dyes used in this operation?

EXPLOSIVES

_____ Are there any explosives (including explosive waste) handled, processed, used, or stored?

_____ Does this operation involve the transportation of explosives or explosive wastes?

THERMAL

_____ Are there any sources of thermal hazards, other than commercially available units such as soldering irons, hot plates, small quantity (<0.5 gal) of cryogenics, etc. that are less than −1 °C (30 °F) or greater than 54 °C (130 °F)?

PRESSURE

_____ Are there any pressure sources to be considered such as compressed gas cylinders, pressure vessels, hydraulic systems, vacuum systems, etc. (excluding house supplied sources) in this operation?

_____ Are any parts of any pressure systems operating in the operation at a positive pressure greater than 15 psig (i.e., 15 psi above local atmospheric pressure) and having a potential stored energy greater than 1.0 kJ?
_____ Are there brittle materials utilized in this system?
_____ Is there a pressure system in the operation intended for both negative (i.e., vacuum) and positive pressures and capable of operating at greater than 1.0 kJ positive stored energy (i.e., under positive pressure) and not equipped with a pressure relief device set at 15 psig (positive) or lower?
_____ Are there brittle materials utilized in this system?
_____ Is there a pressure system present in the operation involving cryogenic system or dewar installation equipped with a pressure relief device set above 15 psig, regardless of the estimated amount of potential stored energy?
_____ Are there brittle materials utilized in this system?
_____ Is there a pressure system present in the operation involving a transportable, commercially available 160–200 L liquid nitrogen dewar with a pressure relief device set above 30 psig?
_____ Does your operation involve a high-hazard pressure system?

NOISE

_____ Are there any sources of excessive noise (e.g., such that you have to shout at a distance of 3 ft. to communicate to a coworker or louder than busy traffic) involved in this operation?

OTHER HAZARDS

_____ Are there any additional hazards, not mentioned above, that should be considered such as biological hazards, firearms, or ergonomics?

BIOLOGICAL

_____ Could a worker be exposed to any biological hazard including handling of human body fluids, human tissues, or mouse droppings?

ERGONOMICS

_____ Do personnel perform functions that involve repetitive motion, excessive force or vibration, lifting, or other ergonomic concerns?

TEMPERATURE EXTREMES

_____ Will personnel be required to perform this operation in extreme climates or temperatures?

MORE HAZARDS

_____ Are you aware of any other hazardous conditions or potential sources of hazards that have not previously been addressed by this system that you feel deserve further consideration?

_____ Does this operation involve any human interfaces critical to safety (e.g., communicating chemical recipe information to the user of an etching solution or communicating maximum load to be used on a pressure cell)?

_____ Does this operation involve the use of equipment, tools, or materials outside of the design specifications or outside of the manufacturer's recommendations or the use of equipment or apparatus built locally?

_____ Will this operation be unattended (operating with or without personnel in attendance) or operated after normal working hours (6 p.m. to 7 a.m.) or operate with a sole attendant (working alone)? Alternatively, require special attention if left unexpectedly for long periods of time (e.g., need cryogen refilled)?

Will this operation be left unattended?

      _____ Will the operation require work outside normal working hours?

      _____ Will this operation require two-person rule?

      _____ Will this operation require special attention in the event it is left unattended unexpectedly for long periods of time?

ENVIRONMENTAL

_____ Are there any environmental concerns, such as wastes, air or wastewater discharges, or any waste generated in a radiological-controlled area, involved with this operation?

_____ Will this operation generate any hazardous wastes, or will personnel be required to handle hazardous waste?

_____ Will any acutely hazardous wastes be generated, handled, or used?

_____ Will this operation generate any radioactive waste, or will personnel be required to handle radioactive waste?

_____ Will this operation generate any mixed waste, or will personnel be required to handle mixed waste?

_____ Will these operations generate any infectious or biohazardous waste, or will personnel be required to handle infectious or biohazardous waste?

_____ Will these operations generate administratively controlled waste?

_____ Will this operation generate air emissions or wastewater discharge?

_____ Will *any* waste (radioactive, hazardous, mixed, sanitary, etc.) be produced in a radiological controlled area as result of this operation?

_____ Is a National Environmental Policy Act (NEPA) evaluation needed for this operation?

CONTROLS

_____ Are there any controls (i.e., ventilation, fume hoods, interlocks, personal protective equipment, HEPA filters or vacuum cleaners, and medical monitoring) associated with this operation?

_____ Is any local ventilation used in this operation?

_____ Are interlocks used in this operation?

_____ Is any personal protective equipment used in this operation?

_____ Are there potential hazards that need to use lockout and tagout procedures?

_____ Are protective gloves necessary to perform this operation?

_____ Are HEPA filters in place and used?

      _____ Are ventilation systems used?

      _____ Are HEPA vacuum cleaners used?

_____ Is any medical monitoring required for this operation?

_____ Will respiratory protection be required for this operation?

OTHER SYSTEMS

Do you rely on any facility systems (listed as subquestions) to provide safety controls for your operations?

_____ Breathable air (BTF only)
_____ Building physical structure (including drainage)
_____ Compressed air
_____ Compressed gas
_____ Chilled water
_____ Deionized or demineralized water
_____ Differential pressure monitors
_____ Electric power (includes grounding)
_____ Fire protection
_____ Hoists and cranes
_____ Heating water
_____ Hydrogen gas (including alarm and monitoring)
_____ Industrial liquid waste (includes radioactive liquid waste)
_____ Lightning protection system
_____ Liquid nitrogen
_____ Nonpotable water
_____ Oxygen monitoring system
_____ Public address
_____ Potable water
_____ Process cooling water (including circulating cooling water)
_____ Sanitary sewer
_____ Steam
_____ Utility gas (natural gas)
_____ Vacuum
_____ Ventilation supply or exhaust

Safety and Health Audit Instrument

These checklists are by no means all-inclusive. You should add to them or delete portions or items that do not apply to your operations; however, carefully consider each item as you come to it and then make your decision. You will also need to refer to Occupational Safety and Health Administration (OSHA) standards for complete and specific standards that may apply to your situation. (Note: These checklists are typical for general industry but not for construction or maritime.) You can find a similar checklist for construction in the *Handbook of OSHA Construction Safety and Health*, 2nd edn., published by CRC Press/Taylor & Francis.

CONTENTS

Means of exit
Medical services and first aid
Nonionizing radiation
Personal protective equipment (PPE)
Pressure vessels
Rigging
Scaffolding
Security
Slips, trips, and falls
Violence
Walking–Working surfaces
Welding and cutting

Biological Hazards

Yes ☐   No ☐   Are there biological organisms present?
Yes ☐   No ☐   Are any biologicals potential pathogens?
Yes ☐   No ☐   Have all biologicals been accounted for?
Yes ☐   No ☐   Have the biological risks been evaluated?
Yes ☐   No ☐   Are there hygiene facilities present?
Yes ☐   No ☐   Is hygiene practiced?
Yes ☐   No ☐   Do specific methods for handling of organisms exist?
Yes ☐   No ☐   Are biohazard procedures in use?
Yes ☐   No ☐   Have workers been trained regarding biological
                        hazards?
Yes ☐   No ☐   Are biological warning signs posted?
Yes ☐   No ☐   Do biohazard warning containers exist?
Yes ☐   No ☐   Are there procedures for disposal of biologicals?
Yes ☐   No ☐   Are laboratory facilities where biologicals
                        exist secure?

Bloodborne Pathogens

Is there a written exposure control plan consisting of the following:
Yes ☐   No ☐   List of employees whose job exposes them to bloodborne diseases.
Yes ☐   No ☐   List of all tasks that present exposure potential.
Yes ☐   No ☐   Procedure for evaluating exposure potential.
Yes ☐   No ☐   Do all personnel take precautions to prevent blood and body fluid contact among
                        employees?
Yes ☐   No ☐   Are hand washing facilities provided that are easily accessible to all employees?
Yes ☐   No ☐   Are contaminated needles bent or recapped by hand?
Yes ☐   No ☐   Are contaminated needles disposed of in approved medical waste containers?
Yes ☐   No ☐   When there is an exposure hazard, is the following allowed eating: drinking,
                        smoking, applying cosmetics or lip balm, or handling contact lenses?
Yes ☐   No ☐   Is food and drink stored in areas where blood or any other body fluid is stored?
Yes ☐   No ☐   Is suction pipetting of any body fluid prohibited?
Yes ☐   No ☐   Are containers specifically designed for the fluid stored in it?
Yes ☐   No ☐   Are containers labeled properly before storage or transfer?
Yes ☐   No ☐   Are labels placed on any container that contains or may contain contaminated
                        items or waste products?

Yes ☐   No ☐   If any contact with blood or infectious material is possible, does the employer provide PPE at No charge?

Yes ☐   No ☐   Is the employer responsible for the maintenance and repair of all PPE?

Yes ☐   No ☐   Is any item that contacts blood or any other body fluid disinfected or discarded?

Yes ☐   No ☐   Do employers ensure that the work site is in a clean and sanitary condition?

Yes ☐   No ☐   Are proper disinfectants that kill pathogens used to clean working surfaces, bins, or other areas where contamination may occur?

Yes ☐   No ☐   Is broken glassware picked up with a device and not using unprotected hands?

Yes ☐   No ☐   Are hepatitis B vaccinations made available to employees who may be exposed to the virus during employment?

Yes ☐   No ☐   Has the employer provided an occupational exposure training program for all employees during working hours?

Yes ☐   No ☐   Are any exposures noted and kept in employees' medical record?

Chemicals

Yes ☐   No ☐   If hazardous substances are used in your processes, do you have a medical or biological monitoring system in operation?

Yes ☐   No ☐   Are you familiar with threshold limit values or permissible exposure limits?

Yes ☐   No ☐   Are limits of airborne contaminants and physical agents used in your workplace?

Yes ☐   No ☐   Have control procedures been instituted for hazardous materials, where appropriate, such as respirators, ventilation systems, and handling practices?

Yes ☐   No ☐   Whenever possible, are hazardous substances handled in properly designed and exhausted booths or similar locations?

Yes ☐   No ☐   Are there written standard operating procedures (SOPs) for the selection and use of respirators where needed?

Yes ☐   No ☐   If you have a respirator protection program, are your employees instructed on the correct usage and limitations of the respirators?

Yes ☐   No ☐   Are the respirators National Institute for Occupational Safety and Health (NIOSH) approved for this particular application?

Yes ☐   No ☐   Are they regularly inspected and cleaned, sanitized, and maintained?

Yes ☐   No ☐   If hazardous substances are used in your processes, do you have a medical or biological monitoring system in operation?

Yes ☐   No ☐   Do you use general dilution or local exhaust ventilation systems to control dusts, vapors, gases, fumes, smoke, solvents, or mists that may be generated in your workplace?

Yes ☐   No ☐   Is ventilation equipment provided for removal of contaminants from such operations as production grinding, buffing, spray painting, and vapor degreasing, and is it operating properly?

Yes ☐   No ☐   Do employees complain about dizziness, headaches, nausea, irritation, or other factors of discomfort when they use solvents or other chemicals?

Yes ☐   No ☐   Is there a dermatitis problem?

Yes ☐   No ☐   Do employees complain about dryness, irritation, or sensitization of the skin?

Yes ☐   No ☐   Are combustion engines used and is carbon monoxide kept within acceptable levels?

Yes ☐   No ☐   Is vacuuming used, rather than blowing or sweeping dusts, whenever possible for cleanup?

Yes ☐   No ☐   Have you considered the use of an industrial hygienist or environmentalist?

Compressed Gas Cylinders

Yes ☐   No ☐   Are CGCs kept away from radiators and other sources of heat?

Yes ☐   No ☐   Are CGCs stored in well-ventilated, dry locations at least 20 ft. away from materials such as oil, grease, excelsior, reserve stocks of carbide, acetylene, or other fuels as they are likely to cause acceleration of fires?

Yes ☐   No ☐   Are CGCs stored only in assigned areas?

Yes ☐   No ☐   Are CGCs stored away from elevators, stairs, and gangways?

Yes ☐   No ☐   Are CGCs stored in areas where they will not be dropped, knocked over, or tampered with?

Yes ☐   No ☐   Are CGCs stored in areas with poor ventilation?

Yes ☐   No ☐   Are storage areas marked with signs such as "OXYGEN, NO SMOKING, or NO OPEN FLAMES?"

Yes ☐   No ☐   Are CGCs stored outside generator houses?

Yes ☐   No ☐   Do storage areas have wood and grass cut back within 15 ft.?

Yes ☐   No ☐   Are CGCs secured to prevent falling?

Yes ☐   No ☐   Are stored CGCs in a vertical position?

Yes ☐   No ☐   Are protective caps in place at all times except when in use?

Yes ☐   No ☐   Are threads on cap or cylinder lubricated?

Yes ☐   No ☐   Are all CGCs legibly marked for the purpose of identifying the gas content with the chemical or trade name of the gas?

Yes ☐   No ☐   Are the markings on CGCs by stenciling, stamping, or labeling?

Yes ☐   No ☐   Are markings located on the slanted area directly below the cap?

Yes ☐   No ☐   Does each employee determine that CGCs are in a safe condition by means of a visual inspection?

Yes ☐   No ☐   Is each portable tank and all piping, valves, and accessories visually inspected at intervals not exceeding 2.5 years?

Yes ☐   No ☐   Are inspections conducted by the owner, agent, or approved agency?

Yes ☐   No ☐   On insulated tanks, is the insulation not removed if, in the opinion of the person performing the visual inspection, external corrosion is likely to be negligible?

Yes ☐   No ☐   If evidence of any unsafe condition is discovered, is the portable tank not returned to service until it meets all corrective standards?

Confined Spaces

Yes ☐   No ☐   Are confined spaces thoroughly emptied of any corrosive or hazardous substances, such as acids or caustics, before entry?

Yes ☐   No ☐   Are all lines to a confined space, containing inert, toxic, flammable, or corrosive materials, valved off and blanked or disconnected and separated before entry?

Yes ☐   No ☐   Are all impellers, agitators, or other moving parts and equipment inside confined spaces locked out if they present a hazard?

Yes ☐   No ☐   Is either natural or mechanical ventilation provided before confined space entry?

Yes ☐   No ☐   Are appropriate atmospheric tests performed to check for oxygen deficiency, toxic substances, and explosive concentrations in the confined space before entry?

Yes ☐   No ☐   Is adequate illumination provided for the work to be performed in the confined space?

Yes ☐   No ☐   Is the atmosphere inside the confined space frequently tested or continuously monitored during conduct of work? Is there an assigned safety standby employee outside of the confined space, when required, whose sole responsibility is to watch the work in progress, sound an alarm if necessary, and render assistance?

Yes ☐ No ☐ Is the standby employee appropriately trained and equipped to handle an emergency?

Yes ☐ No ☐ Is the standby employee or other employees prohibited from entering the confined space without lifelines and respiratory equipment if there is any question as to the cause of an emergency?

Yes ☐ No ☐ Is approved respiratory equipment required if the atmosphere inside the confined space cannot be made acceptable?

Yes ☐ No ☐ Is all portable electrical equipment used inside confined spaces either grounded and insulated, or equipped with ground-fault protection?

Yes ☐ No ☐ Before gas welding or burning is started in a confined space, are hoses checked for leaks, compressed gas bottles forbidden inside of the confined space, torches lighted only outside of the confined area, and the confined area tested for an explosive atmosphere each time before a lighted torch is to be taken into the confined space?

Yes ☐ No ☐ If employees will be using oxygen-consuming equipment, such as salamanders, torches, and furnaces, in a confined space is sufficient air provided to assure combustion without reducing the oxygen concentration of the atmosphere below 19.5% by volume?

Yes ☐ No ☐ Whenever combustion-type equipment is used in a confined space, are provisions made to ensure that the exhaust gases are vented outside of the enclosure?

Yes ☐ No ☐ Is each confined space checked for decaying vegetation or animal matter, which may produce methane?

Yes ☐ No ☐ Is the confined space checked for possible industrial waste that could contain toxic properties?

Yes ☐ No ☐ If the confined space is below the ground and near areas where motor vehicles will be operating, is it possible for vehicle exhaust or carbon monoxide to enter the space?

Crane Safety

Yes ☐ No ☐ Is a wind indicator or wind sock placed on all outside cranes and is it visible to the operator?

Yes ☐ No ☐ Is the rated load capacity of the crane on the crane at all times and visible from the ground?

Yes ☐ No ☐ Is a fire extinguisher of the appropriate size and type on the crane at all times?

Yes ☐ No ☐ Are all walking surfaces of the nonslip type?

Yes ☐ No ☐ Do all ladders, stairs, and railings comply with requirements of the regulations?

Yes ☐ No ☐ Are all moving parts such as gears, set screws, moving components, or anything dangerous guarded?

Yes ☐ No ☐ Is each independent hoisting unit provided with at least one self-setting brake or holding brake?

Yes ☐ No ☐ Do all electrical equipment and wiring components comply with electrical regulations?

Yes ☐ No ☐ Do all ropes, chains, and cables meet the manufacturer's recommendations?

Yes ☐ No ☐ Is any crane that has a power traveling mechanism equipped with a warning signal to include a visual warning light?

Yes ☐ No ☐ Has the employer ensured all operators of cranes are properly trained?

Yes ☐ No ☐ Are the danger areas under the load and any area below where the load will travel marked and blocked off to prevent other employees from walking under suspended loads?

Yes ☐ No ☐ Are all passages and walkways safe from movement of the crane?

Yes ☐ No ☐ Are cones, warning tape, or guards erected?

Yes ☐ No ☐ Does the cab allow the operator to see the load at all times?

Yes ☐ No ☐ Is the cab illuminated to allow operator to see sufficiently?

Yes ☐ No ☐ Is there a preventive maintenance program based on the manufacturer's recommendations?

Yes ☐ No ☐ Is there a location provided to perform maintenance on cranes where it causes the least interference with surrounding operations?

Yes ☐ No ☐ During maintenance are controls in the off position?

Yes ☐ No ☐ Is the main switch locked out and tagged out?

Yes ☐ No ☐ Are signs posted on the crane, and on the hook where it can be seen from the floor, stating "Out of Order?"

Yes ☐ No ☐ Do cranes operating on the same runway as an idle crane have rail stops or suitable means to prevent contact of cranes?

Yes ☐ No ☐ Are all guards in place, safety devices reactivated, and maintenance equipment removed before operating a crane?

Yes ☐ No ☐ Are cranes inspected daily (before each use), monthly, and quarterly?

Yes ☐ No ☐ Are they inspected annually by an outside expert (e.g., manufacturer's representative)?

Yes ☐ No ☐ Is a certificate of the annual inspection retained?

Yes ☐ No ☐ Does the manufacturer's representative inspect cranes annually and retain the certificate?

Before each use, are the following tested:

Yes ☐ No ☐ Hoisting and lowering devices.

Yes ☐ No ☐ Trolley travel.

Yes ☐ No ☐ Bridge travel.

Yes ☐ No ☐ Locking or safety devices.

Yes ☐ No ☐ Are all grooves inspected to detect surface defects that may damage ropes?

Yes ☐ No ☐ Are all ropes inspected at least once a month?

Yes ☐ No ☐ Are ropes, cables, or chains inspected for kinks before lifting?

Yes ☐ No ☐ Has a preventative maintenance program based on the manufacturer's recommendations been established?

Yes ☐ No ☐ Are all adjustments or repairs done by a qualified person?

Driver and Vehicle Safety

Yes ☐ No ☐ Are drivers selected carefully?

Yes ☐ No ☐ Are motor vehicle record checks required for all drivers?

Yes ☐ No ☐ Are all drivers trained?

Yes ☐ No ☐ Are all drivers tested and road tested?

Yes ☐ No ☐ Is seatbelt use mandatory for all drivers?

Yes ☐ No ☐ Are all drivers of age and meet minimum driving experience levels?

Yes ☐ No ☐ Do drivers hold current licenses appropriate to the vehicles they drive?

Yes ☐ No ☐ Have drivers successfully completed a driver training course specifically designed for the vehicle driven (car, van, bus, or truck)?

Yes ☐ No ☐ Does the course include documented classroom curriculum, a written exam, documented minimum of 3 h of driving practice (additional training time for trailers may be required), and a documented driving exam that demonstrates competency in each class of vehicle?

Yes ☐ No ☐ Does the driver possess current certifications in first aid and cardiopulmonary resuscitation (CPR)?

Yes ☐ No ☐ Does the organization have a written policy stating that the maximum speed limit for all vehicles is 55 mph, regardless of posted speed limit?

Yes ☐ No ☐ Is the weight limit of passengers and equipment limited to 75% of the manufacturer's recommended gross carrying weight for passengers and trailers?

Yes ☐ No ☐ Do drivers using prescription medications have written medical approval by a physician (current within the past 12 months) allowing them to drive the designated vehicle?

Yes ☐ No ☐ Is there a designated staff person responsible for transportation program, vehicle safety, and driver selection administration? Responsibilities would include but not be limited to keeping vehicle maintenance records, keeping records of annual vehicle safety inspections (vehicles must be inspected by a certified mechanic), and performing driver training.

Yes ☐ No ☐ Are vehicles inspected daily, and is the inspection documented before operation?

Electrical

Yes ☐ No ☐ Do you specify compliance with OSHA for all contract electrical work?

Yes ☐ No ☐ Are all employees required to report as soon as practicable any obvious hazard to life or property observed in connection with electrical equipment or lines?

Yes ☐ No ☐ Are employees instructed to make preliminary inspections and appropriate tests to determine what conditions exist before starting work on electrical equipment or lines?

Yes ☐ No ☐ When electrical equipment or lines are to be serviced, maintained, or adjusted, are necessary switches opened, locked out, and tagged whenever possible?

Yes ☐ No ☐ Are portable electrical tools and equipment grounded or of the double-insulated type?

Yes ☐ No ☐ Are electrical appliances such as vacuum cleaners, polishers, and vending machines grounded?

Yes ☐ No ☐ Do extension cords used have a grounding conductor?

Yes ☐ No ☐ Are multiple plug adaptors prohibited?

Yes ☐ No ☐ Are ground-fault circuit interrupters installed on each temporary 15 or 20A, 120V AC circuit at locations where construction, demolition, modifications, alterations, or excavations are performed?

Yes ☐ No ☐ Are all temporary circuits protected by suitable disconnecting switches or plug connectors at the junction with permanent wiring?

Yes ☐ No ☐ Do you have electrical installations in hazardous dust or vapor areas? If so, do they meet the National Electrical Code (NEC) for hazardous locations?

Yes ☐ No ☐ Is exposed wiring and cords with frayed or deteriorated insulation repaired or replaced promptly?

Yes ☐ No ☐ Are flexible cords and cables free of splices or taps?

Yes ☐ No ☐ Are clamps or other securing means provided on flexible cords or cables at plugs, receptacles, tools, equipment, etc., and is the cord jacket securely held in place? Are all cord, cable, and raceway connections intact and secure?

Yes ☐ No ☐ In wet or damp locations are electrical tools and equipment appropriate for the use or location or otherwise protected?

Yes ☐ No ☐ Is the location of electrical power lines and cables (overhead, underground, underfloor, other side of walls) determined before digging, drilling, or similar work is begun?

Yes ☐ No ☐ Are metal measuring tapes, ropes, handlines, or similar devices with metallic thread woven into the fabric prohibited where they could come in contact with energized parts of equipment or circuit conductors?

Yes ☐   No ☐   Is the use of metal ladders prohibited in areas where the ladder or the person
                using the ladder could come in contact with energized parts of equipment,
                fixtures, or circuit conductors?

Yes ☐   No ☐   Are all disconnecting switches and circuit breakers labeled to indicate their
                use or equipment served?

Yes ☐   No ☐   Are disconnecting means always opened before fuses are replaced?

Yes ☐   No ☐   Do all interior wiring systems include provisions for grounding metal parts
                of electrical raceways, equipment, and enclosures?

Yes ☐   No ☐   Are all electrical raceways and enclosures securely fastened in place?

Yes ☐   No ☐   Are all energized parts of electrical circuits and equipment guarded against
                accidental contact by approved cabinets or enclosures?

Yes ☐   No ☐   Is sufficient access and working space provided and maintained about all
                electrical equipment to permit ready and safe operations and maintenance?

Yes ☐   No ☐   Are all unused openings (including conduit knockouts) in electrical enclosures
                and fittings closed with appropriate covers, plugs, or plates?

Yes ☐   No ☐   Are electrical enclosures such as switches, receptacles, and junction boxes
                provided with tight fitting covers or plates?

Yes ☐   No ☐   Are disconnecting switches for electrical motors in excess of 2 hp capable of
                opening the circuit when the motor is in a stalled condition, without exploding?
                (Switches must be horsepower rated equal to or in excess of the motor hp rating.)
                Is low-voltage protection provided in the control device of motors driving
                machines or equipment that could cause probable injury from inadvertent starting?

Yes ☐   No ☐   Is each motor disconnecting switch or circuit breaker located within sight of the
                motor control device?

Yes ☐   No ☐   Is each motor located within sight of its controller or the controller disconnecting
                means capable of being locked in the open position or is a separate
                disconnecting means installed in the circuit within sight of the motor?

Yes ☐   No ☐   Is the controller for each motor in excess of 2 hp rated in horsepower equal to or
                in excess of the rating of the motor it serves?

Yes ☐   No ☐   Are employees who regularly work on or around energized electrical equipment
                or lines instructed in CPR methods?

Yes ☐   No ☐   Are employees prohibited from working alone on energized lines or equipment
                over 600 V?

Emergency Response and Planning

Yes ☐   No ☐   Is there a written emergency response planning that is available
                to all employees?

Yes ☐   No ☐   Is there an established procedure specifically outlining the steps to be taken by all
                employees including route of evacuation, place to meet outside building, and
                designation of person responsible for verifying that employees are all accounted for?

Yes ☐   No ☐   Have proper evacuation procedures been communicated to everyone before the
                need for an actual evacuation, and have those procedures been actively practiced
                in a mock evacuation situation?

Yes ☐   No ☐   Is there an established protocol for determining the need for evacuation?

Yes ☐   No ☐   Is there a designated person responsible for making an evacuation decision?

Yes ☐   No ☐   Is the need for evacuation communicated to employees in such a way that
                everyone (other than those designated as the initial contacts) receives the same
                information at the same time?

Yes ☐ No ☐ In the event of electrical failure, is there a backup system for both broadcasting of messages and lighting of escape routes?

Yes ☐ No ☐ Are established escape routes clearly marked, and are maps posted outlining the entire route?

Yes ☐ No ☐ Are escape routes determined to be the shortest safe route possible, allowing adequate room and number of routes for the number of employees?

Yes ☐ No ☐ Are all emergency exits clearly marked and functioning properly?

Yes ☐ No ☐ Are all escape routes free of clutter and tripping hazards?

Yes ☐ No ☐ Is there adequate emergency lighting along the routes?

Yes ☐ No ☐ Is emergency equipment such as fire extinguishers and flashlights located at predetermined sites along escape routes and is this equipment routinely tested for proper operation?

Yes ☐ No ☐ In the event that employees are required to remain within hallways/stairways of escape route for longer than expected, is there adequate ventilation, temperature control, and some type of communication equipment?

Yes ☐ No ☐ Are all established meeting places outside of the building a reasonably safe distance away?

Yes ☐ No ☐ Is there an established method for verification that all employees have left the building and a way to communicate to emergency personnel the identities and possible locations of those who have not?

Ergonomics
Manual Material Handling

Yes ☐ No ☐ Is there lifting of loads, tools, or pans?

Yes ☐ No ☐ Is there lowering of tools, loads, or parts?

Yes ☐ No ☐ Is there overhead reaching for tools, loads, or parts?

Yes ☐ No ☐ Is there bending at the waist to handle tools, loads, or parts?

Yes ☐ No ☐ Is there twisting at the waist to handle tools, loads, or parts?

Physical-Energy Demands

Yes ☐ No ☐ Do tools and parts weigh more than 10 lb?

Yes ☐ No ☐ Is reaching greater than 20 in.?

Yes ☐ No ☐ Is bending, stooping, or squatting a primary task activity?

Yes ☐ No ☐ Is lifting or lowering loads a primary task activity?

Yes ☐ No ☐ Is walking or carrying loads a primary task activity?

Yes ☐ No ☐ Is stair or ladder climbing with loads a primary task activity?

Yes ☐ No ☐ Is pushing or pulling loads a primary task activity?

Yes ☐ No ☐ Is reaching overhead a primary task activity?

Yes ☐ No ☐ Do any of the above tasks require five or more complete work cycles to be done within a minute?

Yes ☐ No ☐ Do workers complain that rest breaks and fatigue allowances are insufficient?

Other Musculoskeletal Demands

Yes ☐ No ☐ Do manual jobs require frequent, repetitive motions?

Yes ☐ No ☐ Do work postures require frequent bending of the neck, shoulder, elbow, wrist, or finger joints?

Yes ☐ No ☐ For seated work, do reaches for tools and materials exceed 15 in. from the worker's position?

Yes ☐ No ☐ Is the worker unable to change his or her position often?
Yes ☐ No ☐ Does the work involve forceful, quick, or sudden motions?
Yes ☐ No ☐ Does the work involve shock or rapid buildup of forces?
Yes ☐ No ☐ Is finger-pinch gripping used?
Yes ☐ No ☐ Do job postures involve sustained muscle contraction of any limb?

Computer Workstation

Yes ☐ No ☐ Do operators use computer workstations for more than 4 h a day?
Yes ☐ No ☐ Are there complaints of discomfort from those working at these stations?
Yes ☐ No ☐ Is the chair or desk nonadjustable?
Yes ☐ No ☐ Is the display monitor, keyboard, or document holder nonadjustable?
Yes ☐ No ☐ Does lighting cause glare or make the monitor screen hard to read?
Yes ☐ No ☐ Is the room temperature too hot or too cold?
Yes ☐ No ☐ Is there irritating vibration or noise?

Environment

Yes ☐ No ☐ Is the temperature too hot or too cold?
Yes ☐ No ☐ Are the worker's hands exposed to temperatures less than 70 °F?
Yes ☐ No ☐ Is the workplace poorly lit?
Yes ☐ No ☐ Is there glare?
Yes ☐ No ☐ Is there excessive noise that is annoying, distracting, or producing hearing loss?
Yes ☐ No ☐ Is there upper extremity or whole-body vibration?
Yes ☐ No ☐ Is air circulation too high or too low?

General Workplace

Yes ☐ No ☐ Are walkways uneven, slippery, or obstructed?
Yes ☐ No ☐ Is housekeeping poor?
Yes ☐ No ☐ Is there inadequate clearance or accessibility for performing tasks?
Yes ☐ No ☐ Are stairs cluttered or lacking railings?
Yes ☐ No ☐ Is proper footwear worn?

Tools

Yes ☐ No ☐ Is the handle too small or too large?
Yes ☐ No ☐ Does the handle shape cause the operator to bend the wrist to use the tool?
Yes ☐ No ☐ Is the tool hard to access?
Yes ☐ No ☐ Does the tool weigh more than 9 lb?
Yes ☐ No ☐ Does the tool vibrate excessively?
Yes ☐ No ☐ Does the tool cause excessive kickback to the operator?
Yes ☐ No ☐ Does the tool become too hot or too cold?

Gloves

Yes ☐ No ☐ Do the gloves require the worker to use more force when performing job tasks?
Yes ☐ No ☐ Do the gloves provide inadequate protection?
Yes ☐ No ☐ Do the gloves present a hazard of catch points on the tool or in the workplace?

Administration

Yes ☐ No ☐ Is there little worker control over the work process?
Yes ☐ No ☐ Is the task highly repetitive and monotonous?

Yes ☐ No ☐   Does the job involve critical tasks with high accountability and little or
              no tolerance for error?
Yes ☐ No ☐   Are work hours and breaks poorly organized?

Excavations/Trenches

Yes ☐ No ☐   Have workers been briefed on excavation hazards related to work?
Yes ☐ No ☐   What is the area to be excavated?
Yes ☐ No ☐   What method is to be used (for example, a backhoe)?
Yes ☐ No ☐   What is the proposed depth of excavation/trench?
Yes ☐ No ☐   Is the excavation deeper than 4 ft.? Trenches deeper than 4 ft. have extra shoring
              requirements.
Yes ☐ No ☐   Does the excavation require shoring/battering?
Yes ☐ No ☐   If shoring is required, is there adequate supply of suitable support?
Yes ☐ No ☐   Is material delivered to the site before trenching commences?
Yes ☐ No ☐   Have all underground services been located, marked, and precautions taken to
              avoid them?
Yes ☐ No ☐   Has all shoring been erected by a competent person?
Yes ☐ No ☐   Have barricades been placed around excavation/trench 3 ft. from the edge?
Yes ☐ No ☐   Have barricades been tagged with a barricade tag? Have all materials and
              equipment been placed no closer than 2 ft. from the side of the excavation?
Yes ☐ No ☐   Have utilities been isolated to preclude contact?
Yes ☐ No ☐   Does the trenching undermine adjacent structures?
Yes ☐ No ☐   Do workers need to enter the trench?
Yes ☐ No ☐   Are means of egress from the trench provided?
Yes ☐ No ☐   Are emergency procedures in place?
Yes ☐ No ☐   Has the atmosphere in the trench been checked?
Yes ☐ No ☐   Have trenches been checked for water seepage?
Yes ☐ No ☐   Is there a means of worker escape (stairway, ladder, ramp) available within 25 ft.
              lateral travel of all workers in the excavation?
Yes ☐ No ☐   Are there walkways or bridges, with standard guardrails, provided where persons
              are required to cross the excavation/trench?

Fall Protection

Yes ☐ No ☐   Are all connectors made of drop forged, pressed, or formed steel or equivalent
              materials?
Yes ☐ No ☐   Do all connectors have a corrosion-resistant finish?
Yes ☐ No ☐   Are all surfaces and edges smooth to prevent damage to other parts of the fall
              arrest system?
Yes ☐ No ☐   Do all lanyards and vertical lifelines that tie-off one employee have a minimum
              breaking strength of 5000 lb?
Yes ☐ No ☐   Do self-retracting lifelines and lanyards that automatically limit free fall distance
              to 2 ft. or less have components capable of sustaining a minimum static tensile
              load of 3000 lb applied to the device with the lifeline or lanyard in the fully
              extended position?
Yes ☐ No ☐   Do self-retracting lifelines and lanyards that do not limit free fall distance to
              2 ft. or less, rip stitch lanyards, tearing and deforming lanyards have
              components capable of sustaining a minimum static tensile load of 5000 lb
              applied to the device with the lifeline or lanyard in the fully extended
              position?

Yes ☐ No ☐   Are all dee-rings and snap-hooks capable of sustaining a minimum tensile load of 5000 lb?

Yes ☐ No ☐   Have all dee-rings and snap-hooks received 100% proof testing to a minimum tensile load of 3600 lb without cracking, breaking, or permanent deformation?

Yes ☐ No ☐   Are all snap-hooks of compatible size with the member to which they are connected, so as to prevent unintentional disengagement of the snap-hook by depression of the snap-hook keeper by the connected member? Are all snap-hooks a designed locking type that is capable of preventing disengagement of the snap-hook by the contact of the snap-hook keeper with the connected member?

Yes ☐ No ☐   When horizontal lifelines are used, are they designed and installed as part of a complete personal fall arrest system, which maintains a safety factor of at least two under the supervision of a qualified person?

Yes ☐ No ☐   Are anchorages to which personal fall arrest equipment is attached capable of holding 5000 lb per employee attached?

Yes ☐ No ☐   Are all ropes and straps (webbing) used in lanyards, lifelines, and strength components of the body harnesses made of synthetic fibers or wire rope?

Yes ☐ No ☐   When stopping a fall while using a body harness is the maximum arresting force on the employee limited to 1800 lb?

Yes ☐ No ☐   When stopping a fall does the fall arrest system bring the employee to a complete stop and limit the maximum deceleration distance an employee travels to 3.5 ft.?

Yes ☐ No ☐   Does the fall arrest system have sufficient strength to withstand twice the potential impact energy of an employee free falling a distance of 6 ft. or the free fall distance permitted by the system, whichever is less?

Yes ☐ No ☐   When appropriate are criteria and protocols modified to provide proper protection for combined body and tool weight when total weight is over 310 lb?

Yes ☐ No ☐   Are snap-hooks, unless of locking type design and used to prevent disengagement, not attached directly to the webbing, rope, or wire rope?

Yes ☐ No ☐   Are snap-hooks unless the locking type not directly attached to each other?

Yes ☐ No ☐   Are snap-hooks unless the locking type not attached to a dee-ring to which another snap-hook or other connector is already attached?

Yes ☐ No ☐   Are snap-hooks unless the locking type not attached to a horizontal lifeline?

Yes ☐ No ☐   Are snap-hooks unless the locking type not attached to any object that is incompatibly shaped or dimensioned in relation to the snap-hook, such that, the connected object could depress the snap-hook keeper with enough force to release itself?

Yes ☐ No ☐   Are devices used to connect to a horizontal lifeline, which may become a vertical lifeline, capable of locking in either direction on the lifeline?

Yes ☐ No ☐   Is the personal fall arrest system rigged to not allow an employee to free-fall more than 6 ft. or contact a lower level?

Yes ☐ No ☐   Is the body harness attachment point located in the center of the wearer's back, near the shoulders, or above the wearer's head?

Yes ☐ No ☐   When vertical lifelines are used, is each employee provided a separate lifeline?

Yes ☐ No ☐   Is use of the personal fall arrest systems or components restricted to use in the fall arrest system only?

Yes ☐ No ☐   When personal fall arrest systems or components are subjected to impact loading, are they immediately removed from service and not used again for employee protection unless inspected and deemed suitable for reuse by a competent person?

Yes ☐ No ☐ Does the employer provide for prompt rescue of employees in the event of a fall or assure the self-rescue capability of employees?

Yes ☐ No ☐ Are all employees who use the personal fall arrest system trained in its inspection care, use, and system performance?

Yes ☐ No ☐ Are personal fall arrest systems inspected before each use for mildew, wear, damage, and other deterioration?

Yes ☐ No ☐ Are defective components found in the personal fall arrest system removed from service if their strength or functions are adversely affected?

Fire Protection and Prevention

Yes ☐ No ☐ Does the employer provide portable fire extinguishers for small fires?

Yes ☐ No ☐ Are all fire extinguishers clearly marked with symbols that distinctly reflect the type of fire hazard for which they are intended?

Yes ☐ No ☐ Are portable fire extinguishers located where they are readily accessible to employees without subjecting them to possible injury?

Yes ☐ No ☐ Are fire extinguishers fully charged and operable at all times?

Yes ☐ No ☐ Are class A and D fire extinguishers no more than 75 ft. apart?

Yes ☐ No ☐ Are class B fire extinguishers no more than 50 ft. apart?

Yes ☐ No ☐ Are class C fire extinguishers patterned among class A and B extinguishers where a class C fire hazard exists?

Yes ☐ No ☐ Are all fire extinguishers clearly marked with symbols that distinctly reflect the type of fire hazard for which they are intended?

Yes ☐ No ☐ Are protective clothing worn to protect the entire body including respiratory, head, hand, foot, leg, eye, and face?

Yes ☐ No ☐ Are fixed extinguishing systems used on specific fire hazards?

Yes ☐ No ☐ Is an alarm with a delay in place to warn employees before a fixed extinguisher is to be discharged?

Yes ☐ No ☐ Are hazard warning or caution signs posted at the entrance to, and inside, areas protected by systems that use agents known to be hazardous to employee safety and health?

Yes ☐ No ☐ Are fire detection systems installed and maintained to assure best detection of a fire?

Yes ☐ No ☐ Is an employee alarm system installed that is capable of warning every employee of an emergency?

Yes ☐ No ☐ Is the alarm system such that it can be heard above the sound level of the work area?

Yes ☐ No ☐ Are warning lights installed, if there are hearing impaired employees?

Yes ☐ No ☐ Is all firefighting equipment inspected at least annually and records kept?

Yes ☐ No ☐ Are portable fire extinguishers inspected at least monthly and records kept?

Yes ☐ No ☐ Is any damaged equipment removed immediately from service and replaced?

Yes ☐ No ☐ Is hydrostatic testing done on each extinguisher at least every 5 years?

Yes ☐ No ☐ Are fixed extinguishing systems inspected annually by a qualified person?

Yes ☐ No ☐ Are fire detection systems tested monthly if they are battery operated?

Yes ☐ No ☐ Is training provided on the use of portable fire extinguishers and records of attending employees kept?

Yes ☐ No ☐ Is training provided to employees designated to inspect, maintain, operate, or repair fixed extinguishing systems?

Yes ☐ No ☐ Is an annual review training required to keep them up to date?

Yes ☐ No ☐ Are all employees trained to recognize the alarm signals for each emergency (fire, tornado, chemical release, etc.)?

Yes ☐ No ☐ Are employees trained in how to report an emergency, where the alarms are, and how to sound them?

Yes ☐ No ☐ Is training provided on evacuation procedures?

Yes ☐ No ☐ Are drills performed periodically to ensure that employees are aware of their duties?

Yes ☐ No ☐ Is all training conducted by a qualified/competent person?

Yes ☐ No ☐ Has the employer established and maintained a written policy that establishes the existence of a fire brigade.

Yes ☐ No ☐ Does the employer use employees who are physically capable of performing the duties as a member of a fire brigade that may be assigned to them during an emergency?

Yes ☐ No ☐ Is training of the duties provided by the employer before the employee is asked to do any emergency response duties?

Yes ☐ No ☐ Are all fire brigade members trained at least annually, and interior structural fire fighters provided with an education session or training at least quarterly?

Yes ☐ No ☐ Did the employer inform the fire brigade members of special hazards, such as storage and use of flammable liquids and gases, toxic chemicals, radioactive sources, and water-reactive substances that they may encounter during an emergency?

Flammable and Combustible Materials

Yes ☐ No ☐ Are combustible scrap, debris, and waste materials (oily rags, etc.) stored in covered metal receptacles and removed from the worksite promptly?

Yes ☐ No ☐ Is proper storage practiced to minimize the risk of fire including spontaneous combustion?

Yes ☐ No ☐ Are approved containers and tanks used for the storage and handling of flammable and combustible liquids?

Yes ☐ No ☐ Are all connections on drums and combustible liquid piping, vapor, and liquid tight?

Yes ☐ No ☐ Are all flammable liquids kept in closed containers when not in use (e.g., parts cleaning tanks, pans, etc.)?

Yes ☐ No ☐ Are bulk drums of flammable liquids grounded and bonded to containers during dispensing?

Yes ☐ No ☐ Do storage rooms for flammable and combustible liquids have explosion-proof lights?

Yes ☐ No ☐ Do storage rooms for flammable and combustible liquids have mechanical or gravity ventilation?

Yes ☐ No ☐ Is liquefied petroleum gas stored, handled, and used in accordance with safe practices and standards?

Yes ☐ No ☐ Are "NO SMOKING" signs posted on liquefied petroleum gas tanks?

Yes ☐ No ☐ Are liquefied petroleum storage tanks guarded to prevent damage from vehicles?

Yes ☐ No ☐ Are all solvent wastes and flammable liquids kept in fire-resistant, covered containers until they are removed from the worksite?

Yes ☐ No ☐ Is vacuuming used whenever possible rather than blowing or sweeping combustible dust? Are firm separators placed between containers of combustibles or flammables, when stacked one upon another, to assure their support and stability?

Yes ☐ No ☐ Are fuel gas cylinders and oxygen cylinders separated by distance and fire-resistant barriers while in storage?

Yes ☐   No ☐   Are fire extinguishers selected and provided for the types of materials in areas where they are to be used?

Yes ☐   No ☐   Are appropriate fire extinguishers mounted within 75 ft. of outside areas containing flammable liquids and within 10 ft. of any inside storage area for such materials?

Yes ☐   No ☐   Are extinguishers free from obstructions or blockage?

Yes ☐   No ☐   Are all extinguishers serviced, maintained, and tagged at intervals not exceeding 1 year?

Yes ☐   No ☐   Are all extinguishers fully charged and in their designated places?

Yes ☐   No ☐   Where sprinkler systems are permanently installed, are the nozzle heads so directed or arranged that water will not be sprayed into operating electrical switchboards and equipment?

Yes ☐   No ☐   Are "NO SMOKING" signs posted where appropriate in areas where flammable or combustible materials are used or stored?

Yes ☐   No ☐   Are safety cans used for dispensing flammable or combustible liquids at a point of use?

Yes ☐   No ☐   Are all spills of flammable or combustible liquids cleaned up promptly?

Yes ☐   No ☐   Are storage tanks adequately vented to prevent the development of excessive vacuum or pressure as a result of filling, emptying, or atmosphere temperature changes?

Yes ☐   No ☐   Are storage tanks equipped with emergency venting that will relieve excessive internal pressure caused by fire exposure?

Yes ☐   No ☐   Are "NO SMOKING" rules enforced in areas involving storage and use of hazardous materials?

Forklifts

Yes ☐   No ☐   Do all new forklifts meet the American National Standards Institute (ANSI) BS6.1-1969?

Yes ☐   No ☐   Are the ANSI label, load ratings, and any plates in place and visible at all times?

Yes ☐   No ☐   Is each forklift examined before each shift and is an operator checklist completed?

Yes ☐   No ☐   If a forklift needs repair, is defective, or unsafe, is it removed from service?

Yes ☐   No ☐   Are all repairs done by trained, authorized personnel?

Yes ☐   No ☐   Is a copy of the maintenance report kept on file?

Yes ☐   No ☐   Are lockout/tagout procedures used during maintenance?

Yes ☐   No ☐   Are only properly licensed operators allowed to operate forklifts?

Yes ☐   No ☐   Is refresher training conducted yearly?

Yes ☐   No ☐   Are new employees tested despite previous experience?

Yes ☐   No ☐   Are special battery changing areas provided for electric trucks?

Yes ☐   No ☐   Is a hoist or crane provided to lift batteries?

Yes ☐   No ☐   Does proper ventilation exist in areas in which exhaust-releasing forklifts will be operated?

Yes ☐   No ☐   Are riders not allowed on forklifts?

Yes ☐   No ☐   Are forklifts turned off, controls in neutral, fork lowered, and brakes set when the driver is not in the driver's seat?

Yes ☐   No ☐   Do all forklifts have an overhead guard in place?

Yes ☐   No ☐   Are traffic regulations posted in forklift areas and compliance ensured?

Yes ☐   No ☐   Are only safely arranged loads lifted with a forklift?
Yes ☐   No ☐   Is the forklift operated within the rated capacity?
Yes ☐   No ☐   Are forklifts never fueled while they are running?
Yes ☐   No ☐   Are safety devices never allowed to be removed from the forklift?
Yes ☐   No ☐   Is the forklift maintained clean at all times?
Yes ☐   No ☐   Are operators trained for the specific machine that they will be operating?
Yes ☐   No ☐   Is training repeated annually and training materials used
                      for training retained?

Hand and Portable Powered Tools

Hand Tools and Equipment

Yes ☐   No ☐   Are all tools and equipment (both company and employee owned) used by
                      employees at their workplace in good condition?
Yes ☐   No ☐   Are hand tools such as chisels and punches, which develop mushroomed heads
                      during use, reconditioned or replaced as necessary?
Yes ☐   No ☐   Are broken or fractured handles on hammers, axes, and similar equipment
                      replaced promptly?
Yes ☐   No ☐   Are worn or bent wrenches replaced regularly?
Yes ☐   No ☐   Are appropriate handles used on files and similar tools?
Yes ☐   No ☐   Are employees made aware of the hazards caused by faulty or improperly used
                      hand tools?
Yes ☐   No ☐   Are appropriate safety glasses, face shields, etc., used while using hand
                      tools or equipment that might produce flying materials or be subject to
                      breakage?
Yes ☐   No ☐   Are jacks checked periodically to ensure that they are in good operating condition?
Yes ☐   No ☐   Are tool handles wedged tightly in the head of all tools?
Yes ☐   No ☐   Are tool cutting edges kept sharp so the tool will move smoothly without binding
                      or skipping?
Yes ☐   No ☐   Are tools stored in dry, secure locations where they would not be
                      tampered with?
Yes ☐   No ☐   Is eye and face protection used when driving hardened or tempered spuds or
                      nails?

Portable (Power-Operated) Tools and Equipment

Yes ☐   No ☐   Are grinders, saws, and similar equipment provided with appropriate safety
                      guards?
Yes ☐   No ☐   Are power tools used with the correct shield, guard, or attachment, recommended
                      by the manufacturer?
Yes ☐   No ☐   Are portable circular saws equipped with guards above and below the base shoe?
                      Are circular saw guards checked to assure they are not wedged up, thus leaving
                      the lower portion of the blade unguarded?
Yes ☐   No ☐   Are rotating or moving parts of equipment guarded to prevent
                      physical contact?
Yes ☐   No ☐   Are all cord-connected, electrically operated tools and equipment effectively
                      grounded or of the approved double-insulated type?
Yes ☐   No ☐   Are effective guards in place over belts, pulleys, chains, sprockets, on equipment
                      such as concrete mixers and air compressors?

Yes ☐   No ☐   Are portable fans provided with full guards or screens with openings ½ in. or
less?

Yes ☐   No ☐   Is hoisting equipment available and used for lifting heavy objects, and are hoist
ratings and characteristics appropriate for the task?

Yes ☐   No ☐   Are ground-fault circuit interrupters provided on all temporary electrical 15 and
20 A circuits used during periods of construction?

Yes ☐   No ☐   Are pneumatic and hydraulic hoses on power-operated tools checked
regularly for deterioration or damage?

Powder-Actuated Tools

Yes ☐   No ☐   Are employees who operate powder-actuated tools trained in their use and carry a
valid operator's card?

Yes ☐   No ☐   Is each powder-actuated tool stored in its own locked container when not used?

Yes ☐   No ☐   Is a sign at least 7 in. by 10 in. with bold face type reading "POWDER-
ACTUATED TOOL IN USE" conspicuously posted when the tool is used?

Yes ☐   No ☐   Are powder-actuated tools left unloaded until they are actually ready
to be used?

Yes ☐   No ☐   Are powder-actuated tools inspected for obstructions or defects each day before use?

Yes ☐   No ☐   Do powder-actuated tool operators have and use appropriate PPE such as hard
hats, safety goggles, safety shoes, and ear protectors?

Walk-Behind and Power Mowers

Yes ☐   No ☐   Are all walk-behind mowers required to meet ANSI guidelines?

Yes ☐   No ☐   Do walk-behind and riding mowers have a deadman's control switch that will cut
off the motor when pressure is released?

Yes ☐   No ☐   Do the blades on all mowers have guards installed with a discharge opening?

Yes ☐   No ☐   Is the discharge opening set up such that discharge is directed away from the user?

Yes ☐   No ☐   Are power riding mowers required to have a stop installed to prevent the steering
wheel from turning to a point that could cause jack knifing or locking?

Jacks

Yes ☐   No ☐   Are jacks inspected once every 6 months for constant or intermittent use?

Yes ☐   No ☐   Is the load weight of each jack marked on the jack and legible at all times?

Yes ☐   No ☐   If jacks are sent out of shop for special work, are they inspected before they leave
and upon return?

Yes ☐   No ☐   If jacks are subjected to abnormal or extreme conditions, are they inspected
before and after use?

Yes ☐   No ☐   If a defect is found, is the jack taken out of service immediately and tagged?

Yes ☐   No ☐   Before using a jack, is the operator sure it can lift the required weight?

Yes ☐   No ☐   When using a jack, is it ensured that it is on a firm foundation?

Yes ☐   No ☐   Is a jack blocked to prevent forward and backward movement before lifting a load?

Yes ☐   No ☐   Is a jack not used to support a lifted load?

Hazard Communication

Yes ☐   No ☐   Is there a list of hazardous substances used in your workplace?

Yes ☐   No ☐   Is there a written hazard communication program dealing with material safety
data sheets (MSDS), labeling, and employee training?

Yes ☐   No ☐   Is each container for a hazardous substance (i.e., vats, bottles, storage tanks, etc.) labeled with product identity and a hazard warning (communication of the specific health hazards and physical hazards)?

Yes ☐   No ☐   Is there an MSDS readily available for each hazardous substance used?

Yes ☐   No ☐   Is there an employee training program for hazardous substances?

Does this Program Include

Yes ☐   No ☐   An explanation of what an MSDS is and how to use and obtain one.

Yes ☐   No ☐   MSDS contents for each hazardous substance or class of substances.

Yes ☐   No ☐   Explanation of "Right to Know."

Yes ☐   No ☐   Identification of where an employee can see the employer's written hazard communication program and where hazardous substances are present in their work areas.

Yes ☐   No ☐   The physical and health hazards of substances in the work area and specific protective measures to be used.

Yes ☐   No ☐   Details of the hazard communication program, including how to use the labeling system and MSDSs.

Are Employees Trained in the Following

Yes ☐   No ☐   How to recognize tasks that might result in occupational exposure?

Yes ☐   No ☐   How to use work practice, engineering controls, and PPE and to know their limitations?

Yes ☐   No ☐   How to obtain information on the type selection, proper use, location, removal handling, decontamination, and disposal of PPE?

Yes ☐   No ☐   Who to contact and what to do in an emergency?

Hazardous Chemical Exposure

Yes ☐   No ☐   Are employees trained in the safe handling practices of hazardous chemicals such as acids, caustics, etc.?

Yes ☐   No ☐   Are employees aware of the potential hazards involving various chemicals stored or used in the workplace such as acids, bases, caustics, epoxies, and phenols?

Yes ☐   No ☐   Is employee exposure to chemicals kept within acceptable levels?

Yes ☐   No ☐   Are eyewash fountains and safety showers provided in areas where corrosive chemicals are handled?

Yes ☐   No ☐   Are all containers, such as vats and storage tanks labeled as to their contents, e.g., "CAUSTICS?"

Yes ☐   No ☐   Are all employees required to use personal protective clothing and equipment when handling chemicals (gloves, eye protection, and respirators)?

Yes ☐   No ☐   Are flammable or toxic chemicals kept in closed containers when not in use?

Yes ☐   No ☐   Are chemical piping systems clearly marked as to their content?

Yes ☐   No ☐   Where corrosive liquids are frequently handled in open containers or drawn from storage vessels or pipelines, are adequate means readily available for neutralizing or disposing of spills or overflows, and performed properly and safely?

Yes ☐   No ☐   Have SOPs been established, and are they followed when cleaning up chemical spills?

Yes ☐   No ☐   Where needed for emergency use, are respirators stored in a convenient, clean, and sanitary location?

Yes ☐ No ☐   Are respirators intended for emergency use adequate for the various uses for which they may be needed?

Yes ☐ No ☐   Are employees prohibited from eating in areas where hazardous chemicals are present?

Yes ☐ No ☐   Is PPE provided, used, and maintained whenever necessary?

Yes ☐ No ☐   Are there written SOPs for the selection and use of respirators where needed?

Yes ☐ No ☐   If you have a respirator protection program, are your employees instructed on the correct usage and limitations of the respirators? Are the respirators NIOSH-approved for this particular application? Are they regularly inspected and cleaned, sanitized, and maintained?

Yes ☐ No ☐   If hazardous substances are used in your processes, do you have a medical or biological monitoring system in operation?

Yes ☐ No ☐   Are you familiar with the threshold limit values or permissible exposure limits of airborne contaminants and physical agents used in your workplace?

Yes ☐ No ☐   Have control procedures been instituted for hazardous materials, where appropriate, such as respirators, ventilation systems, and handling practices?

Yes ☐ No ☐   Whenever possible, are hazardous substances handled in properly designed and exhausted booths or similar locations?

Yes ☐ No ☐   Do you use general dilution or local exhaust ventilation systems to control dusts, vapors, gases, fumes, smoke, solvents, or mists that may be generated in your workplace?

Yes ☐ No ☐   Is ventilation equipment provided for removal of contaminants from such operations as production grinding, buffing, spray painting, or vapor degreasing and is it operating properly?

Yes ☐ No ☐   Do employees complain about dizziness, headaches, nausea, irritation, or other factors of discomfort when they use solvents or other chemicals?

Yes ☐ No ☐   Is there a dermatitis problem? Do employees complain about dryness, irritation, or sensitization of the skin?

Yes ☐ No ☐   Have you considered the use of an industrial hygienist or environmental health specialist to evaluate your operation?

Yes ☐ No ☐   If internal combustion engines are used, is carbon monoxide kept within acceptable levels?

Yes ☐ No ☐   Is vacuuming used, rather than blowing or sweeping dusts, whenever possible for cleanup?

Yes ☐ No ☐   Are materials that give off toxic asphyxiant, suffocating, or anesthetic fumes stored in a remote or isolated location when not in use?

Hazardous Waste Operations and Emergency Response (Hazwoper)

Yes ☐ No ☐   Is a written plan made available to anyone on the site, as well as to federal authorities?

Yes ☐ No ☐   Are all personnel on the site informed of the hazards?

Yes ☐ No ☐   Is all PPE provided at no cost to the employees?

Yes ☐ No ☐   Is a predesignated representative of the company appointed to become the incident commander?

Yes ☐ No ☐   Are written SOPs developed for every process?

Yes ☐ No ☐   Is a written hazard communication program implemented?

Yes ☐ No ☐   Are all excavations during site preparation shored or sloped in a manner that will not allow accidental collapse?

Yes ☐ No ☐ Has a postemergency response plan that involves cleanup, follow-up, and start-up procedures been developed?

Yes ☐ No ☐ Is there a written safety and health program?

Yes ☐ No ☐ Is there an organizational structure chart?

Yes ☐ No ☐ Is there a comprehensive work plan?

Yes ☐ No ☐ Is there a site-specific safety and health plan?

Yes ☐ No ☐ Have the medical surveillance plan requirements been outlined?

Yes ☐ No ☐ Have all personnel on the site been trained in hazardous waste operations before they participate in any activity that could expose them to hazardous substances, or safety or health hazards?

Yes ☐ No ☐ Have general site workers, laborers, and supervisors had a minimum of 40 h off-site instruction and 3 days on-site training under the direct supervision of a trained, experienced supervisor?

Yes ☐ No ☐ Have workers who are on the site occasionally and workers regularly on site received at least 24 h of off-site instruction and 1 day on-site training by a trained, experienced supervisor?

Yes ☐ No ☐ Have regular workers required to wear respirators undergone an additional 16 h of off-site instruction and 2 days on-site training by a trained, experienced supervisor?

Yes ☐ No ☐ Have management and supervisors attended at least 40 h off-site instruction and 3 days of field supervised training and an additional 8 h of specialized training on topics such as PPE, employee training, spill containment, and monitoring techniques?

Yes ☐ No ☐ Have trainers been qualified to instruct employees, completed a trainers' course, and attained certification as a trainer from that course?

Yes ☐ No ☐ Has each certified worker undergone an additional 8 h refresher training course annually?

Health Hazards

Yes ☐ No ☐ Are there physical factors in the workplace that can cause illnesses?

Yes ☐ No ☐ Are there chemical agents in the workplace that can cause illnesses?

Yes ☐ No ☐ Are there ergonomic factors in the workplace that can cause illnesses?

Yes ☐ No ☐ Are there biological agents in the workplace that can cause illnesses?

Yes ☐ No ☐ Are there controls in place to prevent health effects?

Yes ☐ No ☐ Are workers showing any signs and symptoms of illness?

Yes ☐ No ☐ Are workers complaining of symptoms of illness at work?

Yes ☐ No ☐ Can the signs and symptoms be linked to time at work?

Yes ☐ No ☐ Have workers received medical evaluations that identify the work environment as the cause?

Yes ☐ No ☐ Are steps taken to evaluate the work environment for potential causes of illness?

Yes ☐ No ☐ Have any causes of illness been identified?

Yes ☐ No ☐ Are changes in ventilation a possible control if air contamination exists?

Yes ☐ No ☐ Are other controls evaluated?

Yes ☐ No ☐ Is there follow-up evaluation of the workplace?

Yes ☐ No ☐ Have any organisms that could cause health problems been identified?

Yes ☐ No ☐ Have any chemicals that could cause health problems been identified?

Yes ☐ No ☐ Has any physical hazard that could cause health problems been identified?

Yes ☐ No ☐ Has any ergonomic hazard that could cause health problems been identified?
Yes ☐ No ☐ Have steps been taken to control potential health hazards?
Yes ☐ No ☐ Has the appropriate PPE been provided to workers?
Yes ☐ No ☐ Have health-related symptoms been reported by employees?
Yes ☐ No ☐ Has the employee received a medical evaluation?
Yes ☐ No ☐ Are inspections conducted on a regular basis?
Yes ☐ No ☐ Is regular environmental and personal monitoring conducted?
Yes ☐ No ☐ Are health concerns or findings addressed promptly?

Heat Hazards

Yes ☐ No ☐ Are there thermal processes in use?
Yes ☐ No ☐ Are heat sources shielded or guarded?
Yes ☐ No ☐ Do burn risks exist?
Yes ☐ No ☐ Have workers been burned?
Yes ☐ No ☐ Are there extreme heat hazards?
Yes ☐ No ☐ Is heat stress a possible problem?
Yes ☐ No ☐ Does appropriate PPE exist?
Yes ☐ No ☐ Are workers wearing required PPE?
Yes ☐ No ☐ Do heat processes cause eye injuries?
Yes ☐ No ☐ Have workers been trained on avoidance of thermal hazards?

Ionizing Radiation

Yes ☐ No ☐ Are there sources of ionizing radiation?
Yes ☐ No ☐ Are radioisotopes in use?
Yes ☐ No ☐ Are medical sources of radiation in use?
Yes ☐ No ☐ Are x-rays in use?
Yes ☐ No ☐ Are radiation surveys conducted?
Yes ☐ No ☐ Are radiation warning signs posted?
Yes ☐ No ☐ Are workers protected by shielding, time, or distance?
Yes ☐ No ☐ Have workers been trained on the hazards and safeguards?
Yes ☐ No ☐ Is PPE needed?
Yes ☐ No ☐ Have workers been trained in the use of PPE?
Yes ☐ No ☐ Are workers wearing the appropriate PPE?
Yes ☐ No ☐ Are nuclear reactors in use?
Yes ☐ No ☐ Have or do workers exhibit any signs and symptoms of radiation sickness or exposures?

Ladders

Yes ☐ No ☐ Are only type 1 or type 1A industrial ladders used?
Yes ☐ No ☐ Do steps on ladders have a minimum load capacity of 250 lb?
Yes ☐ No ☐ Are all ladders inspected for damage before use?
Yes ☐ No ☐ Are ladders not placed against movable objects?
Yes ☐ No ☐ Are ladders placed to prevent movement by lashing or other means?
Yes ☐ No ☐ Are employees' shoes free of mud, grease, or other substances that could cause a slip or fall?
Yes ☐ No ☐ Are ladders not placed on unstable bases such as boxes or barrels?

Yes ☐ No ☐ Do employees not climb on the top two steps of a stepladder?

Yes ☐ No ☐ Are ladders used to gain access to a roof that extends at least 3 ft. above the point of support, at eave, gutter, or roof line?

Yes ☐ No ☐ Are stepladders fully opened to permit the spreaders to lock?

Yes ☐ No ☐ Are all labels in place and legible on ladders?

Yes ☐ No ☐ Are ladders always moved to prevent and avoid overreaching?

Yes ☐ No ☐ Are single ladders not longer than 30 ft.?

Yes ☐ No ☐ Do extension ladders up to 36 ft. have a 3 ft. overlap between sections?

Yes ☐ No ☐ Do extension ladders over 36 ft. and up to 48 ft. have a 4 ft. overlap between sections?

Yes ☐ No ☐ Do extension ladders over 48 ft. and up to 60 ft. have a 5 ft. overlap between sections?

Yes ☐ No ☐ Do two-section extension ladders not exceed 48 ft. in total length?

Yes ☐ No ☐ Do ladders over two-section not exceed 60 ft. in total length?

Yes ☐ No ☐ Are ladders not used horizontally as scaffolds, runways, or platforms?

Yes ☐ No ☐ Is the area around the top and base of ladders kept free of tripping hazards such as loose materials, trash, cords, hoses, and leaves?

Yes ☐ No ☐ Is the base of a straight or extension ladder set back a safe distance from the vertical or approximately 1/4 of the working length of the ladder?

Yes ☐ No ☐ Are ladders that project into passageways or doorways, where they could be struck by personnel, moving equipment, or materials that are handled, protected by barricades or guards?

Yes ☐ No ☐ Do employees face the ladder when ascending or descending?

Yes ☐ No ☐ Do employees use both hands when going up or down a ladder?

Yes ☐ No ☐ Are materials or equipment raised or lowered by way of lines?

Yes ☐ No ☐ Are employees trained and educated on the proper use of ladders?

Yes ☐ No ☐ Are repairs done professionally?

Yes ☐ No ☐ Are inspections conducted before each use and defective, broken, or damaged ladders pulled from service tagged and marked "Dangerous. Do Not Use?"

Yes ☐ No ☐ Are the rungs tight in the joint of the side rails?

Yes ☐ No ☐ Do all moving parts operate freely without binding?

Yes ☐ No ☐ Are all pulleys, wheels, and bearings lubricated frequently?

Yes ☐ No ☐ Are rungs kept free of grease and oil?

Yes ☐ No ☐ Is rope that is badly worn or frayed replaced immediately?

Yes ☐ No ☐ Are all ladders equipped with slip-resistant feet, free of grease, and in good condition?

Portable Wood Ladders

Yes ☐ No ☐ Are all wood ladders free of splinters, sharp edges, shake, wane, compression failures, decay, and other irregularities?

Yes ☐ No ☐ Are portable stepladders no longer than 20 ft.?

Yes ☐ No ☐ Is the step spacing no more than 12 in. apart?

Yes ☐ No ☐ Are stepladders that have a metal spreader or locking device of sufficient strength and size to hold the front and back when open?

Portable Metal Ladders

Yes ☐ No ☐ Are ladders inspected immediately when dropped or tipped over?

Yes ☐ No ☐ Is the step spacing no more than 12 in. apart?

Yes ☐   No ☐   Are metal ladders not for electrical work or in areas where they could contact energized conductors?

Fixed Ladders

Yes ☐   No ☐   Are the steps no more than 12 in. apart?

Yes ☐   No ☐   Are job-made ladders constructed to conform with the established OSHA standards?

Yes ☐   No ☐   Are all fixed ladders painted or treated to prevent rusting?

Yes ☐   No ☐   Do fixed ladders 20 ft. or higher have a landing every 20 ft. if there is no surrounding cage?

Yes ☐   No ☐   If it has a cage or safety device, is a landing required every 30 ft.?

Lifting Safety

Yes ☐   No ☐   Have all workers been trained on proper lifting techniques?

Yes ☐   No ☐   Was the object to be lifted inspected to decide the best way to grasp it?

Yes ☐   No ☐   Has the load been sized up to ensure that it can be lifted?

Yes ☐   No ☐   Are loads kept small to prevent heavy lifts?

Yes ☐   No ☐   Is the feet placed close to the object?

Yes ☐   No ☐   Can a good grip be made on the load?

Yes ☐   No ☐   Are the knees bent while keeping the back straight?

Yes ☐   No ☐   Is the load held close to the worker's body?

Yes ☐   No ☐   Can the worker see past the load?

Yes ☐   No ☐   Did the worker get help for large or heavy objects?

Yes ☐   No ☐   Is the lift more appropriate for a team lift?

Yes ☐   No ☐   Are harsh jerking movements when pushing, pulling, or lifting a load avoided?

Yes ☐   No ☐   Are materials stacked so that they are positioned between the knees and waist?

Yes ☐   No ☐   Are gloves worn when handling sharp or rough objects?

Yes ☐   No ☐   Is power lifting equipment used when possible to prevent injuries instead of manual lifting?

Yes ☐   No ☐   Is care taken not to drop materials that might hit someone?

Yes ☐   No ☐   Are equipment, carts, or table kept at a proper height to help prevent back injuries?

Lockout/Tagout Procedures

Yes ☐   No ☐   Is all machinery or equipment capable of movement required to be de-energized or disengaged and locked out during cleaning, servicing, adjusting, or setting up operations whenever required?

Yes ☐   No ☐   Where the power disconnecting means for equipment does not also disconnect the electrical control circuit, a Are appropriate electrical enclosures identified?

Yes ☐   No ☐   Are means provided to assure that the control circuit can also be disconnected and locked out?

Yes ☐   No ☐   Is the locking out of control circuits in lieu of locking out main power disconnects prohibited?

Yes ☐   No ☐   Are all equipment control valve handles provided with a means for locking out?

Yes ☐   No ☐   Does the lockout procedure require that stored energy (mechanical, hydraulic, air, etc.) be released or blocked before equipment is locked out for repairs?

Yes ☐   No ☐   Are appropriate employees provided with individually keyed personal safety locks?

Yes ☐   No ☐   Are employees required to keep personal control of their keys while they have safety locks in use?

Yes ☐   No ☐   Is it required that only the employee exposed to the hazard place or remove the safety lock?

Yes ☐   No ☐   Is it required that employees check the safety of the lockout by attempting a start-up after making sure no one is exposed?

Yes ☐   No ☐   Are employees instructed to always push the control circuit stop button immediately after checking the safety of the lockout?

Yes ☐   No ☐   Is there a means provided to identify any or all employees who are working on locked-out equipment by their locks or accompanying tags?

Yes ☐   No ☐   Are a sufficient number of accident preventive signs or tags and safety padlocks provided for any reasonably foreseeable repair or emergency?

Yes ☐   No ☐   When machine operations, configuration, or size requires the operator to leave his or her control station to install tools or perform other operations, and that part of the machine could move if accidentally activated, is that element required to be separately locked or blocked out?

Yes ☐   No ☐   In the event that equipment or lines cannot be shut down, locked out, and tagged, is a safe job procedure established and rigidly followed?

Machine Guarding and Safety

Yes ☐   No ☐   Do the safeguards provided meet the minimum OSHA requirements?

Yes ☐   No ☐   Do the safeguards prevent workers' hands, arms, and other body parts from making contact with dangerous moving parts?

Yes ☐   No ☐   Are the safeguards firmly secured and not easily removable?

Yes ☐   No ☐   Do the safeguards ensure that no objects fall into the moving parts?

Yes ☐   No ☐   Do the safeguards permit safe, comfortable, and relatively easy operation of the machine?

Yes ☐   No ☐   Can the machine be oiled without removing the safeguard?

Yes ☐   No ☐   Is there a system for shutting down the machinery before safeguards are removed?

Yes ☐   No ☐   Can the existing safeguards be improved?

Yes ☐   No ☐   Is there a point-of-operation safeguard provided for the machine?

Yes ☐   No ☐   Does it keep the operator's hands, fingers, and body out of the danger area?

Yes ☐   No ☐   Is there evidence that the safeguards have been tampered with or removed?

Yes ☐   No ☐   Could you suggest a more practical, effective safeguard?

Yes ☐   No ☐   Could changes be made on the machine to eliminate the point-of-operation hazard entirely?

Yes ☐   No ☐   Are there any unguarded gears, sprockets, pulleys, or flywheels on the apparatus?

Yes ☐   No ☐   Are there any exposed belts or chain drives?

Yes ☐   No ☐   Are there any exposed setscrews, key ways, collars, etc.?

Yes ☐   No ☐   Are starting and stopping controls within easy reach of the operator?

Yes ☐   No ☐   If there is more than one operator, are separate controls provided?

Yes ☐   No ☐   Are safeguards provided for all hazardous moving parts of the machine, including auxiliary parts?

Yes ☐   No ☐   Have appropriate measures been taken to safeguard workers against noise hazards?

Yes ☐   No ☐   Have special guards, enclosures, or PPE been provided, where necessary, to protect workers from exposure to harmful substances used in machine operation?

Yes ☐   No ☐   Is the machine installed in accordance with National Fire Protection Association and NEC requirements?

Yes ☐   No ☐   Are there loose conduit fittings?

Yes ☐   No ☐   Is the machine properly grounded?

Yes ☐   No ☐   Is the power supply correctly fused and protected?

Yes ☐   No ☐   Do workers occasionally receive minor shocks while operating any of the machines?

Yes ☐   No ☐   Do operators and maintenance workers have the necessary training in how to use the safeguards and why?

Yes ☐   No ☐   Have operators and maintenance workers been trained in where the safeguards are located, how they provide protection, and what hazards they protect against?

Yes ☐   No ☐   Have operators and maintenance workers been trained in how and under what circumstances guards can be removed?

Yes ☐   No ☐   Have workers been trained in the procedures to follow if they notice guards that are damaged, missing, or inadequate?

Yes ☐   No ☐   Is protective equipment required?

Yes ☐   No ☐   If protective equipment is required, is it appropriate for the job, in good condition, kept clean and sanitary, and stored carefully when not in use?

Yes ☐   No ☐   Is the operator dressed safely for the job (i.e., no loose-fitting clothing or jewelry)?

Yes ☐   No ☐   Have maintenance workers received up-to-date instruction on the machines they service?

Yes ☐   No ☐   Do maintenance workers lock out the machine from its power sources before beginning repairs?

Yes ☐   No ☐   Where several maintenance persons work on the same machine, are multiple lockout devices used?

Yes ☐   No ☐   Do maintenance persons use appropriate and safe equipment in their repair work?

Yes ☐   No ☐   Is the maintenance equipment itself property guarded?

Yes ☐   No ☐   Are maintenance and servicing workers trained in the requirements of 29 CFR 1910.147, lockout/tagout hazard, and do the procedures for lockout/tagout exist before they attempt their tasks?

Material Handling
Material-Handling Equipment

Yes ☐   No ☐   Are all operators of material-handling equipment trained. (hand trucks, cranes, hoists, fork trucks, or any motorized equipment)?

Yes ☐   No ☐   Are all operators of forklifts trained by a certified instructor?

Yes ☐   No ☐   Is all material-handling equipment kept in good repair and maintained by trained personnel?

Yes ☐   No ☐   Is all material-handling equipment inspected before use, daily, monthly, and annually as required?

Yes ☐   No ☐   Is all material-handling equipment properly marked with load ratings?

Yes ☐   No ☐   Are forklifts marked "FLAMMABLE," if they use propane or any other compressed-gas source?

Yes ☐   No ☐   Are railroad cars, heavy equipment, and rolling hoists or cranes chocked or blocked to prevent rolling?

Yes ☐   No ☐   Are grading or ramps installed between two working levels for safe vehicle movement?

Yes ☐   No ☐   Is material-handling equipment that poses a danger to equipment or personnel guarded to prevent access within a safe distance?

Storage Areas

Yes ☐   No ☐   Are maximum safe load limits observed?

Yes ☐   No ☐   Are load limits posted for platforms and floors?

Yes ☐   No ☐   Are storage racks stable and secure?

Yes ☐   No ☐   Are stored material neatly stacked, racked, blocked, or interlocked?

Yes ☐   No ☐   Are height limits set and posted to ensure stability of stacked material?

Yes ☐   No ☐   Do all aisles, loading docks, doorways, turns, and passages have safe clearances for equipment and material?

Yes ☐   No ☐   Are clearance signs posted in a visible place to warn employees of clearance limits?

Yes ☐   No ☐   Are all ramps, open pits, tanks, vats, ditches, and elevated surfaces 4 ft. or more guarded?

Housekeeping

Yes ☐   No ☐   Are storage areas kept clean, dry, and in good condition?

Yes ☐   No ☐   Are storage areas kept free of tripping and slipping hazards?

Yes ☐   No ☐   Are storage areas kept free of fire hazards (trash, paper, oily rags, or empty flammable liquid containers)?

Yes ☐   No ☐   Are storage areas kept free of explosion hazards (unsecured CGCs, flammable vapors, or dusts)?

Yes ☐   No ☐   Are storage areas kept free of pests such as rats, mice, roaches, other vermin?

Means of Exit

Yes ☐   No ☐   Do all exits have an illuminated sign above them that states "EXIT"?

Yes ☐   No ☐   Are there signs that state "NOT AN EXIT" placed over doors if there is the possibility that it could be mistaken for an exit, e.g., closets, stairways, and doors?

Yes ☐   No ☐   Are there procedures to ensure that under no circumstances exits are locked while the building is occupied?

Yes ☐   No ☐   Are all emergency exit doors equipped with panic bars?

Yes ☐   No ☐   Do all emergency exit doors designated for fire escape lead to a safe area of refuge?

Yes ☐   No ☐   Do all emergency exit doors or passageways have emergency illumination, in case of power loss?

Yes ☐   No ☐   Is there access to exits that are unobstructed at all times?

Yes ☐   No ☐   Are all floor areas around exits clean and dry at all times?

Yes ☐   No ☐   Is an inspection from a fire marshal done at least once a year?

Yes ☐   No ☐   Is a general inspection of exit signs, exit doors, exit accesses, and alarm systems conducted by a trained person who has the authority to rectify any problems?

Yes ☐   No ☐   Is training done on the identification of all exits and their locations?

Medical Services and First Aid

Yes ☐ No ☐ Are medical facilities and medically trained personnel on-site if possible?

Yes ☐ No ☐ In the absence of a medical facility that is close and available, are adequately trained personnel readily available to render first aid?

Yes ☐ No ☐ Are physician-approved first-aid supplies readily available?

Yes ☐ No ☐ Are quick drenching or flushing facilities available in work areas where the eyes or body may be exposed to injurious corrosive materials or chemicals?

Yes ☐ No ☐ Is a first-aid log kept on employees?

Yes ☐ No ☐ Is an inventory checklist kept of all first-aid supplies?

Yes ☐ No ☐ Are all employees trained on basic first-aid techniques and procedures?

Yes ☐ No ☐ Are all employees trained on usage of PPE while first aid is performed?

Nonionizing Radiation

Yes ☐ No ☐ Are there radio frequency sources?

Yes ☐ No ☐ Are there ultraviolet sources?

Yes ☐ No ☐ Are there infrared sources?

Yes ☐ No ☐ Are there lasers in use?

Yes ☐ No ☐ Are there microwaves in use?

Yes ☐ No ☐ Is there known risk for nonionizing radiation sources?

Yes ☐ No ☐ Are safeguards in place?

Yes ☐ No ☐ Are interlocks in place?

Yes ☐ No ☐ Are caution, warning, and information signs in place?

Yes ☐ No ☐ If PPE is needed, is it provided?

Yes ☐ No ☐ Are workers wearing the appropriate PPE?

Yes ☐ No ☐ Have workers complained of medically related signs or symptoms from exposure to nonionizing sources?

Yes ☐ No ☐ Does welding take place?

Yes ☐ No ☐ Have flash-related injuries occurred?

PPE

Yes ☐ No ☐ Do employers assess the workplace to determine if hazards that require the use of PPE (e.g., head, eye, face, hand, or foot protection) are present or are likely to be present?

Yes ☐ No ☐ If hazards or the likelihood of hazards are found, do employers select and have affected employees use properly fitted PPE suitable for protection from these hazards?

Yes ☐ No ☐ Have employees been trained on PPE procedures, that is, what PPE is necessary for a job task, when they need it, and how to properly adjust it?

Yes ☐ No ☐ Are protective goggles or face shields provided and worn where there is any danger of flying particles or corrosive materials?

Yes ☐ No ☐ Are approved safety glasses required to be worn at all times in areas where there is a risk of eye injuries such as punctures, abrasions, contusions, or burns?

Yes ☐ No ☐ Are employees who need corrective lenses (glasses or contacts) in working environments with harmful exposures required to wear only approved safety glasses, protective goggles, or use other medically approved precautionary procedures?

Yes ☐ No ☐ Are protective gloves, aprons, shields, or other means provided and required where employees could be cut or where there is reasonably anticipated exposure to corrosive liquids, chemicals, blood, or other potentially infectious materials?

Yes ☐ No ☐ Are hard hats provided and worn where danger of falling objects exists?

Yes ☐ No ☐ Are hard hats inspected periodically for damage to the shell and suspension system?

Yes ☐ No ☐ Is appropriate foot protection required where there is risk of foot injuries from hot, corrosive, or poisonous substances, falling objects, crushing or penetrating actions?

Yes ☐ No ☐ Are approved respirators provided for regular or emergency use where needed?

Yes ☐ No ☐ Is all protective equipment maintained in a sanitary condition and ready for use?

Yes ☐ No ☐ Do you have eyewash facilities and a quick drench shower within the work area where employees are exposed to injurious corrosive materials? Where special equipment is needed for electrical workers, is it available?

Yes ☐ No ☐ Where food or beverages are consumed on the premises, are they consumed in areas where there is no exposure to toxic material, blood, or other potentially infectious materials?

Yes ☐ No ☐ Is protection against the effects of occupational noise exposure provided when sound levels exceed those of the OSHA noise standard?

Yes ☐ No ☐ Are adequate work procedures, protective clothing, and equipment provided and used when cleaning up spilled toxic or otherwise hazardous materials or liquids?

Yes ☐ No ☐ Are there appropriate procedures in place for disposing of or decontaminating PPE contaminated with, or reasonably anticipated to be contaminated with, blood or other potentially infectious materials?

Pressure Vessels

Yes ☐ No ☐ Has the vessel been inspected for leaks or cracks?

Yes ☐ No ☐ Have flue gas leaks been spotted by looking for black dust (soot) around sheet-metal joints?

Yes ☐ No ☐ Are any part leaks of boiler enclosure and breaching, especially in the connection to the stack, observed?

Yes ☐ No ☐ Are electrical wires inspected to ensure that covers are installed on over-limit switches, temperature sensors, and controls.

Yes ☐ No ☐ Are all controls and valves in a normal operating position?

Yes ☐ No ☐ Is the internal pressure of the system monitored?

Yes ☐ No ☐ Are safety valves installed with full-sized discharge piping properly supported and directed to a point of safe discharge and is safety valve set pressure equal to or less than boiler maximum allowable working pressure, because safety valve relieving capacity must be equal to or greater than boiler output?

Yes ☐ No ☐ Have any open panels, covers, and signs of rewiring on the floor or bottom of panels been observed?

Yes ☐ No ☐ Have checks been completed for jumper wires and locked shutoffs?

Yes ☐ No ☐ Has the exterior shell or insulation been inspected for indications of overheating?

Yes ☐ No ☐ Has a search for water leaks on the floor been performed?

Yes ☐ No ☐ Has a check for water or steam escaping from any part of a pressurized system including the boiler, valves, or piping been performed?

Yes ☐ No ☐ Are all temperature and pressure gages operational and located for proper monitoring?

Yes ☐ No ☐ Has piping been checked for proper supports and allowance for expansion and contraction?

Yes ☐ No ☐ Are there ways to shut off the fuel sources to the vessel and have they been checked?

Yes ☐ No ☐ Has isolation relief valve been identified?

Yes ☐ No ☐ Has a checklist to record isolation valve status and operation been developed?

Yes ☐ No ☐ Has approval (by signature) of the person named as responsible on the Statement of Authority been obtained before maintenance?

Yes ☐ No ☐ Has the operating certificate been observed noting the last date of inspection and expiration date when required?

Rigging

Yes ☐ No ☐ Is ANSI-approved equipment used?

Yes ☐ No ☐ Are daily inspections conducted before use by the user or operator?

Yes ☐ No ☐ Are monthly inspections conducted by a person trained to recognize defects and authorized to remove equipment from service?

Yes ☐ No ☐ Are annual inspections by the manufacturer or outside contractor done but not required?

Yes ☐ No ☐ Is only the manufacturer allowed to repair these devices?

Yes ☐ No ☐ Do all chains, slings, and cables have an identification tag attached always that shows their load rating, limitations, etc?

Yes ☐ No ☐ Is the load rating never exceeded for chains, slings, or cables?

Yes ☐ No ☐ Are only alloy steel chains used?

Yes ☐ No ☐ Are chains inspected before use for wear, abrasions, collapse, visible damage, or any damage no matter how insignificant?

Yes ☐ No ☐ Are damaged chains removed from service?

Yes ☐ No ☐ Do hooks, rings, links, or any coupling devices have the same or higher rating as the chain to which they are affixed?

Yes ☐ No ☐ Are wire rope slings and cables inspected before use?

Yes ☐ No ☐ Do all attachments meet the same load standards as the wire rope sling they are attached to?

Yes ☐ No ☐ Are wire rope slings and fiber-core wire ropes operated at temperatures below 200 °F?

Yes ☐ No ☐ Are nonfiber core, wire rope slings only used at temperatures below 400 °F and above 60 °F?

Yes ☐ No ☐ Do the handle's metal mesh slings meet the minimum requirements of the sling?

Yes ☐ No ☐ Are metal mesh slings not impregnated with elastomers used in temperatures not exceeding 500 °F or below 20 °F?

Yes ☐ No ☐ Are metal mesh slings impregnated with polyvinyl chloride or neoprene, used in the temperature range from 0 °F to 200 °F?

Yes ☐ No ☐ Are natural or synthetic fiber rope slings used only in temperature ranges of above 20 °F to 180 °F, unless they are wet or frozen?

Yes ☐ No ☐ Are metal mesh slings never spliced except when the manufacturer makes alterations to slings?

Yes ☐ No ☐ Are natural or synthetic fiber slings removed from service if there is abnormal wear, powdered fibers appear between strands, fibers are broken or cut, variation in size or roundness of strands occurs, discoloration or rotting is detected, or distortion of hardware is detected?

Yes ☐   No ☐   Are synthetic web slings uniform in thickness?

Yes ☐   No ☐   Are polyester and nylon webbedslings not used where fumes, vapors, sprays, mists, liquids of acids, phonetics, or caustics are present?

Yes ☐   No ☐   Are synthetic fiber slings removed from service when the following conditions are present: acids or caustics, burns, melting or charring of any part of the sling, snags, punctures, tears, or cuts, broken or worn stitches, and distortion of any fitting?

Scaffolding

Yes ☐   No ☐   Are scaffolds erected and dismantled under the direction of a competent person?

Yes ☐   No ☐   Are there guardrails and toeboards on all open sides and ends of scaffold platforms 10 ft. or more above the ground?

Yes ☐   No ☐   Are tube and coupler scaffold posts accurately spaced, erected on suitable base, maintained plumb, and brace connections secured?

Yes ☐   No ☐   Are planks secured or not less than 6 in. or more than 12 in. over the end support?

Yes ☐   No ☐   Is a ladder or equivalent means of access provided?

Yes ☐   No ☐   Are scaffolds tied into structures when height and length requirements are met?

Yes ☐   No ☐   Are all footings or anchorages sound, rigid, and can support the intended load?

Yes ☐   No ☐   Are screens placed between toeboard and guardrail where persons pass under scaffolds?

Yes ☐   No ☐   Is overhead protection provided for workers on scaffolds if overhead exposures exist?

Yes ☐   No ☐   Can scaffold components support four times their intended load?

Yes ☐   No ☐   Is planking overlapped 12 in. or secured and planks extend over end supports 6 to 12 in.?

Yes ☐   No ☐   Are workers not permitted on scaffolds during storms, high winds, ice, or snow?

Yes ☐   No ☐   Does the maximum work level height not exceed four times the least base dimension unless outrigger frames are used?

Yes ☐   No ☐   Are scaffold platform widths not less than 18 in.?

Yes ☐   No ☐   Are lifelines and harnesses provided and used by each worker on swing and single-point adjustable suspension scaffolds?

Suspended Scaffolds

Yes ☐   No ☐   Are suspended or hanging scaffold components protected such as wire and fiber ropes from heat, chemicals, or corrosive substances?

Yes ☐   No ☐   Can scaffolding ropes support six times the load?

Aerial Lifts

Yes ☐   No ☐   Are safety harnesses with lanyards attached to the boom or basket and worn by occupants of aerial lifts?

Yes ☐   No ☐   Are safety rails present on all open sides of elevated work platforms?

Security

Yes ☐   No ☐   Is there a plan to provide for security of workers?

Yes ☐   No ☐   Is there a trained security force on duty?

Yes ☐   No ☐   Are workers isolated from the public and outside as much as possible?

Yes ☐   No ☐   Is there a separate sign in or sign-out screening performed for all visitors?

Yes ☐   No ☐   Do surveillance cameras exist?
Yes ☐   No ☐   Are alarms available for emergency situations?
Yes ☐   No ☐   Are emergency telephone numbers posted?
Yes ☐   No ☐   Have workers been trained regarding personal security?
Yes ☐   No ☐   Is there escort service for employees?
Yes ☐   No ☐   Are all areas well lighted?
Yes ☐   No ☐   Are parking areas well lighted?
Yes ☐   No ☐   Does landscaping permit high visibility?

Slips, Trips, and Falls

Yes ☐   No ☐   Are all walking surfaces clear of debris, tools, etc.?
Yes ☐   No ☐   Are travel surfaces even?
Yes ☐   No ☐   Are spills cleaned immediately?
Yes ☐   No ☐   Do workers wear nonslip footwear?
Yes ☐   No ☐   Are materials not stored in passageways and aisles?
Yes ☐   No ☐   Are stairs clear of items such as tools?
Yes ☐   No ☐   Are floors and walking surfaces free of spills, grease, or oil?
Yes ☐   No ☐   Are handrails in place?
Yes ☐   No ☐   Are signs or mirrors used for blind corners?

Violence

Yes ☐   No ☐   Is there zero tolerance for workplace violence?
Yes ☐   No ☐   Are all weapons precluded from the workplace?
Yes ☐   No ☐   Have workers and supervisors been trained related to acts of violence?
Yes ☐   No ☐   Is there a protocol for violence incidents?
Yes ☐   No ☐   Are protective steps taken to prevent violent acts?
Yes ☐   No ☐   Are potentials for violence reported?
Yes ☐   No ☐   Is action taken immediately when potential for violent acts
               is possible?
Yes ☐   No ☐   Are there emergency procedures to handle cases of workplace violence?
Yes ☐   No ☐   Are emergency telephone numbers posted?

Walking–Working Surfaces

Yes ☐   No ☐   Is a documented, functioning housekeeping program in place?
Yes ☐   No ☐   Are all worksites clean, sanitary, and orderly?
Yes ☐   No ☐   Are work surfaces kept dry or is appropriate steps taken to assure that the surfaces
               are slip-resistant?
Yes ☐   No ☐   Are all spilled hazardous materials or liquids, including blood and other
               potentially infectious materials, cleaned up immediately and according to proper
               procedures?
Yes ☐   No ☐   Is combustible scrap, debris, and waste stored safely and removed from the
               worksite properly?
Yes ☐   No ☐   Is all regulated waste, as defined in the OSHA bloodborne pathogens standard
               (1910.1030), discarded according to federal, state, and local regulations?
Yes ☐   No ☐   Are accumulations of combustible dust routinely removed from elevated surfaces
               including the overhead structure of buildings, etc.?
Yes ☐   No ☐   Is combustible dust cleaned up with a vacuum system to prevent the dust from
               going into suspension?

Yes ☐   No ☐   Is metallic or conductive dust prevented from entering or accumulating on or around electrical enclosures or equipment?

Yes ☐   No ☐   Are covered metal waste cans used for oily and paint-soaked waste?

Walkways

Yes ☐   No ☐   Are aisles and passageways kept clear?

Yes ☐   No ☐   Are aisles and walkways marked as appropriate?

Yes ☐   No ☐   Are wet surfaces covered with nonslip materials?

Yes ☐   No ☐   Are holes in the floor, sidewalk, or other walking surfaces repaired properly, covered, or otherwise made safe?

Yes ☐   No ☐   Is there safe clearance for walking in aisles where motorized or mechanical handling equipment is operating?

Yes ☐   No ☐   Are materials or equipment stored in such a way that sharp projectives will not interfere with the walkway?

Yes ☐   No ☐   Are spilled materials cleaned up immediately?

Yes ☐   No ☐   Are changes of direction or elevation readily identifiable?

Yes ☐   No ☐   Are aisles or walkways that pass near moving or operating machinery, welding operations, or similar operations arranged such that employees will not be subjected to potential hazards?

Yes ☐   No ☐   Is adequate headroom provided for the entire length of any aisle or walkway?

Yes ☐   No ☐   Are standard guardrails provided wherever aisle or walkway surfaces are elevated more than 30 in. above any adjacent floor or the ground?

Yes ☐   No ☐   Are bridges provided over conveyors and similar hazards?

Floor and Wall Openings

Yes ☐   No ☐   Are floor openings guarded by a cover, a guardrail, or equivalent on all sides (except at entrance to stairways or ladders)?

Yes ☐   No ☐   Are toeboards installed around the edges of permanent floor openings (where persons may pass below the opening)?

Yes ☐   No ☐   Are skylight screens of such construction and mounting that they will withstand a load of at least 200 lb?

Yes ☐   No ☐   Is the glass in the windows, doors, glass walls, etc., which are subject to human impact, of sufficient thickness and type for the condition of use?

Yes ☐   No ☐   Are grates or similar type covers over floor openings such as floor drains of such design that foot traffic or rolling equipment will not be affected by the grate spacing?

Yes ☐   No ☐   Are unused portions of service pits and pits not actually in use either covered or protected by guardrails or equivalent?

Yes ☐   No ☐   Are manhole covers, trench covers, and similar covers, plus their supports designed to carry a truck rear axle load of at least 20,000 lb when located in roadways and subject to vehicle traffic?

Yes ☐   No ☐   Are floor or wall openings in fire-resistive construction provided with doors or covers compatible with the fire rating of the structure and provided with a self-closing feature when appropriate?

Stairs and Stairways

Yes ☐   No ☐   Are standard stair rails or handrails provided on all stairways with four or more risers?

Yes ☐   No ☐   Are all stairways at least 22 in. wide?

Yes ☐   No ☐   Do stairs have landing platforms not less than 30 in. in the direction of travel and extend 22 in. in width at every 12 ft. or less of vertical rise?

Yes ☐   No ☐   Do stairs angle no more than 50 and no less than 30°?

Yes ☐   No ☐   Are step risers on stairs uniform from top to bottom?

Yes ☐   No ☐   Are steps on stairs and stairways designed or provided with a surface that renders them slip-resistant?

Yes ☐   No ☐   Are stairway handrails located between 30 and 34 in. above the leading edge of stair treads?

Yes ☐   No ☐   Do stairway handrails have at least 3 in. of clearance between the handrails and the wall or surface they are mounted on?

Yes ☐   No ☐   Where doors or gates open directly on a stairway, is there a platform provided so that the swing of the door does not reduce the width of the platform to less than 21 in.?

Yes ☐   No ☐   Where stairs or stairways exit directly into any area where vehicles may be operated, are adequate barriers and warnings provided to prevent employees stepping into the path of traffic?

Yes ☐   No ☐   Do stairway landings have a dimension measured in the direction of travel, at least equal to the width of the stairway?

Elevated Surfaces

Yes ☐   No ☐   Are signs posted, when appropriate, showing the elevated surface load capacity?

Yes ☐   No ☐   Are surfaces elevated more than 30 in. above the floor or ground provided with standard guardrails?

Yes ☐   No ☐   Are all elevated surfaces (beneath which people or machinery could be exposed to falling objects) provided with standard 4-in. toeboards?

Yes ☐   No ☐   Is a permanent means of access and egress provided to elevated storage and work surfaces?

Yes ☐   No ☐   Is required headroom provided where necessary?

Yes ☐   No ☐   Is material on elevated surfaces piled, stacked, or racked in a manner to prevent it from tipping, falling, collapsing, rolling, or spreading?

Yes ☐   No ☐   Are dockboards or bridge plates used when transferring materials between docks and trucks or rail cars?

Welding and Cutting

Yes ☐   No ☐   Are areas established for welding and cutting equipment based on fire potentials?

Yes ☐   No ☐   Are there designated individuals responsible for authorizing cutting or welding in nonwelding areas?

Yes ☐   No ☐   Are all cutters, welders, and supervisors trained in the safe operation and use of equipment and processes?

Yes ☐   No ☐   Are combustible materials removed or protected from ignition?

Yes ☐   No ☐   Are fire protection and extinguishing equipment properly located and available?

Gas Welding and Cutting

Yes ☐   No ☐   Are gas cylinders secured in an upright position?

Yes ☐   No ☐   Are valve caps in place when cylinders are not in use?

Yes ☐   No ☐   Are special wrenches available when required by cylinders?

Yes ☐   No ☐   Are cylinders transported properly?

Yes ☐   No ☐   Is oxygen and acetylene stored at a distance of 20 ft. or with a firewall between them?

Yes ☐   No ☐   Are cylinder valves closed and equipment purged when not in use?

Yes ☐   No ☐   Are cylinders placed so that sparks, hot slag, or flame cannot reach them, or fire-resistant shields provided?

Yes ☐   No ☐   Are cylinders not allowed to become part of electrical circuit?

Yes ☐   No ☐   Do cylinders meet standard requirements for the general industry and maintenance?

Yes ☐   No ☐   Are fuel gas and oxygen manifolds conspicuously and permanently marked according to contents?

Yes ☐   No ☐   Are supply hose connections not interchangeable between fuel gas or oxygen supply headers?

Yes ☐   No ☐   Are hose connections free of grease and oil?

Yes ☐   No ☐   Are fuel gas hoses and oxygen hoses easily distinguishable from each other and not interchangeable?

Yes ☐   No ☐   Are hoses inspected at the beginning of each shift?

Arc Welding

Yes ☐   No ☐   Are welding curtains used where needed?

Yes ☐   No ☐   Are welding cables in good condition and properly insulated?

Yes ☐   No ☐   Do equipment and apparatus comply with UL Standards or other applicable standards?

Yes ☐   No ☐   Are manual electrode holders designed and insulated for arc welding and cutting?

Yes ☐   No ☐   Are welding cables and connectors insulated and capable of handling maximum current?

Yes ☐   No ☐   Are cables free from repair or splices for 10 ft. from the electrode holders unless of equal insulating quality and cable lug connectors insulated?

Yes ☐   No ☐   Are frames of arc welding and cutting machines grounded?

Yes ☐   No ☐   Are workers assigned to arc welding or gas shielded arc welding instructed and qualified?

Yes ☐   No ☐   Are arc welding or cutting operations shielded by flameproof screens or located in bays or booths to protect from direct rays of the arc?

Yes ☐   No ☐   Are operators specially protected from high intensities of ultraviolet radiation by screening or filter lenses?

Yes ☐   No ☐   Is skin protected by clothing or other devices?

Fire Prevention

Yes ☐   No ☐   Are fire extinguishers immediately available during welding and cutting?

Ventilation and Protection in Welding, Cutting, and Heating

Yes ☐   No ☐   Is there proper ventilation or respirators during welding and cutting operations?

Sample and Blank Material Safety Data Sheets

MSDS	**Material Safety Data Sheet**	24 Hour Emergency Telephone: 918-450-2551 CHEMTREC: 1-800-424-9300
		National Response in Canada CANUTEC: 613-996-6666
		Outside U. S. and Canada Chemtrec: 703-527-3887
From: CDR Chemicals, Inc. 3 Reactor Drive Solution, ZT 98765	*CDR*	All non-emergency questions should be direct to Customer Service (1-866-999-1100) for assistance

ACETONE

MSDS Number: A0446 --- *Effective Date: 04/10/01*

1. Product Identification

Synonyms: Dimethylketone; 2-propanone; dimethylketal
CAS No.: 67-64-1
Molecular Weight: 58.08
Chemical Formula: (CH3)2CO
Product Codes:
J.T. Baker: 5356, 5580, 5805, 9001, 9002, 9003, 9004, 9005, 9006, 9007, 9008, 9009,
9010, 9015, 9036, 9125, 9254, 9271, A134, V655
Mallinckrodt: 0018, 2432, 2435, 2437, 2438, 2440, 2443, 2445, 2850, H451, H580,
H981

2. Composition/Information on Ingredients

```
Ingredient                               CAS No        Percent   Hazardous
----------------------------------       ------------  -------   ---------

Acetone                                  67-64-1       99 - 100%   Yes
```

3. Hazards Identification

Emergency Overview

DANGER! EXTREMELY FLAMMABLE LIQUID AND VAPOR. VAPOR MAY CAUSE FLASH FIRE. HARMFUL IF SWALLOWED OR INHALED. CAUSES IRRITATION TO SKIN, EYES AND RESPIRATORY TRACT. AFFECTS CENTRAL NERVOUS SYSTEM.

J.T. Baker SAF-T-DATA^(tm) Ratings (Provided here for your convenience)

--

Health Rating: 1 - Slight
Flammability Rating: 4 - Extreme (Flammable)
Reactivity Rating: 2 - Moderate
Contact Rating: 1 - Slight
Lab Protective Equip: GOGGLES; LAB COAT; VENT HOOD; PROPER GLOVES; CLASS B EXTINGUISHER
Storage Color Code: Red (Flammable)

--

Potential Health Effects

Inhalation:
Inhalation of vapors irritates the respiratory tract. May cause coughing, dizziness, dullness, and headache. Higher concentrations can produce central nervous system depression, narcosis, and unconsciousness.
Ingestion:
Swallowing small amounts is not likely to produce harmful effects. Ingestion of larger amounts may produce abdominal pain, nausea and vomiting. Aspiration into lungs can produce severe lung damage and is a medical emergency. Other symptoms are expected to parallel inhalation.
Skin Contact:
Irritating due to defatting action on skin. Causes redness, pain, drying and cracking of the skin.
Eye Contact:
Vapors are irritating to the eyes. Splashes may cause severe irritation, with stinging, tearing, redness and pain.
Chronic Exposure:
Prolonged or repeated skin contact may produce severe irritation or dermatitis.
Aggravation of Pre-existing Conditions:
Use of alcoholic beverages enhances toxic effects. Exposure may increase the toxic potential of chlorinated hydrocarbons, such as chloroform, trichloroethane.

4. First Aid Measures

Inhalation:
Remove to fresh air. If not breathing, give artificial respiration. If breathing is difficult, give oxygen. Get medical attention.

Ingestion:
Aspiration hazard. If swallowed, vomiting may occur spontaneously, but DO NOT INDUCE. If vomiting occurs, keep head below hips to prevent aspiration into lungs. Never give anything by mouth to an unconscious person. Call a physician immediately.

Skin Contact:
Immediately flush skin with plenty of water for at least 15 minutes. Remove contaminated clothing and shoes. Get medical attention. Wash clothing before reuse. Thoroughly clean shoes before reuse.

Eye Contact:
Immediately flush eyes with plenty of water for at least 15 minutes, lifting upper and lower eyelids occasionally. Get medical attention.

5. Fire Fighting Measures

Fire:
Flash point: -20C (-4F) CC
Autoignition temperature: 465C (869F)
Flammable limits in air % by volume:
lel: 2.5; uel: 12.8
Extremely Flammable Liquid and Vapor! Vapor may cause flash fire.

Explosion:
Above flash point, vapor-air mixtures are explosive within flammable limits noted above. Vapors can flow along surfaces to distant ignition source and flash back. Contact with strong oxidizers may cause fire. Sealed containers may rupture when heated. This material may produce a floating fire hazard. Sensitive to static discharge.

Fire Extinguishing Media:
Dry chemical, alcohol foam or carbon dioxide. Water may be ineffective. Water spray may be used to keep fire exposed containers cool, dilute spills to nonflammable mixtures, protect personnel attempting to stop leak and disperse vapors.

Special Information:
In the event of a fire, wear full protective clothing and NIOSH-approved self-contained breathing apparatus with full facepiece operated in the pressure demand or other positive pressure mode.

6. Accidental Release Measures

Ventilate area of leak or spill. Remove all sources of ignition. Wear appropriate personal protective equipment as specified in Section 8. Isolate hazard area. Keep unnecessary and unprotected personnel from entering. Contain and recover liquid when possible. Use non-sparking tools and equipment. Collect liquid in an appropriate container or absorb with an inert material (e.g., vermiculite, dry sand, earth), and place in a chemical waste container. Do not use combustible materials, such as saw dust. Do not flush to sewer! If a leak or spill has not ignited, use water spray to disperse the vapors, to protect personnel attempting to stop leak, and to flush spills away from exposures. U.S. Regulations (CERCLA) require reporting spills and releases to soil, water and air in excess of reportable quantities. The toll free number for the U.S. Coast Guard National Response Center is (800) 424-8802.

J.T. Baker SOLUSORB(R) solvent adsorbent is recommended for spills of this product.

7. Handling and Storage

Protect against physical damage. Store in a cool, dry well-ventilated location, away from any area where the fire hazard may be acute. Outside or detached storage is preferred. Separate from incompatibles. Containers should be bonded and grounded for transfers to avoid static sparks. Storage and use areas should be No Smoking areas. Use non-sparking type tools and equipment, including explosion proof ventilation. Containers of this material may be hazardous when empty since they retain product residues (vapors, liquid); observe all warnings and precautions listed for the product.

8. Exposure Controls/Personal Protection

Airborne Exposure Limits:
Acetone:
-OSHA Permissible Exposure Limit (PEL):
1000 ppm (TWA)

-ACGIH Threshold Limit Value (TLV):
500 ppm (TWA), 750 ppm (STEL) A4 - not classifiable as a human carcinogen
Ventilation System:
A system of local and/or general exhaust is recommended to keep employee exposures below the Airborne Exposure Limits. Local exhaust ventilation is generally preferred because it can control the emissions of the contaminant at its source, preventing dispersion of it into the general work area. Please refer to the ACGIH document, *Industrial Ventilation, A Manual of Recommended Practices,* most recent edition, for details.
Personal Respirators (NIOSH Approved):
If the exposure limit is exceeded, a half-face organic vapor respirator may be worn for up to ten times the exposure limit or the maximum use concentration specified by the appropriate regulatory agency or respirator supplier, whichever is lowest. A full-face piece organic vapor respirator may be worn up to 50 times the exposure limit or the maximum use concentration specified by the appropriate regulatory agency or respirator supplier, whichever is lowest. For emergencies or instances where the exposure levels are not known, use a full-face piece positive-pressure, air-supplied respirator. WARNING: Air-purifying respirators do not protect workers in oxygen-deficient atmospheres.
Skin Protection:
Wear impervious protective clothing, including boots, gloves, lab coat, apron or coveralls, as appropriate, to prevent skin contact.
Eye Protection:
Use chemical safety goggles and/or a full face shield where splashing is possible. Maintain eye wash fountain and quick-drench facilities in work area.

9. Physical and Chemical Properties

Appearance:
Clear, colorless, volatile liquid.
Odor:
Fragrant, mint-like
Solubility:
Miscible in all proportions in water.
Specific Gravity:
0.79 @ 20C/4C
pH:
No information found.
% Volatiles by volume @ 21C (70F):
100
Boiling Point:
56.5C (133F) @ 760 mm Hg
Melting Point:
-95C (-139F)
Vapor Density (Air=1):
2.0
Vapor Pressure (mm Hg):
400 @ 39.5C (104F)
Evaporation Rate (BuAc=1):
ca. 7.7

10. Stability and Reactivity

Stability:
Stable under ordinary conditions of use and storage.
Hazardous Decomposition Products:
Carbon dioxide and carbon monoxide may form when heated to decomposition.
Hazardous Polymerization:
Will not occur.
Incompatibilities:
Concentrated nitric and sulfuric acid mixtures, oxidizing materials, chloroform, alkalis, chlorine compounds, acids, potassium t-butoxide.
Conditions to Avoid:
Heat, flames, ignition sources and incompatibles.

11. Toxicological Information

Oral rat LD50: 5800 mg/kg; Inhalation rat LC50: 50,100mg/m3; Irritation eye rabbit, Standard Draize, 20 mg severe; investigated as a tumorigen, mutagen, reproductive effector.

```
--------\Cancer Lists\-------------------------------------------------------
                                       ---NTP Carcinogen---
Ingredient                             Known    Anticipated    IARC Category
------------------------------------   -----    -----------    -------------
Acetone (67-64-1)                      No          No          None
```

12. Ecological Information

Environmental Fate:
When released into the soil, this material is expected to readily biodegrade. When released into the soil, this material is expected to leach into groundwater. When released into the soil, this material is expected to quickly evaporate. When released into water, this material is expected to readily biodegrade. When released to water, this material is expected to quickly evaporate. This material has a log octanol-water partition coefficient of less than 3.0. This material is not expected to significantly bioaccumulate. When released into the air, this material may be moderately degraded by reaction with photochemically produced hydroxyl radicals. When released into the air, this material may be moderately degraded by photolysis. When released into the air, this material is expected to be readily removed from the atmosphere by wet deposition.
Environmental Toxicity:
This material is not expected to be toxic to aquatic life. The LC50/96-hour values for fish are over 100 mg/l.

13. Disposal Considerations

Whatever cannot be saved for recovery or recycling should be handled as hazardous waste and sent to a RCRA approved incinerator or disposed in a RCRA approved waste facility. Processing, use or contamination of this product may change the waste management options. State and local disposal regulations may differ from federal disposal regulations. Dispose of container and unused contents in accordance with federal, state and local requirements.

14. Transport Information

Domestic (Land, D.O.T.)

Proper Shipping Name: ACETONE
Hazard Class: 3
UN/NA: UN1090
Packing Group: II
Information reported for product/size: 350LB

International (Water, I.M.O.)

Proper Shipping Name: ACETONE
Hazard Class: 3.1
UN/NA: UN1090
Packing Group: II
Information reported for product/size: 350LB

15. Regulatory Information

```
--------\Chemical Inventory Status - Part 1\-----------------------------------
Ingredient                                    TSCA   EC    Japan  Australia
-----------------------------------------     ----   ---   -----  ---------
Acetone (67-64-1)                             Yes    Yes   Yes    Yes

--------\Chemical Inventory Status - Part 2\-----------------------------------
                                                     --Canada--
Ingredient                                    Korea  DSL   NDSL   Phil.
-----------------------------------------     -----  ---   ----   -----
Acetone (67-64-1)                             Yes    Yes   No     Yes

--------\Federal, State & International Regulations - Part 1\----------------
                                              -SARA 302-      ------SARA 313------
Ingredient                                    RQ    TPQ     List   Chemical Catg.
-----------------------------------------     ---   -----   ----   --------------
Acetone (67-64-1)                             No    No      Yes    No

--------\Federal, State & International Regulations - Part 2\----------------
                                                          -RCRA-    -TSCA-
Ingredient                                    CERCLA      261.33    8(d)
-----------------------------------------     ------      ------    ------
Acetone (67-64-1)                             5000        U002      No
```

```
Chemical Weapons Convention:  No     TSCA 12(b):  Yes    CDTA:  Yes
SARA 311/312:  Acute: Yes      Chronic: No    Fire: Yes  Pressure: No
Reactivity: No            (Pure / Liquid)
```

Australian Hazchem Code: 2[Y]E
Poison Schedule: No information found.
WHMIS:
This MSDS has been prepared according to the hazard criteria of the Controlled Products Regulations (CPR) and the MSDS contains all of the information required by the CPR.

---

16. Other Information

NFPA Ratings: Health: 1 Flammability: 3 Reactivity: 0
Label Hazard Warning:
DANGER! EXTREMELY FLAMMABLE LIQUID AND VAPOR. VAPOR MAY CAUSE FLASH FIRE. HARMFUL IF SWALLOWED OR INHALED. CAUSES IRRITATION TO SKIN, EYES AND RESPIRATORY TRACT. AFFECTS CENTRAL NERVOUS SYSTEM.
Label Precautions:
Keep away from heat, sparks and flame.
Keep container closed.
Use only with adequate ventilation.
Wash thoroughly after handling.
Avoid breathing vapor.
Avoid contact with eyes, skin and clothing.
Label First Aid:
Aspiration hazard. If swallowed, vomiting may occur spontaneously, but DO NOT INDUCE. If vomiting occurs, keep head below hips to prevent aspiration into lungs. Never give anything by mouth to an unconscious person. Call a physician immediately. If inhaled, remove to fresh air. If not breathing, give artificial respiration. If breathing is difficult, give oxygen. In case of contact, immediately flush eyes or skin with plenty of water for at least 15 minutes. Remove contaminated clothing and shoes. Wash clothing before reuse. In all cases, get medical attention.
Product Use:
Laboratory Reagent.
Revision Information:
No changes.

Material Safety Data Sheet	U.S. Department of Labor	

May be used to comply with

OSHA's Hazard Communication Standard, 29 CFR 1910.1200. Standard must be consulted for specific requirements.

Occupational Safety and Health Administration

(Non-Mandatory Form)

Form Approved

OMB No. 1218-0072

IDENTITY *(As Used on Label and List)*	Note: Blank spaces are not permitted. If any item is not applicable, or no information is available, the space must be marked to indicate that.

Section I

Manufacturer's Name	Emergency Telephone Number
Address *(Number, Street, City, State, and ZIP Code)*	Telephone Number for Information
	Date Prepared
	Signature of Preparer *(optional)*

Section II - Hazard Ingredients/Identity Information

Hazardous Components (Specific Chemical Identity; Common Name(s))	OSHA PEL	ACGIH TLV	Other Limits Recommended	%*(optional)*

Section III - Physical/Chemical Characteristics

Boiling Point		Specific Gravity ($H_2O = 1$)	
Vapor Pressure (mm Hg.)		Melting Point	
Vapor Density (AIR = 1)		Evaporation Rate (Butyl Acetate = 1)	
Solubility in Water			
Appearance and Odor			

Section IV - Fire and Explosion Hazard Data

Flash Point (Method Used)		Flammable Limits	LEL	UEL
Extinguishing Media				
Special Fire Fighting Procedures				
Unusual Fire and Explosion Hazards				

(Reproduce locally)   OSHA 174, Sept. 1985

Section V - Reactivity Data

Stability	Unstable		Conditions to Avoid
	Stable		
Incompatibility *(Materials to Avoid)*			
Hazardous Decomposition or Byproducts			
Hazardous Polymerization	May Occur		Conditions to Avoid
	Will Not Occur		

Section VI - Health Hazard Data

Route(s) of Entry:	Inhalation?	Skin?	Ingestion?
Health Hazards *(Acute and Chronic)*			
Carcinogenicity:	NTP?	IARC Monographs?	OSHA Regulated?
Signs and Symptoms of Exposure			
Medical Conditions Generally Aggravated by Exposure			
Emergency and First Aid Procedures			

Section VII - Precautions for Safe Handling and Use

Steps to Be Taken in Case Material is Released or Spilled
Waste Disposal Method
Precautions to Be taken in Handling and Storing
Other Precautions

Section VIII - Control Measures

Respiratory Proctection *(Specify Type)*		
Ventilation	Local Exhaust	Special
	Mechanical *(General)*	Other
Protective Gloves		Eye Protection
Other Protective Clothing or Equipment		
Work/Hygienic Practices		

U.S.G.P.O.: 1986-491-529/45775

Ergonomics Solution Chart

Physical Work Activities and Conditions	Ergonomic Risk Factors That May Be Present	Examples of Controls
1. Exerting considerable physical effort to complete a motion	a. Force	Use powered tools
		Change pinch to power grip
		Use longer handle
		Use powered lift assist
		Use lift tables
	b. Awkward postures	Provide better mechanical advantage such as a longer handle
		Move the items closer to the worker
		Design task for smooth movements
	c. Contact stress	Attach a handle
		Wrap or coat the handle with cushioning and nonslip material
		Wear gloves that improve the grip
2. Doing the same motion over and over again	a. Repetition	Use power tools
	b. Force	Use job enlargement
		Use job rotation
		Reallocate tasks
	c. Awkward postures	Provide wrist rest
		Allow short breaks
	d. Cold temperatures	Take break in a warm area
		Provide heat where the hands are located
3. Performing motions constantly without short pauses or breaks in between	a. Repetition	Use job enlargement
	b. Force	Allow breaks as needed
	c. Awkward postures	
	d. Static postures	
	e. Contact stress	
	f. Vibration	
4. Performing tasks that involve long reaches	a. Awkward postures	Redesign the workplace layout
		Reposition object
		Provide better access to machinery

(continued)

(continued)

Physical Work Activities and Conditions	Ergonomic Risk Factors That May Be Present	Examples of Controls
		Rotate pallet or work surface
		Keep work in front of the worker
		Use a tool to extend the reach
	b. Static postures	Provide adjustability
		Allow short breaks
		Use job enlargement
		Allow tools and items to be set aside periodically
	c. Force	Use lift tables or pallet jacks
5. Working surfaces are too high or too low	a. Awkward postures	Provide adjustability
		Raise/lower the worker
		Use a tool to extend the reach
	b. Static postures	Use job enlargement
	c. Force	Reorient work
		Allow short breaks
		Use lift tables
	d. Contact stress	Ensure round edges
		Pad surfaces
6. Maintaining same position or posture while performing tasks	a. Awkward postures	Use job enlargement
		Reposition object
	b. Static postures	Reduce weight of object
		Use job rotation
		Use job enlargement
		Allow short breaks
		Use sit or stand workstation
		Use antifatigue mats
		Provide foot rest
		Provide cushioned insoles
	c. Force	Use balanced powered hand tools
		Provide lift assist
	d. Cold temperatures	Wear thermal clothing
		Take break in a warm area
		Provide localized heating
7. Sitting for a long time	a. Awkward postures	Stand occasionally
	b. Static postures	Provide lumbar support
	c. Contact stress	Allow short breaks
		Provide chairs with padding on the seat
		Make seat height adjustment
8. Using hand and power tools	a. Force	Support weight of the tool mechanically
	b. Awkward postures	Ensure tool has good balance
	c. Static postures	Use appropriate size handles
	d. Contact stress	Avoid sharp edges and finger slots on the handle
	e. Vibration	Use low-vibration tools

(continued)

Physical Work Activities and Conditions	Ergonomic Risk Factors That May Be Present	Examples of Controls
	f. Cold temperatures	Isolate source of vibration from the worker
		Maintain tools
		Reduce vibration
		Insulate hands
		Eliminate or reduce draft or blow back on the hands
9. Vibrating working surfaces, machinery, or vehicles	a. Vibration	Isolate source of vibration
	b. Force	Use job rotation
	c. Cold temperatures	Use adsorbing material to reduce the magnitude of the vibration
		Provide insulation from the cold
		Allow breaks in a warm area
10. Workstation edges or objects press hard into muscles or tendons	a. Contact stress	Provide round edges
		Enlarge handles
		Pad surfaces and handles
11. Using the hand as a hammer	a. Contact stress	Review design specifications
	b. Force	Use soft mallet
		Provide frequent maintenance
12. Using hands or body as a clamp to hold object while performing tasks	a. Force	Use a fixture, clamp or jig
	b. Static posture	Use job rotation
	c. Awkward posture	Provide round edges
	d. Contact stress	Pad surfaces
13. Gloves are bulky, too large, or too small	a. Force	Provide several sizes and weights of gloves
	b. Contact stress	
Manual Handling (lifting/lowering, pushing/pulling, and carrying)		
14. Objects or people moved are heavy	a. Force	Lighten load
	b. Repetition	Use lift assist
	c. Awkward postures	Use lift table
	d. Static posture	Place package in larger containers that have to be mechanically handled
	e. Contact stress	Use two-people lift team
		Rely on gravity to move the object
		Reduce friction
15. Horizontal reach is long	a. Force	Redesign the workplace layout
	b. Repetition	Reposition object closer to the employee
	c. Awkward postures	Provide pallet, table that can be rotated
	d. Static posture	Provide space so that the employee can walk around to the object
	e. Contact stress	Reduce the size of the object
		Slide the object closer before lifting
		Eliminate unnecessary barriers
16. Vertical reach is below knees or above the shoulders	a. Force	Do not place objects to be lifted on the floor
	b. Repetition	Use adjustable height tables
		Store heavy objects at waist height

(continued)

(continued)

Physical Work Activities and Conditions	Ergonomic Risk Factors That May Be Present	Examples of Controls
	c. Awkward postures	Put employee on a platform
		Store heavy objects at waist height
	d. Static posture	Put handles on the object
	e. Contact stress	Change the workplace layout
17. Objects or people are moved significant distances	a. Force	Modify the process to eliminate or reduce moves over a significant distance
	b. Repetition	Convey the object e.g., conveyor, ball casters, air
	c. Awkward postures	Use forklifts, hand dollies, carts, or chairs (for people)
	d. Static posture	Use appropriate wheels on carts (and maintain the wheels)
	e. Contact stress	Provide handles for pushing, pulling, or carrying
18. Bending or twisting during manual handling	a. Force	Raise work to the appropriate height
	b. Repetition	Lower the employee
	c. Awkward postures	Arrange workstation so that work is done in front of the worker
	d. Static posture	Use conveyors, chutes, slides, or turntables to change direction of the object
19. Object is slippery or has no handles	a. Force	Provide good handles
	b. Repetition	Provide belt with hand holds to assist in moving patients
	c. Awkward postures	Provide gloves that assist in holding slippery objects
	d. Static posture	
20. Floor surfaces are uneven, slippery, or sloped	a. Force	Redesign the handling job to avoid movement over poor surfaces
	b. Repetition	Use surface with treatments or antiskid strips
	c. Awkward postures	Provide footwear that improves friction
	d. Static posture	

OSHA Safety and Health Training Requirements

GENERAL INDUSTRY TRAINING REQUIREMENTS

29 CFR Part 1910

Subpart E	Means of Egress
	Employee Emergency Plans and Fire Prevention Plans—1910.38(a)(5)(i), (ii)(a) through (c), (iii), (b)(4)(i) and (ii)
Subpart F	Powered Platforms, Manlifts, and Vehicle-Mounted Work Platforms
	Powered Platforms for Building Maintenance—Operations-Training 1910.66(i), (ii)(A) through (E) and (iii) through (v)
	Care and use Appendix C, Section 1—1910.66 (e)(9)
Subpart G	Occupational Health and Environmental Control
	Dip Tanks—Personal Protection—1910.94 (d)(9)(I)
	Respirators—1910.94 (d)(9)(vi)
	Inspection, Maintenance, and Installation—1910.94(d)(11)(v)
	Hearing Protection—1910.95 (i)(4)
	Training Program—1910.95 (k)(1)-(3)(i)-(iii)
Subpart H	Hazardous Materials
	Flammable and Combustible Liquids—1910.106(b)(5)(v)(2) and (3)
	Explosives and Blasting Agents—1910.109(d)(3)(i) and (iii)
	Bulk Delivery and Mixing Vehicles—1910.109(h)(3)(d)(iii)
	Storage and Handling of Liquefied Petroleum Gases—1910.110(b)(16) and 1910.110(d)(12)(i)
	Process Safety Management of Highly Hazardous Chemicals—1910.119(g)(1)(i) and (ii)
	Contract Employer Responsibilities—1910.119(h)(3)(i) through (iv)
	Mechanical Integrity—1910.119(j)(3)
	Hazardous Waste Operations and Emergency Response—1910.120(e)(1)(i) and (ii);(2)(i) through (vii);(3)(i) through (iv) and (4) through (9)
	Hazardous Waste Cleanup Workers—1910.120 Appendix C
	New Technology Programs—1910.120(o)(i)
	Hazardous Waste—Emergency Responders—1910.120(p)(8)(iii)(A) through (C)
Subpart I	Personal Protective Equipment
	Personal Protective Equipment—1910.132(f)(1)(i)–(v);(2), (3)(i) through (iii) and (4)
	Respiratory Protection—1910.134(k)(1)(i) through (vii); (2), (3) and (5)(i) through (iii)
	Respiratory Protection for *M. Tuberculosis*—1910.139(a)(3); 1910.139(b)(3)

(continued)

29 CFR Part 1910 (continued)

Subpart J	General Environmental Controls
	Temporary Labor Camps—1910.142(k)(1) and (2)
	Specifications for Accident Prevention Signs and Tags—1910.145(c)(1)(ii),(2)(ii) and (3)
	Permit-Required Confined Spaces—1910.146(g)(1) and (2)(i) through (iv)(3) and (4) and (k) (1)(i) through (iv)
	The Control of Hazardous Energy (Lockout/Tagout) Lockout or Tagout
	Devices Removed—1910.147(a)(3)(ii);(4)(i)(D);(7)(i)(A) through (C);(ii)(A) through (F);(iii)(A) through (C)(iv) and (8)
	Outside Personnel—1910.147(f)(2)(i)
Subpart K	Medical Services and First Aid
	Medical Services and First Aid—1910.151(a) and (b)
Subpart L	Fire Protection
	Fire Protection—1910.155(c)(iv)(41)
	Fire Brigades—1910.156(b)(1)
	Training and Education—1910.156(c)(1)–(4)
	Portable Fire Extinguishers—1910.157(g)(1), (2), and (4)
	Fixed Extinguishing Systems—1910.160(b)(10)
	Fire Detection Systems—1910.164(c)(4)
	Employee Alarm Systems—1910.165(d)(5)
Subpart N	Materials Handling and Storage
	Servicing of Multipiece and Single-Piece Rim Wheels—1910.177(c)(1)(i) through (iii); (2)(i) through (viii), and (3)
	Powered Industrial Trucks—1910.178(1)
	Moving the Load—1910.179(n)(3)(ix)
	Crawler Locomotives and Truck Cranes—1910.180(i)(5)(ii)
Subpart O	Machinery and Machine Guarding
	Mechanical Power Presses—1910.217(e)(3)—1910.217(f)(2)
	Mechanical Power Presses—Instructions to Operators—1910.217(e)(2)
	Training of Maintenance Personnel—1910.217(e)(3)
	Operator Training—1910.217(H)(13)(i)(A) through (E) and (ii)
	Forging Machines—1910.218(a)(2)(iii)
Subpart Q	Welding, Cutting, and Brazing
	General Requirements—1910.252(a)(2)(xiii)(c)
	Oxygen—Fuel Gas Welding and Cutting—1910.253(a)(4)
	Arc Welding and Cutting—1910.254(a)(3)
	Resistance Welding—1910.255(a)(3)
Subpart R	Special Industries
	Pulp, Paper, and Paperboard Mills—1910.261(h)(3)(ii)
	Laundry Machinery and Operating Rules—1910.264(d)(1)(v)
	Sawmills—1910.265(c)(3)(x)
	Logging—1910.266(i) and (2)(i) through (iv); (3)(i) through (vi); (4) and (5)(i) through (iv); (6) and (7)(i) through (iii); (8) and (9)
	Telecommunications—1910.268(b)(2)(i)
	Derrick Trucks—1910.268(j)(4)(iv)(D)
	Cable Fault Locating—1910.268(l)(1)
	Guarding Manholes—1910.268(o)(1)(ii)
	Joint Power and Telecommunication Manholes—1910.268(o)(3)
	Tree Trimming—Electrical Hazards—1910.268(q)(1)(ii)(A) through (D)
	Electric Power Generation, Transmission, and Distribution—1910.269(b)(1)(i) and (ii); (d)(vi) (A) through (C); (vii); (viii)(A) through (C); and (ix)

29 CFR Part 1910 (continued)

	Grain Handling Facilities—1910.272(e)(1)(i) and (ii) and (2)
	Entry Into Bins, Silos, and Tanks—1910.272(g)(5)
	Contractors—1910.272(h)(2)
Subpart S	Electrical Safety-Related Work Practices
	Content of Training—1910.332(b)(1)
Subpart T	Commercial Diving Operations
	Qualifications of Dive Team—1910.410(a)(1); (2)(i)–(iii); (3) and (4)

TRAINING REQUIREMENTS IN OSHA STANDARDS AND TRAINING GUIDELINES FOR TOXIC AND HAZARDOUS MATERIALS

Subpart Z	Toxic Substances
	Asbestos—1910.1001(j)(7)(i) through (iii)(A) through (H)
	4-Nitrobiphenyl—1910.1003(e)(5)(i)(a)–(h)(i) and (ii)
	Alpha-Naphthylamine—1910.1004(e)(5)(i)(a) through (h)(i) and (ii)
	Methyl Chloromethyl Ether—1910.1006(e)(5)(i)(a) through (h)(i) and (ii)
	3,3-Dichlorobenzidine (and its salts)—1910.1007(e)(5)(i)(a) through (h)(i) and (ii)
	Bis-Chloromethyl Ether—1910.1008(e)(5)(i)(a) through (h)(i) and (ii)
	Beta-Naphthylamine—1910.1009(e)(5)(i)(a) through (h)(i) and (ii)
	Benzidine—1910.1010(e)(5)(i)(a) through (h)(i) and (ii)
	4-Aminodiphenyl—1910.1011(e)(5)(i)(a) through (h)(i) and (ii)
	Ethyleneimine—1910.1012(e)(5)(i)(a) through (h)(i) and (ii)
	Beta-Propiolactone—1910.1013(e)(5)(i)(a) through (h)(i) and (ii)
	2-Acetylaminofluorene—1910.1014(e)(5)(i)(a) through (h)(i) and (ii)
	4-Dimethylaminoazobenzene—1910.1015(e)(5)(i)(a)–(h)(i) and (ii)
	N-Nitrosodimethylamine—1910.1016(e)(5)(i)(a) through (h)(i) and (ii)
	Vinyl Chloride—1910.1017(j)(1)(i)–(ix)
	Inorganic Arsenic—1910.1018(o)(1)(i) and (ii)(A) through (F) and (2)(i) and (ii)
	Lead—1910.1025(l)(1)(i)–(v)(A) through (G)(2)(i) through (iii)
	Cadmium—1910.1027(m)(4)(i) through (iii)(A) through (H) and (m)(4)(iv)(A) and (B)
	Benzene—1910.1028(j)(3)(i) through (iii)(A) and (B)
	Coke Oven Emissions—1910.1029(k)(1)(i) through (iv)(a) through (e) and (k)(2)(i) and (ii)
	Bloodborne Pathogens—1910.1030(g)(2)(i); (ii)(A) through (C); (iii) through (vii)(A) through (N); (viii) and (ix)(A) through (C)
	Cotton Dust—1910.1043(i)(1)(i)(A) through (F) and (2)(i) and (ii)
	1,2-Dibromo-3-Chloropropane—1910.1044(n)(1)(i) and (ii)(a) through (e) and (n)(2)(i) and (ii)
	Acrylonitrile (Vinyl Cyanide)—1910.1045(o)(1) and (iii)(A) through (G) and (2)(i) and (ii)
	Ethylene Oxide—1910.1047(j)(3)(i);(ii)(A) through (D) and (iii)(A)–(D)
	Formaldehyde—1910.1048(n)(1) through (3)(i) and (ii)(A) and (B)(iii)–(vii)
	4,4 Methylenedianiline—1910.1050(k)(3)(i) and (ii)(A) and (4)(i)(ii)
	Ionizing Radiation Testing—1910.1096(f)(3)(viii)
	Posting—1910.1096(i)(2)
	Hazard Communication—1910.1200(h)(1),(2)(i) through (iii) and (3)(i) through (iv)
	Occupational Exposure to Hazardous Chemicals in Laboratories—1910.1450(f)(1)(2) and (f)(4)(i)(A)–(C) and (ii)

SHIPYARD EMPLOYMENT TRAINING REQUIREMENTS

29 CFR Part 1915

Subpart A	General Provisions
	Commercial Diving Operations
	Competent Person—1915.7(b)(1)(i) through (iv);(2)(i) though (iii)(A) through (C); and (c)(1) through (7)
Subpart B	Explosive and Other Dangerous Atmospheres
	Confined and Enclosed Spaces—1915.12(d)(1) and (2)(i) through (iii),(3)(i) through (iii),(4)(i) and (ii),(5)(i) and (ii)
	Precautions Before Entering—1915.12(a)(1)(i) through (v)
	Cleaning and Other Cold Work—1915.13(b)(2) and (4)
	Certification before Hot Work Is Begun—1915.14(b)(1)(i) through (v)
	Maintaining Gas Free Conditions, Ship Repairing—1910.15(c)
Subpart C	Surface Preparation and Preservation
	Painting—1915.35(b)(1) and (8)
	Flammable Liquids—1915.36(a)(2) and (5)
Subpart D	Welding, Cutting, and Heating
	Fire Prevention—1915.52(b)(3) and (c)
	Welding, Cutting, and Heating in Way of Preservative Coatings—1915.53(b)
	Welding, Cutting, and Heating of Hollow Metal Containers and Structures Not Covered by 1915.12
	Gas Welding and Cutting—1915.55(d)(1) through (6)
	Arc Welding and Cutting—1915.56(d)(1) through (4)
	Uses of Fissionable Material—1915.57(b)
Subpart E	Scaffolds, Ladders, and Other Working Surfaces
	Scaffolds or Staging—1915.71(b)(7)
Subpart F	General Working Conditions
	Work on or in the Vicinity of Radar and Radio—1915.95(a)
	First-Aid—1915.98(a)
Subpart G	Gear and Equipment for Rigging and Materials Handling
	Ropes, Chains, and Slings—1915.112(c)(5)
	Use of Gear—1915.116(1)
	Qualifications of Operators—1915.117(a) and (b)
Subpart H	Tools and Related Equipment
	Powder-Actuated Fastening Tools—1915.135(a) and (c)(1) through (6)
	Internal Combustion Engines, Other Than Ships' Equipment—1915.136(c)
Subpart I	Personal Protective Equipment
	General Requirements—1915.152(e)(1)(i) through (v); (2), (3)(i) through (iii); and (4)
	Respiratory Protection—1915.152(a)(4)
	Personal Fall Arrest Systems—1915.159(d)
	Positioning Device Systems—1915.160(d)
Subpart K	Portable, Unfired Pressure Vessels, Drums, and Containers, Other Than Ships' Equipment
	Portable Air Receivers and Other Unfired Pressure Vessels—1915.172(b)
Subpart Z	Toxic and Hazardous Substances
	Asbestos—1915.1001(k)(9)(i) through (vi)(A) through (J)
	Carcinogens—1915.1003
	Vinyl Chloride—1915.1017

29 CFR Part 1915 (continued)

Inorganic Arsenic—1915.1018
Lead—1915.1025
Cadmium—1915.1027
Benzene—1915.1028
Bloodborne Pathogens—1915.1030
1,2-Dibromo-3-Chloropropane—1915.1044
Acrylonitrile—1915.1045
Ethylene Oxide—1915.1047
Formaldehyde—1915.1048
Methylenedianiline—1915.1050
Ionizing Radiation—1915.1096
Hazard Communication—1915.1200
Occupational Exposure to Hazardous Chemicals in Laboratories—1915.1450

MARINE TERMINAL TRAINING REQUIREMENTS

29 CFR Part 1917

Subpart A	Scope and Definitions
	Commercial Diving Operations—1917.1(a)(2)(iii)
	Electrical Safety-Related Work Practices—1917.1(a)(2)(iv)
	Grain Handling Facilities—1917.1(a)(2)(v)
	Hazard Communication—1917.1(a)(2)(vi)
	Ionizing Radiation—1917.1(a)(2)(vii)
	Hearing Protection—1917.1(a)(2)(viii)
	Respiratory Protection—1917.1(a)(2)(x)
	Servicing Multipiece and Single-Piece Rim Wheels—1917.1(a)(2)(xii)
	Toxic and Hazardous Substances—1917.1(a)(2)(xiii)
Subpart B	Marine Terminal Operations
	Hazardous Atmospheres and Substances—1917.23(b)(1)
	Fumigants, Pesticides, Insecticides, and Hazardous Preservatives—1917.25(e)(2) and (3)
	Personnel—1917.27(a)(1) and (b)(1) and (2)
	Hazard Communication—1917.28
	Emergency Action Plans—1917.30(a)(5)(i) and (ii)(A) through (C)(iii)
Subpart C	Cargo Handling Gear and Equipment
	General Rules Applicable to Vehicles—1917.44(i) and (ii)(A) through (G)
Subpart D	Specialized Terminals
	Terminal facilities—Handling Menhaden and Similar Species of Fish—1917.73(d)
	Related Terminal Operations and Equipment
	Welding, Cutting, and Heating (Hot Work)—1917.152(c)(4)

LONGSHORING TRAINING REQUIREMENTS

29 CFR Part 1918

Subpart A	Scope and Definitions
	Commercial Diving Operations—1918.1(b)(2)
	Electrical Safety-Related Work Practices—1918.1(b)(3)
	Hazard Communication—1918.1(b)(4)
	Ionizing Radiation—1918.1(b)(5)
	Hearing Protection—1918.1(b)(6)
	Respiratory Protection—1918.1(b)(8)
	Toxic and Hazardous Substances—1918.1(b)(9)
Subpart H	Handling Cargo
	Containerized Cargo Operations—Fall Protection Systems—1918.85(k)(12)
Subpart I	General Working Conditions
	Hazardous Atmospheres and Substances—1918.93(d)(3)
	Ventilation and Atmospheric Conditions and Fumigants—1918.94(b)(v)
	First-Aid and Life Saving Facilities—1918.97(b)
	Qualifications of Machinery Operators—1918.98(a)(1)

CONSTRUCTION INDUSTRY TRAINING REQUIREMENTS

29 CFR Part 1926

Subpart C	General Safety and Health Provisions
	General Safety and Health Provisions—1926.20(b(2) and (4))
	Safety Training and Education—1926.21(a)
	Employee Emergency Action Plans—1926.26.35(e)(1) and (2)(i) through (iii) and (3)
Subpart D	Occupational Health and Environmental Controls
	Medical Services and First Aid—1926.50©
	Ionizing Radiation—1926.53(b)
	Nonionizing Radiation—1926.54(a) and (b)
	Gases, Vapors, Fumes, Dusts, and Mists—1926.55(b)
	Hazard Communication—1926.59
	Methylenedianiline—1926.60(l)(3)(i) and (ii)(A) through (C)
	Lead in Construction—1926.62(l)(1)(i) through (iv); (2)(i) through (viii); and (3)(i) and (ii)
	Process Safety Management of Highly Hazardous Chemicals—1926.64
	Hazardous Waste Operations and Emergency Response—1926.65
Subpart E	Personal Protective and Life Saving Equipment
	Hearing Protection—1926.101(b)
	Respiratory Protection Subpart F Fire Protection and Prevention—1926.103(c)(1)
	Fire Protection Subpart G Signs, Signals, and Barricades—1926.150(a)(5)
	Signaling—1926.201(a)(2)
Subpart I	Tools—Hand and Power
	Power-Operated Hand Tools—1926.302(e)(1) and (12)
	Woodworking Tools—1926.304(f)
Subpart J	Welding and Cutting

29 CFR Part 1926 (continued)

	Gas Welding and Cutting—1926.350(d)(1) through (6)
	Arc Welding and Cutting—1926.351(d)(1) through (5)
	Fire Prevention—1926.352(e)
	Welding, Cutting, and Heating in Way of Preservative Coatings—1926.354(a)
Subpart K	Electrical
	Ground-Fault Protection—1926.404(b)(iii)(B)
Subpart L	Scaffolding
	Scaffolding—Training Requirements—1926.454(a)(1) through (5) and (b)(1) through (4) and (c)(1) through (3)
Subpart M	Fall Protection
	Fall Protection—Training Requirements—1926.503(a)(1) and (2)(ii) through (vii)
Subpart N	Cranes, Derricks, Hoists, Elevators, and Conveyors
	Cranes and Derricks—1926.550(a)(1), (5), and (6)
	Material Hoists, Personnel Hoists, and Elevators—1926.552(a)(1)
Subpart O	Motor Vehicles, Mechanized Equipment, and Marine Operations
	Material-Handling Equipment—1926.602(c)(1)(vi)
	Powered Industrial Trucks (Forklifts)—1926.602(d)
	Site Clearing—1926.604(a)(1)
Subpart P	Excavations
	General Protection—1926.651(c)(1)(i)
Subpart Q	Concrete and Masonry Construction
	Concrete and Masonry Construction—1926.701(a)
Subpart R	Steel Erection
	Bolting, Riveting, Fitting-up, and Plumbing-up—1926.752(d)(4)
Subpart S	Underground Construction, Caissons, Cofferdams, and Compressed Air
	Underground Construction—1926.800(d)
	Compressed Air—1926.803(a)(1) and (2)
Subpart T	Demolition
	Preparatory Operations—1926.850(a)
	Chutes—1926.852(c)
	Mechanical Demolition—1926.859(g)
Subpart U	Blasting and Use of Explosives
	General Provisions—1926.900(a)
	Blaster Qualifications—1926.901(c), (d), and (e)
	Surface Transportation of Explosives—1926.902(b) and (i)
	Firing the Blast—1926.909(a)
Subpart V	Power Transmission and Distribution
	General Requirements—1926.950(d)(1)(ii)(a),(c), (vi) and (vii)
	Overhead Lines—1926.955(b)(3)(i)
	Underground Lines—1926.956(b)(1)
	Construction in Energized Substations—1926.957(a)(1)
Subpart X	Stairways and Ladders
	Ladders—1926.1053(b)(15)
	Training Requirements—1926.1060(a)(i) through (v) and (b)
Subpart Y	Diving
	Commercial Diving Operations
Subpart Z	Toxic and Hazardous Substances—1926.1076

(continued)

29 CFR Part 1926 (continued)

Asbestos—1926.1101(9)(i) through (viii)(A) through (e)(10)
Carcinogens—1926.1103
Vinyl Chloride—1926.1117
Inorganic Arsenic—1926.1118
Cadmium—1926.1127(m)(4)(i) through (iii)(A) through (E)
Benzene—1926.1128
Coke Oven Emissions—1926.1129
1,2-Dibromo-3-Chloropropane—1926.1144
Acrylonitrile—1926.1145
Ethylene Oxide—1926.1147
Formaldehyde—1926.1148
Methylene Chloride—1926.1152

AGRICULTURE TRAINING REQUIREMENTS

29 CFR Part 1928

Subpart B	Applicability of Standards
	Temporary Labor Camps—1928.142
	Logging—1928.266
	Hazard Communication—1928.1200
	Cadmium—1928.1027
Subpart C	Roll-Over Protective Structures
	Roll-Over Protective Structures (ROPS) for Tractors Used in Agricultural Operations—1928.51(d)
Subpart D	Safety for Agricultural Equipment
	Guarding of Farm Field Equipment, Farmstead Equipment, and Cotton Gins—1928.57(a)(6)(i) through (v)
Subpart M	Occupational Health
	Cadmium—1928.1027

FEDERAL EMPLOYEE TRAINING REQUIREMENTS

29 CFR Part 1960

Subpart B	Financial Management
Subpart D	Inspection and Abatement
	Qualifications of Safety and Health Inspectors and Agency Inspections—1960.25(a)
Subpart E	General Services Administration and Other Federal Agencies
	Safety and Health Services—1960.34(e)(1)
Subpart F	Occupational Safety and Health Committees
	Agency Responsibilities—1960.39(b)
Subpart H	Training of
	Top Management—1960.54
	Supervisors—1960.55(a) and (b)
	Safety and Health Specialists—1960.56(a) and (b)
	Safety and Health Inspectors—1960.57
	Collateral Duty, Safety and Health Personnel, and Committee Members—1960.58
	Employees and Employee Representatives—1960.59(a) and (b)
	Training Assistance—1960.60(a) through (d)
Subpart K	Federal Safety and Health Councils
	Role of the Secretary—1960.85(b)
	Objectives of Field Councils—1960.87(d)

OSHA Offices

REGIONAL OFFICES

Region I: Connecticut, Maine, Massachusetts, New Hampshire, Rhode Island, Vermont
JFK Federal Building, Room E340
Boston, MA 02203
Phone: (617) 565-9860
Fax: (617) 565-9827

Region II: New Jersey, New York, Puerto Rico, Virgin Islands
201 Varick Street, Room 670
New York, NY 10014
Phone: (212) 337-2378
Fax: (212) 337-2371

Region III: Delaware, District of Columbia, Maryland, Pennsylvania, Virginia, West Virginia
The Curtis Center, Suite 740 West
170 S. Independence Mall West
Philadelphia, PA 19106-3309
Phone: (215) 861-4900
Fax: (215) 861-4904

Region IV: Alabama, Florida, Georgia, Kentucky, Mississippi, North Carolina,
South Carolina, Tennessee
61 Forsyth Street, SW
Atlanta, GA 30303
Phone: (404) 562-2300
Fax: (404) 562-2295

Region V: Illinois, Indiana, Michigan, Minnesota, Ohio, Wisconsin
230 South Dearborn Street, Room 3244
Chicago, IL 60604
Phone: (312) 353-2220
Fax: (312) 353-7774

Region VI: Arkansas, Louisiana, New Mexico, Oklahoma, Texas
525 Griffin Square Building, Room 602
Dallas, TX 75202
Phone: (972) 850-4145
Fax: (972) 850-4149

Region VII: Iowa, Kansas, Missouri, Nebraska
  Two Pershing Square
  2300 Main Street, Suite 1010
  Kansas City, MO 64108
  Phone: (816) 283-8745
  Fax: (816) 283-0547

Region VIII: Colorado, Montana, North Dakota, South Dakota, Utah, Wyoming
  1999 Broadway, Suite 1690
  Denver, CO 80202
  Phone: (720) 264-6550
  Fax: (720) 264-6585

Region IX: Arizona, California, Hawaii, Nevada, American Samoa, Guam,
Trust Territory of the Pacific Islands
  90 7th Street, Suite 18100
  San Francisco, CA 94103
  Phone: (415) 625-2547
  Fax: (415) 625-2534

Region X: Alaska, Idaho, Oregon, Washington
  1111 Third Avenue, Suite 715
  Seattle, WA 98101-3212
  Phone: (206) 553-5930
  Fax: (206) 553-6499

STATE PLAN OFFICES

Alaska
  Alaska Department of Labor and Workforce Development
  PO Box 11149
  1111 W. 8th Street, Room 304
  Juneau, AK 99801-1149
  Phone: (907) 269-4955
  Fax: (907) 269-4950

Arizona
  Industrial Commission of Arizona
  800 W. Washington
  Phoenix, AZ 85007-2922
  Phone: (602) 542-5795
  Fax: (602) 542-1614

California
  California Department of Industrial Relations
  455 Golden Gate Avenue—10th Floor
  San Francisco, CA 94102
  Phone: (415) 703-5050
  Fax: (415) 703-5059

Connecticut
  Connecticut Department of Labor
  Conn-OSHA
  200 Folly Brook Boulevard
  Wethersfield, CT 06109
  Phone: (860) 263-6900
  Fax: (860) 263-6940

Hawaii
  Hawaii Department of Labor and Industrial Relations
  830 Punchbowl Street
  Honolulu, HI 96813
  Phone: (808) 586-8842
  Fax: (808) 586-9099

Indiana
  Indiana Department of Labor
  State Office Building
  402 West Washington Street, Room W195
  Indianapolis, IN 46204-2751
  Phone: (317) 232-2378
  Fax: (317) 233-3790

Iowa
  Iowa Division of Labor
  1000 E. Grand Avenue
  Des Moines, IA 50319-0209
  Phone: (515) 242-8067
  Fax: (515) 281-7995

Kentucky
  Kentucky Department of Labor
  1047 U.S. Highway 127 South, Suite 4
  Frankfort, KY 40601
  Phone: (502) 564-3070
  Fax: (502) 564-5387

Maryland
  Maryland Division of Labor and Industry
  Department of Labor, Licensing and Regulation
  1100 North Eutaw Street, Room 613
  Baltimore, MD 21201-2206
  Phone: (410) 767-2241
  Fax: (410) 767-2986

Michigan
  Michigan Occupational Safety and Health Administration
  PO Box 30643
  Lansing, MI 48909-8143
  Phone: (517) 322-1814
  Fax: (517) 322-1775

Minnesota
  Minnesota Department of Labor
    and Industry
  443 Lafayette Road
  St. Paul, MN 55155-4307
  Phone: (651) 284-5310
  Fax: (651) 284-5741

Nevada
  Nevada Division of Industrial Relations
  400 West King Street, Suite 200
  Carson City, NV 89073
  Phone: (702) 486-9020
  Fax: (702) 990-0358

New Jersey
  New Jersey Department of Labor
  1 John Fitch Plaza—Labor Building
  Market and Warren Streets
  PO Box 386
  Trenton, NJ 08625-0386
  Phone: (609) 292-2975
  Fax: (609) 633-9271

New Mexico
  New Mexico Environment Department
  1190 St. Francis Drive, Suite 4050
  PO Box 26110
  Santa Fe, NM 87502
  Phone: (505) 827-2850
  Fax: (505) 827-2836

New York
  New York Department of Labor
  W. Averell Harriman State Office
    Building—12, Room 500
  Albany, NY 12240
  Phone: (518) 457-3518
  Fax: (518) 457-1519

North Carolina
  North Carolina Department of Labor
  4 West Edenton Street
  Raleigh, NC 27601-1092
  Phone: (919) 733-0359
  Fax: (919) 733-6197

Oregon
  Oregon Occupational Safety and Health Division
  Department of Consumer & Business Services
  350 Winter Street, NE, Room 430
  Salem, OR 97309-0405
  Phone: (503) 378-3272
  Fax: (503) 947-7461

Puerto Rico
    Puerto Rico Department of Labor and Human Resources
    Prudencio Rivera Martínez Building, 21st Floor
    505 Muñoz Rivera Avenue
    Hato Rey, PR 00918
    Phone: (787) 754-2172
    Fax: (787) 767-6051

South Carolina
    South Carolina Department of Labor, Licensing, and Regulation
    Koger Office Park, Kingstree Building
    110 Centerview Drive
    PO Box 11329
    Columbia, SC 29210
    Phone: (803) 896-4300
    Fax: (803) 896-4393

Tennessee
    Tennessee Department of Labor
    200 French Landing Drive
    Nashville, TN 37243
    Phone: (615) 741-2793
    Fax: (615) 741-3325

Utah
    Utah Labor Commission
    160 East 300 South, 3rd Floor
    PO Box 146650
    Salt Lake City, UT 84114-6650
    Phone: (801) 530-6898
    Fax: (801) 530-6390

Vermont
    Vermont Department of Labor and Industry
    5 Green Mountain Drive
    PO Box 485
    Montpelier, VT 05601-0458
    Phone: (802) 828-2288
    Fax: (802) 828-2748

Virgin Islands
    Virgin Islands Department of Labor
    2203 Church Street
    Christiansted, St. Croix, VI 00820-4660
    Phone: (340) 773-1994
    Fax: (340) 773-1858

Virginia
    Virginia Department of Labor and Industry
    Powers-Taylor Building
    13 South 13th Street
    Richmond, VA 23219
    Phone: (804) 786-2377
    Fax: (804) 371-6524

Washington
  Washington Department of Labor and Industries
  PO Box 44001
  Olympia, WA 98504-4001
  Phone: (360) 902-4200
  Fax: (360) 902-4202

Wyoming
  Wyoming Department of Employment
  Workers' Safety and Compensation Division
  1510 East Pershing Building—West Wing
  Cheyenne, WY 82002
  Phone: (307) 777-7786
  Fax: (307) 777-3646

50 Most Cited Violations by Major Industrial Groups

Listed below are the standards that were cited by Federal OSHA for the specified SIC during the period October 2006 through September 2007. Code of Federal Regulations (CFR) 1910 is the standard for general industry and 1926 is the standard for construction.

DIVISION A—AGRICULTURE, FORESTRY, AND FISHING (1–9)

CFR Standard	# Cited	Description
1910.1200	115	Hazard Communication
1910.132	75	Personal Protective Equipment, General Requirements
5A1	64	General Duty Clause (Section of OSHA Act)
1910.67	55	Vehicle-Mounted Elevating and Rotating Work Platforms
1910.266	54	Logging Operations
1910.133	48	Eye and Face Protection
1910.305	47	Electrical, Wiring Methods, Components and Equipment
1910.303	34	Electrical Systems Design, General Requirements
1910.135	30	Head Protection
1928.57	27	Guarding on Farm Equipment, Farmstead Equipment, and Cotton Gins
1910.212	25	Machines, General Requirements
1910.134	23	Respiratory Protection
1910.157	23	Portable Fire Extinguishers
1910.178	23	Powered Industrial Trucks (Forklifts)
1910.22	21	Walking–Working Surfaces, General Requirements
1910.147	21	The Control of Hazardous Energy, Lockout/Tagout
1910.215	20	Abrasive Wheel Machinery
1910.23	19	Guarding Floor and Wall Openings and Holes
1910.219	17	Mechanical Power-Transmission Apparatus
1904.29	16	Forms
1910.213	16	Woodworking Machinery Requirements
1910.37	12	Maintenance, Safeguards, and Operational Features for Exit Routes
1910.141	11	Sanitation
1910.304	11	Electrical, Wiring Design and Protection
1910.119	10	Process Safety Management, Highly Hazardous Chemicals
1910.146	10	Permit-Required Confined Spaces

(continued)

CFR Standard	# Cited	Description
1910.253	10	Oxygen–Fuel Gas Welding and Cutting
1910.269	10	Electric Power Generation, Transmission, and Distribution
1904.32	9	Annual Summary
1904.39	9	Reporting Fatalities and Multiple Hospitalization Incidents to OSHA
1910.151	9	Medical Services and First Aid
1910.333	9	Electrical, Selection and Use of Work Practices
1910.95	8	Occupational Noise Exposure
1926.451	8	Scaffolds, General Requirements
1926.550	8	Cranes and Derricks
1960.8	8	Agency Responsibilities
1910.138	7	Hand Protection
1910.179	7	Overhead and Gantry Cranes
1910.334	7	Electrical, Use of Equipment
1910.106	6	Flammable and Combustible Liquids
1910.1030	6	Bloodborne Pathogens
1926.453	6	Aerial Lifts
1910.242	5	Hand and Portable Powered Tools and Equipment
1910.1000	5	Air Contaminants
1910.101	4	Compressed Gases, General Requirements
1910.177	4	Servicing Multipiece and Single-Piece Rim Wheels
1926.651	4	Specific Excavation Requirements
1928.51	4	Roll-Over Protective Structures (ROPS) for Tractors Used in Agriculture Operations
1904.41	3	Annual OSHA Injury and Illness Survey for 10 or More Employers
1910.26	3	Portable Metal Ladders

DIVISION B—MINING

CFR Standard	# Cited	Description
1910.23	166	Guarding Floor and Wall Openings and Holes
5A1	156	General Duty Clause (Section of OSHA Act)
1910.305	128	Wiring Methods, Components, and Equipment for General Use
1910.1200	119	Hazard Communication
1910.134	82	Respiratory Protection
1910.132	79	Personal Protective Equipment, General Requirements
1910.303	76	Electrical Systems Design, General Requirements
1910.151	71	Medical Services and First Aid
1910.157	56	Portable Fire Extinguishers
1910.219	55	Mechanical Power-Transmission Apparatus
1910.184	46	Slings
1910.178	35	Powered Industrial Trucks (Forklifts)
1910.212	35	Machines, General Requirements
1910.24	28	Fixed Industrial Stairs
1910.146	26	Permit-Required Confined Spaces
1910.141	24	Sanitation

CFR Standard	# Cited	Description
1910.22	23	Walking–Working Surfaces, General Requirements
1910.242	20	Hand and Portable Powered Tools and Equipment
1910.304	20	Electrical, Wiring Design and Protection
1910.215	19	Abrasive Wheel Machinery
1910.27	15	Fixed Ladders
1910.133	15	Eye and Face Protection
1910.253	15	Oxygen–Fuel Gas Welding and Cutting
1910.307	15	Hazardous (Classified) Locations
1910.1030	15	Bloodborne Pathogens
1910.147	14	The Control of Hazardous Energy, Lockout/Tagout
1904.29	13	Forms
1910.252	13	Welding, Cutting, and Brazing, General Requirements
1910.145	9	Specifications for Accident Prevention Signs and Tags
1910.107	8	Spray Finishing with Flammable and Combustible Materials
1910.332	8	Training
1910.333	7	Electrical, Selection and Use of Work Practices
1904.39	6	Reporting Fatalities and Multiple Hospitalization Incidents to OSHA
1910.26	6	Portable Metal Ladders
1926.651	6	Specific Excavation Requirements
1910.36	5	Design and Construction Requirements for Exit Routes
1910.120	5	Hazardous Waste Operations and Emergency Response
1910.135	5	Head Protection
1910.136	5	Occupational Foot Protection
1910.334	5	Electrical, Use of Equipment
1926.652	5	Requirements for Protective Systems
1910.37	4	Maintenance, Safeguards, and Operational Features for Exit Routes
1910.106	4	Flammable and Combustible Liquids
1910.110	4	Storage and Handling of Liquefied Petroleum Gases
1910.119	4	Process Safety Management of Highly Hazardous Chemicals
1910.335	4	Safeguards for Personnel Protection
1904.32	3	Annual Survey
1904.40	3	Providing Records to Government Representatives
1910.95	3	Occupational Noise Exposure

DIVISION C—CONSTRUCTION (SIC 15–17)

CFR Standard	# Cited	Description
1926.451	10489	General Requirements for all Types of Scaffolding
1926.501	6824	Duty to Have Fall Protection
1926.1053	2931	Ladders
1926.503	1986	Fall Protection Training Requirements
1926.20	1952	Construction, General Safety and Health Provisions
1926.453	1934	Aerial Lifts
1926.651	1886	Specific Excavations Requirements

(continued)

CFR Standard	# Cited	Description
1926.100	1845	Head Protection
1910.1200	1697	Hazard Communication
1926.454	1426	Training Requirements for all Types of Scaffolding
1926.404	1316	Electrical, Wiring Design and Protection
1926.652	1312	Excavations, Requirements for Protective Systems
1926.405	1255	Electrical Wiring Method, Component and Equipment General Use
1926.1052	1062	Stairways
1926.021	1056	Construction, Safety Training and Education
1926.102	1016	Eye and Face Protection
1926.502	989	Fall Protection Systems Criteria and Practices
1926.452	859	Additional Requirements for Specific Scaffolding
1926.403	696	Electrical, General Requirements
1910.134	670	Respiratory Protection
1926.062	526	Lead
1926.1060	494	Stairways and Ladders, Training Requirements
1926.095	432	Criteria for Personal Protective Equipment
1926.1101	419	Asbestos
1926.416	418	Electrical Safety-Related Work Practices, General Requirements
1926.760	398	Fall Protection
5A1	380	General Duty Clause (Section of OSHA Act)
1926.701	330	Concrete/Masonry, General Requirements
1926.25	327	Construction, Housekeeping
1926.602	321	Material-Handling Equipment
1910.178	296	Powered Industrial Trucks (Forklifts)
1926.304	288	Woodworking Tools
1926.150	282	Fire Protection
1926.1051	280	Stairways and Ladders, General Requirements
1926.350	267	Gas Welding and Cutting
1926.550	237	Cranes and Derricks
1926.300	216	Hand and Power Tools, General Requirements
1926.251	156	Rigging Equipment for Material Handling
1926.050	137	Medical Services and First Aid
1926.028	132	Construction, Personal Protective Equipment
1904.29	126	Forms
1926.850	126	Demolition, Preparatory Operations
1926.200	117	Accident Prevention Signs and Tags
1626.761	104	Steel Erection, Training
1926.153	90	Liquefied Petroleum Gas
1903.19	85	Abatement Verification
1926.51	82	Sanitation
1926.059	76	Hazard Communication
1910.146	69	Permit-Required Confined Spaces

DIVISION D—MANUFACTURING (SIC 20–39)

CFR Standard	# Cited	Description
1910.147	3175	The Control of Hazardous Energy, Lockout/Tagout
1910.1200	2694	Hazard Communication
1910.212	2258	Machines, General Requirements
1910.134	2197	Respiratory Protection
1910.305	2011	Electrical, Wiring Methods, Components and Equipment
1910.178	1920	Powered Industrial Trucks (Forklifts)
1910.303	1502	Electrical Systems Design, General Requirements
1910.219	1476	Mechanical Power-Transmission Apparatus
1910.215	1023	Abrasive Wheel Machinery
1910.132	952	Personal Protective Equipment, General Requirements
1910.23	861	Guarding Floor and Wall Openings and Holes
1910.95	861	Occupational Noise Exposure
1910.213	804	Woodworking Machinery Requirements
1910.157	686	Portable Fire Extinguishers
1910.217	659	Mechanical Power Presses
1910.107	637	Spray Finishing with Flammable and Combustible Materials
1910.146	608	Permit-Required Confined Spaces
1910.266	600	Logging Operations
1910.22	577	Walking–Working Surfaces, General Requirements
1910.37	569	Maintenance, Safeguards, and Operational Features for Exit Routes
1910.106	516	Flammable and Combustible Liquids
1904.29	473	Forms
1910.179	454	Overhead and Gantry Cranes
5A1	422	General Duty Clause (Section of OSHA Act)
1910.304	413	Electrical, Wiring Design and Protection
1910.253	403	Oxygen–Fuel Gas Welding and Cutting
1910.242	398	Hand and Portable Powered Tools and Equipment, General
1910.119	374	Process Safety Management, Highly Hazardous Chemicals
1910.151	372	Medical Services and First Aid
1910.133	310	Eye and Face Protection
1910.1000	266	Air Contaminants
1910.184	260	Slings
1904.32	217	Annual Summary
1910.1025	208	Lead
1910.1026	199	Chromium (VI)
1910.36	189	Design and Construction Requirements for Exit Routes
1910.141	184	Sanitation
1910.252	182	Welding, Cutting, and Brazing, General Requirements
19101.30	172	Bloodborne Pathogens
1910.335	170	Safeguards for Personnel Protection
1910.24	165	Fixed Industrial Stairs
1910.176	159	Materials Handling, General
1910.333	157	Electrical, Selection and Use of Work Practices
1910.243	150	Guarding of Portable Power Tools
1910.334	143	Electrical, Use of Equipment
1910.1052	141	Methylene Chloride
1910.120	122	Hazardous Waste Operations and Emergency Response
1910.138	119	Hand Protection
1904.41	118	Annual OSHA Injury and Illness Survey of 10 or More Employees
1910.110	116	Storage and Handling of Liquefied Petroleum Gases

DIVISION E—TRANSPORTATION, COMMUNICATIONS, ELECTRIC, GAS, AND SANITARY SERVICES (SIC 40–49)

CFR Standard	# Cited	Description
1910.178	480	Powered Industrial Trucks (Forklifts)
1910.1200	376	Hazard Communication
1910.305	316	Electrical, Wiring Methods, Components and Equipment
1910.303	241	Electrical Systems Design, General Requirements
1910.37	178	Maintenance, Safeguards, and Operational Features for Exit Routes
1910.132	162	Personal Protective Equipment, General Requirements
1910.147	160	The Control of Hazardous Energy, Lockout/Tagout
1910.157	149	Portable Fire Extinguishers
1910.23	147	Guarding Floor and Wall Openings and Holes
1910.134	145	Respiratory Protection
1904.29	134	Forms
1910.215	122	Abrasive Wheel Machinery
5A1	118	General Duty Clause (Section of OSHA Act)
1910.22	104	Walking–Working Surfaces, General Requirements
1910.212	99	Machines, General Requirements
1910.1030	97	Bloodborne Pathogens
1910.151	88	Medical Services and First Aid
1904.32	73	Annual Summary
1910.253	70	Oxygen–Fuel Gas Welding and Cutting
1910.219	69	Mechanical Power-Transmission Apparatus
1910.36	63	Design and Construction Requirements for Exit Routes
1910.304	59	Electrical, Wiring Design and Protection
1910.146	55	Permit-Required Confined Spaces
1910.106	52	Flammable and Combustible Liquids
1910.176	50	Materials Handling, General
1910.269	37	Electric Power Generation, Transmission, and Distribution
1910.95	36	Occupational Noise Exposure
1910.119	36	Process Safety Management, Highly Hazardous Chemicals
1910.141	34	Sanitation
1910.38	33	Emergency Action Plans
1904.41	31	Annual OSHA Injury and Illness Survey of 10 or More Employers
1910.1025	30	Lead
1910.24	26	Fixed Industrial Stairs
1910.133	26	Eye and Face Protection
1910.1001	26	Asbestos, Tremolite, Anthophyllite, and Actinolite
1910.177	23	Servicing Multipiece and Single-Piece Rim Wheels
1910.334	23	Electrical, Use of Equipment
1910.179	22	Overhead and Gantry Cranes
1910.213	22	Woodworking Machinery Requirements
1917.26	22	First-Aid and Lifesaving Facilities
1917.43	22	Powered Industrial Trucks
1910.101	21	Compressed Gases (General Requirements)
1910.242	21	Hand and Portable Powered Tools and Equipment, General
1910.120	18	Hazardous Waste Operations and Emergency Response
1910.272	18	Grain Handling Facilities
1926.651	18	Specific Excavation Requirements
1904.2	17	Partial Exemptions for Establishments in Certain Industries
1910.107	17	Spray Finishing Using Flammable and Combustible Materials
1917.45	17	Cranes and Derricks
1910.110	16	Storage and Handling of Liquefied Petroleum Gases

DIVISION F—WHOLESALE TRADE (SIC 50–51)

CFR Standard	# Cited	Description
1910.178	461	Powered Industrial Trucks (Forklifts)
1910.1200	345	Hazard Communication
1910.305	255	Electrical, Wiring Methods, Components and Equipment
1910.147	216	The Control of Hazardous Energy, Lockout/Tagout
1910.303	195	Electrical Systems Design, General Requirements
1910.37	172	Maintenance, Safeguards, and Operational Features for Exit Routes
1910.23	170	Guarding Floor and Wall Openings and Holes
1910.157	143	Portable Fire Extinguishers
1910.132	140	Personal Protective Equipment, General Requirements
1910.212	134	Machines, General Requirements
1910.134	133	Respiratory Protection
1910.22	85	Walking–Working Surfaces, General Requirements
1910.219	83	Mechanical Power-Transmission Apparatus
1910.215	77	Abrasive Wheel Machinery
1904.29	75	Forms
1910.151	72	Medical Services and First Aid
5A1	65	General Duty Clause (Section of OSHA Act)
1910.36	61	Design and Construction Requirements for Exit Routes
1910.176	61	Materials Handling, General
1910.304	52	Electrical, Wiring Design and Protection
1910.253	51	Oxygen–Fuel Gas Welding and Cutting
1910.272	49	Grain Handling Facilities
1910.1025	47	Lead
1904.32	41	Annual Summary
1910.110	37	Storage and Handling of Liquefied Petroleum Gases
1910.38	35	Emergency Action Plans
1910.106	34	Flammable and Combustible Liquids
1910.141	33	Sanitation
1910.133	31	Eye and Face Protection
1910.119	29	Process Safety Management, Highly Hazardous Chemicals
1910.146	27	Permit-Required Confined Spaces
1910.24	26	Fixed Industrial Stairs
1910.213	25	Woodworking Machinery Requirements
1910.184	24	Slings
1910.101	23	Compressed Gases, General Requirements
1910.242	22	Hand and Portable Powered Tools and Equipment, General
1910.95	21	Occupational Noise Exposure
1910.334	20	Electrical, Use of Equipment
1910.179	17	Overhead and Gantry Cranes
1904.2	16	Partial Exemption for Establishments in Certain Industries
1910.138	16	Hand Protection
1910.27	15	Fixed Ladders
1910.1027	15	Cadmium
1910.1030	15	Bloodborne Pathogens

(continued)

CFR Standard	# Cited	Description
1910.26	13	Portable Metal Ladders
1910.107	12	Spray Finishing with Flammable and Combustible Materials
1904.41	11	Annual OSHA Injury and Illness Survey of 10 or More Employers
1910.333	11	Electrical, Selection and Use of Work Practices
1926.62	11	Lead
1904.7	10	General Recording Criteria

DIVISION G—RETAIL TRADE (SIC 52–59)

CFR Standard	# Cited	Description
1910.1200	305	Hazard Communication
1910.178	205	Powered Industrial Trucks (Forklifts)
1910.305	181	Electrical, Wiring Methods, Components and Equipment
1910.303	179	Electrical Systems Design, General Requirements
1910.37	169	Maintenance, Safeguards, and Operational Features for Exit Routes
1910.132	127	Personal Protective Equipment, General Requirements
1910.157	108	Portable Fire Extinguishers
1910.22	102	Walking–Working Surfaces, General Requirements
1910.23	95	Guarding Floor and Wall Openings and Holes
1910.134	81	Respiratory Protection
1910.213	67	Woodworking Machinery Requirements
1910.151	60	Medical Services and First Aid
1910.36	56	Design and Construction Requirements for Exit Routes
1910.147	56	The Control of Hazardous Energy, Lockout/Tagout
1904.29	54	Forms
1910.212	53	Machines, General Requirements
1910.176	48	Materials Handling, General
1910.215	43	Abrasive Wheel Machinery
5A1	43	General Duty Clause (Section of OSHA Act)
1910.1030	42	Bloodborne Pathogens
1910.133	39	Eye and Face Protection
1910.304	35	Electrical, Wiring Design and Protection
1910.141	34	Sanitation
1910.219	26	Mechanical Power-Transmission Apparatus
1910.38	21	Emergency Action Plans
1910.138	21	Hand Protection
1910.106	16	Flammable and Combustible Liquids
1910.177	16	Servicing Multipiece and Single-Piece Rim Wheels
1904.32	15	Annual Summary
1910.24	14	Fixed Industrial Stairs
1910.101	13	Compressed Gases, General Requirements
1910.110	13	Storage and Handling of Liquefied Petroleum Gases
1910.253	13	Oxygen–Fuel Gas Welding and Cutting
1910.334	13	Electrical, Use of Equipment
1910.1052	13	Methylene Chloride

CFR Standard	# Cited	Description
1926.501	12	Duty to Have Fall Protection
1910.107	11	Spray Finishing with Flammable and Combustible Materials
1904.40	9	Providing Records to Government Representatives
1910.26	9	Portable Metal Ladders
1910.242	9	Hand and Portable Powered Tools and Equipment, General
1910.95	7	Occupational Noise Exposure
1910.145	7	Specifications, Accident Prevention Signs and Tags
1910.146	7	Permit-Required Confined Spaces
1910.184	7	Slings
1926.503	7	Training Requirements, Fall Protection
1904.41	6	Annual OSHA Survey of Injury and Illness of 10 or More Employers
1910.333	6	Electrical, Selection and Use of Work Practices
1910.1000	6	Air Contaminants
1926.453	6	Aerial Lifts
1903.19	5	Abatement Verification

DIVISION H—FINANCE, INSURANCE, AND REAL ESTATE (SIC 60–67)

CFR Standard	# Cited	Description
1910.1200	35	Hazard Communication
1926.1101	17	Asbestos
1910.132	12	Personal Protective Equipment, General Requirements
1910.1001	12	Asbestos, Tremolite, Anthophyllite, and Actinolite
1910.146	11	Permit-Required Confined Spaces
1910.36	10	Design and Construction Requirements for Exit Routes
1910.134	9	Respiratory Protection
1910.147	7	The Control of Hazardous Energy, Lockout/Tagout
1910.138	5	Hand Protection
1910.303	5	Electrical Systems Design, General Requirements
5A1	5	General Duty Clause (Section of OSHA Act)
1910.22	4	Walking–Working Surfaces, General Requirements
1910.141	4	Sanitation
1910.305	4	Electrical, Wiring Methods, Components and Equipment
1910.1030	4	Bloodborne Pathogens
1910.23	3	Guarding Floor and Wall Openings and Holes
1910.133	3	Eye and Face Protection
1910.219	3	Mechanical Power-Transmission Apparatus
1910.333	3	Electrical, Selection and Use of Work Practices
1910.67	2	Vehicle-Mounted Elevating and Rotating Work Platforms
1910.101	2	Compressed Gases, General Requirements

(continued)

CFR Standard	# Cited	Description
1910.212	2	Machines, General Requirements
1910.334	2	Electrical, Use of Equipment
1910.1025	2	Lead
1926.20	2	General Safety and Health Provisions
1926.453	2	Aerial Lifts
1926.501	2	Duty to Have Fall Protection
1926.850	2	Demolition, Preparatory Operations
1903.19	1	Abatement Verification
1904.29	1	Forms
1910.26	1	Portable Metal Ladders
1910.37	1	Maintenance, Safeguards, and Operational Features for Exit Routes
1910.151	1	Medical Services and First Aid
1910.157	1	Portable Fire Extinguishers
1910.178	1	Powered Industrial Trucks
1910.332	1	Training, Electrical
1910.335	1	Safeguards for Personnel Protection
1926.21	1	Safety Training and Education
1926.100	1	Head Protection
1926.454	1	Scaffolds, Training Requirements
1926.503	1	Fall Protection, Training Requirements
1926.1053	1	Ladders

DIVISION I—SERVICES (SIC 70–80)

CFR Standard	# Cited	Description
1910.1030	1173	Bloodborne Pathogens
1910.1200	882	Hazard Communication
1910.134	523	Respiratory Protection
1910.132	307	Personal Protective Equipment, General Requirements
1910.147	293	The Control of Hazardous Energy, Lockout/Tagout
1910.305	285	Electrical, Wiring Methods, Components and Equipment
1910.303	251	Electrical Systems Design, General Requirements
1904.29	186	Forms
1910.37	171	Maintenance, Safeguards, and Operational Features for Exit Routes
1910.151	159	Medical Services and First Aid
1910.157	126	Portable Fire Extinguishers
1910.215	124	Abrasive Wheel Machinery
1910.23	121	Guarding Floor and Wall Openings and Holes
1910.22	108	Walking–Working Surfaces, General Requirements
1910.212	108	Machines, General Requirements
1910.178	103	Powered Industrial Trucks (Forklifts)
5A1	98	General Duty Clause (Section of OSHA Act)

CFR Standard	# Cited	Description
1926.1101	93	Asbestos
1910.133	84	Eye and Face Protection
1910.304	82	Electrical, Wiring Design and Protection
1910.1025	81	Lead
1904.32	80	Annual Summary
1910.142	79	Temporary Labor Camps
1910.107	77	Spray Finishing with Flammable and Combustible Materials
1910.36	71	Design and Construction Requirements for Exit Routes
1910.146	71	Permit-Required Confined Spaces
1910.219	69	Mechanical Power-Transmission Apparatus
1910.67	58	Vehicle-Mounted Elevating and Rotating Work Platforms
1910.253	57	Oxygen–Fuel Gas Welding and Cutting
1910.106	50	Flammable and Combustible Liquids
1910.138	49	Hand Protection
1910.141	49	Sanitation
1904.41	47	Annual OSHA Injury and Illness Survey of 10 or More Employers
1910.1001	43	Asbestos, Tremolite, Anthophyllite, and Actinolite
1910.1052	37	Methylene Chloride
1910.333	32	Electrical, Selection and Use of Work Practices
1926.453	31	Aerial Lift
1910.184	29	Slings
1910.101	27	Compressed Gases, General Requirements
1910.242	27	Hand and Portable Powered Tools and Equipment, General
1910.95	23	Occupational Noise Exposure
1926.550	22	Cranes and Derricks
1904.8	20	Recording Criteria for Needlestick and Sharps Injuries
1926.451	21	General Requirements for All Types of Scaffolding
1910.252	20	General Requirements, Welding
1910.1000	20	Air Contaminants
1926.501	20	Duty to Have Fall Protection
1910.24	19	Fixed Industrial Stairs

DIVISION J—PUBLIC ADMINISTRATION (SIC 91–99)

CFR Standard	# Cited	Description
1910.303	82	Electrical Systems Design, General Requirements
1910.305	61	Electrical, Wiring Methods, Components and Equipment
1910.37	53	Maintenance, Safeguards, and Operational Features for Exit Routes
1910.215	47	Abrasive Wheel Machinery
1910.157	46	Portable Fire Extinguishers
1910.22	34	Walking–Working Surfaces, General Requirements
1910.32	32	Annual Summary
1910.1200	32	Hazard Communication
1910.213	30	Woodworking Machinery Requirements
1910.304	30	Electrical, Wiring Design and Protection
1910.1030	28	Bloodborne Pathogens

(continued)

CFR Standard	# Cited	Description
1910.23	27	Guarding Floor and Wall Openings and Holes
1910.119	26	Process Safety Management of Highly Hazardous Chemicals
1910.212	25	Machines, General Requirements
1910.1025	23	Lead
1960.55	23	Training of Supervisors
1910.147	22	The Control of Hazardous Energy, Lockout/Tagout
1910.178	21	Powered Industrial Trucks (Forklifts)
1910.134	19	Respiratory Protection
1926.1101	17	Asbestos
1910.36	16	Design and Construction Requirements for Exit Routes
1904.29	15	Forms
1910.132	13	Personal Protective Equipment, General Requirements
1960.8	13	Agency Responsibilities
1910.219	12	Mechanical Power-Transmission Apparatus
1910.107	11	Spray Finishing Using Flammable and Combustible Materials
1910.334	11	Electrical, Use of Equipment
1910.1001	10	Asbestos
1960.59	9	Training Employees and Employee Representatives
1910.39	8	Fire Prevention Plans
1904.30	7	Multiple Business Establishments
1910.24	7	Fixed Industrial Stairs
1910.38	7	Emergency Action Plans
1910.101	7	Compressed Gases (General Requirements)
1910.151	6	Medical Services and First Aid
1910.176	6	Handling Materials (General)
1910.253	6	Oxygen–Fuel Gas Welding and Cutting
1960.67	6	Federal Agency Certification of the Injury and Illness Annual Summary
1910.95	5	Occupational Noise Exposure
1910.146	5	Permit-Required Confined Spaces
1910.333	5	Electrical, Selection and Use of Work Practices
1910.335	5	Safeguards for Personnel Protection
1960.71	5	Agency Annual Report
1904.40	4	Providing Records to Government Representatives
1910.110	4	Storage and Handling of Liquefied Petroleum Gases
1910.141	4	Sanitation
1910.242	4	Hand and Portable Powered Tools and Equipment, General
1910.1047	4	Ethylene Oxide
1960.12	4	Dissemination of Occupational Safety and Health Program Information
1910.67	3	Vehicle-Mounted Elevating and Rotating Work Platforms

Summary of 29 CFR 1910

In this appendix you will find an overview approach to the general industry standard entitled 29 CFR 1910. A paragraph highlights the content of each subpart of this standard. You will also find a listing of the sections that are contained within each subpart. In addition, a checklist is included for each subpart. If you answer "Yes" to any question, then some or all of the subpart would be applicable to your operation. Thus, a yes answer suggests that your workplace needs to be in compliance with the applicable sections of that subpart.

You can find a similar appendix for 29 CFR 1926 in the *Handbook of OSHA Construction Safety and Health*, 2nd edn., 2006, published by CRC Press/Lewis Publishers.

This appendix provides a tool that can be of assistance to any industry or business as they attempt to comply with the regulations that the Occupational Safety and Health Administration (OSHA) have developed, promulgated, and enforced.

PART 1910—OCCUPATIONAL SAFETY AND HEALTH STANDARDS FOR THE GENERAL INDUSTRY

Subpart A—General

This subpart explains the purpose and scope, definitions, petitions for issuance, amendment, and repel of a standard in 29 CFR 1910. The purpose of these standards is to make the workplace safer and healthier. It explains the applicability of the OSHA standards relevant to the workplaces covered, the geographic location covered, and specific entities (i.e., Federal agencies) not covered. It also lists regulations that have been incorporated in this standard by reference into 29 CFR 1910 as well as the requirements for nationally recognized testing laboratories.

Checklist
_____ Do you want to see a standard issued, amended, or repealed?
_____ Does this standard include your operation or business?
_____ Are you interested in the requirements for nationally recognized testing laboratories?
_____ Do you want to order a copy of a standard incorporated by reference into 29 CFR 1910?

Sections of Subpart A
1910.1 Purpose and scope
1910.2 Definitions
1910.3 Petitions for the issuance, amendment, or repeal of a standard
1910.4 Amendments to this part
1910.5 Applicability of standards
1910.6 Incorporation by reference

1910.7 Definition and requirements for a nationally recognized testing laboratory
1910.8 OMB control numbers under the Paperwork Reduction Act

Subpart B—Adoption and Extension of Established Federal Standards

Subpart B adopts and extends the applicability of established federal standards to every employer, employee, and place of employment covered by the Act. Only standards relating to safety or health are adopted into this Act. This also pertains to any facility engaging in construction, alterations, or repair, including painting and decorating.

The Construction Safety Act adopts as occupational safety and health standards under section 6 of the Act the standards that are prescribed in part 1926 of this appendix. Thus, the standards (substantive rules) published in subpart C and the following subparts of part 1926 of this appendix are applied. This section does not incorporate subparts A and B of part 1926 of this appendix.

This subpart covers adoption and extension of established safety and health standards for shipyard employment, and the standards prescribed by part 1915 (formerly parts 1501–1503) of this title and in effect since April 28, 1971 (as revised) are adopted as occupational safety or health standards under section 6(a) of the Act and shall apply, according to the provisions thereof, to every employment and place of employment of every employee engaged in ship repair, shipbreaking, and shipbuilding, or a related employment.

Part 1918 of this chapter shall apply exclusively, according to the provisions thereof, to all employment of every employee engaged in longshoring operations, marine terminals, or related employment aboard any vessel. All cargo transfer accomplished with the use of shore-based material-handling devices shall be governed by part 1917 of this chapter.

Workplaces that expose workers to asbestos, tremolite, anthophyllite, and actinolite dust; vinyl chloride; acrylonitrile; inorganic arsenic; lead; benzene; ethylene oxide; 4,4′-methylenedianiline; formaldehyde; cadmium; 1,3-butadiene; and methylene chloride are covered by appropriate 1910 standards.

Checklist
_____Are construction activities taking place?
_____Is shipyard employment occurring?
_____Is longshoring taking place?
_____Are workers exposed to the hazardous chemicals mentioned here?

Sections of Subpart B
1910.11 Scope and purpose
1910.12 Construction work
1910.15 Shipyard employment
1910.16 Longshoring and marine terminals
1910.17 Effective dates
1910.18 Changes in established federal standards
1910.19 Special provisions for air contaminants

Subpart C—Removed and Reserved

1910.20 [redesignated as 1910.1020]

Subpart D—Walking and Working Surfaces

This subpart addresses the requirements for maintaining walking and working surfaces. Subpart D applies to all permanent places of employment. It contains regulations pertaining to housekeeping, aisles and passageways, guarding wall and floor openings, fixed stairs, portable wood and metal ladders, fixed ladders, scaffolding, and manually propelled mobile ladder stand and scaffolds from frame to suspended types as well as dockboards, forging machine areas, and veneering machine areas.

Checklist

_____Do you use dockboards?

_____Do you have forging machines or veneering machines at your site?

_____Is attention paid to housekeeping?

_____Are there floor and wall openings or holes at your facility?

_____Do you have manually propelled mobile ladder stands and scaffolds?

_____Do your workers use scaffolds in the performance of their work?

_____Do you own scaffolds?

_____Do you enforce housekeeping?

_____Do you erect, tear down, or maintain scaffolds?

_____Are you responsible for training workers regarding scaffolds and their safety?

_____Are there scaffolds on your worksite?

_____Do your workers use ladders in performing their work?

_____Are your workers required to ascend and descend industrial stairs?

_____Does your company own ladders?

_____Do your workers have to climb fixed ladders?

Sections of Subpart D

1910.21 Definitions

1910.22 General requirements

1910.23 Guarding floor and wall openings and holes

1910.24 Fixed industrial stairs

1910.25 Portable wood ladders

1910.26 Portable metal ladders

1910.27 Fixed ladders

1910.28 Safety requirements for scaffolding

1910.29 Manually propelled mobile ladder stands and scaffolds (towers)

1910.30 Other working surfaces

Subpart E—Exit Routes, Emergency Action Plans, and Fire Prevention Plans

This subpart deals specifically with providing a safe, continuous, and unobstructed means of exit and the design, construction, and maintenance to assure an open travelway from any point in a building or structure to a safe exit. The standard addresses exits by describing the makeup of an exit, specific physical requirements for an exit, and the number of exits required. It requires the employer to comply with the National Fire Protection Association's (NFPA) Life Safety Code regarding safe exits. This subpart also contains the requirements essential to providing a safe means of exit from fire and like emergencies. The subpart sets forth the requirements for emergency action plans and fire prevention plans. Emergency action and fire prevention plans are required, which can assure adequate escape procedures, evacuation routes, alarm systems, and other emergency actions.

Checklist

_____Are exits designed to meet the NFPA Life Safety Code?

_____Do exits meet the requirements of the regulation for construction?

_____Are exits maintained to meet the intended purpose?
_____Do you have an emergency action or escape plan or procedure?
_____Do you have a fire prevention plan for your facility?
_____Are all exits unlocked and free from impediments?
_____Is there a safe means of exit for all your workers?
_____Are all exits designed to be visible and allow for a safe exit from your facility?

Sections of Subpart E
1910.33 Table of contents
1910.34 Coverage and definitions
1910.35 Compliance with the NFPA 101-2000, Life Safety Code
1910.36 Design and construction requirements for exit routes
1910.37 Maintenance, safeguards, and operational features for exit routes
1910.38 Emergency action plans
1910.39 Fire prevention plans

Subpart E Appendix—Exit Routes, Emergency Action Plans, and Fire Prevention Plans

Subpart F—Powered Platforms, Manlifts, and Vehicle-Mounted Work Platforms

This subpart covers powered platform installations permanently dedicated to interior or exterior building maintenance of a specific structure or group of structures. It does not apply to suspended self-powered platforms used to service buildings as well as the guidelines for personal fall arrest systems. This subpart applies to all permanent installations completed after July 23, 1990, and contains information on powered platforms for building maintenance. Building maintenance covers a wide array of activities from window cleaning to engineering design of equipment as well as expressing the need to train workers. In addition, this section specifically addresses the requirements for vehicle-mounted elevating and rotating work platforms and manlifts.

Checklist
_____Do you provide fall protection for your workforce?
_____Are powered platforms used for building maintenance?
_____Do you have vehicle-mounted elevated and rotating work platforms?
_____Do you have manlifts?

Sections of Subpart F
1910.66 Powered platforms for building maintenance
1910.67 Vehicle-mounted elevating and rotating work platforms
1910.68 Manlifts

Subpart G—Occupational Health and Environmental Control

The standards in subpart G deal with air quality, noise exposure exceeding 85 dB, and nonionizing radiation exposure in the workplace. Ventilation is specific for facilities that use abrasive blasting; facilities that have spray booths, or open surface tanks used for cleaning, and facilities with grinding, polishing, and buffing operations.

Checklist
_____Do you have ventilation issues caused by abrasive blasting?
_____Do you have ventilation issues caused by spray booths?
_____Do you have ventilation issues caused by open surface tanks?

_____Do you have ventilation issues caused by grinding, polishing, and buffing operations?
_____Do you have noise exposure in excess of the 85 dBA level?
_____Do you have a source of nonionizing radiation?
_____Does your company use any chemicals that could be considered hazardous?
_____Does your company have a medical officer for examinations, advice, or consultation?
_____Have you had injuries or illnesses that require first aid?
_____Have you had to do environmental or air monitoring?
_____Do you provide drinking water to workers?
_____Do you provide for toilets and washing facilities?
_____Do you have high-noise worksites or tasks?
_____Do you have sources of ionizing or nonionizing (lasers) radiation at your worksites?
_____Do you do contracting jobs where chemical processes involving highly hazardous chemicals take place?
_____Do you use some form of ventilation to remove airborne contaminants?
_____Do you do hazardous waste remediation work?
_____Do you do night work or work in areas with limited light?

Sections of Subpart G
1910.94 Ventilation
1910.95 Occupational noise exposure
1910.96 [Redesignated as 1910.1096]
1910.97 Nonionizing radiation
1910.98 Effective dates

Subpart H—Hazardous Materials

Subpart H contains information on compressed gases, acetylene, hydrogen, oxygen, nitrous oxide, flammable and combustible liquids, spray finishing using flammable and combustible materials, dip tanks using flammable or combustible liquids, explosive and blasting agents, storage and handling of liquid petroleum gasses, and storage and handling of anhydrous ammonia. This section also covers process-safety management requirements of highly hazardous chemicals and hazardous waste operations and emergency response. The final part of the regulation deals with dipping and coating processes.

Checklist
_____Do you use compressed gases?
_____Do you have acetylene, hydrogen, oxygen, or nitrous oxide on the premises?
_____Do handle, use, or store flammable or combustible liquids?
_____Do have spray finishing operations with flammable or combustible liquids?
_____Do have dip tanks using flammable or combustible liquids?
_____Do you have highly hazardous chemicals or chemical processes?
_____Do you have workers trained to remediate hazardous chemicals or respond to HAZMAT situations?
_____Do you have dipping and coating operations?
_____Does your company have equipment used for explosives and blasting?
_____Do any of your workers perform explosive handling and blasting operations?
_____Do you have blasting materials on your jobsites?
_____Do you have individuals who are qualified blasters employed by you?
_____Do you have a contract blaster doing your blasting operations?
_____Does your company contract to do blasting activities?
_____Does your company or workers transport explosives or blasting materials?
_____Do blasting activities occur on your jobsites or projects?

Sections of Subpart H
1910.101 Compressed gases (general requirements)
1910.102 Acetylene
1910.103 Hydrogen
1910.104 Oxygen
1910.105 Nitrous oxide
1910.106 Flammable and combustible liquids
1910.107 Spray finishing using flammable and combustible materials
1910.108 Dip tanks containing flammable or combustible liquids
1910.109 Explosives and blasting agents
1910.110 Storage and handling of liquefied petroleum gases
1910.111 Storage and handling of anhydrous ammonia
1910.112 [Reserved]
1910.113 [Reserved]
1910.119 Process safety management of highly hazardous chemicals
1910.120 Hazardous waste operations and emergency response
1910.121 [Reserved]
1910.122 Table of contents
1910.123 Dipping and coating operations: coverage and definitions
1910.124 General requirements for dipping and coating operations
1910.125 Additional requirements for dipping and coating operations that use flammable or combustible liquids
1910.126 Additional requirements for special dipping and coating applications

Subpart I—Personal Protective Equipment (PPE)

Subpart I requires employers to provide employees with proper PPE for the work performed. As part of this requirement, the employer must conduct a hazard survey of the work to determine the control measures to use where hazards cannot be eliminated. This serves as a resource in guiding the selection of appropriate PPE. This includes PPE for eyes, face, head, and extremities. Other types of equipment that may be required are protective clothing and equipment as well as respiratory devices All PPE is to be maintained and in a sanitary condition.

Not only are the employers required to provide needed PPE, but they are also required to train workers how to use and wear their PPE. Equipment for emergency use should be stored in an accessible location known to all workers. The requirements for respirators and their use is the most extensive part of the subpart.

This subpart provides the standard for quality, as well as selection of PPE such as eye and face protection, head protection, respiratory protection, foot protection, and hand and arm protection.

Checklist
_____Do you require PPE to be used?
_____Do you have the potential for falling, flying, or electrical hazards?
_____Do you require head protection?
_____Are there opportunities for heavy material to fall onto the workers' feet?
_____Do you provide hand and arm protection, i.e., gloves?
_____Does your workforce come into contact with electricity where they need protective equipment?
_____Do you have the potential at any time for workers to suffer eye injuries?
_____Do environment or air contaminants require the use of respirators?
_____Do your workers need eye and face protection?
_____Do your workers have the potential to be exposed to tuberculosis?

Sections of Subpart I
1910.132 General requirements
1910.133 Eye and face protection
1910.134 Respiratory protection
1910.135 Head protection
1910.136 Foot protection
1910.137 Electrical protective devices
1910.138 Hand Protection
1910.139 Respiratory protection for *M. tuberculosis*

Subpart J—General Environmental Controls

This section specifically applies to places of employment where such items as sanitary facilities, e.g., toilet facilities, washing facilities, sanitary food storage, and food handling, are present. It also addresses temporary labor camps, safety colors for marking physical hazards, and requirements for accident prevention signs and tags. Two additional items specifically addressed by this section and of considerable importance are permit-required confined spaces and the control of hazardous energy (lockout/tagout).

Checklist
_____Do you provide sanitary facilities for your workforce?
_____Do you have temporary labor camps?
_____Do you have warning or accident prevention signs or tags posted in your workplace?
_____Do you use appropriate colors to mark physical hazards?
_____Do you have confined spaces in your workplace?
_____Does your workforce enter confined spaces where permits are needed?
_____Do you have a lockout/tagout program in place?
_____Do you require lockout/tagout procedures to be followed?

Sections of Subpart J
1910.141 Sanitation
1910.142 Temporary labor camps
1910.143 Nonwater carriage disposal systems [Reserved]
1910.144 Safety color code for marking physical hazards
1910.145 Specifications for accident prevention signs and tags
1910.146 Permit-required confined spaces
1910.147 The control of hazardous energy (lockout/tagout)

Subpart K—Medical and First Aid

The purpose of medical and first aid is to provide the employee with readily available medical consultation. If medical personnel are not readily available, then personnel adequately trained to administer first aid are to be present. These individuals should be provided protection and PPE to prevent exposure to bloodborne pathogens. The employer is required to provide fully equipped first-aid kits and they are to be maintained and in suitable numbers to meet the needs of the workforce.

Checklist
_____Are there qualified medical personnel at the facility?
_____Do you have personnel trained in first aid available?
_____Do you have first-aid kits available?
_____Do you keep first-aid kits adequately stocked?

Sections of Subpart K
1910.151 Medical services and first aid
1910.152 [Reserved]

Subpart L—Fire Protection

Subpart L is concerned with fire protection and fire prevention. This subpart contains requirements for fire brigades, all portable extinguishers, fixed fire suppression systems and fire detection systems, and alarm systems. It contains training requirements for the organization and personnel. It describes requirements for training and protective equipment for fire brigades.

In addition, this subpart establishes the requirements for the placement, use, maintenance, and testing of portable fire extinguishers provided for use by employees, as well as the requirements for all automatic sprinkler systems installed to meet a particular OSHA standard. Firefighting equipment is to be available and readily accessible. Workers are to be trained annually on the use of fire extinguishers.

The fire detection system should be in a labeled specific location. Lastly, a unique alarm system must be established at the worksite that will alert employees to a fire.

Checklist
_____Does your worksite have a fire hazard potential?
_____Do you have a fire prevention program?
_____Do you use fire extinguishers at your site?
_____Do you train workers in fire prevention and firefighting?
_____Do you have a fire brigade?
_____Do you have a fire detection system or fire alarm system?
_____Are your employees expected to fight fires?

Sections of Subpart L
1910.155 Scope, application, and definitions applicable to this subpart
1910.156 Fire brigades

Portable Fire Suppression Equipment

1910.157 Portable fire extinguishers
1910.158 Standpipe and hose systems

Fixed Fire Suppression Equipment

1910.159 Automatic sprinkler systems
1910.160 Fixed extinguishing systems, general
1910.161 Fixed extinguishing systems, dry chemical
1910.162 Fixed extinguishing systems, gaseous agent
1910.163 Fixed extinguishing systems, water spray and foam

Other Fire Protective Systems

1910.164 Fire detection systems
1910.165 Employee alarm systems
Appendices to Subpart L
Appendix A to Subpart L—Fire Protection
Appendix B to Subpart L—National Consensus Standards
Appendix C to Subpart L—Fire Protection References for Further Information

Appendix D to Subpart L—Availability of Publications Incorporated by Reference in Section 1910.156 Fire Brigades

Appendix E to Subpart L—Test Methods for Protective Clothing

Subpart M—Compressed-Gas and Compressed-Air Equipment

This subpart applies to compressed-air receivers and other equipment used in providing and using compressed air for performing operations such as cleaning, drilling, hoisting, and chipping. However, this section does not deal with the special problems created by using compressed air to convey materials, or the problems created when work is performed in compressed-air environments such as in tunnels and caissons. This section is not intended to apply to compressed-air machinery and equipment used for transportation vehicles such as steam railroad cars, electric railway cars, and automotive equipment.

Checklist

_____Do you use a compressed-air receiver?

_____Do you have equipment that provides compressed air?

Sections of Subpart M

1910.166 [Reserved]

1910.167 [Reserved]

1910.168 [Reserved]

1910.169 Air receivers

Subpart N—Materials Handling and Storage

Subpart N details the storage of materials and how to stack, rack, and secure them against falling or sliding. Materials should not create a hazard due to storage in aisles or passageways. Housekeeping is an important component of handling and storing of materials.

Subpart N provides provisions for cranes, derricks, hoists, helicopters, conveyors, and aerial lifts. This subpart delimits many common safety requirements for material-handling equipment and reinforces the need to follow the manufacturer's requirements regarding load capacities, speed limits, special hazards, and unique equipment characteristics. A competent person must inspect all cranes and derricks before daily use and a thorough inspection must be accomplished annually by an OSHA-recognized qualified person. A record must be maintained of that inspection for each piece of hoisting equipment.

The industrial trucks section covers the classifications of trucks and designated areas where a truck can be used. It also describes the required inspections and maintenance actions for those vehicles. Safe operation procedures are also covered in this section.

Procedures for keeping and using slings are also covered in this section. It describes the proper sizes for loads, and safe hook-up procedures and inspection requirements are stated and required markings discussed. The rigging of materials for handling is a critical component of Subpart H. This includes the safe use of slings made from wire rope, chains, synthetic fibers ropes or webs, and natural fiber ropes. Specifications for the use of rigging is found in this subpart regarding carrying capacity, inspection for defects, and safe operating procedures.

This subpart applies to the use of helicopters for lifting purposes. Helicopters must comply with the Federal Aviation Administration regulations. The pilot of the helicopter has the primary responsibility for the load's weight, size, and rigging. Static charge must be eliminated before workers touch the load. Visibility is critical to the pilot in maintaining visual contact with ground crew members so that constant communications can be maintained.

All hoists are to comply with the manufacturer's specifications. If these do not exist, then as with cranes and derricks, the limitations are based on the determination of a professional engineer. In the operation of a hoist there should be a signaling system, a specified line speed, and a sign stating "No Riders." Permanently enclosed hoist cars are to be used to hoist personnel and these cars must be able to stop at any time using safety breaks or a similar system. All hoists are to be tested, inspected, and maintained on an ongoing basis and at least every 3 months. In addition, requirements exist in this subpart for base-mounted drum hoists and overhead hoists.

The servicing of single- and multipiece rims is also covered in this subpart.

Checklist
_____Does your company own or use cranes or derricks?
_____Does your company employ helicopters for lifting purposes?
_____Do you use material hoists on your worksite?
_____Do your workers work around cranes, derricks, helicopters, or hoists?
_____Does your company use cranes or hoists for lifting personnel?
_____Do you rent cranes, derricks, hoists, or other lifting devices?
_____Do you operate powered industrial trucks (forklifts) at your facility?
_____Do you rig loads of handling?
_____Do you use slings for rigging?
_____Do you have single- and multipiece rims at your site?
_____Do you have materials stored on the worksite?
_____Do you have waste materials on the jobsite?
_____Do your workers use rigging to handle materials?
_____Do your workers know the limitations for the use of wire ropes, chains, etc.?
_____Does your company have responsibility for housekeeping?

Sections of Subpart N
1910.176 Handling material—general
1910.177 Servicing multipiece and single-piece rim wheels
1910.178 Powered industrial trucks
1910.179 Overhead and gantry cranes
1910.180 Crawler locomotive and truck cranes
1910.181 Derricks
1910.183 Helicopters
1910.184 Slings
Appendix A to 1910.178—Stability of powered industrial trucks (nonmandatory appendix to paragraph (l) of this section)

Subpart O—Machinery and Machine Guarding

Subpart O covers the machine guarding for any equipment that exposes employees to a hazard during use due to exposed moving or rotating parts; generally, this covers any device that has an exposed point of operation. This subpart covers guards for woodworking machinery, abrasive-wheel machinery, cooperage machinery, mills and calendars, mechanical presses, forging machines, and mechanical power-transmission apparatus.

The woodworking section covers the parts that must be guarded and the type of guards that must be used, while the abrasive wheel section describes the amount of wheel that can be exposed for the various types of abrasive grinding equipment and other precautions to take.

Mechanical-power presses are required to have switches and brakes that protect the operator. Many presses are to be protected mechanically by guarding and by other means. This section describes the actions to be taken to assure safe operations. Many types of guarding systems can be used on presses.

The safe operation, inspection, and maintenance of forging machines in this section as well as the best practices for guarding these pieces of equipment are discussed. Special guarding needs are discussed for certain processes.

The mechanical-power-transmission apparatus in this subpart covers all belts, pulleys, and conveyors that are used in industry. It describes the ones that need to be guarded. For specific applications, guidance is given for preferred operations.

Checklist

_____Do you have power presses?

_____Do your workers use woodworking machinery?

_____Does your facility operate any abrasive wheel machinery?

_____Are you considered to be a rubber or plastics industry that has mill and calendars?

_____Do you operate mechanical-power presses?

_____Do you have a forging or die shop?

_____Are there power-transmission belts, pulley, etc., present?

Sections of Subpart O

1910.211 Definitions

1910.212 General requirements for all machines

1910.213 Woodworking machinery requirements

1910.214 Cooperage machinery

1910.215 Abrasive wheel machinery

1910.216 Mills and calenders in the rubber and plastics industries

1910.217 Mechanical-power presses

1910.218 Forging machines

1910.219 Mechanical-power-transmission apparatus

Subpart P—Hand and Portable Powered Tools and Other Handheld Equipment

The subpart P regulation is dedicated to the safe use of both power and hand tools including employer- and employee-owned tools. This subpart requires that hand tools be safe and free from defects. It also cautions against misuse of tools.

This subpart addresses the need for properly guarded power tools. It discusses the areas where guarding is required and the types of guards that should be used, as well as the proper protective equipment to be used, when tools create such hazards as flying materials. The power tools that are covered by the regulation include electrical, pneumatic, fuel, hydraulic, and powder-actuated powered tools. These tools are to be secured if maintained in a fixed place and all electrically powered equipment must be effectively grounded. Special attention is given to abrasive wheels and tools. Some special requirements exist for powder-actuated tools.

Jacks and their use are covered regarding the blocking and securing of objects that are lifted. This includes jack maintenance and inspection. It also pertains to riding and walk-behind lawn mowers and other internal-combustion-engine-powered machines.

Checklist

_____Do your workers use hand or power tools?

_____Do your workers use woodworking tools?

_____Do your workers use abrasive wheels or tools?

_____Do you supply tools to workers?

_____Do your workers use jacks?

_____Do your workers use walk-behind or riding mowers?

Sections of Subpart P
1910.241 Definitions
1910.242 Hand and portable powered tools and equipment, general
1910.243 Guarding of portable powered tools
1910.244 Other portable tools and equipment

Subpart Q—Welding, Cutting, and Brazing

Subpart Q covers the use and installation of arc or gas welding, cutting, and brazing equipment. It covers the different types of welding and ties the specific safety needs of each. This subpart also regulates the use of oxygen–fuel gas welding and cutting, arc welding and cutting, and resistance welding. Subpart Q covers the procedures and precautions associated with gas welding, cutting, arc welding, fire prevention, compressed-gas cylinders, and welding materials. Special attention is given to transporting, moving, and storing of compressed-gas cylinders, as well as apparatuses such as hoses, torches, and regulators used for welding. Defective gas cylinders should not be used. All cylinders should be marked and labeled with 1 in. letters. Hoses should be identifiable and designed such that they cannot be misconnected to the wrong cylinder regulators. Prework inspections are an important component of this subpart.

Arc welding and its unique precautions are covered by this regulation. This includes grounding, care of cables, and care of electrode holders. As with all welding and cutting operations, appropriate PPE and safety are addressed in this subpart.

Fire prevention is an important part of welding and cutting and such work is not to be performed near flammable vapors, fumes, or heavy dust concentrations. Firefighting equipment must be readily accessible and in good working order.

Checklist
_____Do your workers perform welding and cutting tasks?
_____Do you have compressed-gas cylinders on your jobsite?
_____Do you have adequate firefighting equipment?
_____Is there a need for ventilation?
_____Do your welders wear PPE?
_____Does your company weld or cut in confined spaces?
_____Do your workers have to weld or cut on toxic materials?
_____Do your workers perform resistance welding?

Sections of Subpart Q
1910.251 Definitions
1910.252 General requirements
1910.253 Oxygen–fuel gas welding and cutting
1910.254 Arc welding and cutting
1910.255 Resistance welding

Subpart R—Special Industries

Subpart R deals with industries singled out by OSHA that need to be addressed in industry-specific standards. These industries include pulp, paper and paperboard mills, textile mills, bakeries, laundries, and sawmills. It also includes industries such as pulpwood logging, telecommunications, electric-power generation, transmission, and distribution; and grain-handling facilities.

Checklist
_____Does the work involve the manufacturing of pulp, paper, and paperboard?
_____Does the work involve operation and maintenance of textile mills and machinery?

_____Does the work involve operation and maintenance of machinery and equipment used within a bakery?

_____Does the laundry equipment that is used have point-of-operation hazards?

_____Is the work conducted at a sawmill?

_____Does the work involve the normal operations including logging operations?

_____Does the work involve processes in telecommunications centers and at telecommunications field installations?

_____Do the procedures involve working with the operation and maintenance of electric-power generation, transmission, and distribution lines and equipment?

_____Does the process require the operation of grain elevators, grain storage, and processing facilities?

Sections of Subpart R

1910.261 Pulp, paper, and paperboard mills

1910.262 Textiles

1910.263 Bakery equipment

1910.264 Laundry machinery and operations

1910.265 Sawmills

1910.266 Logging operations

1910.267 [Reserved]

1910.268 Telecommunications

1910.269 Electric-power generation, transmission, and distribution

1910.272 Grain handling facilities

Subpart S—Electrical

Subpart S relates to the installation and use of electrical power on worksites, including both permanent and temporary. The two areas of emphasis within this subpart are installation safety requirements and safety-related work practices.

Installation safety requirements sections of subpart S require that all electrical parts be inspected for durability, quality, and appropriateness. Installation that follows the National Electric Code is considered to be in compliance with OSHA. Grounding is an important part of this regulation and the use of ground fault circuit interrupters (GFCIs) or assured grounding is required. Emphasis is placed on temporary and portable lighting, as well as the use of extension cords. All listed, labeled, and certified equipment must be installed according to instructions from the manufacturer. This subpart includes special purpose equipment installation such as cranes and monorail hoists, electric welders, and x-ray equipment. It discusses work in high-hazard locations as well as special systems such as remote control and power-limited circuits.

Safety-related work practices include workers not working on energized circuits; this includes precautions for working on hidden underground power sources. This subpart addresses the use of barriers to protect workers from electrical sources. In addition, working around electrically energized equipment and power lines is explained as well as the procedures for lockout/tagout of energized circuits to protect workers.

The primary purpose of this subpart is to protect workers from coming into contact with energized electrical power sources.

Checklist:

_____Do you employ electricians?

_____Do your employees perform electrical installations?

_____Do your workers work around energized electrical circuits?

_____Do you follow a lockout/tagout procedure?

_____Do you use temporary lighting and extension cords?

_____Do your workers use GFCIs?

_____Do you have workers working in hazardous environments?
_____Do your workers use electrically powered tools?
_____Are there energized power lines on your jobsite?
_____Do your workers work around energized power lines?
_____Is there special electrically powered equipment on your worksite?

Sections of Subpart S

General

1910.301 Introduction

Design Safety Standards for Electrical Systems

1910.302 Electric utilization systems
1910.303 General requirements
1910.304 Wiring design and protection
1910.305 Wiring methods, components, and equipment for general use
1910.306 Specific purpose equipment and installations
1910.307 Hazardous (classified) locations
1910.308 Special systems
1910.309–1910.330 [Reserved]

Safety-Related Work Practices

1910.331 Scope
1910.332 Training
1910.333 Selection and use of work practices
1910.334 Use of equipment
1910.335 Safeguards for personnel protection
1910.336–1910.360 [Reserved]

Safety-Related Maintenance Requirements

1910.361–1910.380 [Reserved]

Safety Requirements for Special Equipment

1910.381–1910.398 [Reserved]

Definitions

1910.399 Definitions applicable to this subpart
Appendix A to Subpart S—Reference Documents
Appendix B to Subpart S—Explanatory Data [RESERVED]
Appendix C to Subpart S—Tables, Notes, and Charts [RESERVED]

Subpart T—Commercial Diving Operations

Subpart T applies to dives and diving support operations that take place within all waters in the
United States, trust territories, DC, Commonwealth of Puerto Rico, other U.S. protected islands,

etc. It does not apply to instructional diving and search and rescue. This subpart describes requirements, qualifications, and training certifications for divers and dive teams, as well as the need to use specific safe practices for pre-, during-, and postdives. It also includes emergency care procedures such as recompression and evacuation.

This subpart delineates the criteria and procedures for different types of diving operations such as scuba, surface supplied air, and mixed gas diving. The margin for error and risk are high; thus, all diving procedures within this regulation are very precise and require more than superficial knowledge and experience with diving operations.

The care and maintenance of all equipment involved, whether cylinders, decompression chambers, oxygen safety, or other diving equipment, requires a unique expertise. This subpart makes all diving and diving operation procedures very exacting and requires record keeping of all dives and injuries.

Checklist
_____Does your company employ any divers?
_____Does your company oversee any diving operations?
_____Does your company own any diving equipment?
_____Do you have divers or diving operations on or at your workplace that belong to other contractors?

Sections of Subpart T

General

1910.401 Scope and application
1910.402 Definitions

Personnel Requirements

1910.410 Qualifications of dive team

General Operations Procedures

1910.420 Safe practices manual
1910.421 Predive procedures
1910.422 Procedures during dive
1910.423 Postdive procedures

Specific Operations Procedures

1910.424 Scuba diving
1910.425 Surface-supplied air diving
1910.426 Mixed-gas diving
1910.427 Liveboating

Equipment Procedures and Requirements

1910.430 Equipment

Record Keeping

1910.440 Record keeping requirements
1910.441 Effective date

Appendix A to Subpart T—Examples of conditions that may restrict or limit exposure to hyperbaric conditions
Appendix B to Subpart T—Guidelines for scientific diving

Subparts U—Y [RESERVED]

1910.442–1910.999 [Reserved]

Subpart Z—Toxic and Hazardous Substances

Subpart Z provides specific regulations for a select group of toxic or hazardous chemicals. The regulations set specific exposure limits, detail acceptable work procedures, delineate workplace and environmental sampling requirements, set specific PPE requirements, and denote the need for regulated work areas. This subpart also has the permissible exposure limits (PELs) for in excess of 500 hazardous chemicals. Subpart Z discusses, in some detail, working with and around potential cancer-causing chemicals. With many of the chemicals, unique training requirements exist, as well as medical monitoring and surveillance. Requirements exist for posting and labels that warn of the dangers from exposure to specific chemicals. In many cases, precise decontamination is required, along with hygiene procedures to minimize potential contamination to workers or the spread of contamination. These regulations communicate the hazards involved and discuss the target organs, signs, and symptoms that accompany an occupational illness from one of these hazardous or toxic chemicals.

Because each of these chemicals have unique properties, adverse affects, handling procedures, signs and symptoms of overexposure, and regulatory requirements, the regulation specific to each chemical must be consulted and complied with.

This subpart also covers hazard communication, bloodborne pathogens, ionizing radiation, placarding, and laboratories' chemical safety.

Checklist
_____Does your company use any of the chemicals listed in sections 1000 through 1052?
_____Do any of the chemical mixtures that you use on your jobsites contain any chemicals in sections 1000 through 1052?
_____Do your workers do asbestos or lead abatement work?
_____Do you have any sources of ionizing radiation?
_____Do your workers perform hazardous waste remediation work?
_____Do other contractors use any of the chemicals in sections 1000 through 1052, which might expose your own workers inadvertently?
_____Do you have a hazard communication program?
_____Do you provide training to your workers on any of the chemicals listed in section 1000 through 1052?
_____Does any of your work take your workers onto or into worksites where exposure to any of the chemicals in 1000 through 1052 could occur?
_____Do you have laboratories where hazardous chemicals exist or are used?

Sections of Subpart Z
1910.1000 Air contaminants
1910.1001 Asbestos
1910.1002 Coal tar pitch volatiles; interpretation of term
1910.1003 13 Carcinogens (4-nitrobiphenyl, etc.)
1910.1004 alpha-Naphthylamine
1910.1005 [Reserved]

1910.1006 Methyl chloromethyl ether
1910.1007 3,3′-Dichlorobenzidine (and its salts)
1910.1008 Bis-chloromethyl ether
1910.1009 Beta-naphthylamine
1910.1010 Benzidine
1910.1011 4-Aminodiphenyl
1910.1012 Ethyleneimine
1910.1013 Beta-propiolactone
1910.1014 2-Acetylaminofluorene
1910.1015 4-Dimethylaminoazobenzene
1910.1016 *N*-Nitrosodimethylamine
1910.1017 Vinyl chloride
1910.1018 Inorganic arsenic
1910.1020 Access to employee exposure and medical records
1910.1025 Lead
1910.1026 Chromium (VI)
1910.1027 Cadmium
1910.1028 Benzene
1910.1029 Coke oven emissions
1910.1030 Bloodborne pathogens
1910.1043 Cotton dust
1910.1044 1,2-Dibromo-3-chloropropane
1910.1045 Acrylonitrile
1910.1047 Ethylene oxide
1910.1048 Formaldehyde
1910.1050 Methylenedianiline
1910.1051 1,3-Butadiene
1910.1052 Methylene chloride
1910.1096 Ionizing radiation
1910.1200 Hazard communication
1910.1201 Retention of DOT markings, placards, and labels
1910.1450 Occupational exposure to hazardous chemicals in laboratories

Most Common Air Pollutants and Toxic Chemicals

OZONE (GROUND-LEVEL OZONE IS THE PRINCIPAL COMPONENT OF SMOG)

- Source—chemical reaction of pollutants; volatile organic compounds (VOCs), and NO_x.
- Health effects—breathing problems, reduced lung function, asthma, eye irritation, stuffy nose, reduced resistance to colds and other infections, may speed up aging of lung tissue.
- Environmental effects—ozone can damage plants and trees; smog can cause reduced visibility.
- Property damage—damages rubber, fabrics, etc.

VOCS*; SMOG-FORMERS

- Source—VOCs are released from burning fuel (gasoline, oil, wood coal, natural gas, etc.), solvents, paints, glues, and other products used at work or at home. Cars are an important source of VOCs. VOCs include chemicals such as benzene, toluene, methylene chloride, and methyl chloroform.
- Health effects—in addition to ozone (smog) effects, many VOCs can cause serious health problems such as cancer and other effects.
- Environmental effects—in addition to ozone (smog) effects, some VOCs, such as formaldehyde and ethylene, may harm plants.

Nitrogen Dioxide (One of the NO_x); Smog-Forming Chemical

- Source—burning of gasoline, natural gas, coal, oil, etc. Cars are an important source of NO_2.
- Health effects—lung damage, illnesses of breathing passages and lungs (respiratory system).
- Environmental effects—nitrogen dioxide is an ingredient of acid rain (acid aerosols), which can damage trees and lakes. Acid aerosols can reduce visibility.
- Property damage—acid aerosols can eat away stone used on buildings, statues, monuments, etc.

---

* All VOCs contain carbon (C), the basic chemical element found in living beings. Carbon-containing chemicals are called organic. Volatile chemicals escape into the air easily. Many VOCs, such as the chemicals listed in the table, are also hazardous air pollutants, which can cause very serious illnesses. Environmental Protection Agency (EPA) does not list VOCs as criteria air pollutants, but they are included in this list of pollutants because efforts to control smog target VOCs for reduction.

CARBON MONOXIDE (CO)

- Source—burning of gasoline, natural gas, coal, oil, etc.
- Health effects—reduces ability of blood to bring oxygen to body cells and tissues; cells and tissues need oxygen to work. Carbon monoxide may be particularly hazardous to people who have heart or circulatory (blood vessel) problems and people who have damaged lungs or breathing passages.

PARTICULATE MATTER (PM-10) (DUST, SMOKE, SOOT)

- Source—burning of wood, diesel, and other fuels; industrial plants; agriculture (plowing, burning off fields); unpaved roads
- Health effects—nose and throat irritation, lung damage, bronchitis, early death
- Environmental effects—particulates are the main source of haze that reduces visibility
- Property damage—ashes, soots, smokes, and dusts can dirty and discolor structures and other property, including clothes and furniture

SULFUR DIOXIDE

- Source—burning of coal and oil, especially high-sulfur coal from the eastern U.S.; industrial processes (paper, metals).
- Health effects—breathing problems, may cause permanent damage to lungs.
- Environmental effects—O_2 is an ingredient in acid rain (acid aerosols), which can damage trees and lakes. Acid aerosols can also reduce visibility.
- Property damage—acid aerosols can eat away stone used in buildings, statues, monuments, etc.

LEAD

- Source—leaded gasoline (being phased out), paint (houses, cars), smelters (metal refineries); manufacture of lead storage batteries.
- Health effects—brain and other nervous system damage; children are at special risk. Some lead-containing chemicals cause cancer in animals. Lead causes digestive and other health problems.
- Environmental effects—lead can harm wildlife.

TOXIC CHEMICALS

CAS Number	Chemical Name
75070	Acetaldehyde
60355	Acetamide
75058	Acetonitrile
98862	Acetophenone
53963	2-Acetylaminofluorene
107028	Acrolein
79061	Acrylamide
79107	Acrylic acid
107131	Acrylonitrile
107051	Allyl chloride
92674	Aminobiphenyl

(continued)

CAS Number	Chemical Name
62533	Aniline
90040	*o*-Anisidine
1332214	Asbestos
71432	Benzene (including benzene from gasoline)
92875	Benzidine
98077	Benzotrichloride
100447	Benzyl chloride
92524	Biphenyl
117817	Bis(2-ethylhexyl)phthalate (DEHP)
542881	Bis(chloromethyl)ether
75252	Bromoform
106990	1,3-Butadiene
156627	Calcium cyanamide
105602	Caprolactam
133062	Captan
63252	Carbaryl
75150	Carbon disulfide
56235	Carbon tetrachloride
463581	Carbonyl sulfide
120809	Catechol
133904	Chloramben
57749	Chlordane
7782505	Chlorine
79118	Chloroacetic acid
532274	2-Chloroacetophenone
108907	Chlorobenzene
510156	Chlorobenzilate
67663	Chloroform
107302	Chloromethyl methyl ether
126998	Chloroprene
1319773	Cresols/cresylic acid (isomers and mixture)
95487	*o*-Cresol
108394	*m*-Cresol
106445	*p*-Cresol
98828	Cumene
94757	2,4-D, salts and esters
3547044	DDE
334883	Diazomethane
132649	Dibenzofurans
96128	1,2-Dibromo-3-chloropropane
84742	Dibutylphthalate
106467	1,4-Dichlorobenzene(p)
91941	3,3-Dichlorobenzidene
111444	Dichloroethyl ether (bis(2-chloroethyl)ether)
542756	1,3-Dichloropropene
62737	Dichlorvos

(continued)

(continued)

CAS Number	Chemical Name
111422	Diethanolamine
121697	*N,N*-Diethyl aniline (*N,N*-dimethylaniline)
64675	Diethyl sulfate
119904	3,3-Dimethoxybenzidine
60117	Dimethyl aminoazobenzene
119937	3,3-Dimethyl benzidine
79447	Dimethyl carbamoyl chloride
68122	Dimethyl formamide
57147	1,1-Dimethyl hydrazine
131113	Dimethyl phthalate
77781	Dimethyl sulfate
534521	4,6-Dinitro-*o*-cresol, and salts
51285	2,4-Dinitrophenol
121142	2,4-Dinitrotoluene
123911	1,4-Dioxane (1,4-Diethyleneoxide)
122667	1,2-Diphenylhydrazine
106898	Epichlorohydrin (l-Chloro-2, 3-epoxypropane)
106887	1,2-Epoxybutane
140885	Ethyl acrylate
100414	Ethyl benzene
51796	Ethyl carbamate (urethane)
75003	Ethyl chloride (chloroethane)
106934	Ethylene dibromide (dibromoethane)
107062	Ethylene dichloride (1,2-dichloroethane)
107211	Ethylene glycol
151564	Ethylene imine (aziridine)
75218	Ethylene oxide
96457	Ethylene thiourea
75343	Ethylidene dichloride (1,1-dichloroethane)
50000	Formaldehyde
76448	Heptachlor
118741	Hexachlorobenzene
87683	Hexachlorobutadiene
77474	Hexachlorocyclopentadiene
67721	Hexachloroethane
822060	Hexamethylene-1,6-diisocyanate
680319	Hexamethylphosphoramide
110543	Hexane
302012	Hydrazine
7647010	Hydrochloric acid
7664393	Hydrogen fluoride (hydrofluoric acid)
7783064	Hydrogen sulfide
123319	Hydroquinone
78591	Isophorone
58899	Lindane (all isomers)
108316	Maleic anhydride

(continued)

CAS Number	Chemical Name
67561	Methanol
72435	Methoxychlor
74839	Methyl bromide (bromomethane)
74873	Methyl chloride (chloromethane)
71556	Methyl chloroform (1,1,1-trichloroethane)
78933	Methyl ethyl ketone (2-butanone)
60344	Methyl hydrazine
74884	Methyl iodide (iodomethane)
108101	Methyl isobutyl ketone (hexone)
624839	Methyl isocyanate
80626	Methyl methacrylate
1634044	Methyl tert butyl ether
101144	4,4-Methylene bis(2-chloroaniline)
75092	Methylene chloride (dichloromethane)
101688	Methylene diphenyl diisocyanate (MDI)
101779	4,4-Methylenedianiline
91203	Naphthalene
98953	Nitrobenzene
92933	4-Nitrobiphenyl
100027	4-Nitrophenol
79469	2-Nitropropane
684935	*N*-Nitroso-*N*-methylurea
62759	*N*-Nitrosodimethylamine
59892	*N*-Nitrosomorpholine
56382	Parathion
82688	Pentachloronitrobenzene (quintobenzene)
87865	Pentachlorophenol
108952	Phenol
106503	*p*-Phenylenediamine
75445	Phosgene
7803512	Phosphine
7723140	Phosphorus
85449	Phthalic anhydride
1336363	Polychlorinated biphenyls (Aroclors)
1120714	1,3-Propane sultone
57578	beta-Propiolactone
123386	Propionaldehyde
114261	Propoxur (Baygon)
78875	Propylene dichloride (1,2-dichloropropane)
75569	Propylene oxide
75558	1,2-Propylenimine (2-methyl aziridine)
91225	Quinoline
106514	Quinone
100425	Styrene
96093	Styrene oxide
1746016	2,3,7,8-Tetrachlorodibenzo-*p*-dioxin

(*continued*)

(continued)

CAS Number	Chemical Name
79345	1,1,2,2-Tetrachloroethane
127184	Tetrachloroethylene (perchloroethylene)
7550450	Titanium tetrachloride
108883	Toluene
95807	2,4-Toluene diamine
584849	2,4-Toluene diisocyanate
95534	o-Toluidine
8001352	Toxaphene (chlorinated camphene)
120821	1,2,4-Trichlorobenzene
79005	1,1,2-Trichloroethane
79016	Trichloroethylene
95954	2,4,5-Trichlorophenol
88062	2,4,6-Trichlorophenol
121448	Triethylamine
1582098	Trifluralin
540841	2,2,4-Trimethylpentane
108054	Vinyl acetate
593602	Vinyl bromide
75014	Vinyl chloride
75354	Vinylidene chloride (1,1-dichloroethylene)
1330207	Xylenes (isomers and mixture)
95476	o-Xylenes
108383	m-Xylenes
106423	p-Xylenes
0	Antimony compounds
0	Arsenic compounds (inorganic including arsine)
0	Beryllium compounds
0	Cadmium compounds
0	Chromium compounds
0	Cobalt compounds
0	Coke oven emissions
0	Cyanide compounds
0	Glycol ethers
0	Lead compounds
0	Manganese compounds
0	Mercury compounds
0	Fine mineral fibers
0	Nickel compounds
0	Polycyclic organic matter
0	Radionuclides (including radon)
0	Selenium compounds

Workplace Security Program

Our establishment's program for workplace security addresses the hazards known to be associated with the three major types of workplace violence. Type I workplace violence involves a violent act by an assailant with no legitimate relationship to the workplace who enters the workplace to commit a robbery or other criminal act. Type II involves a violent act or threat of violence by a recipient of a service provided by our establishment, such as a client, patient, customer, passenger, a criminal suspect, or a prisoner. Type III involves a violent act or threat of violence by a current or former worker, supervisor or manager, or another person who has some employment-related involvement with our establishment, such as a worker's spouse or lover, a worker's relative or friend, or another person who has a dispute with one of our workers.

RESPONSIBILITY

We have decided to assign responsibility for security in our workplace. The Security Program administrator for workplace security is_____who has the authority and responsibility for implementing the provisions of this program for_____.

All managers and supervisors are responsible for implementing and maintaining this Security Program in their work areas and for answering worker questions about it. A copy of this Security Program is available from each manager and supervisor.

COMPLIANCE

We have established the following policy to ensure compliance with our rules on workplace security. The management of our establishment is committed to ensuring that all safety and health policies and procedures involving workplace security are clearly communicated and understood by all workers. All workers are responsible for using safe work practices, for following all directives, policies and procedures, and for assisting in maintaining a safe and secure work environment. Our system of ensuring that all workers, including supervisors and managers, comply with work practices that are designed to make the workplace more secure, and do not engage in threats or physical actions that create a security hazard for others in the workplace, includes

1. Informing workers, supervisors, and managers of the provisions of our Security Program
2. Evaluating the performance of all workers in complying with our establishment's workplace security measures
3. Recognizing workers who perform work practices that promote security in the workplace
4. Providing training and counseling to workers whose performance is deficient in complying with work practices designed to ensure workplace security
5. Disciplining workers for failure to comply with workplace security practices

The following practices shall ensure worker compliance with workplace security directives, policies, and procedures:_____

COMMUNICATION

At our establishment, we recognize that to maintain a safe, healthy, and secure workplace we must have open, two-way communication between all workers, including managers and supervisors, on all workplace safety, health, and security issues. Our establishment has a communication system designed to encourage a continuous flow of safety, health, and security information between management and our workers without fear of reprisal and in a form that is readily understandable. Our communication system consists of the following checked items:

- New-worker orientation on our establishment's workplace security policies, procedures, and work practices
- Periodic review of our Security Program with all personnel
- Training programs designed to address specific aspects of workplace security unique to our establishment
- Regularly scheduled safety meetings with all personnel that include workplace security discussions
- A system to ensure that all workers, including managers and supervisors, understand the workplace security policies
- Posted or distributed workplace security information
- A system for workers to inform management about workplace security hazards or threats of violence
- Procedures for protecting workers who report threats from retaliation by the person making the threats
- Addressing security issues at our workplace security team meetings
- Our establishment has less than 10 workers and communicates with, and instructs, workers orally about general safe work practices with respect to workplace security
- Other: _____
  _____
  _____

HAZARD ASSESSMENT

We will be performing workplace hazard assessment for workplace security in the form of periodic inspections. Periodic inspections to identify and evaluate workplace security hazards and threats of workplace violence are performed by the following observers in the following areas of our workplace:

Observer   Area

Periodic inspections are performed according to the following schedule:

1. _____. Frequency (daily, weekly, monthly, etc.)
2. When we initially establish our Security Program
3. When new, previously unidentified security hazards are recognized
4. When occupational injuries or threats of injury occur
5. Whenever workplace security conditions warrant an inspection

Periodic inspections for security hazards consist of identification and evaluation of workplace security hazards and changes in worker work practices and may require assessing for more than one type of workplace violence. Our establishment performs inspections for each type of workplace violence using the methods specified below to identify and evaluate workplace security hazards.

Inspections for Type I workplace security hazards include assessing

1. The exterior and interior of the workplace for its attractiveness to robbers
2. The need for security surveillance measures, such as mirrors or cameras
3. Posting of signs notifying the public that limited cash is kept on the premises
4. Procedures for worker response during a robbery or other criminal act
5. Procedures for reporting suspicious persons or activities
6. Posting of emergency telephone numbers for law enforcement, fire, and medical services where workers have access to a telephone with an outside line
7. Limiting the amount of cash on hand and using time access safes for large bills
8. Other: _____
   _____
   _____

Inspections for Type II workplace security hazards include assessing

1. Access to, and freedom of movement within, the workplace
2. Adequacy of workplace security systems, such as door locks, security windows, physical barriers, and restraint systems
3. Frequency and severity of threatening or hostile situations that may lead to violent acts by persons who are service recipients of our establishment
4. Workers' skill in safely handling threatening or hostile service recipients
5. Effectiveness of systems and procedures to warn others of a security danger or to summon assistance, e.g., alarms or panic buttons
6. Use of work practices such as "buddy" systems for specified emergency events
7. Availability of worker escape routes
8. Other: _____
   _____
   _____

Inspections for Type III workplace security hazards include assessing

1. How well our establishment's antiviolence policy has been communicated to workers, supervisors, or managers
2. How well our establishment's management and workers communicate with each other
3. Our workers', supervisors', and managers' knowledge of the warning signs of potential workplace violence
4. Access to, and freedom of movement within, the workplace by nonworkers, including recently discharged workers or persons with whom one of our workers has a dispute
5. Frequency and severity of worker reports of threats of physical or verbal abuse by managers, supervisors, or other workers
6. Any prior violent acts, threats of physical violence, verbal abuse, property damage, or other signs of strain or pressure in the workplace
7. Worker disciplinary and discharge procedures
8. Other: _____
   _____
   _____

INCIDENT INVESTIGATIONS

We have established the following policy for investigating incidents of workplace violence. Our procedures for investigating incidents of workplace violence, which includes threats and physical injury, include

1. Reviewing all previous incidents
2. Visiting the scene of an incident as soon as possible
3. Interviewing threatened or injured workers and witnesses
4. Examining the workplace for security risk factors associated with the incident, including any previous reports of inappropriate behavior by the perpetrator
5. Determining the cause of the incident
6. Taking corrective action to prevent the incident from recurring
7. Recording the findings and corrective actions taken
8. Other: _____

_____

_____

HAZARD CORRECTION

Hazards that threaten the security of workers shall be corrected in a timely manner based on severity when they are first observed or discovered. Corrective measures for Type I workplace security hazards can include

1. Making the workplace unattractive to robbers
2. Using surveillance measures, such as cameras or mirrors, to provide information as to what is going on outside and inside the workplace
3. Procedures for reporting suspicious persons or activities
4. Posting of emergency telephone numbers for law enforcement, fire, and medical services where workers have access to a telephone with an outside line
5. Posting of signs notifying the public that limited cash is kept on the premises
6. Limiting the amount of cash on hand and using time access safes for large bills
7. Worker, supervisor, and management training on emergency action procedures
8. Other: _____

_____

_____

Corrective measures for Type II workplace security hazards can include

1. Controlling access to the workplace and freedom of movement within it, consistent with business necessity
2. Ensuring the adequacy of workplace security systems, such as door locks, security windows, physical barriers, and restraint systems
3. Providing worker training in recognizing and handling threatening or hostile situations that may lead to violent acts by persons who are service recipients of our establishment
4. Placing effective systems to warn others of a security danger or to summon assistance, e.g., alarms or panic buttons
5. Providing procedures for a "buddy" system for specified emergency events
6. Ensuring adequate worker escape routes
7. Other: _____

_____

_____

Corrective measures for Type III workplace security hazards include

1. Effectively communicating our establishment's antiviolence policy to all workers, supervisors, or managers
2. Improving how well our establishment's management and workers communicate with each other

3. Increasing awareness by workers, supervisors, and managers of the warning signs of potential workplace violence
4. Controlling access to, and freedom of movement within, the workplace by nonworkers including recently discharged workers or persons with whom a worker has a dispute
5. Providing counseling to workers, supervisors, or managers who exhibit behavior that represents strain or pressure, which may lead to physical or verbal abuse of coworkers
6. Ensure that all reports of violent acts, threats of physical violence, verbal abuse, property damage, or other signs of strain or pressure in the workplace are handled effectively by the management and that the person making the report is not subject to retaliation by the person making the threat
7. Ensure that worker disciplinary and discharge procedures address the potential for workplace violence
8. Other: _____
   _____
   _____

TRAINING AND INSTRUCTION

We have established the following policy on training all workers with respect to workplace security. All workers, including managers and supervisors, shall have training and instruction on general and job-specific workplace security practices. Training and instruction shall be provided when the Security Program is first established and periodically thereafter. Training shall also be provided to all new workers and to other workers for whom training has not been provided previously and to all workers, supervisors, and managers given new job assignments for which specific workplace security training for that job assignment has not been provided previously. Additional training and instruction will be provided to all personnel whenever the employer is made aware of new or previously unrecognized security hazards. General workplace security training and instruction include, but are not limited to, the following:

1. Explanation of the Security Program including measures for reporting any violent acts or threats of violence
2. Recognition of workplace security hazards including the risk factors associated with the three types of workplace violence
3. Measures to prevent workplace violence, including procedures for reporting workplace security hazards or threats to managers and supervisors
4. Ways to defuse hostile or threatening situations
5. Measures to summon others for assistance
6. Worker routes of escape
7. Notification of law enforcement authorities when a criminal act may have occurred
8. Emergency medical care provided in the event of any violent act upon a worker
9. Postevent trauma counseling for workers desiring such assistance

In addition, we provide specific instructions to all workers regarding workplace security hazards unique to their job assignment to the extent that such information is not already covered in other training.

We have chosen the following checked items for Type I training and instruction for managers, supervisors, and workers:

_____ Crime awareness
_____ Location and operation of alarm systems
_____ Communication procedures
_____ Proper work practices for specific workplace activities, occupations, or assignments, such as late night retail sales, taxicab driver, or security guard

_____ Other: _____
_____
_____

We have chosen the following checked items for Type II training and instruction for managers, supervisors, and workers:

_____ Self-protection
_____ Dealing with angry, hostile, or threatening individuals
_____ Location, operation, care, and maintenance of alarm systems and other protective devices
_____ Communication procedures
_____ Determination of when to use the "buddy" system or other assistance from coworkers
_____ Awareness of indicators that lead to violent acts by service recipients
_____ Other: _____
_____
_____

We have chosen the following checked items for Type III training and instruction for managers, supervisors, and workers:

_____ Preemployment screening practices
_____ Worker assistance programs
_____ Awareness of situational indicators that lead to violent acts
_____ Managing with respect and consideration for worker well-being
_____ Review of antiviolence policy and procedures
_____ Other: _____
_____
_____

Occupational Safety and Health Resources

All the potential sources of information on occupational safety and health cannot possibly be provided in one appendix. The listings here are some of the most recent and useful ones.

ACCIDENT/HAZARD ANALYSIS

Reese, C.D. *Accident/Incident Prevention Techniques*. New York, NY: Taylor & Francis, 2001.
Reese, C.D. and J.V. Eidson. *Handbook of OSHA Construction Safety & Health*, 2nd edn. Boca Raton, FL: CRC/Taylor & Francis, 2006.

ACCIDENT INVESTIGATION

Reese, C.D. *Accident/Incident Prevention Techniques*. New York, NY: Taylor & Francis, 2001.
Reese, C.D. and J.V. Eidson. *Handbook of OSHA Construction Safety & Health*, 2nd edn. Boca Raton, FL: CRC/Taylor & Francis, 2006.
U.S. Department of Energy. *Accident/Incident Investigation Manual (SSDC 27, DOE/SSDC 76-45/27)*, 2nd edn. Washington, DC: November 1985.
Vincoli, J.W. *Basic Guide to Accident Investigation and Loss Control*. New York, NY: John Wiley & Sons, 1994.

ACCIDENT PREVENTION

Michaud, P.A. *Accident Prevention and OSHA Compliance*. Boca Raton, FL: CRC/Lewis Publishers, 1995.
Reese, C.D. *Accident/Incident Prevention Techniques*. New York, NY: Taylor & Francis, 2001.
Reese, C.D. and J.V. Eidson. *Handbook of OSHA Construction Safety & Health*, 2nd edn. Boca Raton, FL: CRC/Taylor & Francis, 2006.
U.S. Department of Labor, National Mine Health and Safety Academy. *Accident Prevention Techniques*. Beckley, WV: 1984.
U.S. Department of Labor, Mine Safety and Health Administration. *Accident Prevention, Safety Manual No. 4*. Beckley, WV: revised 1990.

CONSTRUCTION SAFETY AND HEALTH

Hess, K. *Construction Safety Auditing Made Easy: A Checklist Approach to OSHA Compliance*. Rockville, MD: Government Institutes, 1998.

Moran, M.M. *Construction Safety Handbook: A Practical Guide to OSHA Compliance and Injury Prevention.* Rockville, MD: Government Institutes, 1996.

Reese, C.D. *Annotated Dictionary of Construction Safety and Health.* Boca Raton, FL: CRC/Lewis Publishers, 2000.

Reese, C.D. and J.V. Eidson. *Handbook of OSHA Construction Safety & Health*, 2nd edn. Boca Raton, FL: CRC/Taylor & Francis, 2006.

CONSULTANTS

Reese, C.D. and J.V. Eidson. *Handbook of OSHA Construction Safety & Health.* Boca Raton, FL: CRC/Lewis Publishers, 1999.

Reese, C.D. *Accident/Incident Prevention Techniques.* New York, NY: Taylor & Francis, 2001.

ENVIRONMENTAL

Arms, K. *Environmental Science*, 2nd edn. Saddle Brook, NJ: HBJ College and School Division, 1994.

Henry, J.G. and G.W. Heinke. *Environmental Science and Engineering*, 2nd edn. New York, NY: Prentice Hall, 1995.

Jackson, A.R. and J.M. Jackson. *Environmental Science: The Natural Environment and Human Impact.* New York, NY: Longman, 1996.

Koren, H. and M. Bisesi. *Handbook of Environmental Health and Safety: Principles and Practices* (Vols. I and II), 3rd edn. Boca Raton, FL: Lewis Publishers, 1996.

Lynn, L. *Environmental Biology.* North Port, FL: Kendall-Hunt, 1995.

Manahan, S.E. *Fundamentals of Environmental Chemistry.* Boca Raton, FL: CRC Press/Lewis Publishers, 1993.

Moron, J.M. et al. *Introduction to Environmental Science.* New York, NY: W.H. Freeman and Company, 1986.

Que Hee, S.S. *Hazardous Waste Analysis.* Rockville, MD: Government Institutes, 1999.

Schell, D.J. *What Environmental Managers Really Need to Know.* Rockville, MD: Government Institutes, 1999.

Spellman, F.R. and N.E. Whiting. *Environmental Science and Technology: Concepts and Applications.* Rockville, MD: Government Institutes, 1999.

Sullivan, T.F.P. *Environmental Law Book*, 16th edn. Rockville, MD: Government Institutes, 2001.

Wentz, C.A. *Hazardous Waste Management.* New York, NY: McGraw-Hill, 1990.

ERGONOMICS

Erdil, M. and O.B. Dickerson. *Cumulative Trauma Disorders: Prevention, Evaluation, and Treatment.* New York, NY: Van Nostrand Reinhold, 1997.

Eastman Kodak Company, *Ergonomic Design for People at Work: Volumes 1 and 2.* New York, NY: Van Nostrand Reinhold, 1983.

Kromer, K.H.E. *Ergonomics Design of Material Handling Systems.* Boca Raton, FL: CRC/Lewis Publishers, 1997.

Kromer, K.H.E. and E. Grandjean. *Fitting the Task to the Human.* New York, NY: Taylor & Francis, 1997.

Kromer, K., H. Kromer and K. Kromer-Elbert. *Ergonomics: How to Design for Ease and Efficiency.* Englewood Cliffs, NJ: Prentice Hall, 1994.

Laing, P.M., *Ergonomics: A Practical Guide*, 2nd edn. Itasca, IL: National Safety Council, 1993.

MacLeod, D. *The Ergonomics Edge.* New York, NY: Van Nostrand Reinhold, 1995.

Putz-Anderson, V. *Cumulative Trauma Disorders: A Manual for Musculoskeletal Disease of the Upper Limbs.* New York, NY: Taylor & Francis, 1994.

Reese, C.D. and J.V. Eidson. *Handbook of OSHA Construction Safety & Health*. Boca Raton, FL: CRC/Lewis Publishers, 1999.

FLEET SAFETY

National Safety Council. *Motor Fleet Safety Manual*, 3rd edn. Itasca, IL: NSC, 1986.
Reese, C.D. *Accident/Incident Prevention Techniques*. New York, NY: Taylor & Francis, 2001.
Reese, C.D. and J.V. Eidson. *Handbook of OSHA Construction Safety & Health*, 2nd edn. Boca Raton, FL: CRC/Taylor & Francis, 2006.

HAZARD IDENTIFICATION

National Safety Council. *Supervisors' Safety Manual*, 9th edn. Itasca, IL: NSC, 1997.
Reese, C.D. *Accident/Incident Prevention Techniques*. New York, NY: Taylor & Francis, 2001.
Reese, C.D. and J.V. Eidson. *Handbook of OSHA Construction Safety & Health*. Boca Raton, FL: CRC/Lewis Publishers, 1999.
U.S. Department of Labor, Mine Health and Safety Administration. *Hazard Recognition and Avoidance: Training Manual (MSHA 0105)*. Beckley, WV: revised May 1996.

HEALTH HAZARDS

Levy, B.S. and D.H. Wegman. *Occupational Health: Recognizing and Preventing Work-Related Disease*, 3rd edn. Boston, MA: Little, Brown and Company, 1995.
Reese, C.D. *Accident/Incident Prevention Techniques*. New York, NY: Taylor & Francis, 2001.

INDUSTRIAL HYGIENE

Hathway, G.J., N.H. Proctor and J.P. Hughes. *Proctor & Hughes' Chemical Hazards of the Workplace*, 4th edn. New York, NY: John Wiley & Sons, 1996.
Kamrin, M. *Toxicology*. Boca Raton, FL: CRC/Lewis Publishers, 1988.
Plog, B.A. *Fundamentals of Industrial Hygiene*, 5th edn. Itasca, IL: National Safety Council, 2001.
Reese, C.D. and J.V. Eidson. *Handbook of OSHA Construction Safety & Health*, 2nd edn. Boca Raton, FL: CRC/Taylor & Francis, 2006.
Scott, R. *Basic Concepts of Industrial Hygiene*. Boca Raton, FL: CRC/Lewis Publishers, 1997.

JOB HAZARD ANALYSIS

Reese, C.D. *Accident/Incident Prevention Techniques*. New York, NY: Taylor & Francis, 2001.
Reese, C.D. and J.V. Eidson. *Handbook of OSHA Construction Safety & Health*, 2nd edn. Boca Raton, FL: CRC/Taylor & Francis, 2006.
U.S. Department of Labor, Occupational Safety and Health Administration. *Job Hazard Analysis (OSHA 3071)*. Washington, DC: 1992.
U.S. Department of Labor, Mine Safety and Health Administration. *Job Safety Analysis: A Practical Approach (Instruction Guide No. 83)*. Beckley, WV: 1990.
U.S. Department of Labor, Mine Safety and Health Administration. *Job Safety Analysis (Safety Manual No. 5)*. Beckley, WV: revised 1990.
U.S. Department of Labor, National Mine Health and Safety Academy. *Accident Prevention Techniques: Job Safety Analysis*. Beckley, WV: 1984.

JOB SAFETY OBSERVATION

Reese, C.D. *Accident/Incident Prevention Techniques*. New York, NY: Taylor & Francis, 2001.

Reese, C.D. and J.V. Eidson. *Handbook of OSHA Construction Safety & Health*, 2nd edn. Boca Raton, FL: CRC/Taylor & Francis, 2006.

U.S. Department of Labor, Mine Safety and Health Administration. *Safety Observation (MSHA IG 84)*. Beckley, WV: revised 1991.

OFFICE SAFETY AND HEALTH

Reese, C.D. *Office Building Safety and Health*. Boca Raton, FL: CRC Press, 2004.

OSHA COMPLIANCE

Blosser, F. *Primer on Occupational Safety and Health*. Washington, DC: The Bureau of National Affairs, 1992.

Reese, C.D. *Accident/Incident Prevention Techniques*. New York, NY: Taylor & Francis, 2001.

Reese, C.D. and J.V. Eidson. *Handbook of OSHA Construction Safety & Health*, 2nd edn. Boca Raton, FL: CRC/Taylor & Francis, 2006.

U.S. Department of Labor, Occupational Safety and Health Administration. *Field Inspection Reference Manual (FIRM) (OSHA Instruction CPL 2.103)*. Washington, DC: September 26, 1994.

U.S. Department of Labor, Occupational Safety and Health Administration, Office of Training and Education. *OSHA Voluntary Compliance Outreach Program: Instructors Reference Manual*. Des Plaines, IL: 1993.

U.S. Department of Labor, Occupational Safety and Health Administration. *OSHA 10- and 30-Hour Construction Safety and Health Outreach Training Manual*. Washington, DC: 1991.

U.S. Department of Labor, Occupational Safety and Health Administration, General Industry. Code of Federal Regulations *(Title 29, Part 1910)*. Washington, DC: U.S. Government Printing Office, 1998.

U.S. Department of Labor, Occupational Safety and Health Administration, Construction. Code of Federal Regulations *(Title 29, Part 1926)*. Washington, DC: U.S. Government Printing Office, 1998.

PSYCHOLOGY OF SAFETY

Brown, P.L. and R.J. Presbie. *Behavior Modification in Business, Industry and Government*. Paltz, NY: Behavior Improvement Associates, 1976.

Geller, E.S. *The Psychology of Safety Handbook*. Boca Raton, FL: CRC/Lewis Publishers, 2001.

Herzberg, F. One more time: How do you motivate employees? *Harvard Business Review* (January–February, 1968): 53–62.

Mager, R.F. *Analyzing Performance Problems*. Belmont, CA: Fearson Publishers, Inc., 1970.

Riggio, R.E. *Introduction to Industrial/Organizational Psychology*, 3rd edn. Upper Saddle River, NJ: Prentice Hall, 2000.

Reese, C.D. *Accident/Incident Prevention Techniques*. New York, NY: Taylor & Francis, 2001.

Reese, C.D. and J.V. Eidson. *Handbook of OSHA Construction Safety & Health*, 2nd edn. Boca Raton, FL: CRC/Taylor & Francis, 2006.

REGULATIONS

Reese, C.D. *Accident/Incident Prevention Techniques*. New York, NY: Taylor & Francis, 2001.

Reese, C.D. and J.V. Eidson. *Handbook of OSHA Construction Safety & Health*, 2nd edn. Boca Raton, FL: CRC/Taylor & Francis, 2006.

U.S. Department of Labor, Occupational Safety and Health Administration. *General Industry Digest (OSHA 2201)*. Washington, DC: U.S. Government Printing Office, 1995.

U.S. Department of Labor, Occupational Safety and Health Administration. *29* Code of Federal Regulations *1910*. Washington, DC: U.S. Government Printing Office, 1999.

U.S. Department of Labor, Occupational Safety and Health Administration. *29* Code of Federal Regulations *1926*. Washington, DC: U.S. Government Printing Office, 1999.

SAFETY HAZARDS

Reese, C.D. *Accident/Incident Prevention Techniques*. New York, NY: Taylor & Francis, 2001.

Reese, C.D. and J.V. Eidson. *Handbook of OSHA Construction Safety & Health*. Boca Raton, FL: CRC/ Lewis Publishers, 1999.

U.S. Department of Energy, *OSHA Technical Reference Manual*. Washington, DC: 1993.

SAFETY AND HEALTH HAZARDS

Goetsch, D.L. *Occupational Safety and Health for Technologists, Engineers, and Managers*, 3rd edn. Upper Saddle River, NJ: Prentice Hall, 1999.

Hagan, P.E., J.F. Montgomery and J.T. O'Reilly. *Accident Prevention Manual for Business and Industry: Engineering and Technology*, 12th edn. Itasca, IL: National Safety Council, 2001.

Spellman, F.R. and N.E. Whiting. *Safety Engineering: Principles and Practices*. Rockville, MD: Government Institutes, 1999.

SAFETY AND HEALTH MANAGEMENT

Dougherty, J.E. *Industrial Safety Management: A Practical Approach*. Rockville, MD: Government Institutes, 1999.

Hagan, P.E., J.F. Montgomery and J.T. O'Reilly. *Accident Prevention Manual for Business and Industry: Administration & Programs*, 12th edn. Itasca, IL: National Safety Council, 2001.

Kohn, J.P. and T.S. Ferry. *Safety and Health Management Planning*. Rockville, MD: Government Institutes, 1999.

Lack, R.W. *Essentials of Safety and Health Management*. Boca Raton, FL: CRC/ Lewis Publishers, 1996.

Lack, R.W. *Safety, Health, and Asset Protection: Management Essentials*, 2nd edn. Boca Raton, FL: Lewis Publishers, 2002.

Petersen, D. *Human Error Reduction and Safety Management*, 3rd edn. New York, NY: Van Nostrand Reinhold, 1996.

Petersen, D. *Safety Management: A Human Approach*, 2nd edn. Goshen, IN: Aloray Inc., 1988.

Petersen, D. *Techniques of Safety Management: A Systems Approach*, 3rd edn. Goshen, IN: Aloray Inc., 1989.

Reese, C.D. *Occupational Health and Safety Management: A Practical Approach*. Boca Raton, FL: Lewis Publishers, 2003.

U.S. Department of Labor, Occupational Safety and Health Administration. *Federal Register: Safety and Health Program Management Guidelines (Vol. 54, No. 16)*. Washington, DC: January 26, 1989, pp. 3904–3916.

SERVICE INDUSTRY SAFETY AND HEALTH

Reese, C.D. *Handbook of Safety and Health for the Service Industry, Volume 1: Industrial Safety and Health for Goods and Material Services*. Boca Raton, FL: CRC/Taylor & Francis, 2008.

Reese, C.D. *Handbook of Safety and Health for the Service Industry, Volume 2: Industrial Safety and Health for Infrastructure Services*. Boca Raton, FL: CRC/Taylor & Francis, 2008.

Reese, C.D. *Handbook of Safety and Health for the Service Industry, Volume 3: Industrial Safety and Health for Administrative Services*. Boca Raton, FL: CRC/Taylor & Francis, 2008.

Reese, C.D., *Handbook of Safety and Health for the Service Industry, Volume 4: Industrial Safety and Health for People Oriented Services*. Boca Raton, FL: CRC/Taylor & Francis, 2008.

SYSTEM SAFETY

ABS Group. *Root Cause Analysis Handbook: A Guide to Effective Incident Investigation*. Rockville, MD: Government Institutes, 1999.

Bahr, N.J. *System Safety Engineering, and Risk Assessment: A Practical Approach*. New York, NY: Taylor & Francis, 1997.

Reese, C.D. *Accident/Incident Prevention Techniques*. New York, NY: Taylor & Francis, 2001.

Kavianian, H.R. and C.A. Wentz, Jr. *Occupational and Environmental Safety Engineering and Management*. New York, NY: Van Nostrand Reinhold, 1990.

U.S. Department of Energy, Office of Nuclear Energy. *Root Cause Analysis Guidance Document*. Washington, DC: February 1992.

TRAINING

Reese, C.D. *Accident/Incident Prevention Techniques*. New York, NY: Taylor & Francis, 2001.

Reese, C.D. and J.V. Eidson. *Handbook of OSHA Construction Safety & Health*, 2nd edn. Boca Raton, FL: CRC/Taylor & Francis, 2006.

Reese, C.D. *Occupational Health and Safety Management: A Practical Approach*. Boca Raton, FL: Lewis Publishers, 2003.

U.S. Department of Labor. *Training Requirements in OSHA Standards and Training Guidelines (OSHA 2254)*. Washington, DC: 1998.

WORKPLACE VIOLENCE

U.S. Department of Health and Human Services, National Institute for Occupational Safety and Health. *NIOSH Current Intelligence Bulletin 57. Violence in the Workplace: Risk Factors and Prevention Strategies*. Washington, DC: 1996.

U.S. Department of Labor, Bureau of Labor Statistics. *National Census of Fatal Occupational Injuries*. Washington, DC: 1998.

Warchol, G. *Workplace Violence, 1992–96*. National Crime Victimization Survey (Report No. NCJ-168634). Washington, DC: 1998.

PROFESSIONAL ORGANIZATIONS AND AGENCIES

These are national organizations that specialize in the many aspects of occupational safety and health. They have a wide range of resources as well as unique materials that have been developed by individuals and organizations with special expertise in occupational safety and health. Some key organizations and agencies are as follows:

HEALTH AND ENVIRONMENTAL ASSISTANCE

American Board of Industrial Hygiene (ABIH)
4600 West Saginaw, Suite 101
Lansing, MI 48917
(517) 321-2638

American Conference of Governmental Industrial Hygienists (ACGIH)
Building D-7
6500 Glenway Avenue
Cincinnati, OH 45211
(513) 661-7881

American Industrial Hygiene Association (AIHA)
PO Box 8390
475 White Pond Drive
Akron, OH 44311
(216) 873-3300

SAFETY AND ENGINEERING CONSENSUS STANDARDS

American National Standards Institute (ANSI)
11 West 42nd Street
New York, NY 10038
(212) 354-3300

American Society of Mechanical Engineers (ASME)
345 East 47th Street
New York, NY 10017
(212) 705-7722

American Society for Testing and Materials (ASTM)
655 15th Street NW
Washington, DC 20005
(202) 639-4025

National Safety Management Society (NSMS)
12 Pickens Lane
Weaverville, NC 28787
(800) 321-2910

PROFESSIONAL SAFETY ORGANIZATIONS

American Society for Safety Engineers (ASSE)
1800 East Oakton Street
Des Plaines, IL 60016
(847) 699-2929

Board of Certified Safety Professionals (BCSP)
208 Burwash Avenue
Savoy, IL 61874
(312) 359-9263

Industrial Safety Equipment Association (ISEA)
1901 North Moore Street
Arlington, VA 22209
(703) 525-1695
Fax: (703) 528-2148

National Safety Council (NSC)
1121 Spring Lake Drive
Itasca, IL 60143-3201
(708) 285-1121

Human Factors Society (HFS)
PO Box 1369
Santa Monica, CA 90406
(310) 394-1811

SPECIALTY ASSOCIATIONS (WITH SPECIFIC EXPERTISE)

American Welding Society (AWS)
PO Box 351040
550 LeJeune Road, NW
Miami, FL 33135
(305) 443-9353

American Gas Association (AGA)
1515 Wilson Blvd.
Arlington, VA 22209
(703) 841-8400

American Petroleum Institute (API)
1220 L Street, NW
Washington, DC 20005
(202) 682-8000
Fax: (202) 682-8159

American Society of Heating, Refrigerating, and Air Conditioning
  Engineers (ASHRAE)
1791 Tullie Circle, NE
Atlanta, GA 30329
(404) 636-8400

American Society for Training and Development (ASTD)
1640 King Street
PO Box 1443
Alexandria, VA 22313-2043
(703) 683-8129

Compressed Gas Association (CGA)
1235 Jefferson Davis Highway
Arlington, VA 22202
(703) 979-0900

Illuminating Engineering Society of North America
120 Wall Street, 17th Floor
New York, NY 10005
(212) 248-5000

Institute of Makers of Explosives
1120 19th Street, NW
Washington, DC 20036
(202) 429-9280

Laser Institute of America
12424 Research Parkway, Suite 130
Orlando, FL 32826
(407) 380-1553

National Fire Protection Association (NFPA)
1 Batterymarch Park
Quincy, MA 02269
(800) 344-3555

National Propane Gas Association
1150 176th Street, NW
Washington, DC 20036
(202) 466-7200

The Chlorine Institute
2001 L Street
Washington, DC 20036
(202) 775-2790
Fax: (202) 223-7225

FEDERAL GOVERNMENT SOURCES

The federal government is not the enemy, as many individuals surmise. It is a great resource for all types of information such as publications, training materials, compliance assistance, audiovisuals, access to experts, and other assorted occupational safety and health aids. In most cases, resources offered by the federal government are current and the response time is very reasonable. Asking for information does not act as a trigger for your company to become a target for inspections or audits. The federal government would prefer to assist you in solving your safety and health issues before they become problems. You will be pleasantly surprised by the help that you receive. All you need to do is ask. A list of government agencies that have information regarding occupational safety and health is as follows:

Bureau of Labor Statistics (BLS)
U.S. Department of Labor
Occupational Safety and Health Statistics
441 G Street, NW
Washington, DC 20212
(202) 523-1382

Center for Disease Control (CDC)
U.S. Department of Health and Human Services
1600 Clifton Avenue, NE
Atlanta, GA 30333
(404) 329-3311

Environmental Protection Agency (EPA)
410 M Street, SW
Washington, DC 20460
(202) 382-4361

U.S. Government Printing Office (GPO)
Superintendent of Documents
732 N. Capitol Street, NW
Washington, DC 20402
(202) 512-1800

Mine Safety and Health Administration (MSHA)
U.S. Department of Labor
4015 Wilson Blvd.
Arlington, VA 22203
(703) 235-1452

National Audio Visual Center (NAC)
National Archives and Records Administration
Customer Services Section CL
8700 Edgewood Drive
Capitol Heights, MD 20743-3701
(301) 763-1896

National Institute of Standards and Technology
U.S. Department of Commerce
National Engineering Laboratory
Route I-270 and Quince Orchard Road
Gaithersburg, MD 20899
(310) 921-3434

National Institutes of Health (NIH)
U.S. Department of Health and Human Services
9000 Rockville Pike
Bethesda, MD 20205
(310) 496-5787

National Institute for Occupational Safety and Health (NIOSH)
U.S. Department of Health and Human Services
Publications Dissemination
4676 Columbia Parkway
Cincinnati, OH 45226
(513) 533-8287 or (800) 35-NIOSH

National Technical Information Services (NTIS)
U.S. Department of Commerce
5285 Port Royal Road
Springfield, VA 22161
(703) 487-4636

Occupational Safety and Health Administration—National Office (OSHA)
U.S. Department of Labor
200 Constitution Avenue, NW
Washington, DC 20210
(202) 523-8151

OSHA (after Hours), National Hotline—(800) 321-OSHA
OSHA's Training Institute
1555 Times Drive
Des Plaines, IL 60018
(708) 297-4913

OSHA Publications Office
Room N3101
Washington, DC 20210
(202) 219-9667

Occupational Safety and Health Review Commission (OSHRC)
1825 K Street, NW
Washington, DC 20006
(202) 643-7943

ELECTRONIC SOURCES (INTERNET)

To access electronic sources, you must be connected to the Internet; this means you must select an Internet provider. This provider may be your local or long-distance phone company, your cable television, or it may be commercial services such as AOL, YAHOO, etc. (just to name a few). It goes without saying that you will need a computer with a reasonably fast modem (the faster the better), a telephone line, and some software such as Microsoft Explorer or Netscape Navigator (your Internet provider usually provides this), which allows you to browse the Internet.

Once you have access to the Internet, there are several good search engines that are helpful in finding information (Internet sites). These have names such as Google, which is the most common. These search engines allow you to find the sites or locations of the information that you are interested in (i.e., machine guarding, fire safety). Internet addresses are constantly changing, so the ability to search is critical.

The Internet sites have names that help you understand what they are. Some of the most common abbreviated names are:

http this is a transfer protocol, a standard web programming language
www means World Wide Web and is a connective or networking component of the Internet
com means commercial
edu stands for education
gov means government
org stands for organization

Most Internet sites start with http://www. followed by an abbreviation for entity (company or institution) and other numbers, symbols, or abbreviations that seem to make no sense. The ending is usually com, gov, edu, or org. You can use established site addresses, such as the ones that follow, to access these specific locations.

Government

Addresses of Government Agencies, http://www.fedworld.gov
Agency for Toxic Substances and Disease Registry, http://atsdrl.atsdr.cdc.gov.8080/atsdrhome.html
ATSDR Hazardous Substance Release/Health Effects Database, http://atsdrl.atsdr.cdc.gov.8080/hazdat.html
Building and Fire Research Laboratory, http://www.bfrl.nist.gov/
Bureau of Labor Statistics, http://stats.bls.gov/
California Department of Industrial Relations, http://www.dir.ca.gov/
Centers for Disease Control and Prevention, http://www.cdc.gov/
Consumer Product Safety Commission, http://www.cpsc.gov/
Emerging Infectious Diseases Home Page, http://www.cdc.gov/ncidod/EID/eid.html
Federal Emergency Management Agency, http://www.fema.gov/
Mine Safety and Health Administration, http://www.msha.gov/
Mining Accident and Injury Information, http://www.msha.gov/STATINFO.HTML
National Agriculture Safety Database, http://www.cdc.gov/niosh/nasd/nasdhome.html
National Highway Traffic Safety Administration, http://www.nhtsa.dot.gov/

National Institute of Environmental Health Sciences, http://heww.niehs.nih.gov/
National Institutes of Health, http://www.nih.gov/
National Institute for Occupational Safety and Health, http://www.cdc.gov/niosh/homepage.html
Occupational Safety and Health Administration, http:/www.osha.gov/and http:www.osha.gov/STLC
OSHA Ergonomics, http://www.osha.gov/ergo
U.S. Department of Transportation, http://www.dot.gov/
U.S. Department of Energy Chemical Safety Program, http://tis-hq.eh.doe.gov/web/chem_safety /
U.S. Environmental Protection Agency, http://www.epa.gov/
U.S. Department of Health and Human Services, http://www.dhhs.gov/
U.S. Department of Labor Office of Inspector General, http://gatekeeper.dol.gov/dol/oig/

Associations and Societies

American Association of Occupational Health Nurses, http://www.aaohn.org/
American Chemical Society, http://www.acs.org/
American College of Occupational and Environmental Medicine, http://www.acoem.org/
American Conference of Governmental Industrial Hygienists, http://www.acgih.org/
American Industrial Hygiene Association, http://www.aiha.org/
American National Standards Institute, http://web.ansi.org/default.htm
American Society of Heating, Refrigerating and Air-Conditioning Engineers, http://www.ashrae.org/
American Society of Safety Engineers, http://www.asse.org/
American Society for Testing and Materials, http://www.astm.org/
American Speech-Language-Hearing Association, http://www.asha.org/
Board of Certified Safety Professionals, http://www.bcsp.com/
British Safety Council, http://www.britishsafetycouncil.co.uk/
Building Owners and Managers Association International, http://www.boma.org/
Canada Safety Council, http://www.safety-council.org/english/index.htm
Canadian Society of Safety Engineering, http://www.csse.org/
Chemical Manufacturers Association, http://www.cmahq.corn/index.html
Industrial Safety Equipment Association, http://www.safetycentral.org/isea/
National Association of Demolition Contractors, http://Iwww.voicenet.corn/-NAOC
National Association of Tower Erectors, http://www.daknet.corn/nate/
National Fire Protection Association, http://www.wpi.edu/-fpe/nfpa.html
National Hearing Conservation Association, http://www.globaldialag.corn/-nhca/
National Safety Council, http://www.nsc.org/

Many other electronic sites for information are available and more are being posted with lightning speed. The electronic information system is very fluid and new avenues of information are constantly emerging.

Bibliography

Allison, W.W. *Profitable Risk Control*. Des Plaines, IL: American Society of Safety Engineers, 1986.

Anton, T.J. *Occupational Safety and Health Management*, 2nd edn. New York, NY: McGraw Hill, 1989.

Australian Health & Safety Commission. Violence and Bullying at Work. Available at www.workplacevio-lence911.com, December 2003.

Bell, C.L. *et al. Environmental Law Book*, 16th edn. Rockville, MD: Government Institutes, 2001.

Bertinuson, J. and S. Weinstein. *Occupational Hazards of Construction: A Manual for Building Trades' Apprentices*. Berkeley, CA: Labor Occupational Health Program, University of California, 1978.

Blake, R.R. and J. Srygley. Principles of behavior for sound management, *Training and Development Journal* (October 1979): 26–28.

Blanchard, K. How to get better feedback, *Success* (June 1991): 6.

Blosser, F. *Primer on Occupational Safety and Health*. Washington, DC: The Bureau of National Affairs, Inc., 1992.

Brigham, E.F. *Fundamentals of Financial Management*. Chicago, IL: Dryden, 1989.

Brown, P.L. and R.J. Presbie. *Behavior Modification in Business, Industry and Government*. Paltz, NY: Behavior Improvement Associates, Inc., 1976.

California Department of Industrial Relations (Cal/OSHA). *Easy Ergonomics: A Practical Approach for Improving the Workplace*. 1999.

California Department of Labor, *Guidelines for Security and Safety of Health Care and Community Service Workers*. Available at http://www.ca.gov. Sacramento, CA: March 1998.

Canadian Safety Council. Bullying in the Workplace. Available at www.safety-council.org, 2004.

Chiu, C. *A Comprehensive Course in Root Cause Analysis and Corrective Action for Nuclear Power Plants, Workshop Manual*. San Juan Capistrano, CA: Failure Prevention, Inc., 1988.

Environmental Statutes (2001 Edition). Rockville, MD: Government Institutes, 2001.

Federal Aviation Administration. *System Safety Handbook*, Chap. 3. Washington, DC: 2000.

Feiberg, P. Bullying in the Workplace is a Violence Warning Sign. Available at www.workplaceviolence911. com, July 1998.

Fine, W. Mathematical evaluations for controlling hazards. In J. Widner (Ed.), *Selected Readings in Safety*. Macon, GA: Academy Press, 1973.

Gano, D.L. Root cause and how to find it, *Nuclear News*, August 1987.

Geller, E.S. *The Psychology of Safety Handbook*. Boca Raton, FL: CRC Press/Lewis Publishers, 2001.

Goetsch, D.L. Checking your culture. *Occupational Health and Safety*. October, 2007.

Heinrich, H.W. *Industrial Accident Prevention: A Scientific Approach*. New York, NY: McGraw-Hill, 1959.

Herzberg, F. One more time: how do you motivate employees? *Harvard Business Review* (January–February 1968): 53–62.

Johnson, L.F. *Choosing a Safety Consultant*. Dallas, TX: Occupational Safety and Health, July 1999.

Kohn, J.P. and T.S. Ferry. *Safety and Health Management Planning*. Rockville, MD: Government Institutes, 1999.

Maslow, A.H. *Motivation and Personality*. New York, NY: Harper and Brothers, 1954.

Murphy, W.C. and J.R. Hanson. *A Maine Guide to Employment Law*. Orono, ME: The University of Maine, 1995.

Namie, G. US Hostility Workplace Survey. Bellingham, WA: The Workplace Bullying & Trauma Institute, September 2000.

Namie, G. Report on Abusive Workplaces. Bellingham, WA: The Workplace Bullying & Trauma Institute, October 2003.

National Safety Council, *Motor Fleet Safety Manual*, 3rd edn. Itasca, IL: 1986.

National Safety Council, *Injury Facts 2007 Edition*. Itasca, IL: 2007.

Occupational Safety and Health Administration. Available at http://www.osha.gov, Washington, DC: 2001.

Petersen, D. *Techniques of Safety Management: A Systems Approach*, 3rd edn. Goshen, IN: Aloray, Inc., 1989.

Powers, R. *Military Ethics and Conflicts of Interest*. Available at http://usmilitary.about.com/es generalinfo/a/ stanconduct.htm, 2006.

Protecting Workers Lives: A Safety and Health Guide for Unions, 2nd edn. Itasca, IL: National Safety Council, 1992.

Reese, C.D. *Joint Labor/Management Committees: A Guide for Committee Members*. Storrs, CT: UConn Press, 1996.

Reese, C.D. *Material Handling Systems: Designing for Safety and Health*. New York: Taylor & Francis, 2000.

Reese, C.D. *Accident/Incident Prevention Techniques*. New York, NY: Taylor & Francis, 2001.

Reese, C.D. *Occupational Health and Safety Management: A Practical Approach*. Boca Raton: Lewis Publishers, 2003.

Reese, C.D. and J.V. Eidson. *Handbook of OSHA Construction Safety & Health*, 2nd edn. Boca Raton, FL: CRC/Lewis Publishers, 1999.

Reese, C.D. and J.V. Eidson. *Handbook of OSHA Construction Safety & Health*, 2nd edn. Boca Raton, FL: CRC/Taylor & Francis, 2006.

Reese, C.D., J.B. Moran, and K. Lapping. *Model Construction Safety and Health Program*. Laborers' Health and Safety Fund of North America. Washington, DC: 1993.

Swartz, G. (Ed.), *Safety Culture and Effective Safety Management*. Itasca, IL: National Safety Council, 2000.

U.S. Bureau of Labor Statistics, *Workplace Injuries and Illnesses in 2006*. Washington, DC: 2000.

U.S. Bureau of Labor Statistics, *National Census of Fatal Occupational Injuries in 2000*. Washington, DC: 2006.

U.S. Bureau of Labor Statistics. Available at http://www.bls.gov, Washington, DC: 2007.

U.S. Department of Energy. *OSH Technical Reference Manual*. Washington, DC: 1993.

U.S. Department of Energy, Office of Nuclear Energy. *Root Cause Analysis Guidance Document*. Washington, DC: February 1992.

U.S. Department of Health and Human Services: National Institute for Occupational Safety and Health. *Violence in the Workplace: Risk Factors and Prevention Strategies (CIB 57)*. Washington, DC: June 1996.

U.S. Department of Health and Human Services: National Institute for Occupational Safety and Health. *Elements of Ergonomics Programs (DHHS-97-117)*. 1997.

U.S. Department of Justice's Bureau of Justice Statistics, *Annual National Crime Victimization Survey for 1992–1996*. Washington, DC: 1998.

U.S. Department of Labor, *Training Requirements in OSHA Standards and Training Guidelines (OSHA 2254)*. Washington, DC: 1998.

U.S. Department of Labor/OSHA, *All About OSHA (OSHA 2056)*. Washington, DC: 1985.

U.S. Department of Labor/OSHA, *Record keeping Guidelines for Occupational Injuries and Illnesses (OMB No. 1220-0029)*. Washington, DC: 1986.

U.S. Department of Labor/OSHA, *Access to Medical Records and Exposure Records (OSHA 3110)*. Washington, DC: 1988.

U.S. Department of Labor/OSHA, *OSHA: Employee Workplace Rights (OSHA 3021)*. Washington, DC: 1991.

U.S. Department of Labor/OSHA, *Training Course in OSHA for the Construction Industry (Course #500)*. Des Plaines, IL: OSHA Training Institute, 1997.

U.S. Department of Labor, Mine Safety and Health Administration. *Accident Prevention (Safety Manual No. 4)*. Beckley, WV: Revised 1990.

U.S. Department of Labor, Mine Safety and Health Administration. *Job Safety Analysis (Safety Manual No. 5)*. Beckley, WV: Revised 1990.

U.S. Department of Labor, Mine Safety and Health Administration. *Job Safety Analysis: A Practical Approach (Instruction Guide No. 83)*. Beckley, WV: 1990.

U.S. Department of Labor, Mine Safety and Health Administration. *Safety Observation (MSHA IG 84)*. Beckley, WV: Revised 1991.

U.S. Department of Labor, National Mine Health and Safety Academy. *Accident Prevention Techniques*. Beckley, WV: 1984.

U.S. Department of Labor, National Mine Health and Safety Academy. *Accident Prevention Techniques: Job Safety Analysis*. Beckley, WV: 1984.

U.S. Department of Labor, Occupational Safety and Health Administration. *Job Hazard Analysis, (OSHA 3071)*. Washington, DC: 1992.

U.S. Department of Labor, Occupational Safety and Health Administration, *Office of Training and Education. OSHA Voluntary Compliance Outreach Program: Instructors Reference Manual.* Des Plaines, IL: 1993.

U.S. Department of Labor, Occupational Safety and Health Administration. *General Industry Digest (OSHA 2201).* Washington, DC: U.S. Government Printing Office, 1995.

U.S. Department of Labor, Occupational Safety and Health Administration. *Training Manual OSHA #500 for the Construction Industry.* Des Plaines, IL: OSHA Training Institute, 1997.

U.S. Department of Labor, Occupational Safety and Health Administration. *29* Code of Federal Regulations *1910.* Washington, DC: U.S. Government Printing Office, 1999.

U.S. Department of Labor, Occupational Safety and Health Administration. *29* Code of Federal Regulations *1926.* Washington, DC: 1999.

U.S. Department of Labor, Occupational Safety and Health Administration. *Subject Index.* "Internet." April 1999. Available at http://www.osha.gov.

U.S. Department of Labor, Occupational Safety and Health Administration, *Office of Training and Education. Manual for Trainer Course in OSHA Standards for the General Industry.* Des Plaines, IL: 2001.

U.S. Department of Labor, Occupational Safety and Health Administration. *Trainer Manual OSHA # 501 for the General Industry.* Des Plaines, IL: OSHA Training Institute, 2001.

U.S. Department of Labor, Occupational Safety and Health Administration. *Job Hazard Analysis and Control (1910.917–922)*, Subject Index. "Internet" 2001. Available at http://www.osha.gov.

U.S. Department of Labor, Occupational Safety and Health Administration. *OSHA Handbook for Small Businesses (OSHA 2209).* Washington, DC: 2006.

U.S. Environmental Protection Agency's Website at http://www.epa.gov.

U.S. Office of Personnel Management, *Dealing with Workplace Violence: A Guide for Agency Planners.* Washington, DC: February 1998.

Waxman, K.T. *A Practical Guide to Finance & Budget: Skills for Nurse Managers.* Marblehead, MA: HCPro, Inc., 2004.

Weisinger, H. and N. Lobsewz. *Nobody's Perfect: How To Give Criticism and Get Results.* New York, NY: Stratford Press, 1981.

Wieneke, R.E. et al. *Success in Behavior-Based Safety at Los Alamos National Laboratory's Plutonium Facility.* Los Alamos, NM: Los Alamos National Laboratory, 2002.

Yandrick, R.M. Lurking in the shadows. *HR Magazine, 44,* October 1999.

Zhang, H. et al. Proceedings of the 46th Annual Meeting of the Human Factors and Ergonomics Society. Safety Culture: A Concept in Chaos? Santa Monica, CA: 2002.

Index